D0609296

# The Fruited Plain

Walter Ebeling

# The Fruited Plain

## The Story of American Agriculture

UNIVERSITY OF CALIFORNIA PRESS
BERKELEY • LOS ANGELES • LONDON

FRONTISPIECE: Pear and apple orchards in bloom, Hood River Valley, Oregon. Courtesy of Nick Bielemeier.

University of California Press
Berkeley and Los Angeles, California

University of California Press, Ltd.
London, England


ISBN 0-520-03751-0
Library of Congress Catalog Card Number 78-62837

Designed by A. Marshall Licht

Printed in the United States of America

1 2 3 4 5 6 7 8 9

# Contents

# 4 The South

# Preface

When we at long last escape from the bonds of routine and specialization, some of us choose to retrace cherished but long-forsaken paths to resume a "sentimental journey." For me, with predilections derived from boyhood on a farm, the belated journey led to where early settlers followed Daniel Boone's Wilderness Trace to Kentucky's rich but perilous Bluegrass country; to where sturdy boatmen, having floated cargoes of farm produce down the Mississippi to New Orleans, returned through four hundred and fifty miles of wilderness along the Natchez Trace; to where Texas cowboys drove great herds of half-wild Longhorn cattle up the Chisholm Trail to Kansas railheads; to where Spanish *conquistadores* carved a transient empire out of the great Southwest for the king of Spain; to colorful high mesas where peaceful corn-planting Zuñis and Hopis watched from their cliff dwellings for the approach of fierce nomadic Apache and Navajo raiders; to where hardy, fearless pioneers followed hazardous trails across the immense American prairie and over mountain passes discovered by bold, reckless mountain men, then across fearsome deserts to the fabled lands of Oregon and California. It was as great a thrill to visit the routes of early migration and the sites of pioneer agricultural settlements as to see the impressive panorama of modern agriculture.

In research of the literature I have followed the course of American agriculture from Neolithic beginnings that led to the corn-bean-squash culture of the Indians, from small clearings of the early white settlers in the eastern forest to the plantation system of the South, and then to the immense monocultures of the Midwest and Great Plains and to the diversified irrigated agriculture of the arid lands of the West.

This book attempts to weave the story of agriculture into the general fabric of our nation's history. It also purports to evaluate our failures and successes as to the utilization and care of the precious and fragile layer of living soil that tenuously covers an otherwise lifeless globe and supports the plant and animal products of the land, wild or domesticated, that have sustained us and shaped our history. Brief statements concerning climate, physical features, and native vegetation have been included to give the reader a feeling for the varied and distinctive environments that influenced the course of agricultural development in each of the seven regions discussed.

Some sections of the book are intentionally somewhat anecdotal. There are certain anecdotes that illustrate with an ageless flavor of human interest what is unique in the American experience. An attempt was made to reveal some of the exciting, colorful drama that accompanied the westward march of our nation. Is there anyone "with soul so dead" that he or she is not moved to patriotic fervor by the incomparable drama starring the indomitable people who founded this nation and extended its boundaries to the Pacific, all the while wresting their sustenance from the perilous, stubborn, but bountiful wilderness? And on what a stage was the drama played!—the great trackless forests, the vast prairies, the sky-piercing mountains, the colorful mesas, deep canyons, and searing deserts—a romantic setting for drama unsurpassed elsewhere on this planet.

This country is well supplied with professional agricultural historians and I make no pretense to being one of them. I consider this book to be the *story,* rather than the history, of American agriculture. To geographically delimit the story, it was confined to the coterminous forty-eight states. No doubt much history has been omitted, but probably agricultural historians would have to face a similar dilemma in any single volume. I have placed more than the usual casual emphasis on certain ethnocentric features of the American agricultural experience: the rich legacy of crops and crop cultures the white man inherited from pre-Columbian native Americans, as well as the vital influences of the various immigrant ethnic groups. Perhaps I have also placed more emphasis on the socioeconomic and environmental impact of agriculture as it has evolved in this country, and as currently practiced, than would be justified in a purely historical account.

## ACKNOWLEDGMENTS

The many persons who kindly helped to make my explorations of our nation's "fruited plain" both rewarding and enjoyable are too numerous to mention here. They include many agricultural scientists, extension personnel, foresters, naturalists, conservationists, farmers, and ranchers. Special thanks are due to certain particularly kind and hospitable hosts. Mr. and Mrs. J. C. Redd, prominent citizens of Jackson, Mississippi, took my wife and me on a tour of that state's farmlands and the historic Natchez Trace (chapter 4). (My wife, Ora May, was an enthusiastic and helpful companion on much of my "sentimental journey.") We were also guests of Mr. and Mrs. James W. Evans of Windsor, Missouri, and received a wealth of firsthand information on the day-to-day operation of the Evans's large, technologically progressive cattle ranch. Mr. and Mrs. Roy H. Norman of Riverton, Wyoming, were our gracious hosts for several days, facilitating informative conversations with prominent farmers and ranchers of the area and a trip to the unforgettable "Wyoming Shangri-la" described in chapter 6.

The entire manuscript was read by irrigation engineer Paul F. Keim, botanist Robert Ornduff, food industry consultant Arthur N. Prater, agricultural historian Theodore Saloutos, and plant physiologist Vernon T. Stoutemyer. (Keim and Stoutemyer are also owners of Midwest farms.) Chapter 1, 2, 3, and 8 were read by plant physiologist and soil scientist O. Raynal Lunt, and chapter 8 by Eugene L. MacFarlane, M.D., amateur archaeologist and anthropologist. I greatly appreciate the criticisms, corrections, and valuable insights provided by these distinguished scientists. However, they did not unanimously agree with all opinions expressed by the author. Neither they nor any other persons consulted are in any way responsible for such errors, omissions, or inadequacies that may appear in this book. For these I take sole responsibility.

I wish to express appreciation to Professor Rainer Berger, of UCLA's geology department for the radiocarbon dating of an ancient ear of corn (chapter 1), and to United Airlines and Federal Aviation Administration for permission to ride in the flight deck of a transcontinental airliner for an overview of the nation's agricultural resource base (chapter 2).

Search for literature was obviously vital. In this respect I am greatly indebted to the courteous and efficient work of library personnel operating under the terms of the Interlibrary Loan Code adopted by cooperating libraries in the United States and Canada, thus supplementing in important ways the considerable volume of literature in the libraries of the UCLA campus.

I greatly appreciate the excellent editorial assistance and most cordial and efficient working relationship with the University of California Press throughout the preparation of this book.

W. E.

# 1 How It All Began

*A house and yoke of oxen first provide,*
*A maid to guard your herds, and then a bride.*
—Hesiod

Ancient literature commonly reflects a nostalgic racial memory of a preagricultural Eden-like Golden Age, when, as the Greek poet Hesiod (eighth to seventh century B.C.) proclaimed, "The life-sustaining soil/Yielded its copious fruits/Unbribed by human toil" (Elton, 1815). According to Hesiod, next came the Silver Age, when mankind lost certain virtues possessed in the previous age. It was symbolized by the "yoke of oxen" and "herds" in his quaint verse. It was a time that had some degree of bucolic charm but was characterized by a new ingredient in human experience—work! Mankind was next to descend to the Bronze and then to the Iron Age, with increasing increments of evil—a predicament from which presumably we have not yet emerged.

Unknown in Hesiod's day was the fact that the Golden Age—the age of the hunter-gatherer—extended back two million years to the dawn of man's cultural evolution. It was a period long enough for our species to have acquired certain physical, physiological, and behavioral characteristics and emotional needs that we now find to be not entirely compatible with the artificiality of civilized life. Hunting and gathering are activities modern man, practically Cro-Magnon genetically, now widely considers to be recreation. Goldsmith (1974) points out that no word for "work" is ever found in the various languages of hunting-gathering societies. He believes that they probably never consumed more than about a third of the available food supplies and spent only a few hours a day in satisfying their needs. Lee (1968, 1969) observed that even during a long drought, the food supply of !Kung bushmen in the Kalahari Desert in Botswana is surprisingly dependable, unlike that among many African farming groups. Native wild plants on which hunter-gatherers subsist are well adapted to their environment.

Occasionally we moderns escape to the wilderness to wander about like our primitive ancestors, in wild abandon—in freedom. We revive dormant racial memories that still haunt the human psyche. Thus we obtain some measure of physical and spiritual restoration. Although fallen from Eden, basically we are not far removed from the hunter-gatherer.

## TRANSITION FROM HUNTING-GATHERING TO AGRICULTURE

During the long period of man's experience as a hunter-gatherer, the human population increased very slowly. Estimates of Pleistocene population growth rates vary from 0.0007 to 0.0020 percent per annum, compared to 0.1 percent in the Neolithic (late Stone Age) period (Cohen, 1977) and from 1 to 3 percent in modern times. But slow population growth was apparently not the result of an inadequate food supply. Modern anthropologists generally do not attribute the slow growth to the harshness of preagricultural life, but to the exigencies of the nomadic life-style of the hunter-gatherer, necessitating various forms of birth control practice, as may be noted in hunting-gathering societies existing today (Deevy, 1960, 1968; Birdsell, 1968; Lee, 1972; Shepard, 1973; Dumond, 1975; Pfeiffer, 1977).

Among the many foods of the gatherers were the wild grass seeds. Remarkably little effort was required for gathering them. In one hour a person might harvest the equivalent of 1 kilogram (2.2 lb) of clean grain of wild wheat (einkorn) in Turkey (Harlan, 1967). It contains 57 percent more protein than modern wheat.[1] In 3.5 hours, an 11-day supply of "wild corn" (teosinte) could be gathered in Mexico (Harlan, 1975). The Ojibwa Indians of Wisconsin knocked the kernels of wild rice (*Zizania aquatica*) into their canoes until, in a few hours, the conveyances were fully loaded and low in the water (Stickney, 1896). The supply was inexhaustible (Bakeless, 1950). It is reasonable to suppose that modern hunter-gatherers have been pushed out to what farmers would consider to be marginal lands and that preagricultural hunter-gatherers had even more productive areas at their disposal.

There is no human group on earth so primitive as to be unaware of the fact that most plants grow from seeds (Flannery, 1968). It is not ignorance, but lack of need, that prevents more groups from practicing horticulture. Hunter-gatherers had in some instances increased the productivity of land by such measures as using fire to keep the forests more open and thereby increase productivity for fruits, berries, and game. Fire also improved prairies for large game animals such as buffalo. The Paiute Indians of Owens Valley, California, are known to have irrigated their wild seed plots to increase their yield (Steward, 1933). Agriculture was probably a de facto accumulation of new habits rather than a conceptual breakthrough (Cohen, 1977).

But why, after such a long sojourn in a veritable Eden, so blissful in retrospect,[2] did mankind accept the restrictions and burdens of farming, simultaneously and on a worldwide basis, about 10,000 years ago? Mark Cohen (1977) believes the most plausible explanation is that, despite the factors favoring population stability among nomads, there was nevertheless continuous population growth, along with the mechanisms that were effective in distributing population pressure evenly from region to region. Unlike other species, man was able to adapt to population pressure with technological change, that is, by the adoption of farming, albeit reluctantly, as a way of life. Hunting and gathering, while very successful for small human groups, did not suffice to support large or dense populations. Agriculture was only one of a long series of adaptations to increased populations, for example, use of fire in the mid-Pleistocene period in northern and temperate regions and various measures to increase land productivity, as previously related.

[1]All wild and primitive wheat types contain more protein than many commercial wheats. Apparently wheat was selected for high yield and good semolina or bread-making quality, whereby starch content was increased relative to that of protein (Johnson and Waines, 1977).

[2]But at the higher latitudes primitive tribes might suffer heat and drought in summer and cold and hunger in winter. In areas of harsh climate and meager food resource, life could be somewhat less than idyllic (Malthus, 1806; Fehrenbach, 1974).

By 8000 B.C., human populations had already expanded from tropical to temperate and finally arctic latitudes in both hemispheres, for *modern* man has been in the New World as long as in the Old World. Big game animals had become scarce, and increasing quantities of small animals and vegetable foods, including those involving complex preparation, were being eaten. There was decreased opportunity for the selection of the wide variety of foods that characterizes hunting-gathering societies—even the few such societies now remaining.[3] Also, fire became more important for creating conditions favorable for growth of cereals and other crops, and animals, on which human populations were placing increased reliance. Hunter-gatherers will turn to agriculture only as a last resort when population pressure has decreed that still greater productivity *per unit of space* is required.

## AGRICULTURE, ENERGY, AND CIVILIZATION

Civilization was spawned by agriculture. Civilized life could not have developed if it were not for the "agricultural revolution" in the Neolithic period of human prehistory. By domesticating plants and animals, man passed from savagery to barbarism and, hopefully, on to civilization; man himself became domesticated. Although some sedentism occurred in preagricultural societies, farming was the basis for the expansion of the first sedentary communities into sophisticated cities where prehistoric civilizations originated.

The Neolithic farmer could provide food not only for his own family, but for several other people. Those freed by surpluses from the task of providing food for themselves and their families were able to devote their time to the establishment of permanent communities and the expansion of knowledge. A division of labor was required; merchants and craftsmen appeared on the scene and public works and community functions were begun.

The first agricultural revolution and the first cultural revolution developed symbiotically. Agriculture enabled man to more efficiently utilize solar energy, through the medium of photosynthesis, and thereby support a much greater population than was possible by hunting-and-gathering tribes. The maximum number of people the earth's surface could sustain if man had never progressed beyond the hunting and food-gathering stage has been estimated to be only twenty to thirty million (Dimbleby, 1967). The new energy input was also essential for the advancement of culture, which could not advance beyond the limits set by available energy resources (White, 1959). The next great increment in energy available to man, thousands of years later, was provided by fossil fuels—the sun's energy stored in the earth—ushering in the industrial revolution and energizing another quantum jump in culture and a new level of civilization. Cultural growth depended on more than new ideas and aspirations; these would have been stillborn without new sources of energy. However, the fact that vast new sources of energy have freed mankind from much drudgery and misery and have energized advancing civilization does not imply that "more is better" ad infinitum. The time has at last arrived when, for environmental and ecological reasons, we must reconsider what we mean by civilization.

The empathy expressed by most authors for the Neolithic farmer is not shared by

[3]The high quality of the diet of the hunter-gatherer is shown by the disastrous physical deterioration of primitive peoples who have changed to the modern diet. The most conspicuous visual signs are decayed teeth and malformations of the dental arches (Price, 1939). The principal weakness of modern foods from a nutritional standpoint results from their refinement and devitalization in processing and the alarming quantity and ubiquity of sugar (Price, 1939; Yudkin, 1975).

everyone. For example, Nigel Calder (1967) and Paul Shepard (1973) express their preference for the leisurely, fun-filled life of the Stone Age hunter, to which they believe *Homo sapiens* is anatomically and psychologically better adapted than to the unremitting, monotonous, backbreaking toil, much of it excruciating "stoop labor," that characterized agriculture until, among developed countries, mechanized equipment and fossil-fuel energy eventually provided some relief, though at the price of environmental degradation and social collapse.

Anticipating a world population of at least eight billion, Calder and Shepard suggest that human society nevertheless try to develop a "new cynegetics." This would be accomplished by means of a sophisticated technology to rid the earth of environmentally destructive agriculture, including its industrial extensions and mental correlates. These authors believe the "wretchedness" foisted on mankind by agriculture may eventually be alleviated by the development of unlimited sources of energy (solar or nuclear fusion energy), automation, miniaturization, computers, and a food technology based on one-celled organisms (bacteria, yeasts, protozoa, and algae) and green leaves, the latter known to be high in protein, or based on food produced synthetically or from animals subsisting on synthetically manufactured feed. In the vast areas released from agriculture and returned to wilderness, man would regain his million-year-old cynegetic legacy. Men would do the hunting, and only with hand weapons (no guns). Women's wilderness experience would be fulfilled by gathering, reenacting their role in prehistory. (This scenario fails to reckon with the Women's Lib movement.)

Even primitive unmechanized agriculture requires a greater input of energy per unit of food produced than does hunting and gathering in an uncrowded environment. (Modern mechanized agriculture requires an energy input far exceeding the food energy produced [Pimental et al., 1973; Steinhart and Steinhart, 1974].) The farmer found that he had to work from dawn to dusk. And most of those he liberated from three-hour seed-gathering days or wild-game hunting eventually found themselves slaving twelve-hour days or more in the sweatshops of the Industrial Revolution or mining coal in the hazardous bowels of the earth. For thousands of years leisure and affluence came to relatively few. Harlan (1975) reminds us that at long last there is some indication that the industrial age may permit us to regain the leisure we lost with the advent of agriculture.

Agriculture saddled us with a much greater problem than loss of leisure. In agriculture societies children were found to be an economic asset. They could work in the crop fields at a relatively early age and later provided security for their aged parents. To this day, in agrarian societies childless couples may become impoverished to the point of starvation. Agricultural peoples generally believe that the more children they have the better. Improved public health measures greatly reduced the hazard of losing children prematurely, while high birth rates continued. More people require more food, which in turn requires more intensive farming, more energy input per unit of food, and more environmental degradation, particularly in the form of soil erosion. Population has continuously pressed closely on the available food supply as Malthus (1806), with uncommon prescience for his time, predicted. Crop failure invariably results in starvation and death. In times past, at least there were fertile lands to which surplus people could migrate, but new lands are now increasingly marginal in quality or, in many areas, nonexistent.

Demographers hope for social programs that will provide poor people the economic security they find now only through large families, as well as information on family planning. Only then, they believe, will the poor be convinced that having fewer children is both rational and feasible. With rapidly rising populations and shrinking resources, it remains to be seen if the "demographic transition"[4] experienced by the developed nations will occur soon enough in the overcrowded developing nations to avert wide-scale catastrophe. Even those who convincingly argue that a worldwide new political, socioeconomic, and ecological approach to agriculture could put an end to food deficiencies (e.g., Lappé and Collins, 1977) do not minimize the importance of stopping population growth as soon as possible. Otherwise, political and socioeconomic gains and improved technology, which could be great liberating forces, become no more than temporary holding actions.

For over 99 percent of the estimated two million years that Cultural Man has been on earth he has sustained himself solely by hunting and gathering. The environment of the hunter-gatherer was "secure and abundant" (Pfeiffer, 1977). Hunting-gathering was a system that was "stable, reliable, permanent, and basic" (Harlan, 1975). It entailed only minimal environmental disruption. Will agricultural-industrial society endure that long? At a symposium on "Two Centuries of American Agriculture," agricultural economist Don Paarlberg (1976), who takes a dim view of portrayers of gloom and doom and ranges himself "firmly on the side of the optimists," expressed his opinion: "It seems well within reason to project the existence of the earth, of mankind, and of agriculture for another couple of centuries." But he based his hopes for this rather modest projection on a proviso: "I assume that the human race is essentially reasonable and is likely to stop an adverse trend somewhat short of disaster." Among the blessings that technology and good common sense could bestow upon mankind, Paarlberg considered the most important to be "advances in family planning and in greater acceptance of the replacement sized family so that mankind might move out from under the Malthusian shadow."

With the agricultural revolution man lost his innocence. He was the only species with the necessary mental and physical endowment, along with the advantage of a cultural evolution, that did not have to accept the environment, but could change it. Only in recent decades has an awareness developed that the resulting ecological imbalance to which man has become accustomed has caused a degree of stress that may be greater than the planet's biological systems can tolerate. To return to the "Golden Age" is now no longer even physically possible. In any case, most of us would not contemplate with equanimity the loss of many of the cultural and material amenities that civilization has made available to us, particularly the considerable inroads against ignorance and superstition and the near conquest of infectious diseases. Possibly with a more sophisticated and benevolent technology the imbalance that agricultural-industrial civilization has created can in some way be managed and sustained. But this remains to be seen.

No doubt the agricultural-industrial revolution has resulted in life-styles often at odds with the legacy of instinct, emotion, and impulse derived from countless millennia of hunting and gathering, as various authors have suggested. But given our unique mental endowment, cultural innovations such as toolmaking, the discovery of the use of fire, and the development of agriculture and subsequent technologies were inevitable for better or worse. Assuming an end to population growth, a continually improved agriculture and improved social and political institutions offer our only hope for substantial improvement in the lot of the world's teeming billions.

[4]Demographers and sociologists placing their bets on demographic transition must have been chilled by the words of a Philippine farmer, father of ten, at an international conference on the "War on Hunger," who praised a high-yielding rice variety for providing food enough for himself and his neighbors so that they could continue to see their women in the condition in which they were most beautiful—pregnant (Paddock and Paddock, 1973).

## ORIGIN OF VEGECULTURE

Agriculture probably had its beginning in the tropics. There man could manipulate the wide variety of wild plants and animals that characterize the rain forest, the species-rich habitats of upland-lowland margins, and where forest gives way to more open country or abuts on grassland, swamp, river, lake, or coast. In such surroundings man could most readily combine the gathering of wild plants with hunting and fishing. These complex ecosystems provided the best opportunities for the transition from simple gathering to planting and harvesting—the full domestication of plants.[5] Progressive fishermen living along rivers in a mild climate may have been the progenitors of the earliest agriculturists. Plants containing rotenoids, saponins, and alkaloids were probably used to stun fish, just as they are used by primitive tribes today. Unlike other hunting-gathering activities, fishing permitted a settled life, for one needed only to go to the water's edge to catch fish. The settled life, in turn, favored agricultural activities (Sauer, 1952; Binford, 1968; Harris, 1972).

In humid tropical lowlands, *vegeculture* became the dominant indigenous mode of cultivation. Starch-rich crops such as manioc (yuca or cassava), sweet potato, taro, and the yams were vegetatively produced by planting tubers, corms, and rhizomes. They were grown in swiddens—areas incompletely cleared and burned in the forest. Stem-cuttings were planted among tangled and rotting debris. Soil nutrients were not readily depleted and soil erosion was minimal. At signs of soil exhaustion, a second swidden could be cleared and planted while the soil was naturally restored in the first. Many man-made plants or *cultigens* were eventually developed that lost their ability to bear viable seeds and are now dependent on man for their reproduction. Fishermen may have begun the cultivation and alteration of plants as sources of fiber and ceremonial dyes, as well as food, sometimes all contained in one plant; food may not have always been the most important objective (Sauer, 1952).

Geographer Carl O. Sauer (1952) presented evidence for the dispersal of plant cultigens—rice, taro, yam, sugarcane, coconut, breadfruit, banana, citrus fruits, persimmon, ginger, bamboo, and *Derris* (a source of rotenone)—and domesticated animals (pig, dog, chicken, duck, and goose), from a Southeast Asian tropical hearth, and such cultigens as manioc, sweet potato, peach palm or pejibaye, arrowroot, tobacco, and *Lonchocarpus* (cube root, a source of rotenone), from a tropical hearth in northwestern South America. The idea of centers or hearths of origin and dispersal of food crops had its origin in the investigations of the brilliant Russian geneticist and agronomist N. I. Vavilov (1951). On the basis of hundreds of thousands of plant collections, he proposed eight such centers, accounting for most of the world's cultivated crops. Modern investigators conclude that some of Vavilov's centers can still be so considered, for in those regions a number of plants and animals were domesticated within a relatively small area and diffused out from a center. Harlan (1971, 1976), for example, retains Vavilov's concept of narrowly circumscribed centers of plant crop origin in the Near East, North China, and Mesoamerica, but has designated large areas of scattered, independent origins and distributions in Africa, Southeast Asia and the South Pacific, and South America as "noncenters" on his map. The increasing tendency of anthropolo-

[5]Anthropologist John Pfeiffer (1977) tells of seeing a boy carefully pulling out weeds in a tropical "garden" in the Amazon Basin, in which it seemed that nothing but weeds were growing. Pfeiffer found that every one of the more than thirty different species left was a source of food, seasoning, dye, medicine, poison, perfume, fiber, or other useful product. Such tropical gardens were primitive but effective agricultural experiment stations where people learned about new plants and ways of growing them.

gists and archeologists in recent years to recognize widespread independent origins of agriculture fits in well with the more or less simultaneous and worldwide beginning of agriculture at around 8000 B.C. (Cohen, 1977).

## SEED CULTURE IN THE OLD WORLD

The second major type of Neolithic agricultural system to develop was *seed culture,* the propagation of plants by planting their seeds rather than tubers, corms, rhizomes, and cuttings. Seed culture was the characteristic mode of cultivation in the drier subtropics of both hemispheres. Approximately 8,000-10,000 years ago, in the valleys of the Tigris and Euphrates rivers in the Near East's Fertile Crescent, the Indus River of northern India and Pakistan, the Huang Ho (Yellow River) of northern China, the Tehuacán Valley of southern Mexico, and a few other hearths, Neolithic man began to scratch the soil and plant seed. A higher degree of social organization was required in dry climates, where irrigation was necessary and foods had to be stored for long periods during the rainless or cold seasons when crops could not grow, than in regions where harvests extended throughout the year. The development of civilization was favored where food was potentially abundant but where great effort was required to obtain it. World agriculture obtained its greatest impetus, and the major civilizations found their origins, in regions of seed culture.

Particularly destined for exploitation were the cereals. Areas where seed culture developed were characterized by very marked wet and dry seasons. Most of the ancestors of what are now our cultivated cereals grew in soil that dried out completely in summer drought. (In Tehuacán, winter is the dry season.) They had to germinate and grow quickly when the rains came in the autumn and spring, and they had to mature their seeds before the ground dried out in summer. Even before selection by man, such plants had developed large seeds with ample food reserves. Large seeds could resist drought and support rapid growth during the brief rainy season. Neither plants with small seeds nor perennial herbs, shrubs, and trees could survive under such conditions (Hawkes, 1970).

The ancestral cereal plants were attractive weeds that tended to colonize the bare ground around man's dwellings. Weeds are favored by disturbed earth such as found around human habitations. It was necessary to eliminate other plant competitors in order to protect and encourage these promising food plants. After domestication, not only cereals but all domesticated crop plants had decreasing ability to compete with wild varieties, as artificial selection progressed. Man modified plants (and animals) to suit his own needs, rather than their unique ability to compete under natural conditions. Today herbicides (weed killers) eliminate competition from wild plants and are probably the most economically important of all pesticides.

Seed culture developed most rapidly in regions with diverse ecological niches provided, within a reasonably limited area, by a diversity of soils and climates ranging from rich alluvial river bottoms to foothills and on to high mountains, as, for example, the Zagros area of western Iran. In the highlands of this region, in the earliest indisputable village-farming communities so far excavated (Jarmo and Tepe Sarab), archaeologists found the remains of two-row barley (cultivated barley now has six rows of grain to a spike) and two forms of domesticated wheat. Among domesticated animals were goats, dogs, and possibly sheep. The first farmers who grew both wheat and barley must have lived in the highlands, for wild wheat was native to the highlands of the Fertile Crescent from 2,000 to 4,300 feet elevation. (Wild barley was endemic from central Asia to the Atlantic, but the earliest farmers never cultivated barley alone.)

In wild grasses, especially those related to wheat, the mature axis or rachis is brittle and shatters, scattering its seeds to the ground. The principal contribution of Neolithic farmers to agriculture was the selection of nonshattering grains. Wherever stone sickles were used, grain had already been domesticated, for wild grain could not be harvested with sickles. Nonshattering grains could not be spread by wind; man and most grains had become interdependent.

Neolithic farmers must have found it advantageous to move their wheat cultivation down from the mountain slopes to more level ground, to where the water supply was more dependable, and better accommodations for human habitation were present. Plants brought from foothills to valleys were then subjected to tillage, irrigation, and variety selection (Braidwood, 1960).

Irrigation had not been necessary in the uplands, but was necessary in the lower and more arid alluvial river valleys. At first farmers made do with small-scale irrigation systems involving breaches in the natural embankments of the rivers and with uncontrolled local flooding. In the preurban society of southern Mesopotamia, small communities were scattered along natural watercourses. Fish were obtained from rivers and marshes and the latter also supplied reeds used as building materials. The date palm was cultivated and yielded large and dependable supplies of fruit, as well as building material. Large-scale canal networks were begun only after the advent of fully established cities (Adams, 1960).

The Zagros area provided a rich diversity of plant species, a fact that has been amply demonstrated by archeological findings. Such important field crops as wheat, barley, rye, oats, several millets, pea, chick-pea or garbanzo, lentil, vetch, horse bean, and wine grape, were cultivated. Cultivation of grains antedated cultivation of fruits (olives, grapes, dates, figs, and pomegranates) in the Near East by several millennia. The earliest fruits were those that could be readily propagated vegetatively, as by cuttings or offshoots (in the case of the date), which obviated sophisticated techniques such as budding and grafting (Zohary and Spiegel-Roy, 1975). Since fruit trees do not start bearing fruit for several years, they would be one of the factors favoring a settled type of life.

There were also important Neolithic agricultural developments in the Huang Ho Valley and the southern part of northeastern China, where the soybean was the dominant agricultural contribution (Li, 1970). Likewise in tropical southeast Asia a seed culture (rice cultivation) progressively replaced an indigenous form of vegeculture (yam and taro), utilizing irrigation where necessary (Sauer, 1952; Harris, 1972).

## SEED CULTURE IN THE NEW WORLD

The area of earliest Neolithic seed culture in the Western Hemisphere was the Tehuacán Valley of southern Mexico, where maize (corn), beans, squash, pumpkins, green peppers, avocados, and other indigenous crops were grown. As in the Fertile Crescent of the Near East, and contemporaneously in human prehistory, agriculture began to be practiced in a semiarid area that was surrounded by foothills and mountains that had increasingly greater rainfall as elevation increased, providing a multitude of ecological niches and a correspondingly varied plant life. The Tehuacán type of agriculture was to be duplicated in many areas of Mexico, Central America, and South America along the entire axis of the Andes and the river valleys along the Pacific Coast. Everywhere it led to prosperous pre-Columbian communities. Successful agriculture was then, and still is today, the forerunner of economic development, a fact that is too often ignored by political leaders throughout the world.

Unlike the situation in Southwest Asia, in the areas of the New World where agriculture originated, no food plants grew in extensive stands and few animals were amenable to domestication. In Tehuacán, for example, wild maize (teosinte) was a low-yielding plant that required thousands of years to begin to show its ultimate amazing potential. Favorable genetic changes required considerable crossing and backcrossing, a slow development as it occurs in nature. Maize is wind-pollinated and favorable mutations obtained constant infusions of wild pollen, retarding their development and spread. Maize was deficient in protein, edible legumes were not abundant, and there were no herd animals to make up for the protein deficiencies. Mobile hunting-gathering populations of low density were the rule until the maize-bean-squash *milpa* was developed on an adequate scale to support sedentary groups around 1500 B.C. (MacNeish, 1964; Bender, 1975).

The peoples of the New World had the same potential for cultural evolution as those in the Old World. Their civilizations were going through the same evolutionary processes, but cultural evolution was retarded by about four millennia by the long time required to bring corn to its full potential as an agricultural crop. This was in sharp contrast to the ease with which wheat and other small grains could be domesticated in the Old World. Peoples of the Eastern Hemisphere also had the advantage of effective draft animals, as well as herd animals for meat and milk. This probably accounts for the cultural gap between the peoples of the two hemispheres when they made their first contacts from the late fifteenth to the early seventeenth century, and consequently, in the rapid demise of the New World cultures. In a conflict between two peoples of approximately equally advanced culture, the vanquished can usually sustain their civilization and sometimes even succeed in imposing their culture on the conquerors. However, there was too much difference in the degree of cultural development between Europeans and Amerindians to allow for this kind of an accommodation (Weatherwax, 1954; Fehrenbach, 1974; Harris, 1977; Pfeiffer, 1977).

The regions of the New World to which we now refer as "Latin America" supplied us with such important crops as corn, potato, sweet potato, manioc, kidney and other beans, peanut, cashew nut, sunflower (also native to the United States), tomato, green pepper, squash, pumpkin, Jerusalem artichoke (also native to the States), avocado, pineapple, papaya, cherimoya, guava, cotton (*Gossypium hirsutum* and *G. barbadense*),[6] and tobacco, as well as gourds, agave, prickly pear, arrowroot, maté or Paraguay tea, quinoa, chayote, jicama, soursop, sapota, and custard apple. Twined cotton fabrics from the El Riego cultural phase of the Tehuacán valley, dated between 7200 and 5000 B.C., may represent the beginning of textile fabrication in Mexico (Smith, 1965). The area now comprising the United States contributed the Concord grape, cranberry, blueberry, and pecan.

In sharp contrast to the eventual abundance of its food crops and fiber plants, the New World contributed only the turkey, the llama, and the alpaca among useful domesticated animals! The turkey was domesticated by the Aztecs, then was taken to Spain by the early explorers. In Europe it was confused with the African guinea fowl, which had arrived in Europe via Turkey (Kramer, 1973). In a sense, the New World can also claim the horse, for the horse family (Equidae) originated in North America.

[6]Although cotton is indigenous to the Old and New Worlds, the commercial cottons now grown throughout the world were derived principally from the species and varieties cultivated by prehistoric peoples of the Western Hemisphere.

About a million years ago the horse emigrated from North America to Asia across the "land bridge" then connecting the two continents during one of the periods of widespread glaciation, when much of the earth's water was stored in enlarged polar icecaps and continental glaciers, lowering the level of the ocean by about 300 feet. The horse then dispersed to all continents except Australia and Antarctica. About 400,000 years later the horse again returned to North America and became abundant in the late Pleistocene, but inexplicably disappeared as late as 6,000 years ago. Its bones can be found in dwelling sites of North American aborigines. The horse was then reintroduced by man into North and South America and introduced into Australia.

Proteins of cereals lack certain amino acids essential to human nutrition. In order to sustain themselves, the members of an agrarian culture must be able to obtain animal or plant protein that complements cereal protein in such a way that the total diet contains all required amino acids. The principal plant protein supplements to cereals among prehistoric peoples were derived from legumes: peas, chick-peas, lentils, horsebeans, and soybeans in the Old World and navy beans, kidney beans, lima beans, and peanuts in the New World.

Sauer (1952) considered southern Mexico and western Central America to be the hearth for the dispersal of New World seed culture. Maize was the principal source of calories. It was generally grown in a symbiotic "maize-beans-squash complex." The three crops were commonly grown in the same milpa, thereby ensuring a diet with good protein balance. Corn plants grew tall and had first claim on sunlight and moisture. Bean vines climbed up the corn stalks for their share of light and their roots supported colonies of nitrogen-fixing bacteria. Squashes or pumpkins grew prone on the ground and completed the ground cover. Presumably they minimized weed growth. These three crops were grown together as far north as the lower St. Lawrence River and in such adverse conditions as the margins of deserts inhabited by the Hopi Indians, where rains were scant and late, and the summers short with cold nights. This implies an amazing degree of crop selection practiced by the pre-Columbian peoples of North America. The time span involved, of course, can be measured in millennia.

Corn culture also spread south as far as Peru and Brazil. Figure 1 shows an ear of corn excavated in 1957 from an ancient coastal-Indian grave in a hacienda near Huaral, Peru. The grave was 10 feet below the present ground surface. The ear was 2.5 inches long. Radiocarbon dating of the corn and other organic material (cloth, spindles, hair, etc.) indicated that the material was approximately 500 years old.

The protein in corn not only is not abundant, but, as in other cereals, it is also deficient in certain amino acids, particularly lysine. Beans are deficient in methionine. Corn and beans, however, complement each other to provide an adequate supply of amino acids. Likewise, cooking techniques for corn in which alkali is used in making the cornmeal (as in preparing tortillas) enhance the balance of essential amino acids and free the otherwise almost unavailable niacin, thereby minimizing the risk of pellagra.[7] This would be a vital factor in any civilization in which corn is a major component in the diet (Katz et al., 1974).

The aboriginal farmers of the New World perfected four species of beans, the two most important being the common kidney or navy bean (*Phaseolus vulgaris*) and the lima bean (*P. lunatus*). They had a high protein and oil content and were so superior to the pulses (leguminous crops) of the Old World that, after the discovery of America, they soon became widely used from western Europe to the Orient.

[7]Pellagra became a serious medical-nutritional problem in areas of the Old World where maize was introduced as a single food rather than as a part of a maize-bean-squash complex (Grivetti, 1978).

*Fig. 1. Ancient corn from an Inca grave near Huaral, Peru.*

Squash was originally domesticated for its seed. Eventually cultivated variants were selected for their starchy and sugary flesh and large forms were developed, but the seeds were always eaten. The large-fruited squash races are to this day a principal staple food and their seeds are a source of protein and oil in many native and mestizo communities of Mexico and Central America. Roasted seeds may be eaten out of hand or mixed into meal or sauces. Squashes have separate female and male flowers and the latter, rich in carotene, were picked and used in stews, soups, and salads—a valuable source of vitamin A. Proper thinning of squashes resulted in maximum size for the fruits that were left to mature, and the young fruits served as green cooked vegetables. Many New World seed plants, such as corn, beans, amaranth, and chenopods, are used in various stages of their growth as vegetables, as well as for their seeds. Something may be taken out of the traditional Latin American milpa during the entire growth season, unlike with Old World seed culture, in which no harvesting of a crop takes place until some traditional time of harvest. Many New World crops, such as corn and squash, can await the convenience of the farmer. Therefore harvest festivals are less important in the New World than in the Old, and have a less conspicuous position in the agricultural calendar.

During the same period of human prehistory, man in both the Old and New Worlds learned to recognize plants and animals of special value to him, and he began planting or feeding them and discouraging their competitors. By selection of superior strains he eventually changed plants to the point at which in some cases their botanical antecedents were no longer traceable. Man's early success in this important enterprise is attested to by the fact that there are today very few economically important plants or animals for the origin of which we have an historical record.

The advanced development of maize (corn) as first seen by white men in America in the fifteenth century was probably the most amazing of all prehistoric contributions to agriculture. Among domesticated animals, sheep probably present the most striking example of what was apparently prolonged conscious selection of livestock for human needs. The ancestors of our domestic sheep had no wool, or at least none that could be used for making cloth (Lowie, 1940). To the extent that it is sustained by agriculture, civilization rests on a foundation laid by prehistoric peoples; they deserve our admiration and respect. No domesticated plant or animal of major importance has been added to the bountiful legacy handed down to us by peoples largely unknown to history.

## DOMESTICATION OF ANIMALS

As with plants, the domestication of animals played a vital and fascinating role in the development of civilization. It happens that cattle, sheep, and goats were indigenous to the same region (Asia Minor) as wheat and barley. Their bones are found in early village sites of seed farmers. Herd animals were domesticated by sedentary seed planters, not nomadic hunters as formerly believed. Just as household and dooryard animals (dogs, pigs, chickens, ducks, and geese) may be associated with tropical vegetative planters, herd animals, with the exception of the caribou and possibly the horse, may be associated with seed planters (Sauer, 1952). Domestication of animals is most viable when practiced in conjunction with plant cultivation (Bender, 1975).

The original reasons for keeping animals may not have been practical ones. When first domesticated, sheep gave little if any indication of the luxuriant fleece that would someday be of such great value to man. Likewise it is unlikely that Neolithic man, who domesticated the cow, would have foreseen the day when the cow would yield abundant milk after suckling its calf (Lowie, 1940). Yet when they became domesticated, animals provided man with wool, hides, milk, and meat.

Man has an inborn proclivity for keeping pets and this may account for the beginnings of animal domestication. During a critically impressionable period in its infancy, a young animal has a tendency to become attached to people as a result of "imprinting," the tendency to follow the first living thing it sees and hears. Some animals may have had human wet nurses (Braidwood, 1975).

Further insight into basic requirements for domestication might be found in Alistair Graham's *The Gardeners of Eden* (1973). He notes that the generally aggressive African buffalo, when living in and around the Waukwar Ranger Post in Uganda's Murchison Falls National Park, show no fear or aggression toward humans. The inhabitants of this small village are not allowed to molest wild animals or cultivate crops that could be destroyed by them. Women drive the buffalo away with broom sticks and children ride the animals. Graham believes that such a state of mutual indifference between man and wild animals was required before the latter could be domesticated. One factor that could have encouraged the mutual indifference that allowed wild animals to become tame was totemism—a recognition of a certain object, plant, or animal as a totem—in which case it would not be killed. Milking such animals might have been the first example of exploitation of some types of domesticated animals by humans.

Cattle are now the most important of all livestock animals, supplying about 50 percent of the world's meat, 95 percent of the world's milk, and 80 percent of the hides used for leather. They are also important draft animals. There are seven genera in the cattle family, including the bison, yak, and the Indian and African buffalo, as well as the domestic cow. Nothing is known of the initial phases of the domestication of cattle, but they probably originated in Central Asia and spread to Europe, China, and Africa. Friezes found by archeologists show that cattle were domesticated in the Mideast not later than 9000 B.C. Vedic hymns and Sanskrit writings mentioned milk and some dairy products, indicating that they were consumed in India as early as 6000 B.C. According to Simoons (1978), sheep were domesticated in the Near East about 9000 B.C.

The horse provides us with an example of those occasional lapses in human ingenuity and resourcefulness that defy explanation. In the ninth century A.D., iron shoes, nailed to the hoof, were invented. But a garrote strap formed of a large band of pliant leather "cravatted" the neck of the horse, without contact with the long structure of the shoulder (Prentice, 1939; White, 1962). Not until the tenth century was the rigid shoulder collar invented, probably in France. The broad breast collar did not appear until the twelfth century and was probably of English origin (Prentice, 1939). Previously the horse could pull only four times more than a man (and ate four times more food), but after the introduction of the harness and collar the horse was found to be able to pull fifteen times more than a man.

Turning to the Western Hemisphere, it is equally difficult to understand why the Aztecs, Mayas, and Incas, relatively advanced in many of the arts and sciences, never utilized the wheel in transportation, the plow for turning the soil, or the arch in construction. Failure to invent the wheel and plow may be the result of failure to domesticate a draft animal that would make these inventions useful (Bronowski, 1973). Possibly no animal was available in the New World that could have been as remarkably exploitable for this purpose as the Old World horse and bullock. Without the concept of the wheel and axle, New World man also lacked a model for other inventions that depend on motions of rotation that could be propelled by inanimate forces (water and wind), such as instruments for grinding corn and pulleys for drawing water from underground sources.

## IRRIGATION

Irrigation appears to have had an important influence on the development of incipient civilizations. The rivers that provided the rich alluvial soils in which Neolithic civilizations had their roots also provided water to sustain them. A river flowed through the biblical Eden "to water the garden." Mother Nile, with her annual autumn floods, fed by the monsoon rains of the Abyssinian plateau, left water and mineral nutrients in the fields of all of arable Egypt, thereby nurturing a series of great civilizations extending back to antiquity. Silt deposited by the Nile amounted to only about a twentieth of an inch (1.3 mm) per year, providing sufficient renewal of soil nutrients (and leaching of salts) without excessive siltation. Some authorities believe this to be the reason for the durability (6,000 years) of the successive Egyptian civilizations of native and transplanted cultures (Dale and Carter, 1955). Irrigation engineer Paul Keim informed me (personal communication) that probably the greatest value of the annual flood was in leaching away excessive salts, a function now lost since the building of the Aswan Dam. That dam has brought many other adverse environmental consequences, some as yet to be fully assessed. The basic human dilemma for all the world to ponder is this: the dam made possible a 5 percent increase in cultivated land during a period when Egypt's population increased by one third!

Extension of irrigation beyond the reach of the annual river flood led to salt intrusion and waterlogging. Lack of adequate drainage and siltation of reservoirs, canals, and harbors led to the disappearance of most of the civilizations developed through irrigation. The average life-span of these civilizations was forty to sixty generations (1,000 to 1,500 years). Egypt was a conspicuous exception.

The transient nature of most irrigated communities did not detract from their importance in the origin of civilized life. Because so many of the ancient civilizations developed in arid or semiarid regions, irrigation must be included among the most vital of their accomplishments in both the Old and the New worlds. The Incas, for example, constructed canal systems more extensive than any that have since been developed in the same region. The Hohokam Indians of Arizona likewise constructed a remarkable canal system. Areas that required extensive irrigation and drainage canals, such as Mesopotamia and the Nile Delta, also developed the greatest early civilizations. Annual deposits of mud replaced any soil that may have been lost by erosion, and made "shifting agriculture" unnecessary. So it was desirable to extend the canal network and to enlarge the community.

Even when relatively short lived, civilizations supported by irrigated agriculture had an irreversible effect on the cultural evolution of mankind. Diverting river water for irrigation purposes, sometimes for great distances, and draining marshes or building dykes, encouraged community action and also led to the development of considerable engineering skills. Likewise, mathematics and astronomy are said to have developed in agricultural areas to provide the basis for measurement of land, time, and season.

When the life of a community became dependent on plant crops, a calendar was necessary to ensure the success of the farmer in preparing his ground, sowing his seed, and harvesting his crop at the right time. Some prehistoric Egyptian genius used the sun and the heliacal rising of a star (Sirius) to measure the length of the year in days.[8] The Egyptian year of 365 days was six hours short, an error we now correct with our leap years, but this slight error made possible our calculation of the probable year of the introduction of the Egyptian calendar—4236 B.C. (Childe, 1956).

## AGRARIAN PEOPLES IN NORTHERN EUROPE

Receding glaciation farther north, which resulted in gradually increasing aridity in subtropical Asia Minor and consequently less bountiful hunting and gathering and greater incentive for irrigated seed culture, simultaneously caused northern Europe to become more conducive to hunting and gathering. In any case, wild ancestors of the principal cereals were not indigenous there. North Europeans, with no necessity for cities, were still living in small camps, and a young couple preferred to leave the camp, clear a plot in the forest, and "do its own thing," free from restraints and supervision of its elders.

By the time agrarian peoples had drifted west and north into northern Europe their habits had been changed by the shorter and cooler summers and by the limitations set on their traditional seed culture by forest and woodland. By the time they reached the Baltic and North Sea lowlands they were no longer principally cultivators; they had become stockmen. Later, wheat and barley appeared and with them, as weeds, came rye and oats. Primitive harvesting and winnowing did not separate weed seeds from grain. This may have been a fortunate circumstance, for as the growing of cereals

[8]Every year at about the time the Nile flood reached Cairo, the last star to appear on the horizon before the light of dawn obscured all stars was Sirius. This occurred roughly every 365 days and was adopted as the starting point of a year of that length (Childe, 1956).

spread northward into lands of acid soils and with colder and wetter summers, volunteer rye and oats did relatively better than barley and wheat and eventually became cultivated crops. It is perhaps significant that the Scandinavians became bakers of rye biscuits rather than wheat bread.

Dairying became important in the North Atlantic lands and, with dairying, calves increased the meat supply. Agriculture's limiting factor was the amount of grazing available during inclement weather and the amount of hay that could be stored for winter. Land was cleared more for pastures and meadows than for planting seeds. Stable manure was returned to the fields, maintaining fertility. Topsoil was not lost as it is in seed culture and cultivation. This was the agricultural system maintained in Europe until the seventeenth and eighteenth centuries, when the potato was introduced, along with stock beets, field turnips, and clovers (Sauer, 1952).

## MEDIEVAL FARMING IN ENGLAND

The medieval farming system of England is the one best known and of greatest interest to English-speaking peoples. It revolved around the nature and operation of the *manor,* a unit of English social, economic, and administrative organization in those times that consisted of an estate under a lord enjoying a variety of rights over land and tenants, including the right to hold court. Usually there were tenants of varying degrees of freedom and servitude. The manor was largely economically self-sufficient. Other European nations had similar units of social, economic, and administrative organization, varying in specific features from region to region (Fussell, 1976). The lord of the manor usually had land in the open fields as well as his home farm inside the fence in which the manor house was located. Along the village road were the homes of the freemen, bondsmen, craftsmen, cottagers, and serfs. All held land in "strips" in the open field and under different conditions of tenure. Even the best of the homes were humble, and those of the serfs were only earth-floored, chimneyless sheds of one room, divided in two by a wattle—a partition of poles interwoven by twigs and reeds. The partition separated the family from its cows, pigs, and poultry. Serfs were at the bottom of the hierarchy of wretchedness, even worse off than the bondsmen. They had no legal rights, worked for a pittance or nothing at all, depending on the generosity of the lord, and could not leave the manor. All tenants of the manor paid rents in labor or grain and honey. (In the absence of sugar, honey was the sweetener and mead—fermented honey and water—was the favorite drink [Franklin, 1948]).

Farmers knew that grain could not be grown successfully year after year in the same soil. As soon as one field was sown with wheat or rye, the plow team (eight oxen to a plow) prepared a second field for barley, oats, beans, and peas, and then a third was plowed and allowed to lie fallow for a year. The fallow field was grazed by livestock all winter and spring and the resulting manure was plowed under during the summer in preparation for planting in the fall.

Farmers did not fence in their strips in this sytem of open-field agriculture; they were required only to put up a fence where their strip abutted a road or the edge of the field. The farmer's fence, made of posts and interwoven brush, joined with his neighbor's along road or field edge, but there was no fence between the strips belonging to the different farmers. The manor's fenced-in area was called a *commons.* At harvest time the farmer and his family cut the grain with sickles, bound the sheaves and set them up into shocks, and, on a day of great rejoicing, they carted the grain to the home barn. The grain was threshed with flails. There was a brief period of harvesttime festivity; then the church bells rang to announce that all farmers could release their animals onto the stubble of the harvested strips and the grass of the fallow strips. (Uncultivated strips in hilly or marginal land formed a commons for grazing only.)

The results of open-field agriculture were described by Franklin (1948), who was himself a farmer and was descended from a long line of English farmers:

> For the general appearance of the animals was pitiful, and in their half-starved condition diseases spread through the flocks and herds of the whole village without any possible remedy by the individual. No one attempted to improve the grazing or drainage of the commons, to do so simply amounted to making a present of the improvement to the whole village with little benefit to the improver himself. What was everybody's responsibility was nobody's job.

This is a problem that has universal application, extending beyond agriculture, and is generally referred to as "the tragedy of the commons" (Hardin, 1968).

The British recognized their agricultural problem and began enclosing individual properties about A.D. 1200, and by the year 1800 enclosure became compulsory by an act of Parliament. Farmers were then able to introduce turnips, potatoes, and clover as farm crops, providing winter food for themselves and their animals. They could also plant crops in their strips that did not mature at the same time as those of their neighbors. Franklin points out that "inside his own fences the enterprising farmer could do all the things he wanted to do regardless of his neighbors, and his success or failure was dependent on his own efforts and ingenuity." Franklin considered this to be "the greatest of all changes in the history of farming." It marked the beginning of "agrarian individualism" in farming. There was incentive for improving livestock breeds. The British gained universal recognition as breeders of livestock and attracted buyers from all over the world who paid thousands of pounds for animals they wanted to improve their own stock.

The end of open-field (commons) farming resulted in landlords and yeoman farmers making fortunes, but many "small holders," not able to buy enclosed land, lost their rights of commons, and became farm laborers without the profits of cattle, sheep, hens, and geese they were previously able to sell.[9]

The life of the farm laborer or indentured husbandman in the seventeenth century is described in Gervase Markam's *Farewell to Husbandry,* first published in 1623, and should be read by those who believe that mechanization has led to a dehumanization of working people in farms and factories. The tasks that the lord of the manor expected his employee to accomplish during a day were outlined in great detail, along with the time that should be required for each task. For example, he was left in no doubt as to how many acres of different types of soil he should be able to plow in a day, or how many acres of grain he should be able to cut. According to this schedule, he should rise before 4 A.M., and, after giving thanks to God for his rest, he should clean the stables, rub down the cattle, curry the horses, feed all animals, and prepare everything for the day's work. Then he should spend a half-hour for breakfast and begin plowing, harrowing, harvesting, or whatever the seasonal activity might be. He was allowed a half-hour for dinner (lunch) and then should finish his field work, rub down his oxen or horses, and feed them.

[9]On the other hand, some authorities believe that changes unfavorable to the petty peasantry may have been localized. They point out that enclosure greatly expanded the area under cultivation and also extended the labor-absorbing crops such as turnips and legumes, thereby increasing the numbers employed in farming (Mingay, 1963).

> . . . and by this time it will draw past six of the clocke, at what time he shall come to supper, and after supper he shall either by the fire side, mend shooes both himselfe and their family, or beat and knocke hempe, or flaxe, or pick and stamp apples, or crabs for cider or verdjuice, or else grind malt on the quermes, picke candle-rushes, or do some husbandly office within doores till it be full eight a clocke: Then shall he take his Lanthorne and candle, and goe to his cattle; and having cleansed the stalles and plankes, litter them downe, looke that they be safely tyed, and then fodder and give them meat for all night, then giving God thankes for benefits received that day, let him and the whole household goe to their rest till the next morning.

One should not infer that the lord of the manor was insensitive to the personal needs of his tenants and laborers. Apparently the sixteen-hour workday was not enough to keep them in shape; exercise, physic, and an occasional bleeding were recommended.

## INFLUENCE OF THE INDUSTRIAL REVOLUTION

With the beginning of the Industrial Revolution, small-cottage industries were destroyed in England and rural areas became depressed. Many rural young people moved to the cities or to the colonies. In some parishes, 70 percent of the villagers received parish relief. The end of the Napoleonic Wars in 1815 brought an end to high prices for farm produce. Many large and small farmers lost their property. Finally, in 1834, wages were raised by an act of Parliament and able-bodied men who chose to live on the parish rates of pay had to enter a workhouse, called "Poverty Prison," and submit to its harsh regulations. For example, husbands and wives were separated and lived in different buildings. This ruthless policy reduced taxes for parish relief, enabled farmers to employ labor again, and restored confidence in agriculture.

Development of agricultural colleges and experiment stations began a period of intensive application of science to agriculture. More practical drainage methods were instituted by the invention of cylindrical baked clay pipe. Based on the discoveries of the German chemist Justus von Liebig regarding plant nutrition, the British began to grind certain rocks and treat them with sulfuric acid to form phosphatic fertilizer, cheaper and better than ground bones. Sodium nitrate began to be imported from Chile and guano from Peru. The farmers learned that they still had to use farmyard manure to maintain soil quality, but the addition of cheap artificial fertilizers resulted in greater production.

Of the sixteen elements known to be required by plants, hydrogen, oxygen, and carbon are obtained from air and water. From soil and fertilizers, plants obtain "primary" nutrients (nitrogen, phosphorus, and potassium), "secondary" nutrients (magnesium, calcium, and sulfur), and "minor" or "trace" elements (iron, zinc, manganese, boron, copper, molybdenum, and chlorine (Donahue, 1970). Scientists discovered that none of the soil nutrients can be assumed, a priori, to be present in sufficient quantity for maximum productivity, even in virgin soils. All soils require treatment with chemical fertilizers if they are to be used intensively, even though they may not need them when first cultivated. At least nitrogen and phosphorus are required for economic yields, with potassium next in usual order of need. It is now known that in some poor, sandy soils, even some or all of the minor elements must be added regularly.

Scientists learned that leguminous crops increased fertility because their roots had nodules containing bacteria that extracted nitrogen from the air and turned it into plant food. Improved farm implements were invented, including the reaper and binder. The new agrarian structure, with large farms and landowners with sufficient wealth for investment in new technologies, made British farming of the middle of the nineteenth century the most advanced in the world. Great Britain had entered its "Golden Age" of agriculture (Franklin, 1948).

Scientific technology ushered in an agricultural revolution as momentous in its implications for mankind as the one brought about by our Neolithic forebears in ancient river valleys—before the dawn of history. But history repeatedly tells us that no matter how great the gains in agricultural productive capacity, population gains have been commensurate, so the balance between food sufficiency and famine has remained precarious. Surely the long record of this agonizing dilemma must have taught us by this time that "the quality of life for our descendants will be determined by how many of them there will be" (Handler, 1975). Yet again and again the ghost of Malthus comes back to haunt us. Will mankind ever be allowed to reap the full rewards of its genius?

## GROWTH IN AGRICULTURE ESSENTIAL

Sociologists, psychologists, and economists who consider bumper stickers to be good indicators of contemporary *Zeitgeist* might find considerable philosophical content in the succinct but somewhat exaggerated message one sees on some automobiles these days, reading "AGRICULTURE—THE ONLY ESSENTIAL INDUSTRY." Many thoughtful people in "developed" countries are becoming increasingly disenchanted with growth in gross national product (GNP), which includes all the bads as well as the goods we produce, and the meaningless personal accumulation of material goods as criteria of human progress and well-being. But growth in agriculture, forestry, and their related industries is actively encouraged by people of all political, ideological, sociological, and philosophical persuasions. An increasing number of economists are taking another look at the old physiocratic doctrine that only agriculture and forestry are able to create a significant net product of value. All other so-called "production" merely utilizes finite natural resources and ultimately must cease except on a recycling basis. Most of the energy for the production of genuinely *new* wealth is supplied free by the sun. According to Heichel (1976), American agriculture, which is relatively energy-intensive, nevertheless requires only 2.6 percent of the nation's energy budget that is supplied by man.[10]

Even the eschatological ruminations of the Club of Rome—widely known champions of Zero Growth, that is, zero undifferentiated[11] growth—call for an increase in agricultural production and suggest the diversion of industrial capital to provide for enrichment and preservation of the soil. This follows naturally from the fact that world population will continue to increase, particularly in those countries where the specter of famine already looms ominously whenever drought, floods, or plant pests and diseases result in less-than-average crop production. A minimum of 4 percent annual world increase in food production is believed to be necessary to feed the present world population of 4 billion and the 74 million new mouths added each year. This would leave a

[10]The added energy for processing, packaging, transporting, marketing, shopping for and preparing food for consumption, brings the total per-capita energy annually expended for food up to 13-15 percent of the nation's total energy budget. It is in these categories that energy savings could be most readily effected (Pimentel, 1976).

[11]The Club of Rome distinguishes between *undifferentiated* growth, i.e., the exponential growth exemplified by cancer cells, or by the human population whenever the birth rate is greater than the death rate, and *organic* growth, i.e., the logistic growth that characterizes organic systems, such as plants and animals, as they reach maturity. Organic growth is never static; for example, the human body renovates itself approximately every seven years (Mesarovic and Pestel, 1974).

substantial area of the economy for the "growth" that occupies the attention of traditional economists, for the production of food is a gigantic basic industry, involving about 60 percent of the world's population. In the United States, agribusiness, from farm to consumer, employs some 25 percent of the nation's labor force, and 25 percent of consumer expenditures goes for food and clothing derived from farm products.

The farms of the United States comprise the nation's largest single industry, with total assets estimated to be $731 billion, about three-fifths of the capital assets of all manufacturing. In 1977, gross receipts from marketing crops were $95 billion, more than the combined sales of General Motors and Ford Motor Company, which ranked first and third, respectively, among the nation's companies in terms of sales (Green, 1978).

In the United States, mechanization is the most conspicuous technology input in the "systems approach" that characterizes modern farming. However, of great importance are improved seeds and breeding stock, commonly involving hybridization; improved and environmentally sound tillage; improved water usage; fertilizer; chemicals to control weeds and fungi, and, in a pest-management framework, to control unwanted insects; and the balanced feeding of livestock. Advanced technology from other disciplines is increasingly utilized, a tendency perhaps most spectacularly illustrated by the eradication of screwworms and fruit flies by means of the sterility principle, "one of the most original scientific ideas of this century" (Gall, 1968).

The result of this unprecedented outpouring of advanced technology resource has become known as the American "agricultural miracle." The yield of corn, our number one crop, after remaining at from 22 to 26 bushels per acre for 140 years—up to about 1930—had increased to 80 bushels per acre by 1968. The yield of potatoes had approximately quadrupled, and the yield of wheat and soybeans had doubled. There are similar records of increased production efficiency for broilers, turkeys, beef, and pork. Production of milk per cow and eggs per hen had almost doubled in this period (Wittwer, 1970). Yield increase continued. In 1976 the average yield of corn in the United States was 5.5 metric tons per hectare.[12] Production (metric tons per hectare) in some other countries was as follows: Italy, 5.6; Canada, 5.4; France, 3.9; Hungary, 3.9; Yugoslavia, 3.8; Egypt, 3.7; Poland, 3.5, Rumania, 3.4; U.S.S.R., 3.1; South Africa, 2.3; Argentina, 2.2; People's Republic of China, 2.2; Brazil, 1.6; Kenya, 1.5; Mexico, 1.2; Indonesia, 1.1; Malawi, 1.1; India, 1.0; Nigeria, 0.8; and Philippines, 0.8 (USDA Agr. Stat., 1977). The great range in productivity points up a corresponding potential for increased yields in much of the world, particularly in the developing countries.

Americans spend, on the average, about 17 percent of their income on food, despite unprecedented worldwide inflation, compared with 24 percent in the 1930s. About 4.4 percent of the nation's population[13] produces the food and fiber for the rest of the nation, as well as much for export, bringing to its most bountiful fruition the trend begun by our Neolithic ancestors in various parts of the world, about ten millennia ago, when they produced the first agricultural surpluses. It was these surpluses that set man free to think and dream, to reflect on his origin and destiny, to chart the positions of stars and the movements of the moon and the planets, and finally to set his own feet on a heavenly body, sent aloft in our own generation and from our own blessed land. Now

[12]A metric ton is 2,204 pounds and a hectare (ha) is equivalent to 2.471 acres. In this book "ton" will refer to the metric unit.

[13]Only 5 percent of the labor force in the United States, and only 4.4 percent of the population, is employed in agriculture. This may be compared with the following percentages of the labor force on the land in other parts of the world: Western Europe, 13; U.S.S.R., 32; Latin America, 40; Asia, 58; and Africa, 73 (Heady, 1976). This technological triumph was not achieved without certain adverse socioeconomic consequences, however, as will be recounted in the following chapter.

he is reaching a step further, seeking to harness the eternal fire of the sun—nuclear fusion. Will he be punished for this awesome transgression—like Prometheus, who stole fire from the gods for the benefit of mankind? Greek mythology relates that Prometheus was then chained to a rocky crag, to endure agony for years, but was finally released—having learned humility—"with intelligence and power, human hopes and the will of heaven reconciled at last." The answer must lie in the use we make of knowledge, which increases exponentially and is our most valuable resource, and good judgment and wisdom, born of experience.

In this time of widespread pessimism regarding the human prospect, one nevertheless finds a source of optimism in the seemingly boundless human ingenuity for solving problems. The electronics and miniaturization revolution of the past sixty years, requiring very little energy and materials, is moving at an ever-increasing tempo (Abelson and Hammond, 1977). It promises to be of more enduring consequence than the now obsolete Industrial Revolution, based on profligate use of fossil-fuel energy and nonrenewable resources. It is one of the greatest intellectual achievements of mankind.

# 2 Above the Fruited Plain

*I know of no pursuit in which more real important service can be rendered to any country than by improving its agriculture.*
—George Washington

In this chapter we will observe, from a continental airliner, the topography, climate, soil, natural and human resources and technologies that produced and now sustain the "agricultural miracle." After this brief overview, in subsequent chapters the history and present status of agriculture in particular will be discussed in greater detail for each geographical region from east to west.

We start our transcontinental flight on August 27, 1976, under circumstances that are routine except for the fact that on this flight permission was obtained from the Federal Aviation Administration and from Flight Operations, United Airlines, for me to occupy a seat in the flight deck of a Boeing 747. Thereby I am able to take advantage of the years of observations of the flight crew to add to the fund of knowledge already obtained regarding agriculture, forestry, and natural resource features of the landscape from many transcontinental trips made by air and otherwise, as well as from extensive reading. The flight crew consists of Captain Jack Hanson, First Officer Wally White, and Flight Engineer Harry Metz. Courteous, friendly, and helpful in every possible way, they make the flight especially rewarding for me.

In no single flight can one see all the features to be discussed in this chapter, for some tend to be hidden from view by clouds, particularly in the East. Thus the observations recorded are a composite from a number of transcontinental flights. The recorded chronology, in Eastern time throughout, reads as if all features discussed were seen on this one flight. However, the flight path varies on different flights, depending on weather conditions. The path taken on this trip is shown in figure 2, which also delineates the geographic regions of the United States to be discussed in following chapters. The regions east of 98° longitude are as delineated by McMillen (1966).

12:15 P.M. Leaving Kennedy Airport we fly out to sea and circle back south of the amazing concrete labyrinth of Manhattan, situated on an elongated island purchased from the Indians by the Dutch, three and a half centuries ago, for the equivalent of

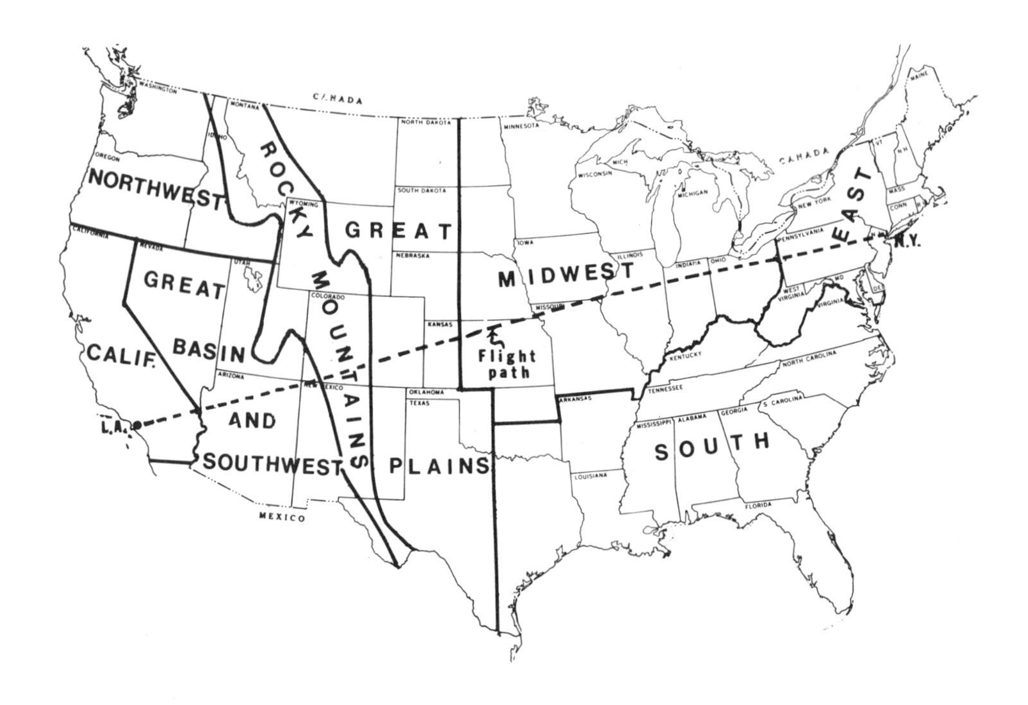


*Fig. 2. Flight path "From Sea to Shining Sea."*

about twenty-four dollars. We soon reach the foothills of the Appalachians and fly over a considerable distance of mountainous country because of an east-west turn in the range that begins in northern New Jersey and extends through most of Pennsylvania. The mountains are heavily forested and the valleys are devoted to general (diversified) farming. Although we will be flying at an altitude of 30,000 feet today, I have with me a Bausch & Lomb 7 x 50 binocular which enables me to clearly distinguish, for example, haystacks in fields in which they are not visible to the unaided eye.

As in my previous transcontinental flights, appropriate lines of Katherine Lee Bates's popular song, *America the Beautiful,* come to mind from time to time as they fit the scenes that unfold below "from sea to shining sea." From the words of that inspiring song I took the title for this book and the heading for this chapter.

## THE EASTERN FOREST

The reader may find figure 3 useful as a guide to the land resource potential of the regions along the route of the flight. The map will also be referred to in later chapters when the crop and forest resources of the various regions of the country are discussed.

On east-west flights, one is soon impressed by the immensity of the forested areas still remaining in the populous eastern states. Although by no means primeval, the forests of the region we are now flying over, and much of the eastern United States as seen from above, appear much as they would have to a pre-Columbian traveler if he could have availed himself of the magic of flight as we do today. A large portion of the eastern forest land has been cleared for farms and towns and a considerable amount has been culled of its valuable timber, devastated by fire, or turned into almost useless brushland. Nevertheless, there are about a quarter of a million square miles of mer-

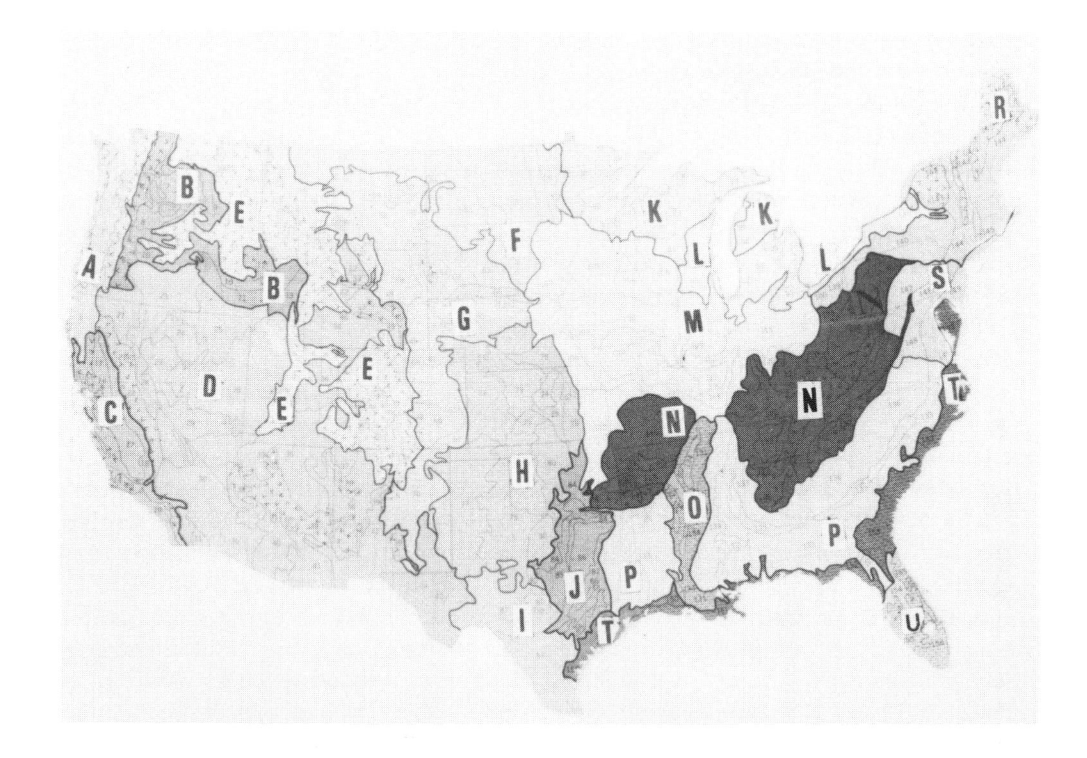


*Fig. 3. Land resource regions of the United States: A, northwestern forage and specialty crop; B, northwestern wheat and range; C, California subtropical fruit, truck, and specialty crop; D, western range and irrigated; E, Rocky Mountain range and forest; F, northern Great Plains spring wheat; G, western Great Plains range and irrigated; H, central Great Plains winter wheat and ranges; I, southwestern plateau and plains range and cotton; J, southwestern prairies cotton and forage; K, northern Lake States forest and forage; L, Lake States fruit truck and dairy; M, central feed grains and livestock; N, east and central general farming and forest; O, Mississippi Delta cotton and feed grain; P, South Atlantic and Gulf Slope cash crop forest and livestock; R, northeastern forage and forest; S, northern Atlantic Slope truck, fruit, and poultry; T, Atlantic and Gulf Coast lowland forest and truck crop; U, Florida subtropical fruit truck crop and range (from Austin, 1965).*

chantable forest lands in the eastern states, about half consisting predominantly of conifers in northern Maine, the Adirondacks of northern New York, the Lake States, and the southern states, and the remainder consisting predominantly of broadleafed (hardwood) species.

Nothing is a better indicator of soil and climate than the composition of vegetation. The eastern forest is not broken by great mountain ranges as in the West; the Appalachian range never reaches great heights. The region consists mainly of rolling country with good soil. The winds that sweep in from the Gulf of Mexico are laden with moisture and precipitate most of it during the growing season, unimpeded in their course by high mountain ranges as are the moisture-laden air masses in the West. Where good soil, favorable temperature, and favorable rainfall occur, the hardwood species predominate and the conifers are almost completely excluded (Shantz and Zon, 1924).

Our flight first takes us over forests that were originally predominantly a mixture of chestnut, chestnut-oak, and yellow poplar until we reach central Ohio, where the oak-hickory forests predominated up to the margin of the tallgrass prairie country, which first appears in small areas of Ohio and Indiana, then in a considerable portion of central Illinois, and finally in the enormous plains west of the Mississippi. The chestnut, chestnut-oak, and yellow poplar forest extended south into northern Mississippi and Alabama. Although these three species predominated, there were many others, probably more than in any other forest areas in North America. The American elm and the chestnut, once so prominent in the species mixtures, have suffered the greatest losses, sharing with other species the conversion of forest to farmland but, in recent decades, suffering additional losses from the ravages of disease.

The forest is a factory working for mankind, using energy supplied by the sun. In transpiration alone, the plants of the earth use five hundred times more energy every year than man obtains from all the fossil fuels and six thousand times more energy than he makes available in the form of electricity. In addition, enormous energy is used to create plant materials such as wood and to perform the carbon dioxide-to-oxygen transformation. It is beyond the technological capacity of man to duplicate more than a small fraction of such prodigious activity. A young, healthy tree can have a net cooling effect, through transpiration, equivalent to ten room-sized air conditioners operating twenty hours a day. Under the canopy of a large tree the air may be 20°F (11.1°C) cooler than elsewhere. Although most of the forest has been cut over two or three times, it is nevertheless beautiful, and young trees provide certain ecological and economic advantages. Whereas an old forest, filled with overmature trees, consumes as much oxygen as it creates, a forest of vigorously growing young trees each year consumes five to six tons of carbon dioxide and gives off four tons of oxygen per acre,[1] besides producing four tons of new wood, about twice as much as an old forest (AFI, 1976).

The forest not only is a major air conditioner; it also supplies us with valuable building material through a relatively clean manufacturing process. Besides the lumber cut from tree trunks, almost all wood residue is now converted into building panels, pulp, and paper products. Plastics, metals, and other materials useful for construction are derived from natural resources that are not renewable, whereas wood is renewable in quantity possibly three or four times greater than any that has been heretofore achieved. And when the usefulness of wood is over, it is biodegradable into its basic elements.

Paradoxically, the eastern forests, which today are held in such high esteem, were a formidable obstacle to the western movement of American civilization, which required two centuries to pass through them. Clearing the forest with ax, fire, and a yoke of oxen was slow, backbreaking toil, discouraging to all but the strongest and most determined people, but such was the nature of the early American backwoods farmer, generally known as the "settler."

## DEMOCRACY FORGED ON THE LAND

Even as late as the last quarter of the eighteenth century, 90 percent of all Americans were still farmers. The frontier, slowly moving westward through the seemingly endless forest, was as wild and hazardous as ever. The self-sufficient trans-Appalachian settlers, with only their bare hands, the ax, the plow, and a yoke of oxen, drew their suste-

[1]Some intensively grown agricultural crops rival a young forest in this respect. An acre of corn producing 100 bushels (5,600 lb; 2540 kg) of grain has been estimated to remove the equivalent of 7-8 tons of carbon dioxide and add 5-7 tons of oxygen to the atmosphere (PINM, 1972).

nance from the rich virgin soil. Although the frontier had its crooks and shysters, the settlers generally lived on a basis of true equality, where men and women rose to prominence in their community on the basis of merit alone. Democracy was being forged on the land. Even statesmen and other civic and intellectual leaders of the time—men like Washington, Jefferson, Franklin, Daniel Webster, and Henry Clay—had strong ties to the land. Jefferson, in particular, considered himself to be a farmer by occupation during his entire life, believing agriculture to be "the first and the most precious of all the arts." The founding fathers adapted the aristocratic and fashionable academic interest in agriculture, science, and commerce that existed in Europe in those days to the needs of Colonial and Revolutionary America, forming societies for their promotion and developing experimental plots on their own lands. They laid the political and social basis in the New World for the flowering of the eighteenth-century ideas of progress, human perfectability, rationality, and scientific improvements (Rasmussen, 1975*b*).

> The land made America. It wasn't so much what people took from it, but what they had to put into it. The land molded American character. It was the lure that drew settlers to America long after the dream of easy gold had been dashed. It was the motivation of the men and women who built this nation. The land sustained them and gave them hope. It also made them tough. It freed them from the bonds of European class society and swallowed up Old World traditions. . . . Those who found quick wealth took it and ran. Those who didn't stayed and looked for deeper value. It was the American farmer who found it, built a nation and made it free. [McLeod, 1975]

## HUMANIZED NATURE

As we leave the lovely Allegheny highlands behind, we fly over perhaps a fifty-mile stretch of what was originally a birch-beech-maple-hemlock forest that extended down like a finger from New York and northern Pennsylvania, where it was the predominant type of forest; then we return again to what was a mixture of chestnut, chestnut-oak, and yellow poplar forest, which will be in evidence until we reach central Ohio. Closer inspection would reveal that, where it has not given way to agriculture, much of the original composition of the forest remains. It becomes increasingly interspersed with rich farmland until, in the upper Ohio River valley, farmland predominates. Yet many farms retain a woodlot, and the uncultivated hills are clothed with the original tree species of the region.

Mile after mile of well-tended fields of green, gold, and brown unfold below us—a panorama of American agriculture at its luxuriant best—dotted at spacious intervals with the comfortable homes and large barns of farmers, each home surrounded by shade trees and often with a woodlot and pond nearby—a broad rural landscape of great beauty, punctuated with wooded hills and laced with tree-lined brooks and streams that eventually find their way to the mighty Ohio River beyond the southern horizon.

Most of us are confined to the city for the major portion of our lives and are apt to forget that, of the nearly two billion acres of the coterminous forty-eight states, about a third consist of pasture and grazing land and nearly a fourth are cropland (Clawson, 1972). To all but the occasional environmental purist, most of this agricultural land presents a pleasing impression of rural America. In our current understandable preoccupation with environmental despoliation by man, have we given enough thought to the beauty he has wrought? This man-made beauty is not necessarily obtained at the cost of ecological impoverishment. One of the most convincing examples of this fact is the captivating charm of rural Britain, which presents an ecological diversity far in excess of the original. Not notably endowed by nature either as to soil or climate, the land was made to bloom through the loving care of its people. The beauty of the British countryside has not been attained at the cost of agricultural productivity. For example, wheat yields increased from about 500 pounds per acre in A.D. 1100 to about 1,500 pounds in 1900. Yields almost doubled in the period 1930 to 1960, reaching 3,000 pounds per acre (Jacks, 1962).

In the mythical Garden of Eden, Paradise was lost with the advent of knowledge, thereby setting mankind on an irrevocable course, for better or for worse. It is idle to yearn for ecological "paradise" lost. In the real world it is necessary to consider not only the threats posed by human intervention to nature, but also its amazing recuperative powers and the extent to which they can be guided by intelligent effort:[2] "nature, although it can tolerate only so much abuse, is remarkably resilient and forgiving" (Carter, 1974).

The rural charm that so pleases the tourist everywhere, even in the most densely populated countries, is based on a special type of beauty that the French-American bacteriologist-ecologist-philosopher René Dubos (1972 *a,b;* 1973) calls "humanized nature." He says of his native French province:

> Ever since the primeval forest was cleared by Neolithic settlers and medieval farmers, the province of Île de France has retained a humanized charm that transcends its natural endowments. To this day, its land has remained very fertile, even though part of it has been in continuous use for more than 2000 years. Far from being exhausted by intensive agriculture over long periods of time, the land still supports a large population and a great variety of human settlements.
>
> What I have just stated about the Île de France is, of course, applicable to many other parts of the world. Ever since the beginning of the agricultural revolution, during the Neolithic period, settlers and farmers have been engaged all over the world in a transformation of the wilderness. Their prodigious labors have progressively generated an astonishing diversity of man-made environments, which have constituted the settings of most of human life. A typical landscape consists of forested mountains and hills serving as a backdrop for pastures and arable lands, villages with their greens, their dwellings, their houses of worship, and their public buildings. People now refer to such a humanized landscape as "nature," even though most of its vegetation has been introduced by man and its environmental quality can be maintained only by individual ecological management. [Dubos, 1973]

This is not meant to disparage the beauty of wilderness. It has a beauty and a meaning of its own that cannot be duplicated in humanized environments. Europe is a beautiful continent, but even its "wild" areas are no more than well-tended parks, easy to keep beautiful because of summer rains. The North American continent has the great good fortune of possessing many national parks, national monuments, and "wilderness areas" that are to remain wild in perpetuity.

1:22 P.M. We are now over central Ohio and have entered the eastern extremity of Region M shown in figure 3, the famous American Corn Belt. The portion of Region M east of the Mississippi was originally largely covered with a forest in which oak and hickory were the principal species. The forest extended westward to the tallgrass prai-

[2]In the Orient, many fields that have been cultivated for 4,000 years are still productive, in some cases yielding more than ever because of increasing knowledge of soil care and genetically superior plant crops (Love, 1971).

rie. Most of the trees have since been felled to make way for farmland. Ohio has 25 percent of its area still covered with forest, Indiana 17 percent, and Illinois 11 percent.

*Fig. 4. Farmland in central Ohio (courtesy Aerial Section of the Ohio Dept. of Transportation).*

Figure 4 shows the typical attractive landscape of central Ohio. Farther south and east are more hills and trees and fewer large farms. To the north and west there is a larger percentage of farmlands, but the land is very flat and there are few woods. The figure shows the usual dispersal of farmsteads wherever family farming predominates in the United States. Farmers tend to live on the farm rather than in rural villages as they do in Europe.

The early settlers could hardly have imagined the sight we see today. Every parcel of land not devoted to cities, towns, roads, and woods, is planted to crops. Most are field crops—corn, soybeans, and winter wheat—but there is also much land planted to truck crops and orchards. The availability of farmland increases as we fly west toward the Mississippi. In 1976, Ohio had 3,920,000 acres of corn, Indiana 6,300,000, and Illinois 11,690000; Ohio had 2,900,000 acres of soybeans, Indiana 3,300,000, and Illinois

7,600,000; Ohio had 1,750,000 acres of winter wheat, Indiana 1,670,000, and Illinois 1,900,000 (USDA Agr. Stat., 1977).

## AMERICA'S LAND PRODIGALITY

The European visitor flying over this route, and familiar with the small parcels of land he sees from the air in his own homeland, cannot fail to be impressed with the large size of our farms. Their large size is the result of the prodigality of the early federal land policy. The states over which we are now flying were originally western extensions of the original states along the Atlantic. These latter were glad to sell their "western lands" to the federal government, for they considered them to be worthless wilderness. The government acquired them through cessions, purchases, and treaties, and they were cheaply disposed of to American citizens. The states ceded 233 million acres of their western lands between 1781 and 1802 at an average cost to the government of 2.7 cents per acre. In the succeeding half-century the following federal acquisitions were made: in 1803, the Louisiana Purchase, 552 million acres; 1819, the Florida Purchase, 43 million acres; 1846, the Oregon Compromise, 181 million acres; 1848, the Mexican Cession, 335 million acres; 1850, purchased from Texas, 79 million acres; and 1853, the Gadsden Purchase, 19 million acres. Payments were made for all these acquisitions except for the Oregon Compromise (obtained by treaty with Great Britain), but the cost per acre for the two largest acquisitions—the Louisiana Purchase and the Mexican Cession—was only 4.4 and 4.9 cents per acre, respectively.[3] The most "expensive" land acquired was the Gadsden Purchase at about 53 cents per acre. In 1867 Alaska, "Seward's Folly," was purchased from Russia—365 million acres at 2 cents per acre (Dana, 1956).

Originally the policy of the federal government was to transfer the land to private owners as rapidly as this huge task could be accomplished, consistent with orderly development of the land. This policy was followed for nearly a century before the modern practice of combining federal ownership and private utilization was seriously advocated. The Continental Congress enacted an ordinance in 1785 that established the rectangular system of public-land surveys. Townships 6 miles square were divided into 36 sections of 640 acres each. Except for sections reserved for common-school and other purposes, lands were sold at auction for cash to the highest bidder at not less than a dollar per acre, but sales were slow. A section was the minimum acreage sold, and few immigrants had the minimal $640 in cash that was needed. Many were "squatting" on the land without payment or legal title. Subsequent acts of Congress raised the minimum price per acre to two dollars, reduced the minimum area that could be sold to 160 acres, and provided for payment in installments over a four-year period. Most of the settlers were delinquent in their payments, however, and others made the first payment just to legalize their presence long enough to exhaust the fertility of the soil or harvest the timber. Then they abandoned the land and moved on to repeat the process elsewhere. In 1820 both the minimum price and the minimum area were reduced. Public lands were sold to the highest bidder in half-quarter sections of 80 acres at $1.25 per acre, with full payment at time of sale. In 1832 the minimum area in private sales was further reduced to 40 acres, and $1.25 per acre remained the standard base price until

[3]The $15,000,000 asked by Napoleon for Louisiana was thought by many Americans to be excessive, only partly mitigated by the stipulation that the United States government should retain $3,750,000 to pay claims of Americans against the French government. Jefferson was widely ridiculed and condemned by friends and enemies alike, for having spent so much for "worthless territory."

1934. No restriction was placed on the maximum area that one person might purchase. This was a fortunate provision for western cattlemen, who required extensive ranges for successful cattle-raising (Dana, 1956).

The early Colonial economy was the exact opposite of what we have today. Land and natural resources were abundant and cheap; labor and capital were scarce. The latter were used sparingly and intensively, while land and natural resources were used as liberally and extensively as possible. This was before the era of contour plowing, strip cropping, and modern methods of fertilization. Soil depletion and erosion became sufficiently serious to cause concern to the more farsighted political leaders of the times. However, in those days a farmer could always leave depleted land and move west.

## THE AMERICAN AGRICULTURAL MIRACLE

The former forest land over which we are now flying in Ohio, Indiana, and Illinois, as well as the deep, dark soil of the former tallgrass prairie farther west, is some of the finest agricultural land on earth. If the European farmers could start with what was mostly relatively poor land and improve it over the centuries to become the most productive on earth, the American farmer of the far richer lands of the Midwest should do at least as well. In the technologically advanced countries, hand labor is one of the most costly inputs in farming, so capital, machines, and agricultural chemicals have been progressively employed, thus freeing former food producers for other walks of life in the industrial society. However, the less-mechanized but highly scientific agriculture of Europe and parts of the Orient, where small holdings of land are meticulously tended with "customized human care" (Schery, 1972), can also be highly productive. In such areas a greater percentage of the population resides on the land. There is a much higher input of agricultural chemicals than in the United States. Land availability, rather than manpower, may be the decisive factor affecting the ratio of production per unit of farm labor in some densely populated countries, even among countries of advanced technology. Labor-intensive agriculture, when well managed, can be highly productive. This kind lends itself to intercropping and an annual productivity per acre exceeding that which can be attained by the most highly mechanized type. It is particularly suitable for developing nations.

It is in the great quantity of food crop produced *per unit of labor* and his ability to feed the nation in abundance and provide for export, that the American farmer is supreme. In 1840, when our nation's population was 17.1 million, 4.4 million (25.7 percent) were farm workers. The average farm worker was supplying food for 3.7 persons (including himself) at home and 0.2 person abroad. In 1969, with a population of 202.2 million, there were still only 4.6 million farm workers—little change in a period of 129 years—but the average farm worker was supplying food for 39.1 persons at home and 6.2 abroad (USDA, 1970*a*). In 1975, with a population of 212.4 million, there were 4.3 million farm workers. The average farm worker was supplying food for 55.8 persons—42.2 at home and 13.6 abroad (D. D. Durost, correspondence). The changes in productivity per worker were principally due to advanced mechanization.

American consumers obtain their food for a lower percentage of their income than do consumers in any other country. The average American eats about three-quarters of a ton of food per year; 165 million tons of food are required to feed us all. Our farmers and commercial fishermen provide about 90 percent of it and the rest is imported. Located in the temperate zone, we have been favored by a wide range of climate and many types of land, making for a great variety of crops and livestock. We have a rich supply of natural resources and good workers, and have been financially able to expend vast sums for research and for applying research findings in agricultural practice. And incredible as it may seem in some parts of the world, there is no central agency to plan the production of all this food. Three million farmers make three million sets of decisions as to what crops and how many acres of each crop will be produced. A constantly increasing population has been amply fed by a declining number of farmers; yet a great amount of agricultural produce is sold abroad every year, constituting our greatest source of foreign exchange. This is the nature of the "American agricultural miracle."

## AN OHIO FARM

Our flight takes us over the flatlands of northwest Ohio and, at 1:29 P.M., just a little north of the small town of Ada in Hardin County, at 83°50′ longitude, near which is situated 115 acres immortalized in Wheeler McMillen's *Ohio Farm.* McMillen (1974) relates the farm life and activities on his family's farm during the early years of this century. He tells the story with a nostalgic sentiment and beauty that is bound to strike a responsive chord in those of rural background who share such precious memories.

The McMillen farm was about average in size for the Midwest general farm in those days. All the vegetables, butter, milk, and eggs consumed on the McMillen farm, practically all the meat, and some of the fruit, were produced on the farm. The farm kept the father and a hired hand busy and, in summer, a team of horses and a team of mules. Some families appeared to live comfortably on as little as 40 acres. But today at least 350 to 400 acres are required in the Midwest for a profitable farm unit, or even a full 640-acre section or more, depending on the crop or animal program followed. The manure produced by animals fed on the McMillen farm, along with their straw- and cornstalk bedding, was the source of almost all the fertilizer. The first commercial fertilizer applied was 2,500 pounds of acid phosphate in 1912.

McMillen describes in detail the threshing bees of the early twentieth century, when families helped one another in the labor-intensive job of threshing their grain, using an old-fashioned steam traction engine, fueled by wood or coal, and a "grain separator." Each family in the neighborhood cooperative project furnished a team and wagon and one or more men. Neighbors who had little time to visit during the busy spring and summer months were brought together at threshing time. Opportunity was afforded to catch up on all the neighborhood gossip and for "much gaiety and banter." The gaiety presumably did not extend to the mother in the family whose grain was being threshed, for upon her fell the awesome task of feeding the threshermen and feeding them well. She had to make preparations days in advance. It was before the days of refrigeration, and if the threshing outfit did not arrive when expected, the meat had to be canned or cold-packed and a new supply obtained.

## THE FAMILY FARM

Although there has been a trend toward very large farms of a thousand acres or more owned by individuals, cooperatives, or corporations, farming is still the only major industry where families make up the largest share of the labor force, constituting about two-thirds (NAC, 1975).[4] From the standpoint of human values, it is unfortunate that

[4]However, 5.5 percent of all farms control more than 50 percent of all farmland (USDA, 1974).

this arrangement has undergone rapid change, for the farm family is a rich social resource. Whereas formerly a father expected his son to eventually take over his farm, more recently sons and daughters have sought more rewarding jobs in the city and generally sold the farm they inherited to someone who needed more land. The small farmer, who loved farm life and tried to stay on the farm, but was not financially able to increase his acreage or properly farm it because of the rapid rise in the cost of equipment, became progressively more impoverished and discouraged. There emerged a continuous increase in larger-than-family-size farms that had an almost factory-type[5] system of operation, particularly in California, Arizona, southern Texas, and Florida.

Gaylord Nelson, senator from Wisconsin and a member of the Senate Finance Committee, informs us that food prices, often blamed for runaway inflation, have risen far less than the prices of most other products. Many consumers believe that farm-support programs are an indefensible expenditure of taxpayer money. Actually, it is the farmers who are subsidizing the consumer, which accounts for the fact that the average American spends far less of his take-home pay for food than people anywhere else in the world. The average American farmer has 390 acres of farmland in which he has an investment of about $200,000. Although efficiently producing a good product, he gets a low return for his labor and often little or no return for investment or management. He would be better off to place his $200,000 in a savings account and free himself of the hard work, financial risk, and dependence on the vagaries of weather and economic factors beyond his control. Better times for farmers would be a guarantee that Americans will continue to enjoy a plentiful supply of quality food at reasonable prices (Nelson, 1977).

But the long exodus from the rural areas may have run its course. Rand Corporation announced that of the twenty-five largest metropolitan areas in the United States, ten experienced no growth in the seventies compared with only one in the sixties. On the other hand, counties outside the metro complexes are suddenly experiencing population increases (Seidenbaum, 1975). According to a Gallup sampling in 1977, nearly 90 percent of Americans would prefer to live in a small city, town, or village, or in a rural area (Tunley, 1977). Improved transportation, communications, and other modern conveniences have eliminated the image of a "backward" rural America. They have also enabled some types of industry to become established in small towns. Some of the small-town inhabitants do part-time farming and may live on a small farm. Conversely, some people living on farms obtain a part of their income from jobs they have in the towns. Perhaps such people are enjoying the best part of two worlds. They currently contribute only a small fraction of the nation's total agricultural production, but they are rapidly increasing in numbers. The American agrarian mystique shows no signs of declining.

More high school students than ever before are taking vocational agriculture courses and enrollment in agricultural colleges is also increasing. The number of female applicants in Cornell's College of Agriculture has more than tripled in the last decade, and the 1978 freshman class will contain nearly equal numbers of men and women. Nationwide, women comprise more than 28 percent of the enrollment in colleges of agricul-

[5]It has been forecast that by the twenty-first century the American farmer will be "sitting in an air-conditioned farm office, scanning a printout from a computer center. This computer will help him decide how many acres to plant, what kinds of seeds to sow, what kind and how much fertilizer to apply, exactly what his soil composition is, and even what day to harvest each crop" (Durost, 1969). Among other developments foreseen are automated machinery directed by tape-controlled programs and supervised by television scanners mounted on towers, and robot harvesters to carry out high-speed picking, grading, packaging, and freezing.

ture. The number of college graduates returning to the farm in the mid-1970s was double what it was ten years previously (Memolo, 1976). One Midwest farm, owned by one of the nation's outstanding young farmers, has three Ph.D.'s on its staff (Drache, 1976). The percentage of young men engaged in agriculture has greatly increased in recent years—21 percent under thirty-five years of age in 1976 compared with 14 percent in 1970 (Thomson, 1976). It is reassuring that young people are at last heading for "where the action is" when one considers the awesome fact that it has taken about ten thousand years to reach the present level of food production, but that level must be doubled in about thirty years to meet the projected world demand (Winstanley et al., 1975).

Many not so fortunate, but with the same yearning, are participating in the burgeoning nationwide community-garden movement, seeking personal reinvolvement in a fulfilling relationship with nature that was lost in the course of urbanization. Shoveling and hoeing with enthusiasm, they capture some degree of that satisfaction sought by Tolstoy's character, the wealthy landowner Konstantin Levin in *Anna Karenina,* who could not refrain from occasionally mingling with the peasants and, scythe in hand, spending an entire day mowing his meadow grasses, despite the risk of ridicule and social ostracism.[6] Or what erstwhile Missouri farmer Harry Truman referred to as "the rare sense of accomplishment that one had at the end of a day of plowing."

The number of jobs in small towns is increasing. Although some corporate giants are planning to acquire more land, others have left the production end of farming, after suffering great financial reverses. Some of these corporations are entering the packaging and marketing end of the business but, in response, the old cooperative movement is reviving around family farms. Some legislatures, as in Iowa (Sidey, 1976) and Minnesota (Salisbury, 1976), are setting up corporate-ownership limits to help preserve family units. In Kansas, a corporation can have no more than ten shareholders and cannot own or control more than 5,000 acres of land. The incorporators must be residents of Kansas. Oklahoma also has a ten-shareholder limit. Furthermore, no more than 20 percent of farm-corporation gross receipts can come from a source other than farming, ranching, or mineral rights.

On the other hand, incorporating the family farm is one of the strategies that farm families can adopt to help them to hand their farms down to their descendants. There are some provisions of the Federal Estate Tax Law that make corporations a rather desirable way of owning property. Also, the Tax Reform Act of 1976 does much to ease the tax burden of family farms being transferred from one generation to the next. The amount of the tax has been reduced for most family-size operations. Further, the law now contains a provision whereby family farms may be assessed for real estate tax evaluation purposes at their current value rather than the customary highest and best uses or market value. In many cases, particularly where farmers are located in areas where there is a high developmental value on their land, this could result in considerable tax savings.

The high cost of equipment for a large and fully modernized farm can be a serious obstacle, but on the other hand, farmers can custom hire all or a part of the field operations required for growing their crops. Probably the first to custom hire all their farming were the "gentlemen farmers" of the southern California citrus industry, starting many decades ago. A farmer may also rent his land and a rancher may rent his cattle.

[6]At the age of sixty-seven, Tolstoy was still swinging a scythe for three hours a day, and going for long swims in cold water, for exercise (Pritchard, 1976). Apparently he used the fictional Levin to reflect the belief in the dignity of labor that Tolstoy demonstrated on his own large estate.

Family farmers also have the advantage of certain types of services that are ordinarily associated with corporate vertical integration: a giant complex of research organizations; information networks (extension); and price and production coordination, most of it supplied by state or federal agencies. Corporations that commenced operating after about 1965 began too late to obtain big operating gains from rising efficiency and just in time to encounter inflated land costs, antipollution crusades, and the rising militancy of organized farm workers (Cordtz, 1972; Raup, 1973).

A USDA study revealed that maximum crop-production efficiency is reached in a relatively small size of operation and does not gain in efficiency with greater acreage. For most crops the family farm, a modern and fully mechanized one- or two-man operation, is the optimum size. It is comparable and often superior in efficiency to the large units that operate with hired management. This conclusion is based on 138 studies on the production costs of different-sized farms, published by the Economic Research Service of the U.S. Department of Agriculture (McDonald, 1975). The studies were based on analyses of the costs and gross profits of many types of farming, including fruit, grain, livestock, cotton, vegetables, alfalfa, and dairy. Crop records are set on farms of moderate size, not on the gigantic landholdings of the conglomerates (Hightower, 1973). (Corporate farms obtain advantages, however, from tax laws, easier access to credit, and vertical integration.) Moreover, the "optimum one- or two-man operation" in the USDA study was found to be about three times larger than the average in the Midwest at that time (1973). The most efficient one-man unit corn farm was determined to be 800 acres compared with the average size of 263 acres, and optimum size for a grain farm was 1,950 acres compared with the average size of 694.

Hiram Drache (1976), a farmer and professor of history at Concordia College, Moorpark, Minnesota, in the period 1971-1975 interviewed 135 farmers in Iowa, Minnesota, North and South Dakota, Montana, and the Province of Manitoba. He concluded from these interviews that modern agricultural equipment, ranging from tractors with four-wheeled articulated power units to two-way radios and intercoms, are going to continue to increase farm size and provide the industrial power to make the farmer's income potential equal to the best in other industries.

Regarding optimum acreages, Drache states:

> All of the corn farmers indicated that the greatest economy was reached between 800 and 1,000 acres; sugar-beet farmers suggested an average of 600 acres; potato men 800 acres. Small-grain farmers in the humid areas (east of the 98th meridian) suggested that 3,400 to 4,000 acres was a very sound economic unit in contrast to the farmers farther west whose suggested average was 2,000 acres of grain and 2,000 acres of fallow. Most of the western farmers hedged somewhat, as they felt that a livestock operation was absolutely essential if the farm were at the 2,000-acre crop level.

Women's lib advocates will be encouraged by these words:

> . . . in the great majority of these interviews the farmer's wife participated and in most cases answered part of every question. Many times the husband asked the wife for her opinion before he gave the answer. Many times the wife had a better grasp of why some things took place than the husband did. It was obvious that in most cases few major changes or expansions took place without consultation with the wife. This point is specifically made so the reader can appreciate that nearly all the farms researched were family-centered enterprises even though the size might give another impression.

One of the questions asked in Drache's interviews concerned the farmers' attitude toward further expansion:

> The answer most often received was the lack of desire to grow larger because the owner preferred to spend his time with his family, in civic or church activities, or in other businesses. The great majority of these people are very active in community, civic, religious or political activities. Most of them were good managers of their time and their labor force, making it possible for them to be involved off the farm.

Drache feels comfortable in contemplating continued increase in farm size.

> If the average of these farms suddenly became the basis for our standard-size commercial farm operation America would need less than 100,000 farmers. (This would be exclusive of the two million small-scale residential and part-time farmers.) I am convinced that it is possible. The social implications are great, however, but as has been true in the past, people will adjust and the end result will be a better life style.

Not all observers of the current agricultural scene, however, share Drache's sanguine conclusion. Anthropologist Walter Goldschmidt (1946, 1947, 1975, 1978), for example, has observed that communities where small farms prevail support more people per unit of agricultural land, have a larger percentage of independent entrepreneurs, do a larger retail business, have more and better community facilities and organizations, and their citizens participate much more actively in democratic decision-making and live a richer social life. Throughout the world, wherever democratic conditions prevail, agriculture is pursued predominantly by farmers who themselves till the soil and control their farm enterprise. Corporate agriculture tends to accelerate rural decline beyond that which is caused by ever-increasing size of farm units. The problem has become so acute that its social and moral implications became the subject of extensive hearings before the Subcommittee on Monopoly of the Select Committee on Small Businesses of the United States Senate in 1971 and 1972. Testimony was principally from family farmers who had witnessed the devastating effect of corporate agriculture on their communities. Those living entirely from farming or ranching generally have little or no taxable nonfarm income against which to offset farming losses. Therefore it is almost entirely the outside investors who obtain tax-loss advantages. How extensively "tax-loss farming" is practiced is indicated by IRS figures revealing that, of the 3,000,000 farm income tax returns filed in 1965, 680,000 had farm losses offsetting nonfarm incomes (SCSB, 1973).

British economist E. F. Schumacher (1973) has concluded from wide-ranging studies that the size of an enterprise must satisfy criteria that are not purely economic. He believes that there is growing awareness of the need, not only in agriculture but in all industries, to sufficiently decentralize or reduce the size of an enterprise to ensure maximum personal involvement and self-reliance and to give free reign to basic and deep-rooted human aspirations and thereby reverse the present trend toward social demoralization.

In fairness to the large corporate farms, it should be pointed out that some family farms are likewise very large and that in any case a decline in the size and quality of the small towns would have occurred just in response to new technology. More change has occurred in farming during the last 40 years than in the preceding 4,000. Farms must become larger to profitably utilize the new technologies, particularly the ever-increasing use of farm machinery. This results in fewer farmers, but with bigger farms. Also, with less diversified agriculture, there are fewer chores to do and large families are no longer needed. Improved transportation and roads facilitate shopping in larger communities at a considerable distance from the farms. Today rural towns that are maintaining or increasing their populations are generally those within about fifty miles of a large city. Or two or three small, deteriorating rural towns may combine to gain new and improved water and sewage facilities, community colleges, and in general more of the amenities expected in modern life. Such new towns are attracting certain types of industry and are inhabited by industrial and farm workers and retired persons seeking refuge from the congestion, pollution, and crime of large cities. Some may engage in small-scale farming to keep themselves busy (Bohlen et al., 1975).

According to USDA estimates, two out of three American farmers receive over half of their annual income from off-farm sources. Their off-farm incomes doubled in a decade, from $8.5 billion in 1960 to $17 billion in 1970, when for the first time it exceeded net farm income (USDA, 1971). It is generally believed that diversity in the size and structure of operating farm units of all sizes will continue for a long time (Kyle et al., 1975).

The foregoing changes in rural America have not diminished interest in who controls agriculture. Even in urban areas, opinion polls indicate that the family farm has a much higher standing than does the large corporate type. Moreover, people everywhere, particularly among the youth, are reevaluating their previous acceptance of corporate dominance in industry. Particularly the many people who are now moving to rural towns do not favor a system of land ownership that tends to displace the local suppliers of agricultural equipment, fertilizer, and pesticides. They are interested in their community's economic survival (Rhodes and Kyle, 1973).

The University of Illinois College of Agriculture has published a group of leaflets (Special Publication 27) written by agricultural scientists from the USDA and various universities who discuss five alternatives for control of agriculture: (1) a dispersed, independent farmer, open-market system; (2) a corporate agriculture; (3) a cooperative agriculture; (4) a government-administered agriculture; and (5) a combination, with a role for each system. In the latter, "individual farmers or land owners, corporations, or cooperatives would not be limited as to the amount of land each could hold, provided the land would not be held primarily to take advantage of tax provisions or for reasons not connected with profitable farming." The purpose of the leaflets is to present and discuss alternatives, not to advocate or predict a particular method of control. They comprise a comprehensive and dispassionate symposium on the American farm problem.

In recent years a new threat to ownership of farms by American farmers, large or small, has arisen because of massive purchase of farmland by foreign buyers from Japan, West Germany, Italy, Iran, Saudi Arabia, and elsewhere. Land that is not worth more than $1,000 per acre for the growing of crops, is purchased at two or three times that amount purely for investment purposes and as a hedge against inflation. Some of the most successful American farmers are now being forced out of agribusiness by the foreign investors. What remains of the tradition of handing farms down from one generation to the next is then ended. The problem is nationwide and on such an alarming scale that legislation is being urged to put a stop to it at the earliest possible date.

1:43 P.M. Below us are the farmlands of north-central Indiana, the first large areas of the eastern intrusions of the great tall grass prairie that once dominated the plains west of the Mississippi.

## THE AMERICAN PRAIRIE

Just as the first settlers to cross the Alleghenies saw before them one of the world's greatest and richest areas of hardwood forest, those who first emerged from the western margins of that forest saw before them one of the world's greatest prairies—700 million acres. To find similar expanses of grassland elsewhere, one would have to visit the steppes of Eurasia, the African veldt, or the Argentine pampas. The eastern portion —the "tallgrass prairie"—contained grasses so tall and dense that cattle could be found only by the tinkling of cowbells and the waving of grass. The edge of the prairie, with considerable penetration into Illinois and even some penetration into Indiana and Ohio, had probably been progressively shifted eastward by repeated fires set by lightning or by the Indians to improve the area for hunting. The burned-over areas were favored by buffalo when the first grasses of spring appeared. Ironically, clearings made by Indians in forest lands along the Atlantic were a decisive factor in the establishment of the incipient white settlements, while their more extensive land-clearing activities in the Midwest ensured rapid settlement by the much greater hordes of whites that poured across the Alleghenies in the latter half of the nineteenth century. In either case, the efforts of the Indians hastened their own demise.

Despite extensive burning, in pioneer days there was a forest mosaic, with the forest extending into the grassland along the rivers and on the uplands. Native tree species thrive today at least as far west as Manhattan, Kansas, if given an opportunity (Gleason and Cronquist, 1964).

The eastern or tallgrass portion of the grassland region is referred to by some authors as "the prairie" and the western shortgrass or mixed-grass portion as "the plains," the two regions being roughly divided, depending on the authority, from anywhere between 98° and 100° longitude; the latter, as indicated in figure 2, is also being used in this book as the approximate dividing line between the Midwest and the Great Plains. Protrusions of the two types of grassland extend east or west of this line and vary with the seasons. Some authors refer to the entire region as "the prairie."

Among the most common grasses in the original tallgrass prairie were big bluestem (*Andropogon geraldi*), little bluestem (*A. scoparius*), and Indian grass (*Sorghastrum nictans*). These grasses are almost entirely replaced and the deep, dark, fertile soil now comprises the famous Corn Belt, devoted mostly to the raising of corn, but also millions of acres of wheat, sorghum, soybeans, and cotton. The only extensive native tallgrass grazing area remaining is in the Flint Hills area of eastern Kansas.

Although grass is the dominant native vegetation in the prairies and plains, there is also a succession of flowers in bloom, as the season advances, to brighten the spring and summer landscape. They are mainly composites: butterweeds and blanketflowers in the spring; sunflowers and coneflowers in midsummer; and asters and goldenrods in late summer and fall. Species of the pea family are also abundant in many places. The larger streams are bordered with cottonwood trees. The bur oak and other small trees grow in the moist spots of the northern half of the plains, as in Nebraska (Gleason and Cronquist, 1964). The bur oak has thick, fire-resistant bark.

The portion of the Midwest where corn was eventually to be grown most intensively was covered by the glaciers of the Ice Age, which, in the central United States, extended from central Ohio to eastern Nebraska and south approximately to what are now the Ohio and Missouri rivers. As the glaciers moved over the Midwest they met little resistance in the present Corn Belt and laid down enormous quantities of ground rock powder known as *till*, filling the erosional wrinkles on the surface of both the level and undulating portions of the plain. This pulverized rock, formed from uplifted strata of erosional and marine depositions consolidated with sandstones, shales, and limestone, was the base for the deep, rich soil of the prairie. The significance of this geological good fortune can be realized by comparing the physical features and soil of the Corn Belt with those of New England, where most of the preglacial land was hard igneous and metamorphic rock that did not pulverize easily. Consequently the till deposits left by glaciers generally have more stones and boulders than those of the Corn Belt. Also it was not sufficiently abundant to fill in and smooth the hilly landscape (Higbee, 1958).

Although the glaciers deposited fewer surface boulders in the Midwest than in New England, there were still more than the farmers could tolerate in their fields. They hauled the boulders from the fields and piled them into long windrows where they now mark the edges of the fields. These boulders became overgrown with shrubs and trees that eventually formed continuous hedgerows between the fields. They can be distinguished from the airplane window with the aid of a good field binocular, and are a characteristic feature of the Midwest roughly north of the Ohio and Missouri rivers. South of this glaciated area the checkered pattern of fields may be seen, but without stone walls. There the hedgerows are formed along fences where the seeds of cedar and osage orange, from the droppings of birds resting on the fences, grow into large trees, if allowed to. These rows of trees are sought by cattle for their shade and are also kept as windbreaks in many Midwest areas.

The deepest and most fertile soils on the North American continent were formed under the vegetative growth of the prairies. Thousands of years of growth and decay of heavy-sodding native grasses resulted in deep, rich, black soil. Such soil does not develop under trees. Where forest trees are removed there are a few inches of dark topsoil under the surface litter of the forest floor, but below this relatively thin layer the soil is light in color and contains only a few tree roots. In contrast, the dark humus-rich topsoil of the tallgrass prairie is one to two feet deep and is a mass of fibrous roots. The tallgrasses are spread by underground stems and form dense sods. The first 4 inches of soil have been estimated to contain 4.5 tons of grass roots per acre. The roots reach depths of 12 feet or more and bring minerals to the growing plant from all intervening levels (Allen, 1967).

Originally farmers had difficulty in breaking the tough sod with their plows. Although cast-iron plows had been successful in the New England and Middle Atlantic states, in the heavy, sticky prairie loam the soil would cling to the moldboard instead of sliding by and turning over. Every plowman carried a paddle to clean the moldboards. Often farmers used six to eight oxen to pull the plow through the prairie sod, a man to guide the plow and a man to drive. And they plowed less than an acre per day. Many settlers gave up their homesteads in disgust and returned east. The problem was solved in 1837 by an Illinois blacksmith, John Deere, who designed a plow with a wrought iron moldboard and with a cutting edge of steel carefully welded to the moldboard. This was a "clean-scouring" plow; the heavy black loam of the prairie would not cling to it. An increasing proportion of steel was used as it became less expensive and more available in the 1850s. By 1857 Deere and his partner were producing about 10,000 plows a year, and by the time of the Civil War the steel plow had largely replaced the cast-iron one on the prairies. The steel plow not only reduced toil and frustration, but also improved the quality of the plowing and reduced by at least a third the animal power needed to turn the soil. It was one of the truly great milestones in farm mechanization.

The John Deere plow conquered the prairie sod, but of equal importance was the gangplow that appeared in the 1860s. These were mostly two-bottom sulky types which had wheels and a seat for the driver. They enabled the farmer to double the amount of land he could handle. Later there were multiple-bladed plows drawn by teams of eight or more horses or mules (Gittins, 1950; Rasmussen, 1960*b*; Schlebecker, 1975). Today giant tractors pulling strings of plows and working day and night can plow 100 acres in a 24-hour day. Some gang equipment plows, drills seed, and applies one dressing of fertilizer at the same time.

The sticky-soil problem was largely solved by steel plows. It was not completely solved, however, until roller bearings, notched disks, and high-grade steel made the use of the disk plow practical for certain types of soil. Sticky soils are scraped off as the disk revolves. Another advantage is that the disk rolls over stones instead of catching on them. The disk plow requires more traction per unit of soil surface than do other plows, but sufficient power is provided by modern tractors (Schlebecker, 1975).

2:02 P.M. We are now over the agriculturally rich lands of central Illinois. Throughout our flight over the Midwest we are impressed by the way monoculture has become the dominant feature of the American agricultural landscape.

## MONOCULTURE AND MECHANIZATION

Monoculture is the cultivation of a single crop to the exclusion of other possible uses of the land. For example, thousands of contiguous acres of corn may be grown in one area because climatic and soil factors are optimum for the crop in that particular area. In other areas the same will be found to be true of wheat, rice, sorghum, soybeans, sugarcane, cotton, sugar beets, beans, potatoes, lettuce, apples, oranges, and other crops. Sometimes a crop is cultivated extensively in an area because it can be economically grown there while other possibly more valuable crops would not thrive. Monoculture tends to maximize the productivity of an area.

Conceding the great initial economic advantages of crop specialization and monoculture, the fact remains that what may at first sight seem brilliant technological advances are often seen in hindsight to have had adverse side effects. Sustained productivity depends not only on the supply of chemicals that can temporarily stimulate a plant to maximum production but also on the long-term maintenance of good structure, aggregation, texture, and mellowness of the soil and the encouragement of its microorganic life. The latter transforms dead plant tissue into humus, necessary for mellow soil of proper structure. Erosion is generally aggravated by inadequate replacement of organic material and improper tillage and it leads to a decrease in soil productivity.

When other factors are equal, crop productivity is directly related to depth of topsoil (Donahue, 1970). Improved plant strains and greatly increased use of fertilizers and pesticides have steadily increased crop yields and have masked the adverse effect of the ever-decreasing depth of topsoil. With steadily increasing crop production per acre and complacency on the part of most farmers, it is understandable that soil erosion has not received as much attention from environmentalists, and from the public in general, as one would expect from the long-range importance of the problem. Yet erosion not only destroys vital cropland but also produces billions of tons of sediment, the nation's largest single water pollutant and which also can be a carrier of other pollutants such as pesticides, toxic metals, and plant nutrients.

The farmer is not entirely to blame, for the sharply rising costs for tractor fuel, fertilizer, pesticides, labor, and equipment, far out of proportion to increases in the prices of farm produce, have tended to put short-term profits ahead of conservation (Carter, 1977*a*).

The short growing season and the high valuation of land force farmers of the rich Corn Belt to grow the most profitable crops—mainly corn and to a lesser extent soybeans. These are clean-cultivated crops and the soil is consequently particularly susceptible to erosion, even though the slope may be very small. Under certain Corn Belt cropping systems some formerly superb prairie loams have been either eroded away or have suffered a serious decline in soil structure and humus content because of constant emphasis on annual plants without winter cover or green manure crops (Higbee, 1958). On the other hand, soil-conservation measures developed in recent years, discussed in greater length in later chapters, encourage us to believe that the battle against soil erosion can be won if these precautions are universally observed. It is a problem that affects us all. "In the large sense every one of us is a farmer," said Cornell's Liberty Hyde Bailey, "for the keeping of the earth is given to the human race."

Monoculture also tends to increase pest and disease problems. Food supply for pests and plant diseases is virtually unlimited, and in the form of winter crop residue, provides pest harborage from one year to the next. (Crop rotation would break the continuity of the pest's life cycle.) The ability of pests to reproduce and expand their loci of infestation is likewise unlimited. Indigenous natural enemies usually are not able to thrive in the area after the native or other plants that previously provided their ecological niches are eliminated. Generally a varied and complex ecosystem supports the maximum number of predators. Such ecosystems are eliminated, of course, in making way for monocultures. Pest and disease control provide some of the most severe challenges in the monoculture of the principal food crops of the world. Entomologists, plant pathologists, and plant breeders are among the shock troops in the global struggle to provide food and fiber for our burgeoning human populations. Along with plant physiologists, soil chemists, agronomists, agricultural engineers, food technologists, and extension specialists, they have provided the scientific knowledge that the farmer can utilize, with his invaluable background of experience, to sustain the agricultural miracle.

Among the scientific advancements of recent decades in the United States can be found a nearly complete mechanization of cultural and management practices for plant crops, livestock, and poultry.[7] One of the mechanical advancements lay in harvesting, affecting many plant crops and ultimately perhaps attainable for all of them. This required the development of new varieties amenable to mechanical harvesting as well as changes in plant densities, spacings, and in water and fertilizer requirements. Also required were new storage and handling methods and new processing plants. Increased use of fertilizers, micronutrients, insecticides, miticides, fungicides, herbicides, nematicides, antibiotics, bioregulants for plant crops, and feed additives, growth factors, and protective agents for livestock and poultry played an important role in the increase in agricultural production. The development of new strains and varieties by plant and animal breeders resulted not only in increased productivity, but also, in the case of plants, sometimes promoted improved protein quality. Finally, agriculture has become a computerized "agribusiness," utilizing high-speed computers and systems analysis in new systems of management and farm accounting (Wittwer, 1970).

For every dollar the consumer paid for food in 1973, 62 cents went to marketers for assembling, processing, packaging, and distributing it, while 38 cents went to farmers who produced it (USDA, 1973*b*). The U.S. Department of Agriculture estimates that the actual cost of wheat in a loaf of bread is only three cents. Farmers in turn invested a large share of their incomes in fuel, fertilizers, pesticides, farm equipment, and labor.

Agriculture is this country's largest industry. If one includes the agriculture-related activities, and then includes the production and marketing of timber products, the total agricultural and timber complex accounts for more than 25 percent of the gross

[7]The same degree or type of mechanization is not necessarily appropriate throughout the world, particularly in poor and overpopulated countries, where small, inexpensive, and labor-intensive equipment may be more suitable. The "A.T." (Appropriate Technology) revolution is already well underway (Ellis, 1977), testing on a universal scale the principles espoused in E. F. Schumacher's (1973) bestseller, *Small Is Beautiful.*

national product (GNP) of the United States. About a third of the American work force is engaged in industries based on our farms and forests[8] (Anonymous, 1966).

As previously related, agriculture has become an important factor in our balance-of-payments position. In 1974 the United States exported two-thirds of its wheat crop, half of its crop of soybeans, of which it is the world's only exporter, and two-fifths of its corn crop, for an export total exceeding $21 billion. This is a sum far exceeding that obtained from either of our major "high-technology" exports—computers and jet aircraft (Handler, 1975). In 1975 American farmers produced 5.8 billion bushels of corn and exported 25 percent of it. They produced 1.52 billion bushels of soybeans and exported 30 percent of it. In a nation priding itself on efficiency it is of interest to note that in the 20-year period 1952 to 1972 agricultural efficiency had increased 330 percent, compared with an increase of 160 percent in manufacturing industries. This despite the fact that the cost of farming always rises at a far greater rate than prices received for farm produce.

Revenue received by selling grain and soybeans to Europe or Japan is used to purchase the 400 million metric tons of oil annually imported. It is tempting to view this as a clever exchange of what is principally inexhaustible solar energy, trapped in plants, for nonrenewable petroleum. (Our investment in fossil energy is only 0.5 quad (quadrillion *Btus*) for agricultural machines, fuel, and fertilizers, which we exchange for 16 quads of oil energy.) But as C. H. M. van Bavel (1977) of Texas A&M University has pointed out, for every ton of grain we export, "we export several tons of topsoil to the Gulf of Mexico." We are trading "soil for oil." Topsoil is replaceable through geological processes, but it is being lost much more rapidly than it is being replaced, on most of our land. For the international trade-off, the Arabs are losing their strictly nonreplaceable oil. However, per unit of land surface, they have more potential for solar energy—the earth's greatest and safest source of energy—than any other of the world's peoples. There are indications that they are not unaware of that fact (Hayes, 1975).

It would be irresponsible to leave the impression that ever-increasing mechanization is an unmixed blessing. As pointed out by the President's National Advisory Commission on Rural Poverty in 1967, a distressing degree of urban poverty and discontent resulted in large part from the fact that in a brief period of 15 years, from 1950 to 1965, new agricultural technology increased the nation's farm output by 45 percent while reducing farm employment by 45 percent. This resulted in a flood of people to the cities, but with no better skills to cope with the new technologies of industry than with those of agriculture. However, 14 million of those dispossessed by the new agricultural technology still remained in rural areas. Of the 14 million rural poor, 11 million were whites and 3 million blacks. Accompanying low incomes were low levels of formal schooling, a condition that tends to be self-perpetuating. The commission found malnutrition and hunger to be widespread.

Unlike the urban poor, those in rural areas were not well organized and had few spokesmen to bring the nation's attention to their plight or to gain benefits from antipoverty programs. Farmers and farm workers were excluded from the Social Security Act until the mid-1950s. At the time of the commission's report, they and workers in agriculturally related occupations were still excluded from other major labor legislation, including the unemployment insurance program and most state workmen's compensation acts (NACRP, 1967).

[8]One seldom-recognized outlet for farm products is the automobile industry. The annual agricultural needs of just one major automobile company consist of 900,000 bushels of corn, 736,000 bushels of flax, 74,000 bushels of cotton, products derived from 364,000 sheep and 36,000 cattle, and many other items such as hog bristles and beeswax (Johnston and Dean, 1969).

Among the twelve recommendations of the commission was "a national policy designed to give the residents of rural America equality of opportunity with all other citizens. This must include equal access to jobs, medical care, housing, education, welfare, and all other public services, without regard to race, religion, or place of residence."

Another recommendation of the commission worthy of special note because of its universal application, was the "development and expansion of family planning programs for the rural poor. Low income families are burdened with relatively numerous children to feed, clothe, and house. They are prepared psychologically to accept family planning. As a matter of principle, they are entitled to facilities and services to help them plan the number and spacing of their children."

The U.S. Department of Agriculture and the Land-Grant College and Experiment Station systems have been justly praised for their productive programs of basic and applied research that have so successfully increased this nation's and the world's capacity to produce more food, more clothing, more shelter, and more of the things that make life comfortable.[9] Therefore it was particularly significant that a land-grant college president, James H. Hilton of Iowa State University, should say at the centennial celebration of the establishment of the USDA and Land-Grant College system:

> ...many of the new problems which are troubling Americans today are a combination of socio-political and economic factors, the complex which in real life does not break down neatly into problems which are either scientific, economic, sociological, or political. The difficulties which confront farm and urban families are no respecters of academic disciplines. And their solutions will often require the special knowledge and competence of a variety of disciplines. (Knoblauch et al., 1962)

Despite our commendable food-production successes, it is well to bear in mind that the United States can continue to export foods principally because its population has not grown as rapidly as that of certain other nations, particularly in Latin America, that were once net food exporters but now must import food to feed their exploding populations. In the developing world, population is increasing at a rate of 2 to 3.5 percent annually. A 2-percent growth rate results in a 7-fold increase and a 3.5-percent rate results in a 32-fold increase in a century. Philip Handler, president of the National Academy of Sciences, stated in no uncertain terms, "The growth of human populations is the principal threat to the survival of our species" (Handler, 1975).

Our own annual growth rate, variously estimated to be between 0.6 and 0.8 percent, plus 400,000 legal and probably twice as many illegal immigrants, results in a far greater rate of increase than most Americans are aware of. Most of the immigrants, legal or otherwise, come from countries where overpopulation has been a prime factor in a gloomy economic outlook for the majority. Yet upon arriving in the United States, these refugees from despair and hopelessness usually continue their large-family tradition, often boasting of it. The government provides no disincentives; on the contrary, it provides tax and welfare incentives for having large families. We could ourselves eventually become deficient in foods, but there would be no place left from which to import. Additionally there is the mounting problem of environmental degradation.

[9]Yet out of the projected $7.7 billion allotted the Department of Agriculture for 1976, $5.4 billion was for the various relief programs. Food stamps, which cost the federal government $1.9 billion in 1970, cost $3.1 billion in 1972, and $8.2 billion in 1975 (Shover, 1976).

2:10 P.M. We fly over the Mississippi, "Father of Waters," at Burlington, Iowa, and will have only a brief glimpse of the Hawkeye State before dipping into northern Missouri and then northern Kansas.

## IOWA AGRICULTURE

Iowa is the only state that can be said to be practically entirely within the tallgrass prairie; it was originally 85 percent covered by the tallgrasses. The tallgrass prairie extends from eastern Texas northward through the eastern portions of Oklahoma, Kansas, Nebraska, the Dakotas, and into southern Manitoba. Likewise southern Minnesota, central Illinois, and northern Missouri share sizable sections of this rich geologic bounty. But Iowa lies entirely within it except for a small part of its northeastern corner. Understandably then, it boasts the greatest agricultural production, for its size, in the nation. It is the center of America's famous Corn Belt. It is logical that our richest lands should be devoted principally to the growing of corn, for corn produces more digestible nutrients per unit area of cropland than any other food crop. Iowa has 13.8 million acres planted to corn, more than any other state. Since 1975, Illinois has surpassed Iowa in production of corn for grain, but Iowa produces more corn for silage. Iowa is first in the production of hogs, with 24 percent of the nation's total of 76 million marketed in 1976. It is second only to Texas in the number of cattle produced (USDA Agr. Stat., 1977).

Of the approximately 60 million acres of corn harvested annually in the United States, amounting to half of the world's supply, 80 percent is earmarked for grain and the remainder for forage and silage. Most of the production of beef, pork, dairy products, and poultry products depends on corn as the basic feed. Less than 10 percent of the American corn crop is industrially processed for such products as breakfast foods, corn meal, flour, syrup, starch, sugar, oil, alcohol, acetone, and other industrial products (Hughes and Metcalfe, 1972). This contrasts strikingly with our neighboring country, Mexico, where 75 percent of the grain corn is used as human food (Wittwer, 1970).

Flying over the Corn Belt from central Ohio to eastern Kansas, we are impressed with the fine homes, huge barns, and towering silos of the farmers. This relatively high degree of rural prosperity is based principally on the fabulous richness of the earth in this region and the utilization of advanced technology by the farmers. One might expect that the short growing season in these northern latitudes would be a severe handicap to agriculture, but the genetically adapted corn varieties grow rapidly and mature in the 90-120 days between late and early frosts, producing as much food per acre as many other plants in tropical climates produce in a year.

It is evident that farmers have had profitable times, during that rare period of real peacetime prosperity from 1898 to 1914, the "Golden Age" of American agriculture, or during the war years, or in 1973, when the nation's net farm income was $33 billion. But prosperity can quickly vanish; in 1977 net farm income dropped to $20 billion and farmers were losing money. Yet prices farmers must pay for machines, energy, and fertilizer rise continuously. In the fall of 1977 there was talk of a farmers' strike, but by then about 80 percent of the winter wheat had been planted by those who hoped others would do the striking.

The disparity between the farmer's contribution to the nation's welfare and his reward has always been a characteristic of his occupation, even during the late Golden Age, when the farmer's economic situation was believed to be so satisfactory that the calculation of "parity" prices for his crops has ever since used that period as a base.

Thus a return to 100 percent of parity would be a situation in which the federal support of prices of farm products would yield farmers the same purchasing power per bushel that they enjoyed in the last four years of the Golden Age—between 1910 and 1914. A bushel of wheat would buy 10 gallons of gasoline, as it did in 1913.

In a massive nationwide survey of rural life by President Theodore Roosevelt's Commission on Country Life in the mid-Golden Age year of 1909, based on 94,000 answers to a questionnaire, one of the questions asked was "Do the farmers in your neighborhood get the return they reasonably should from the sale of their products?" The percentages of farmers by region answering affirmatively averaged as follows: North Central, 54; North Atlantic, 46; Western, 40; South Atlantic, 34; and South Central, 25. Members of a wide variety of other occupations, when asked the same question, substantially supported the farmers' opinion of their condition, albeit somewhat more optimistically. Their corresponding percentages of affirmative answers averaged 65, 56, 46, 40, and 33 (Larson and Jones, 1976).

2:38 P.M. After flying about fifty miles into Kansas we reach the western border of Region M (Fig. 3) and enter Region H, the winter wheat region, and a minute later we are over the northern part of the Flint Hills section of eastern Kansas.

## THE FLINT HILLS PRAIRIE-GRASS REFUGE

The Flint Hills form a north-south band that comprises an ecotone, a region where many habitats and flora meet. The Flint Hills contain the largest remaining area of tallgrass in the world, but shortgrass and midgrass are also found. In the lowlands, big bluestem, Indian grass, and switch grass may reach eight feet tall; yet on top of a ridge there may be buffalo grass only a few inches in height. In areas of intermediate moisture one finds midgrass species: little bluestem, western wheatgrass, June grass, and side oats grama. Under ideal conditions there would be thirty-six species of grasses and about three hundred flowering plants, so that some groups of species would be in bloom throughout most of the year, providing a changing picture of colors, from the scarlet of April's paintbrush to the yellow of October's goldenrod and compass plant.

The Kansas-based Save the Tallgrass Prairie, Inc. (4101 W. Fifty-fourth Terrace, Shawnee Mission, Kansas 66205), is engaged in a campaign to have 60,000 acres (a 10-mile square) of the Flint Hills south of the town of Emporia set aside as a tallgrass Prairie National Park. The hills are mostly too steep and rocky for plowing and are used for the grazing of cattle. Supporters of the national park movement claim that the cattle population in the area is too high to allow the full growth and flowering of plants. They propose allowing limited numbers of buffalo (*Bison*), elk, deer, and antelope in the park in numbers not large enough to continuously graze the whole area. The different wild species would eat generally different plants, keeping a better balance than is maintained by the grazing of one species, such as cattle. Fire would be used as a management tool to burn off shrubs and trees. Fires and periodic droughts, to which the grasses are adapted, are adverse to tree growth and are the factors that have maintained the North American prairie since it first spread across the midlands about twenty million years ago.

Predictably, the arguments of the national park supporters are being challenged by some Kansas citizens, including ranchers, who have organized the Kansas Grassroots Association, Inc. They argue that the grassland of Flint Hills is not disappearing and could continue indefinitely if left in the hands of private owners. They point out that the grassland acreage of the Flint Hills has not diminished during the past century, but on the contrary large tracts of land that were once tilled have been returned to native grass. They propose that the existing Prairie Parkway through the grasslands be improved, with rest and observation areas added and with the parkway extended to a circle drive. They argue that the Flint Hills represent not only a sea of grass, but also a culture and tradition, almost unique in the United States, of independent, self-reliant ranchers and cattlemen who have lived in harmony with the tallgrass prairie and preserved the best aspects of their environment for more than a century.

One can only hope that both the national park advocates and their opponents have as their goal the preservation of a priceless American heritage, found today on a considerable scale[10] only in the Flint Hills of Kansas, in as good condition as possible under the protection and care of man. Economic considerations have been put forth, but they are almost profane in view of the fact that the proposed park would amount to only about 0.3 percent of Kansas's grassland.

2:53 P.M. About halfway through Kansas, between 98 ° and 99° longitude, we enter the shortgrass prairie, but only gradually; even the broad dividing line can shift east or west from year to year, depending on rainfall. As one passes westward, the soil becomes more deficient in moisture, the little bluestem (bunchgrass) predominates, and the shortgrasses of the high plains begin to appear. This has been called the "mixed-grass" or "midgrass" prairie. Here the vegetation of the eastern prairie is mixed with that of the western at its best. The average annual precipitation is twenty to thirty inches compared with thirty to forty inches in the tallgrass prairie. The surface layer of moist soil is from two to four feet in depth. We are now flying over the greatest winter-wheat area of the United States. Farther north, in Nebraska, Montana, and the Dakotas, the native needlegrass (*Stipa spartea*) and slender wheatgrass (*Agropyron tenerum*) become relatively more common in uncultivated areas. They form a dense sod cover but are not as vigorous as the bluestems farther south. Rainfall is from eighteen to thirty inches, evaporation is less than it is farther south, and the soil is dark and moist to a depth of three feet or more. This is the nation's great spring-wheat area (Shantz and Zon, 1924).

## KANSAS WHEAT

Just as corn dominated the landscape as we flew over Iowa and northern Missouri, wheat is king in Kansas. The ripening wheat covers the earth in "amber waves of grain" that roll on and on, rising and dipping in long undulations until the sky's blue vault rests on them on all horizons. Kansas does not present to the air traveler the checkerboard appearance of the field crops of the Midwest, for the area is almost entirely in wheat and each wheat field is very large. Although mixed-crop agriculture, with here and there a bit of woodland and pasture with grazing livestock, presents a more pleasing scene, it is interesting to what extent Americans have found plant monoculture to be aesthetically satisfying and to be a pleasant aspect of humanized nature. No such accommodation has been made to animal monoculture, which would mean for example, tens of thousands of cattle crowded in miles of feedlots.

The Great Plains is a good place to notice the effects of the original land layout sys-

[10]There are a number of small plots of original prairie that give the viewer a good idea of its composition. For example, Rossou Prairie in Iowa is a forty-acre patch of virgin land preserved by the Rossou family since frontier times. The grass grows higher than one's head and the rich, black earth is as springy as a mattress.

tem on roads. Many roads run long and straight, directly north, south, east, or west, crossing one another at right angles at mile intervals. These are usually the principal roads, but there are many minor roads that mark off quarter-sections (160 acres), the original units of land allocation. Streams and other natural boundaries occasionally cause deviations in the road pattern.

Kansas grows about 16 percent of the nation's wheat crop, which in 1975 amounted to 2.13 billion bushels, bringing a record $7.4 billion to American farmers. (A bushel of wheat weighs 60 pounds.) Exports required 60 percent of the American wheat crop of 1975.

With the aid of binoculars, the combines that are now harvesting Kansas wheat can be distinguished. In the United States wheat is harvested almost entirely by means of these giant machines, utilizing the same principles as when wheat was harvested by hand, that is, trampling, beating, or flailing the grain free from the straw and hulls and removing the latter by winnowing—tossing the grain into the air to let the lighter hulls blow away. Combines and combine crews start in the early summer on the winter-wheat crop in Texas and Oklahoma and gradually move northward as the wheat matures. Eventually, in the autumn, they harvest the spring wheat of the northern United States and Canada (Schery, 1972).

3:06 P.M. Our plane happens to be flying directly over the tiny town of Quinter, at 100°15′, in Gove County, northwest Kansas. From an altitude of 30,000 feet the plains in this area look reasonably level, and we can hardly imagine the magnitude of the task that confronted the pioneer wagon trains in crossing the innumerable ditches, gullies, stream beds, live streams, and rivers. But at Quinter the headquarters for Wagons Ho, Inc., is located; it provides its customers with scheduled 3-day, 65-mile, authentic wagon-train trips in summer. The wagon train follows the Smoky Hill Trail, known in the 1800s as the shortest but also the most dangerous of the routes west. On this trip one rides in safety and comfort, but needs little imagination to appreciate the dangers and hardships endured by the pioneers. Persons interested in vegetation can become intimately acquainted with prairie grasses at approximately the juncture of the mixed-grass and shortgrass prairie, as well as myriads of wild flowers and the wild animals of the prairie.

## CIRCULAR FIELDS

At the western border of Kansas, which almost exactly coincides with 102° longitude, we see at 3:16 P.M. a group of 16 large, circular fields. (The first fields of this type we saw at 101° longitude.) These fields (see fig. 25) are irrigated by huge, automatically controlled, center pivot sprinkler systems. Each consists of a pipe to which sprinklers are attached. The pipe and sprinklers are mounted on wheels and self-propelled, moving laterally and pivoting about a point (see figs. 23 and 27). The system is propelled by water pressure or electric motors.

The circular fields are easily seen from the plane for the pipes are usually about a quarter mile long and, revolving through a complete circle, irrigate an area of about 133 acres, which is a quarter section (160 acres) minus the corners.

At this point the western border of Kansas happens to coincide with the eastern border of Region G (see fig. 3). We will pass through a subsection of Region G that the Soil Conservation Service designates as the "Upper Arkansas Valley Rolling Plains." The fields, circular or otherwise, are not green, of course, if the land has been just recently planted to winter wheat or is lying fallow.

## THE FATE OF THE PLAINS INDIANS AND BUFFALO

With the exception of some green irrigated fields along rivers, particularly the Arkansas River in Colorado, the country we fly over is typical of the semiarid High Plains. Shortgrasses dominate the flora, reaching no more than sixteen inches high and competing well under dry conditions and grazing. Blue grama (*Bouteloua gracilis*) forms a shallow but very dense sod. Often mixed with blue grama is buffalo grass (*Buchloë dactyloides*), which grows only four or five inches above the ground, hiding its seed heads within the foliage. As one proceeds westward, the shortgrass gives way to tufts of bunchgrass, because the climate is too dry to support a continuous growth.

The cured stands of blue, black, hairy, and sideoats grama and the buffalo grass, growing on heavily mineralized soil, served the immense herds of buffalo and antelope well, and were later excellent feed for cattle. But the buffalo and the Indians who hunted them are gone. We recall with great pride the hardiness, indomitable will, and great capacity for heroic self-sacrifice of our pioneer ancestors, amply recounted in song and story. But as has happened throughout history where Stone Age cultures have stood in the way of civilization's advance, the heady songs of conquest cannot hide the melancholy undertones of tragedy to the native inhabitants.

The relationship of the first English settlers and the American Indians was amiable. The small bands of settlers in a strange and inhospitable environment needed the help and cooperation of the Indians. Even under the best of circumstances the settlers were just barely able to survive, and "respect for Indian rights was the better part of wisdom" (Udall, 1963). But it was not long before the friendship of the Indians became of less value than their land. According to the Indian way of life, the vast area now represented by the forty-eight contiguous states supported probably less than a million people (USDI, 1968) after tens of thousands of years of occupation by its aboriginal inhabitants. The superior technology of the whites would in three and a half centuries provide for more than two hundred times that number of people in the same land area. It was soon obvious that the two cultures were not going to coexist. This is not to say that the problem could not have been solved in a more just and humane manner. Perhaps the best that could have been expected would have been the proposal of the artist George Catlin, who suggested that large areas be set aside where Indians and buffalo could live much as they did since time immemorial. But that was not to be.

There is no doubt that the western plains could not have been developed in any way acceptable to white man in coexistence with the original vast herds of buffalo, or perhaps not even with numbers exceeding those actually in existence today. Yet one cannot reflect without a profound sadness, on the loss of the irreplaceable heritage with which nature so bountifully blessed our land. A melancholy verse in Omar Khayyam's *Rubaiyat* comes to mind:

> The Moving Finger writes; and, having writ,
> Moves on; nor all your Piety nor Wit
>    Shall lure it back to cancel half a Line,
> Nor all your Tears wash out a Word of it.

## CATTLEMEN AND NESTERS

To most devotees of Americana, the most colorful and exciting period of the Old West was during the years when the immense herds of Longhorn cattle were trailed north from Texas. Longhorns sometimes traveled hundreds of miles in search of more verdant pastures and this often resulted in the mixing of herds. At the "roundup," held at least once a year, all the animals of a region were driven to a central point. Calves were branded with a hot iron, to leave the same mark on their hides as possessed by their mother. Also the animals to be sold at market were selected. "Cowboys and Indians" of those stirring times were later to be the theme of thousands of books, pulp-magazine stories, and movie "westerns" that captivated millions of readers and viewers in America—and worldwide—with undiminished success to the present day. The demise of the great "cattle kings" of the Old West came with the disastrous blizzards of 1885-1886 and 1886-1887, in which about 84 percent of the cattle were lost throughout the West. This resulted in the end of the open-range cattle industry. Buffalo can push away the snow with their massive heads; horses can find grass by pawing through the snow; sheep can eat snow if water is lacking; but cattle can perform none of these survival tricks. It became obvious that the ranges would have to be fenced in and cattle would have to be fed some in winter. This increased the desirability of new, more productive breeds of cattle and was the beginning of the end for the Longhorns.

Free land provided by the Homestead Act of 1862 had lured many an adventurous "nester" to apply farming methods developed in the humid East to the dry lands of the High Plains where, during years of above-average rainfall, fortunes had been made. But rainfall on the High Plains, scanty at best, is extremely variable. Many a young couple, generally with meager financial assets, could be financially, physically, and spiritually broken by a year or two of drought on the vast, empty, lonely expanses of the plains. Combined with drought, the fierce winds of the plains brought the ultimate disaster, the Dust Bowl tragedy of the 1930s and the pathetic trek of refugees from the devastated area to California, a migration immortalized in John Steinbeck's *Grapes of Wrath* and Dorothea Lange's *American Exodus.*

In the 1930s, conservation programs were initiated to preserve and improve the remaining soil and retain the rainfall. Measures known to prevent wind and water erosion were not universally practiced, but enough so that when drought conditions returned in the 1950s, serious dust storms again darkened the sky and some crops were ruined, though erosion was not as severe as in the 1930s. From the vantage point of the flight deck of our airliner, Captain Hanson points to wisps of dust arising in certain places on this windy day, but in general the land appears peaceful and serene today, at least in this area.

## THE ROCKY MOUNTAINS

Having flown about halfway between Pueblo and Walsenburg, Colorado, at 3:36 P.M. we approach the Rocky Mountains at the impressive Sangre de Cristo Range, with some peaks only a short distance to the north and south of us rising to over 14,000 feet in elevation. We have entered Region E (fig. 3). With the aid of binoculars, one has a few moments to observe the grassland merge into sunny open woods of ponderosa (yellow) pine. In the Rockies the ponderosa pine merges with the Douglas-fir. The latter is much smaller than in Oregon and Washington, where it is the principal lumber tree.

When compared with the coniferous forests of the Sierra Nevada, the trees in the Rockies are generally smaller and grow in dense stands. In the Sierra Nevada the heavy snowfall of winter assures a good supply of moisture to great depths, thus supporting larger trees, but the dry summers do not favor heavy reproduction. In the Rockies there is less annual precipitation, but there are many summer rains, promoting abundant reproduction and therefore dense stands of trees, except in the lower ponderosa pine belt. In contrast, in the Sierra Nevada the forests are more open and parklike at all elevations.

Above the Douglas-fir belt are the forests of alpine fir, Colorado blue (silver) spruce, and Engelmann spruce, as well as some small, dwarfed, five-needled pines. With the aid of a binocular, the firs and spruces can readily be distinguished from this height because of their narrow, spire-topped shape. In fact, all details of the landscape are clear here because when passing over mountains the plane is much closer to the earth and also the air is sparkling clear. Blue spruce commonly occurs in groups and small groves along streams and at the foot of most slopes. Until middle age, it is one of the most beautiful of conifers in the Rockies, with regular whorls of densely foliaged branches and often with perfectly conical form. The stiff, sharply pointed needles are about an inch long and vary in color from bright green to a clear silvery bluish tone, especially on the new growth of the season. The blue spruce is the state tree of Colorado and is often used as an ornamental throughout this country and in Europe. It can endure the heat and dry air of the plains and the cold and moisture of such areas as New England.

Common in the middle altitudes throughout the Rockies, as well as the Sierra of California and the Cascades of the Northwest, are the lodgepole pines, so-called because the trunks are very slender and have live branches only at the top. This characteristic made them the favorite of the Indians as poles for their lodges. These trees are often found in dense stands following logging or fire.

A characteristic and striking feature throughout the Rockies at middle and upper elevations are the many groves of quaking aspen along the margins of streams, in swampy meadows, or on rocky drifts. These are deciduous trees related to the poplar and distributed over a greater geographic range than any other tree in North America. The common name is based on the tremulant or quivering motion of the foliage in response to a breeze. Later in the year the many aspen groves in their brilliant fall colors provide a spectacular panorama against the background of somber green of the coniferous trees of the western mountains, where conifers are dominant.

There are many perennial wild flowers in the Rockies where the forest is not too dense. They include asters, penstemons, buttercups, daisies, larkspurs, columbines (Colorado's state flower), lupines, paintbrushes, and shooting stars (*Dodecatheon*). There are also shrubs, other herbs, and grasses wherever sufficient light gets through the forest canopy. This vegetation and the grasses of the meadows sustain cattle and sheep (fig. 5) during summer months, as well as browsing and grazing wild animals. Upon the approach of cold weather, the livestock are driven down to lowland pastures to feed on vegetation kept green by winter rains. As between cattle and sheep—John Muir's "hoofed locusts"—the latter are more destructive to vegetation, for they pull up grasses by their roots whereas cattle merely nibble off the tops. Also sheep travel in close formation, trampling all vegetation in their path.

Having crossed such an imposing mountain range as Sangre de Cristo, one expects to see miles of wild mountain landscape and is surprised to find instead an agricultural valley at the headwaters of the Rio Grande in the San Juan Valley between the Sangre de Cristo and the San Juan Mountains. The valley is three times the size of the state of Delaware and lies at an altitude of 7,500 feet. It receives an average of less than 8 inches of moisture a year. Particularly impressive is the great concentration of the large circular green areas, irrigated by automatically controlled center pivot sprinklers, first noticed in our flight over Kansas. They are located in the area between Del Norte and Alamosa. Most passengers with window seats look down at the scenery when the plane

*Fig. 5. Cattle and sheep in Rocky Mountain meadows (courtesy U.S. Forest Service and USDA Soil Conservation Service, respectively).*

begins to cross the Rockies, and the many green circles are invariably objects of great interest.

3:44 P.M. We are now crossing the San Juan Mountains, with the highest visible peaks about 13,000 feet in altitude. For hundreds of miles along the Continental Divide in a lofty realm of crystal clear beauty, in countless shallow basins scooped out of granite by ancient glaciers are *ciénegas* where mariposa lilies and columbines abound. Here tiny rivulets of clean, cold water begin their missions of mercy, to be joined by a thousand others to form sparkling streams cascading down through groves of fir, spruce, pine, and aspen, then to flow in muddy streams past colorful mesas and over vast plains and deserts, southeast to the Land of Enchantment, west to the awesome chasms of the Colorado, and southwest to the Valley of the Sun. Everywhere water brings the spark of life to arid but potentially productive land. The San Juan River and its tributaries were intensively trapped for beaver by the "mountain men." Looking down on the wild, rugged terrain, one gets a feeling for the magnitude of the tasks so fearlessly undertaken by these hardy, reckless men.

## THE ARID SOUTHWEST

At the latitude at which we crossed the Rockies, ponderosa pines on their western approaches merge into sagebrush or piñon-juniper forests. We have entered Region D (fig. 3), which includes the Great Basin, extending from the Sierra Nevada and the Cascade Mountains in the west to the Rocky Mountains in the east and from southeastern Oregon, southern Idaho, and southwestern Wyoming down through southwestern Utah. Here there is seldom a cloud cover to hide from airline passengers the colorful buttes, mesas, and canyons, sharply delineated by sun and shadow, and vast vistas of green juniper and gray-green sage reaching to far horizons under flawless sky blue. Intriguing as the aerial view may seem, it gives only a tantalizing hint of the region's multitude of colorful scenic treasures—flaming gorges and box canyons carved out of the edges of the plateaus by millions of years of erosion as the Colorado River and its major tributaries cut their way through the red earth, then left further sculpturing to the scouring sands of desert windstorms.

South of the Great Basin and extending far into northern Mexico is the desert country of the Sonoran life zone. The Great Basin and Sonoran regions comprise what is generally known as the "arid Southwest." We will be flying through the southern limits of the Great Basin in approximately the area where it merges with the Sonoran desert and where Great Basin and Sonoran flora interfinger extensively.

Leaving the state of Colorado, the San Juan River flows through the rugged and picturesque Aztec Canyon National Monument in New Mexico. Here prehistoric Indians (but not Aztecs) irrigated their lands with water from the Animas River, which flows into the San Juan. Making a generous loop into northwestern New Mexico, the San Juan brings the state its most abundant supply of water. A series of small community canals built by industrious Mormons bring over 30,000 acres under irrigated cultivation for fine crops of fruit, vegetables, grains, and hay, supporting the towns of Fruitland, Bloomfield, Farmington, Waterflow, and Flora Vista. Dams at Fruitland and Hog Back provide water for about 10,000 acres on the Navajo Reservation.

The San Juan River turns northwest and flows into Utah at Four Corners, plainly in view, the only place in the country where four states join. Then it passes north of Monument Valley and through the Goosenecks of the San Juan to empty into Lake Powell, to which First Officer White calls our attention at 4:08 P.M., as we fly over the colorful Painted Desert of northern Arizona. Lake Powell is a long section of the Colorado River impounded by the Glen Canyon Dam near Page, Arizona. The lake is about fifty miles north of our flight path, but plainly visible in the clear air of the high plateau.

Glen Canyon Dam's hydroelectric plant sells power to cities and industries throughout the West, the revenues paying for the dam and other projects. The dam also permits long-term storage of water for use as it is needed downstream. Glen Canyon Dam rises 583 feet and backs water up for 186 miles in canyons of the Colorado and its tributaries. Lake Powell is the basis for the Glen Canyon National Recreation Area in southern Utah and northern Arizona, popular for fishing and water sports. Subsequent proposals to build dams in this scenic area have met with great opposition from environmentalists.

## THE GRAND CANYON

Crossing the Painted Desert, we fly over a section of the Grand Canyon. "My regrets," said Soviet Party Chairman Leonid Breshnev when he saw the Grand Canyon from an airplane window, "we all have these problems." Breshnev preferred the rectangles of Midwest agribusiness to rugged western scenery (Salisbury, 1976). The Grand Canyon is an immense gorge cut by the Colorado River into the high plateaus of northern Arizona. It is unrivaled in size, majesty, ornate sculpture, and wealth of color. A large part of geologic history, going back half a billion years, is revealed more clearly here than anywhere else on earth. From primitive algae in the lower strata to trilobites, dinosaurs, camels, horses, ground sloths, and elephants, fossils tell the story of evolving life through the ages.

## IRRIGATION IN THE ARID SOUTHWEST

Irrigated land produces 81 percent of the nation's sugar beets, 70 percent of the fruits and vegetables, 40 percent of the cotton and sorghum, 30 percent of the alfalfa, 25 percent of the barley, and 10 percent of the corn and wheat (Wittwer, 1976). Most of the irrigated land is west of approximately 100° longitude, where the Great Plains area begins. Confining our attention now to the Great Basin and Sonoran life zones, it is evident that there is enough rainfall to provide for a varied and sometimes remarkably abundant vegetation, but it is generally not sufficient to produce crops planted by man. Therefore it is not surprising that irrigation technology in this region is highly developed and has been the object of study by irrigation specialists from all parts of the world with similar climatic conditions.

There could be little irrigated agriculture in the Southwest without the impounding reservoirs created by dams. Some transcontinental flights provide a good view of Hoover (Boulder) Dam, completed in 1936, and Lake Mead on the Colorado River where Arizona and Nevada join. Lake Mead can be seen to the north of our flight path today. Hoover is one of the world's highest dams (726 ft.) but it is greatly exceeded in size by Grand Coulee Dam in Washington and Shasta Dam in California.

The system of impounding reservoirs and conveyance canals that supplies the water to make the Southwest desert bloom depends on the Colorado River and its tributaries. We cross the river a few miles north of Needles, California, at 4:36 P.M. One cannot judge the importance of the Colorado from the meager trickle of water in its riverbed at this point. The Colorado is one of the great rivers of North America, rising amid the snowcapped peaks of north central Colorado and flowing 1,400 miles southwest into the Gulf of California and draining about one-fifteenth of the continental United States (excluding Alaska). For about a thousand miles of its way to the sea, the Colorado River travels through deep canyons, the most famous of which is the Grand Canyon in Arizona.

The Colorado River was destined to become a major factor in the agricultural development of the Southwest. Only two years after the completion of Hoover Dam, Parker Dam near Parker, Arizona, and Imperial Dam just north of Yuma, were completed. From Parker Dam the Colorado River Aqueduct carries water 242 miles to supply Los Angeles, San Diego, and other cities in the Metropolitan Water District of Southern California. From Imperial Dam the All-American Canal carries water 80 miles to the Imperial Valley in the southeast corner of California. Then, in this valley, through 2,900 miles of lateral canals, Colorado River water transforms a formerly barren desert into the most productive land in the nation, producing cotton, forage, and vegetable crops throughout the year. A branch canal, 126 miles long, carries water to California's Coachella Valley.

Water for the Palo Verde Valley, where Blythe, California, is located, is diverted by the Palo Verde Diversion Dam, which is a stabilizing weir. The canals of the Palo Verde Irrigation District are supplied by gravity flow. An atomic power-generating plant, planned for a nearby mesa, will require 34,000 acre-feet of water per year to cool its two units.

Between Hoover and Parker dam is Davis Dam, completed in 1949 and supplying power to Arizona. The last of the Colorado River dams to be constructed was the Glen Canyon Dam, completed in 1964. Arizona's share of Colorado River water is 2.8 million acre-feet. About 1.2 million acre-feet of this will not be available until the completion in 1985 of the Central Arizona Project (CAP), which will draw water from Havasu Lake, behind Parker Dam.

## THE MOJAVE DESERT

West of the Colorado River the high plateau of the Mojave Desert extends to the mountains that ring the Los Angeles basin. The only break in native terrain and flora is where water from those mountains supplies an occasional town and the green fields of alfalfa that surround it. The Joshua tree (*Yucca brevifolia*) is the dominant native plant and can be easily distinguished from above. This grotesque tree, commonly sixteen to thirty feet high and with an open crown of armlike branches, forms extensive groves on the Mojave Desert and on mesas north and east as far as Utah.

4:52 P.M. As we fly over the desert, to our left Big Bear Lake can be clearly seen, nestled in a basin of the San Bernardino Mountains. Southeast of the lake southern California's highest peak, Mount San Gorgonio or "Grayback," rises to an elevation of 11,502 feet.

## A NOSTALGIC INTERLUDE

Nostalgically I recall the many boyhood days I spent climbing Grayback and surrounding peaks, following the streams, and exploring the forests of ponderosa pine and white fir in the canyons and on the north slopes of the San Bernardino Mountains. Today the hazards of hiking alone in the mountains are considered unacceptable, but in those days I thought of myself as a latter-day John Muir exploring the enchanting forests of the "Sierra of the South." Muir had nearly always traveled alone in his extensive mountain explorations. My parents, European immigrants who had been practically "on their own" since childhood, did not seem to feel that this was strange behavior.

My parents had purchased five acres of virgin land north of the little town of Beaumont and at the very edge of the foothills of the San Bernardino Mountains. They planted it to apple trees, starting the precarious life of small-scale, hardscrabble farming. The only way we could survive, of course, was by working for absentee owners of orchards, as well as taking care of our own orchard. It was always early to bed, early to rise, six or seven days a week, depending on the season. Our frugality was beyond anything imaginable even among the poorest of present-day families. The constant struggle of the family unit, not only for survival, but to "put some money in the bank," left no room for a generation gap between my parents and their two sons. There were serious setbacks for us and other farmers in the valley. Apple trees planted by naive and misguided beginners had to be removed, with shovel and ax, for the codling moth was too formidable a foe in the relatively mild climate. (Apples are still grown profitably at a much higher elevation—5,000 feet—at Oak Glen.) Apples were replaced by other deciduous fruit trees, mostly cherries. Five acres grew to ten and finally to thirty; the workdays never became shorter.

Yet there remain precious memories, such as sunny spring days in a cherry-blossom wonderland. Our home was on a small knoll at the north edge of Cherry Valley, providing a splendid view. To the south was a mosaic of deciduous fruit orchards, some in the full flush of bloom, others already adorned with spring's soft beauty of tender green. To the southeast, Mount San Jacinto rose majestically above the valley floor, its lofty north face still holding the last patches of winter's snow in deep shady ravines. To the north the tall blue escarpment of the Yucaipa Range rose brightly from somber foothills, culminating in massive, snowcapped San Gorgonio, challenging and beckoning—enriching boyhood dreams with beauty and enchantment.

A similar story could be told for millions of immigrant families, hungry for land—and liberty and hope—that the Old World could not provide. Poor, but ambitious, they came with boundless and almost childlike faith that, in a fabled land across the sea, hard work, thrift, and patience would be rewarded—a faith that did not remain unrequited.

## THE LOS ANGELES BASIN

The San Bernardino and San Gabriel mountains form an east-west range that separates the Mojave Desert from the Los Angeles Basin, which is a level plain extending eastward about seventy-five miles from the Pacific Ocean to Redlands and Riverside. These mountains form the last range we will have to fly over, of the countless ranges we have passed to reach the "promised land." "But over that final range," said Carey McWilliams, "is paradise" (McWilliams, 1946). Below the conifers of these mountains is the chaparral belt. *Chaparral* derives from the Spanish *chaparro,* a dark evergreen oak. In southern California the term refers to a community of plants dominated by evergreen shrubs commonly three to ten feet tall and with firm, mostly rather small, thick, hard leaves resistant to water loss during the long, rainless summer season. Chamiso, various species of manzanita, wild lilac, and a species of scrub oak are characteristic shrubs of the usually dense thicket.

Every part of the country has its own type of natural disaster, such as earthquakes, blizzards, floods, hurricanes, and tornados. Besides earthquakes, southern California has its annual "brush fires" in the resinous chaparral, abetted by a dry north wind called the "Santa Ana." Following a fire, the shrubs reproduce themselves from their underground portions, reestablishing a mature plant community in fifteen to twenty years. Many homes are built in the chaparral-covered hills. Some may be destroyed by brush fires or damaged when the winter rains cause erosion and dangerous mudslides on the burned-over slopes no longer protected by the dense plant cover. Certain areas in the plains below can be severely flooded.

Experience has shown that man cannot prevent chaparral fires. The longer the period between fires the greater the severity of the holocaust, as the quantity of combustible fuel continuously increases. Ironically, from an ecological standpoint chaparral fires are desirable. Chaparral is well adapted to the recurrent fires it has experienced for countless millennia. In southern California there is very little decay of vegetation and fire serves to return nutrients to the soil. The regrown area supports a larger wildlife population than the heavy, impenetrable growth artificially nurtured by man's efforts at fire prevention. The best solution for those who wish to build their homes in the chaparral-covered foothills of southern California would seem to be to build them of adobe, concrete, brick, rock, or other noncombustible materials. Particularly hazardous are shake or shingle roofs.

Southern California's greatest disaster today is not brush fires or even earthquakes. It is a disaster principally issuing from the exhaust pipes of its millions of automobiles —smog! Reaching the rim of the basin we note that on this warm summer day the stratum of smog is so deep and dense that we can barely discern the vast expanse of urban sprawl below us. On the rim itself high above the floor of the basin, smog has caused considerable damage to ponderosa pine trees.

Smog has become a serious problem in many of the world's heavily populated and industrialized areas, especially where motor vehicles are used extensively and where temperature inversions are common. The Los Angeles Basin is such an area. Motor vehicle travel is particularly heavy. The cool moist air from the ocean becomes trapped below a layer of warm dry air, where it remains stationary and becomes laden with pollutants. Finally the warm air rises, allowing the polluted cool sea air to escape over the mountains and become dispersed over the desert. Sometime during the day, smog reaches its peak intensity and then subsides, first in coastal areas and then progressively later as one proceeds east. Coastal areas are generally free of smog by late afternoon.

Smog consists of particulate matter, carbon monoxide and other gaseous hydrocarbons, sulfur oxides, nitrogen oxides and, most importantly, photochemical oxidants. The latter form when nitrogen oxides combine with gaseous hydrocarbons in the presence of sunlight. They include ozone, nitrogen dioxide, peroxyactyl nitrate, aldehydes, and acrolein.

Smog damages materials, home furnishings, and plant life. Some have incriminated smog as the cause of a number of respiratory and other diseases (Middleton, 1973). However, correlation of lung cancer with areas of high concentration of polycyclic aromatic hydrocarbons, once considered to be a strong indictment against smog as a human health hazard, is now believed to be a mere coincidence. Workers in such areas happen to be engaged in occupations in which there is a greater degree of contact with carcinogens than in other areas (Gillette, 1977). All agree, however, that smog causes at least short-term respiratory effects and eye irritation, forcing some people to leave for smog-free areas. An adverse effect on plant life can be clearly demonstrated. Statewide annual loss to agriculture from smog has been estimated at $35 million (Heller, 1972).

Leaving the desert behind near Cajon Pass, we descend into the Los Angeles Basin. As a native of California's southland, I can of course remember when orange and lemon orchards occupied much of this area, forming a nearly continuous carpet of green from the San Fernando Valley in the west to Highland, Redlands, and Riverside in the east. Particularly in winter, tourists from the frozen East stared in amazement at row after row of dark green trees, laden with gold or yellow fruit, continuing in ordered ranks to the foothills of a range of snow-covered mountains (fig. 6). McWilliams could still write, as late as 1946, "Just how important the citrus belt has been in changing the physical appearance of the land can only be sensed by trying to imagine what Southern California would be like were these green belts removed" (McWilliams, 1946). Just three decades later we can sadly say, "amen!" The formerly dreaded north winds are now considered to be a blessing, blowing away the smog and giving an occasional glimpse of the landmarks of a paradise lost. But in place of the humanized beauty of a fabulously productive land, we see the blight of urban sprawl. A bit of the captivating beauty of rural southern California as some once knew it may still be seen in the Santa Clara Valley of Ventura County, about fifty miles northwest of Los Angeles.

If the unconscionable greed of southern California's speculators, investors, promoters, and boomers and the intoxicating hyperbole of its well-intentioned boosters could have been in some way abated, possibly the development of the region could have been planned to preserve more of its early charm and beauty.

5:13 P.M. (2:13 P.M. Pacific Standard Time). We arrive at Los Angeles International Airport and are soon hustled over a network of futuristic freeways, cloverleaves, and ramps at breakneck speed to our home.

## FUTURE SHOCK

There are residents of California whose grandparents spent months of toil and suffering on harsh and dangerous trails to reach this state. Then, after the completion of transcontinental railways, the duration of the trip could be measured in terms of days. Today our trip required five hours. Would there be an advantage in reducing the time to five minutes? When I was about five years old our family traveled from Los Angeles to our home near Beaumont in a wagon drawn by two horses, covering about ninety miles in three days. Today we flew an equivalent distance in nine minutes. Would there be an advantage in reducing the time to nine seconds? Obviously not, yet this degree of acceleration is in principle what is happening to all human activity involving technology. Technology feeds on itself, making more technology possible.[11] But is the possible necessarily desirable? The ecological, social, and psychological price is enormous, as pointed out in Alvin Toffler's best seller, *Future Shock* (1970).

He defines "future shock" as "the human response to overstimulation." The pace of change, diversity, transience, and novelty increases exponentially, rocketing society toward an historical crisis of adaptation. "We create an environment," he says, "so ephemeral, unfamiliar and complex as to threaten millions with adaptive breakdown." (Toffler recognizes, of course, that some level of change is as vital to health as excessive change is damaging.) As one of the antidotes to future shock, Toffler suggests "we cut down on change and stimulation by consciously maintaining longer-term relationships with the various elements of our physical environment."

So my wife and I will take the edge off today's phantasmagoric experience in our humble backyard garden in Westwood, seeking "continuity, order and regularity in the environment," as Toffler recommends. Located only five miles from the ocean, Westwood is relatively free of smog, particularly in the late afternoons and evenings. In our

[11]Fortunately there are resource limits. The stratosphere-polluting Concorde SST, for example, uses five times more fuel per passenger than the Boeing 747 and supersonic development now seems hopelessly uneconomic.

*Fig. 6. Rural scenes in southern California before the years of urban sprawl. Top, a section of the east Los Angeles Basin in the early 1930s, with Mt. San Gorgonio in the distant background (courtesy of Sunkist Growers). Bottom, a winter scene in the mid-1940s, looking toward Mt. Baldy (courtesy Frasher Collection, Pomona Public Library).*

garden we have trees that bear avocados, figs, and citrus fruits much like those of the mission padres and the Spanish dons, linking us with California's romantic past. Roses grow everywhere, filling the garden with a riot of color. A giant elm and other lesser trees comprise a bird haven: robins, finches, goldfinches, sparrows, starlings, shrikes, blackbirds, jays, mockingbirds, mourning doves, and others—about forty-three species during the year if both residents and migrants are included. They fill the day with chatter and song. But their songs will subside as the mantle of night softly falls over a quiet, peaceful scene. Soon the sun will again blaze forth on our land, first sending shafts of light through the canopies of eastern forests, then bringing life to the green and gold of prairies and plains and bathing the high peaks and deep canyons of the West with the warm glow of morning light, and finally calling forth the first burst of song from the birds in our garden as hibiscus petals unfurl to greet the new day. Tomorrow we can expect the same, and a year hence, and for eons of time, linking past, present, and future in an ageless pattern.

Paarlberg (1976) reminds us of an ancient long-term agricultural outlook statement: "While the earth remaineth, seedtime and harvest, and cold and heat, and summer and winter, and day and night shall not cease" (Genesis 8:22).

There are certain unchanging verities in the universe—certain immutable cosmic forces in which humanity has found continuity, order, and beauty. To contemplate them seriously is a cure for future shock.

# 3 The East

*Cultivators of the earth are the most valuable citizens. They are the most vigorous, the most independent, the most virtuous, and they are tied to their country and wedded to its interests by the most lasting bonds.*
—Thomas Jefferson

On December 21, 1620, 102 Pilgrims—73 males and 29 females—went ashore from the *Mayflower.* The vessel had been headed for lands farther south, belonging to the Virginia Company of London, established at Jamestown, Virginia, in 1607. This company was one of the royal monopolies with which England furthered its colonial interests and developed trade and settlements. But foul weather prevented accurate calculation of direction and the emigrants first sighted land in the neighborhood of Cape Cod on November 19, 1620. Most of them stayed on the ship for weeks while Captain Miles Standish and a small company of men reconnoitered the area extensively and finally recommended a landing site nearly opposite the point of Cape Cod. There the Pilgrims established their first settlement, which they called Plymouth, and their colony became the Plymouth Colony. In 1628 another group of English settlers, the Puritans, established the Massachusetts Bay Colony at Salem. Both the Plymouth and the Massachusetts Bay colonies eventually sold out to the colonists, who obtained the necessary revenue for the transaction from the sale of furs, timber, and codfish.

The Pilgrims had expected to barter for corn raised by Indians, based on reports of Captain John Smith of the Virginia Colony, who had explored the New England coast in 1614. But it just so happened that the Indians along the coast where the Pilgrims landed had been exterminated a few years previously by virulent diseases. These had been introduced by white men who previously penetrated the area, and to which the Indians proved to have no resistance. The fact that the land was uninhabited was a temporary misfortune for the colonists but afterward proved to their advantage. They were able to build their crude habitations, and later plant their crops, on land that had been cleared by the Indians and which was penetrated by a stream of fresh water. On Cape Cod they found about ten bushels of corn and beans that had been cached in baskets, in the ground, by the Indians. This served as seed for spring planting. The early settlers of Rhode Island had the same experience, as did those of New Netherland and New

Sweden. In New England many towns were built on open spaces formerly cultivated by the Indians (Bidwell and Falconer, 1925).

As early as 1623, emigrants from Massachusetts were attracted to the fertile lands along the Piscatagua River in New Hampshire. This valley was especially suitable for growing hay and forages, and livestock thrived. It was not until 1636 that a successful English settlement was made in the Connecticut Valley, the largest single area of fertile land in New England and which had already been settled by the Dutch. It was also in 1636 that the first settlement was begun in Rhode Island. The Narraganset Indians of that region were probably second only to the Iroquois of western New York in their agricultural attainments and had done pioneer work in subduing the land for cultivation, clearing it of timber for a distance of eight to ten miles back from the shores of Narragansett Bay (Updike, 1847). Much of the land was purchased from the Indians and corn was obtainable by trade. Rhode Island was noted for its excellent dairy products. Boycotted by the strict and puritanical Pilgrims in the north, the Rhode Islanders developed a lively trade with the southern colonies.

It was fortunate for the early European immigrants that the Indians inhabiting much of the North Atlantic region from the Carolinas to Canada practiced both hunting and crop-growing, for this was not true of all Indians. The Indians the French encountered along the lower Saint Lawrence River lived almost entirely by hunting, and those along the Gulf of Mexico mainly by gathering fruits, nuts, herbs, and fleshy roots of plants growing in the swamps. Where Indians practiced agriculture early white settlers were not only able to trade with them for corn and beans, but they also learned much about growing the indigenous crops. On the other hand, the Indians obtained various plant crop and livestock varieties from the whites. As early as 1687, the commanding officer of a French expedition, after a campaign of destruction of corn crops and livestock of the Iroquois of western New York, wrote that in all the villages were found "plenty of horses, black cattle, fowl, and hogs." Apple orchards were also found by the first white explorers to penetrate the area (Pinkerton, 1819; from Carrier, 1923).

Unlike the Spaniards, who generally sought rapid exploitation of native human and mineral resources and succeeded in this goal, Englishmen found no mineral riches but were satisfied to be able to build a new life in a new land, while doing their work themselves. They were to be joined by other Europeans with a similar goal. The early settler's life was a race against time to clear a farm, build a house, and rear a family, while his settlement might lose half its inhabitants from Indian attack, epidemics, accidents, starvation, and the long, severe winter.

The Pilgrims suffered severely the first winter from exposure, hunger, and disease; half of them perished. The next summer their crops were generally poor, but twenty acres of corn planted under the direction of the Indian Squanto had thrived. In preparation for a three-day festival of thanksgiving in November 1621, the Pilgrims hunted and gathered ducks, geese, clams and other shellfish, eels, leeks, watercress and other greens, wild plums, and dried berries. The Indian Chief Massasoit and some ninety "brightly painted braves" joined the feast, bringing five deer and "other good things," the latter presumably including turkeys, although they were not specifically mentioned. Days of thanksgiving were celebrated sporadically by Americans in later years, but not until 1863 was a Thanksgiving Day formalized, through a proclamation by President Abraham Lincoln.

## THE PILGRIMS RESCUED BY INDIAN CORN

Most of the early settlers had no background of practical agriculture. There was wild game, such as deer, partridges, turkeys, and passenger pigeons, the latter in incredible numbers, in forested areas partially burned out by the Indians, but the Pilgrims were not good hunters; hunting was not a sport of middle-class Englishmen in those days. And few if any had ever caught a fish (Channing, 1905). There was an important resource, however, that the Pilgrims, with the help of the Indians, were able to exploit with great success—Indian corn.

"Indian corn"[1] was a contribution from the American Indians during the few peaceful years following the Pilgrims' landing. In evaluating the great contribution of Indian corn to the survival of the early English settlers at Jamestown and Plymouth, and eventually to all of mankind, bear in mind that corn had long since become so domesticated by the Indians for efficient food production that it could no longer survive without human care in its propagation. The ancient wild grass from which corn developed probably had a seed that could be wind-borne, but this facility was lost thousands of years previously as selection continued for increasingly greater productivity. If the modern ear of corn falls to the ground under conditions in which sprouting is possible, the numerous kernels rob one another of soil nutrients and moisture. There is no record of modern corn ever having maintained itself in nature.

The amount of selection of corn varieties the Indians had undertaken is amazing. They had varieties in southern Canada only 4 or 5 feet tall that matured in less than 90 days and others in the South that grew 10 or 12 feet tall and matured in 5 months. They had varieties for special purposes: meal-making, parching, popping, and sweet corn for roasting ears. They also had early- and late-maturing varieties. The number of ears to the stalk, number of rows of kernels per ear, and the number of kernels in a row were almost the same as for our best modern varieties. Kernels could be white, yellow, red, blue, or variegated, but in all cases produced a white flour. Yield of Indian corn per acre in virgin soil was very good, considering the wide spacing of the plants in those days (Carrier, 1923).

Indian women soaked the corn seed for a few days before planting four seeds to a hill. As in every other detail of corn-planting, the English colonists followed this Indian custom:

> One for the squirrel, one for the crow,
> One for the cutworm and one to grow.

For the Indians the custom derived from the ancient Maya, who attached religious significance to the four colors of their corn—white, red, yellow, and blue—and to the four points of the compass and the four gods who held up the corners of the earth and so influenced the wind and the rain (Giles, 1940).

Other crops, such as beans and squash or pumpkins, were often interplanted with the corn. As related in chapter 1, the corn-beans-squash culture, originating in Mexico and Central America, spread throughout pre-Columbian North America. The white settler, after clearing a few acres in the forest, planted these crops in the manner learned from the Indians. These three crops, along with fish and wild game, sustained the pioneer family until gristmills were at hand and wheat could be grown and ground. To this day in New England one can see fields with cornstalks supporting bean vines and with squash or pumpkin vines covering the remaining ground.

[1] "Corn" was a generic term for all four of the cereal grasses (grains) known to Europeans (wheat, rye, oats, and barley). Therefore, when the first colonists became acquainted with the new cereal they distinguished it from other "corn" with which they were familiar by calling it "Indian corn." Eventually Americans shortened the term to "corn." The colonists called the ground product "Indian meal." An important insect pest of Indian meal and other stored food products was called "Indian meal moth," which has remained its common name to this day.

Corn was an important factor in ensuring the survival of the Pilgrims during their first difficult years at Plymouth, as it had been for the settlers at Jamestown and as it was to be for the Dutch and Swedish settlers—in fact, for all early settlers anywhere in eastern America. If insufficient corn was raised by the settlers it could be obtained by trade with the Indians, who always seemed to have plenty. It gave a bountiful return for the few seeds and labor invested, and, most important, it was ideal for storage. When kept dry and protected from rodents and insects, corn remained unchanged in storage from one harvest season to the next.

## The Story of Squanto

According to the story taught to schoolchildren, and fondly retained in memory as a romantic episode of the early years of our history, on March 16, 1621, an Indian by the name of Squanto strode into the tiny settlement at Plymouth with the greeting, "Welcome, Englishmen," and proceeded to inform its inhabitants of the merits of placing a fish under every hill of corn to supply the necessary soil nutrients. I must admit that the fact that Squanto knew how to speak English never intruded upon my belief that fertilization of soil for plant crops was a part of Indian agriculture. That is, not until reading an article by anthropologist Lynn Ceci (1975), who deromanticized the story by pointing out that in 1614 Squanto was kidnapped[2] and sold into slavery in Malaga, Spain, where he stayed two years, then escaped to England, where he learned English. After living in England for two years, he was taken back across the Atlantic to Newfoundland. Acting as a pilot and guide, he was eventually brought to Cape Cod, Massachusetts.

Thus, Squanto had four years of contact with Europeans, who had long known the use of fertilizers, including fish, in farming practice. Fish fertilizer had probably been used by European settlers on the North American continent before the Pilgrims arrived. Neither the Pilgrims nor anyone else ever saw Indians use a fertilizer of any kind (Hariot, 1588; Smith, 1608). According to Ceci, Amerindians did not use it even after years of exposure to European farming technology. Apparently fish heads, as well as guano (bird excrement) from offshore islands, were used as fertilizer, however, along the coast of Peru in prehistoric times (Weatherwax, 1954; Mason, 1964).

Indians practiced "shifting agriculture," moving their settlements when yield diminished, either to new, more fertile soils, or soils that had lain fallow sufficiently long to have a restored fertility. This was the characteristic practice of early agriculturists over much of the world (Curwen and Hatt, 1953). At first the white settlers had plenty of land and followed the Indian custom of planting their crops on new land as the soil became exhausted. It was not long before the regions settled by the whites became too crowded for the continuance of this Stone Age type of agriculture and the settlers had to find ways of cropping the same land continuously (Clark, 1945). Various means were found of restoring the nutrients that were annually extracted from the soil by the crops.

## How Corn Was Prepared for Eating

Nevertheless, the early colonists learned much from the Indians regarding ways of

[2]Squanto, along with a number of other Indians, was kidnapped by Thomas Hunt, a sea captain with Captain John Smith's previously mentioned exploring expedition of 1614. During his two years in England, Squanto learned to like Englishmen and this was of great value to the colonists. Squanto assured his chief, Massasoit, that the English would be valuable to him in his struggle with the Narragansett Indians. Squanto died in 1622 while on an expedition with the colonists.

raising and storing corn and preparing it for eating. This was "flint corn," the kernels of which consisted of hard starch, difficult to grind. It was also eaten in the green state as "roasting ears." Sweet corn was not known to the colonists until after the Revolution.[3]

The Indians often parched or popped corn before storing it in pots or small caves dug in canyon walls. They also soaked it in wood-ash lye; corn preserved in this way became known to the colonists as "hominy." Much corn was ground into meal by the Indians with a mortar and pestle fashioned from stones. This process left some fine sand in the meal. Because cornmeal was such an important part of the diet, the teeth of some Indians were worn down nearly to the gum line by continual abrasion (Schery, 1972). The Indian also sucked the sweet sap from the green stalk of the corn plant (Gray, 1933).

## THE DUTCH IN NEW YORK

A Dutch trading post was located on the present site of the city of New York by about 1613, but little farming was at first attempted. To buttress their claim to the region, the Dutch in 1625 sent an expedition consisting of two ships with 103 head of cattle and horses, as well as many hogs and sheep, seeds, and agricultural implements. A third ship with hay and water was sent along in case of a delay in the voyage. Only two animals died en route—a remarkable record in sharp contrast to the great losses in animals suffered by the English (O'Callaghan, 1850).

The livestock was landed on the Island of "Manhattes." It was purchased from the Indians for 60 guilders and provided good pasturage. The same year another ship loaded with sheep, hogs, wagons, and plows was sent over. As had the Pilgrims at Plymouth, the Dutch on Manhattan found much land that had been cleared by the Indians. The Dutch West India Company furnished the farmers with a house, barn, farming implements, and tools, four horses, four cows, sheep, and pigs, which, along with the animal progeny, the farmer could "own" for six years, after which he was expected to repay in kind the number of animals received. The farmers paid a yearly rental of 100 guilders and 80 pounds of butter. The Dutch sent over the best foundation stock of horses and cattle to be found in America at that time and good varieties of fruit trees. By 1650 all the wagons and plows needed by the Dutch farmers were constructed in New Netherland (New York). Thus the Dutch settlers were far better treated by the sponsors of their emigration than were those of any other nation. They were excellent farmers and very industrious. They were the first to grow European grain profitably in America, shipping it to the West Indies as early as 1645.

Large Dutch landholdings or manors were developed along the Hudson River as far north as Albany. The owners of the manors were called "patroons." They were given title to eight linear miles on either side of the Hudson River or sixteen miles on one side, and for an indefinite distance into the interior, provided they could bring over fifty families within a period of four years. The patroons were required to buy the land from the Indians. The largess of the Dutch West India Company was little appreciated by the patroons. They soon began to quarrel over payment of rents, contributing to the disorganization of the colony. Cruelty and injustice to the Indians by the incompetent governors sent over by the company provoked Indian hostility. All these factors

[3]In 1790 a soldier with General John Sullivan on his raid on the Indians' Six Nations had seen miles of tall corn. He returned to Plymouth with a pocketful of sweet corn. The seed was planted and yielded roasting ears with larger, fuller, and sweeter kernels than the whites had previously seen, and much superior to the kernels of the immature cobs which they had been using as a summer vegetable (Giles, 1940).

resulted in tame submission of the Dutch colonists to the English, ending for all time Dutch rule in North America. English settlers soon joined the Dutch at Long Island and along the Hudson River.

Unfortunately, the English continued the policy of granting large estates in the New York province, three of them containing over a million acres each. The tracts of land were granted by royal governors largely in consideration of support, friendship, family connections, social position, and wealth. Many large tracts remained uninhabited for decades, or if inhabited, the tenants were ruled by the lord of the manor almost as in the days of feudalism in Europe. Understandably, the manors were not settled as rapidly as had been expected, for although they were choice tracts of land, most immigrants were not keen about exchanging New World serfdom for the serfdom from which they had escaped. The liberal homestead laws, predicated on the democratic theory that the preparation of wild lands for the plow was a contribution to the country's welfare and should be rewarded by a gift of the land to the settler, were not formulated until the middle of the nineteenth century. They greatly aided in the settlement of the Mississippi Valley and the West (Hedrick, 1933).

## THE FATE OF "RED INFIDELS"

Farming was not too difficult where there were plenty of open meadows and fields that had been cleared by the Indians, whom the "Divine Hand" had a way of removing through internecine wars and disease, for the benefit of the white settlers (Carrier, 1923). Among white settlers, the Puritans in particular were noted for their tendency to attribute any Indian disaster to divine intervention. Among the few Indian tribes not decimated by white man's diseases were the Pequots of the Connecticut Valley. When their village was finally surrounded and burned to the ground, one of the Puritan captains said it was "as though the finger of God had touched both match and flint." A minister called upon his congregation to thank God "that on this day we have sent six hundred heathen souls to hell" (Hagan, 1961).

This attitude prevailed despite the fact that Indians tended to be friendly, generous and helpful in their first contacts with whites. Generosity was an Indian cultural trait, along with truthfulness and honesty, at least before they had dealings with whites and succumbed to the ravages of liquor (Vogel, 1972). The establishment of the first white settlements, difficult at best, would have been a far more excruciating and precarious ordeal without the help and cooperation of the Indian. In part the attitude of the Puritans can be explained, but not excused, by the fact that they considered the "red infidels" to be people without a religion, and therefore inferior. The same attitude prevailed among most Euroamericans to greater or lesser degrees during their conquest of the American continent. Actually, Indians were deeply religious, worshiping deities representing the natural elements and forces of the cosmos. In sharp contrast, whites thought of these same elements and forces only as resources to be exploited. Only recently has an appreciable part of white society, aware of its ecological blunders, begun to modify its concepts in this regard (Terrell, 1971).

## THE FUR TRADE

New Englanders were soon to learn that some of the world's richest fishing banks lay offshore, and with timber near at hand, a large fleet of vessels was eventually built to carry fish and lumber to ports throughout the British Empire. However, at first the fishing fleets of Europe almost monopolized the industry.

There remained the fur trade. (It was in anticipation of shipments of furs from the Pilgrims that English merchants agreed to advance them money for their first trip to America.) Furs were a natural resource that was an important item of economic support not only for the first settlers in the East but at every step of the way across the vast North American continent. The fur trade was a prime factor in stimulating continual penetrations westward (Moloney, 1967).

By far the most important fur-bearing animal was the beaver. Although the beaver-fur trade is popularly associated with the West, it also played an important role in the exploration, settlement, and foreign relations of the eastern states. Competition for control of the fur trade resulted in the British seizure of Manhattan from the Dutch, thereby ensuring control of the Hudson River and access to the inland sources of furs. The fur trade helped greatly to sustain the first settlements on the coast, including Plymouth Colony. Beaver pelts were small in bulk, high in value, and well suited for overseas trade. In exchange for furs, the early colonists traded the Indians cheap trinkets, guns, liquor, and corn, the latter having been introduced to them by the Indians only a few years earlier. Beavers were rapidly exterminated in New England, but the fur trade was of great importance in the economy of the region for two or three decades, the critical years of the founding of the Massachusetts Bay towns.

## LIVING HISTORICAL FARMS

For two centuries the forward march of the pioneer from the Atlantic to the grassy plains beyond the Appalachians was marked by the hearth smoke of log cabins in forest clearings. The cabins were so sturdily built that some may still be seen in New England and many in the mountains of Appalachia. In later years the colonists could use bricks for the construction of chimneys for their fireplaces, which rested on foundations of fieldstone. Fireplaces served both for heating the house and for cooking, until iron stoves began to be common about 1850. The early settlers had no matches; starting a fire was difficult and time-consuming. Therefore live embers were kept overnight by burying them in the ashes. The hearth was never cold.

The settler's family sat on wooden benches, ate at a wood-slab table, and had mostly wooden trenchers and gourds for dishes. Wood pegs and deer horns held surplus clothing and the woodsman's rifle with powder horn and bullet pouch.

Tourists may now visit faithfully re-created early American villages that have been made into permanent historical museums in such places as Old Sturbridge, Massachusetts; Cooperstown, New York; and Williamsburg, Virginia. All household articles and farm implements of the early eastern settlers may be seen there (Lent, 1968). Throughout the country there are "living historical farms," which are farmed in the same way as during specific times in the past, in the areas in which the farms occur, including the same types of plants and livestock, and using the same types of tools and equipment. There is an Association for Living Historical Farms and Agricultural Museums to which individuals and institutions may belong. Most states have living historical farms or sites being developed by private nonprofit corporations or the National Park Service. *Living Historical Farms Handbook,* by J. T. Schlebecker and G. E. Peterson (1972) provides those interested in creating such farms with useful information based on experience with existing living historical farms. The handbook also describes one or more such projects in each of forty-one states. (The paperback is for sale by the Superintendent of Documents, Government Printing Office, Washington, D.C. 20402, for 65 cents.)

Space does not permit a description of more than a few of the most thoroughly developed and best-known projects. One of the first to be laid out was the Freeman Farm at

Old Sturbridge Village, Massachusetts. There one can see a large number of animals, including oxen, milch cows, hogs, sheep, and barnyard fowl. Several acres under cultivation are surrounded by split-rail fencing. At appropriate seasons one may see plowing with oxen, mowing of oats with a scythe, fence-building, or dyeing of wool.

At McLean, Virginia, the Turkey Run Living Historical Farm portrays a poor settler's farm as of 1776. It is being developed year by year as the settlers would have done. Eventually the cabin will be improved.

At Lincoln City in southern Indiana the National Park Service's Lincoln Boyhood National Memorial has developed a living historical farm where a visitor can "walk back into history" to see how the Tom Lincoln family, typical Indiana pioneers, probably lived in the years 1827-1830. The site includes a cabin, small smokehouse, toolshed, corncrib, barn, chicken house, a fairly large garden and a growing orchard nearby, and a ten-acre farm with tobacco, cotton, flax, and corn. If the staff is not too busy it may allow visitors to walk behind a plow to experience "participatory history." In striking contrast to the humble cabins at Turkey Run and the Lincoln farm is the manor house on the living historical farm at George Washington's Birthplace, near Fredricksburg, Virginia, but crops are hoed in a similar manner. In Washington, D.C., the agricultural wing of the Smithsonian's Museum of History and Technology traces agricultural developments from colonial times to the present. Wooden harrows, hayforks, portable steam engines, and early gasoline tractors may be seen there (Schlebecker and Peterson, 1972; Rasmussen, 1975*a*).

## THE SETTLERS' DIET

The colonists observed the Indians covering unhusked ears of corn with hot ashes or holding them over a bed of coals on the end of a pointed stick. The Indians' succotash was a mixture of green corn kernels and shell beans boiled together, sometimes with pumpkin and chopped venison added. The colonists enthusiastically adopted this dish. They also learned from the Indians how to prepare a mixture of charred corn, chopped venison, and hard fat. Rolled in a skin case, it sufficed to sustain hunters or fur traders for many days, even if no game was encountered on their journey. The Indians also showed the whites how to make maple sugar. This could be mixed with cornmeal and water to prepare "Indian pudding."

To the wives of the white settlers goes credit for a corn bread made by kneading cornmeal, water, and fat into a dough and baking it on flat stones before a fire. Eventually corn bread was improved by the addition of yeast and maple sugar. This product became known as johnnycake in the North and corn pone in the South and is still a staple farm food. Evidently the white settlers' wives also invented the New England dish of baked beans with salt pork and molasses, the latter obtained after trade with the West Indies was initiated. It is unlikely that Indian forest dwellers had pottery strong enough for baking beans. The Indians probably just boiled them (Clark, 1945).

In some areas the summer diet of the early white settlers could be adequate because, in addition to the crops they grew, they could obtain products of the forest: wild game —usually deer, but including also that unique Thanksgiving symbol, the turkey—fish, small, barely palatable wild fruits (plums, cherries, grapes, and crab apples), berries (blueberries, blackberries, raspberries, gooseberries, strawberries, and cranberries), edible roots, and greens.[4]

[4]O. P. Medsger, in *Edible Wild Plants* (1939), describes 82 fruits and berries, 18 nuts, 12 edible seeds and seedpods, 75 salad plants and potherbs, 35 edible roots and tubers, 28 beverage and flavoring plants, and 9 sugars and gums from northeastern United States and eastern Canada, west to the Mississippi River.

The heavily forested regions of the East never provided the settlers bountifully with game except where the Indians had thinned it by burning or where forest was interspersed with meadows. If the forest had teemed with wildlife there would not have been so many instances of near famine. The dense climax forest with a high proportion of mature trees shading out any understory was a "biological desert," shunned even by Indians. Those few Indians who inhabited the dense forests were the "bark eaters," despised by other members of their race.

The relative inability of densely forested areas to support human populations might be indicated by the fact that east of the Mississippi River the aboriginal population was estimated (Fey and McNickle, 1959) to have been 6.95 per 100 square kilometers, not greatly above that of the barren Arctic Coast (4.02), below that of the arid Southwest (10.7), and only a small fraction of that of California (43.3), which had by far the highest population density in what are now the coterminous United States.

Clearings for planting crops, whether made by Indians or white man, were a step beyond the burning of undergrowth often practiced by the Indians to increase useful plant and animal life and improve hunting. Clearing land for agriculture increases many species of wildlife. For example, for every land bird that is now less abundant than it was before the Pilgrims landed, there are believed to be five or six that are more abundant (Jones, 1974).

The partly abandoned fields of later years provided more feed for game, particularly deer, than did the original forest. Also there were fewer predators. Despite enormous numbers killed by hunters every fall, there are now probably more deer in the East than there were when white men first penetrated the wilderness. Likewise the familiar greens or "potherbs" of the East thrive best in old fields and gardens; the pioneer settler's wife could probably find very few to satisfy the early spring need for vitamins. Many a settler's family had to allay its hunger with the bulbs of the malodorous wild onion or leek. In certain areas of rich alluvial valleys and riverbanks the groundnut (*Apios tuberosa*) was a more nutritious and palatable native food. Many of the native berries, such as the raspberry, blackberry, huckleberry, or blueberry, and the tiny but delicious strawberry, were found mainly in newly cleared lands and abandoned pastures, but were rare in the forest.

The diet of the white settlers during the long winter months of the first few years would not appeal to the modern American's palate, accustomed to endless variety, including meats, fruits, and vegetables from cold storage, canned or frozen, or shipped in from the winter gardens of southern Florida, southern California's Imperial Valley, or the tropics. But it could not have been much worse than the year-round diet of cabbages, turnips, and bread to which the colonists had been accustomed in England. In any case, corn and beans were soon supplemented by wheat, rye, barley, buckwheat, and peas as staple crops, as well as a few vegetables.

The typical diet eventually became considerably diversified beyond the bleak survival regimen of earlier years. In the fireplace, iron or copper pots with venison stews, cornmeal mush, and other foods dangled from S hooks hung on iron cranes. From ceiling rafters hung hams, sides of bacon, festoons of sausages, ropes of dried apples, and bunches of rice, bergamot, boneset, mint, and other herbs used to season meals, brew "teas," blend salves, and make "blood purifiers" and other medicines. Pantries contained cornmeal, beans, wheat flour (when it became available), and other foods. The cellar contained fruits and vegetables, barrels of cider, and crocks of mincemeat, salted beef, and pork (Clark, 1945).

Cane sugar was prohibitively expensive; fortunately the colonists had two natural sources of sugar: maple sugar and honey. Maple sap was obtained by the Indians through a destructive partial girdling of the trees. The whites used an auger to bore a hole and a wooden spout directed the dripping sap into a wooden trough. Later, wood spouts and buckets were replaced by metal. The sap acquires the maple flavor only by boiling.

Honey bees were apparently introduced into Virginia in 1621 (Neill, 1669) and in New England in 1678. Early writers were impressed by their rapid rate of increase. Possibly in those days bees were free of foulbrood, a disease that destroys the bee larvae, and which appears to have been first noticed in America in the 1870s in New York State.

## THE DOMESTIC ECONOMY

Whatever might be said about the pros and cons of New World farming, the most important fact was that the colonist farmer had attained his highest goal: he had become the owner of a piece of land and the freedom, independence, and security that only land could provide. In this respect the white man differed sharply from the Indian, to whom land was merely a limitless hunting ground and a communal area for a temporary, slash-and-burn type of shifting agriculture. This type of agriculture was suitable for Stone Age peoples because they were never able to densely populate an area. The basic irreconcilable difference in the meaning of land to white men and Indians was to become one of the barriers to the coexistence of the two races. Known to the Indians as a "dirt eater," many a luckless pioneer was found scalped and with his mouth stuffed with dirt (Meyer, 1975). There was no way that whites could be dissuaded from settling in what to them appeared to be empty unused land.

The energy, skill, enterprise, and self-reliance of the colonial farmer was remarkable. He usually made and repaired his own farm implements—plows, sleds, wagons, and hoes. The only limiting factor was the availability of metal. He made shoes, hats, and caps out of furs and skins. Furniture, churns, spinning wheels, and looms were made out of the ever-abundant wood. When wrought iron was available, even nails, shovels, and chains were manufactured on farms and plantations. Women made clothes, rugs, soap, candles, bedding, and garments. Some operated processing plants for baking bread, packing meat, preserving or dehydrating fruits and vegetables, or churning butter. Using flax or wool, women spun and wove in their homes, usually producing cloth that was coarse in quality and used mainly for "working clothes," but eventually their finer linens and woolens compared well with imports from Britain (Beard and Beard, 1944).

## CROP PESTS A SEVERE PROBLEM

The colonists suffered terrible crop losses from pests, diseases, and weeds. As early as 1642, passenger pigeons descended upon grain fields around Massachusetts villages in such numbers as to beat the grain to the ground. They continued their depredations for a couple of centuries. In the nineteenth century passenger pigeons sometimes consumed so much mast (beechnuts, acorns, and chestnuts) as to cause the free-running hogs in the forest to starve to death. (The first underground seeder was said to have been built in Wisconsin in 1860 to prevent passenger pigeons from eating grain as fast as it was sown.) Grainfields were also devastated by armies of squirrels; bounties were paid for the destruction of these animals. Squirrels and crows pulled up corn plants as soon as they appeared, in order to obtain the grains of corn attached to the young plants. Among the serious insect pests were grasshoppers, wireworms, Angoumois grain moths, chinch bugs, and the Hessian fly, the latter a pest long before the arrival of the Hessian soldiers. Such aggressive weeds as plantain, cockle, Saint-John's-wort, couch grass, and charlock plagued the farmer everywhere. They had been accidentally introduced from Europe, where in turn they were immigrants from southwest Asia, everywhere following in the wake of man-made clearings of forest and other native vegetation. The worst plant immigrant was barberry, deliberately introduced from Europe because of its berries, good for jams and jellies. Barberry was the host for the black stem rust of wheat, a fact not known by the early settlers (Jones, 1974).

This was all long before the era of the much maligned chemical pesticides, when all farming was "organic." Losses such as suffered by the colonists, sparsely populating a seemingly limitless virgin land before many of the worst foreign pest invaders had arrived, would today be a national and worldwide calamity. Of course, chemical pesticides do not deserve all the credit for improvement in the pest situation. Much has been accomplished with what are now called "bioenvironmental controls." Black stem rust of wheat was controlled by the eradication of the barberry plant. The Hessian fly is kept at low levels by the use of resistant wheat varieties or by delayed planting. Nearly all the small-grain and alfalfa varieties are disease resistant. Cabbage root rot is controlled by rotating cabbage with some noncole crop, causing the pathogen to decline and the soil to be reusable. The spotted alfalfa aphid is controlled by a combination of resistant varieties and natural enemies (parasites and predators). The screwworm, a pest of cattle, sheep, goats, and hogs, has been effectively controlled by the release of screwworm fly males sterilized by X rays or gamma rays. The Klamath weed had already infested 4.5 million acres of range and farmland in the West when leaf beetles were introduced from Australia to control it. These are only a few examples of the many modern bioenvironmental controls (Pimentel, 1976).

Nevertheless, chemical pesticides still play an important role. There is no substitute for them when a pest threatens a crop and no other control procedure would work rapidly enough to save it. Despite the fact that insects and mites become resistant to one pesticide after another, chemists seem to be infintely versatile in developing new pesticides that are for a while effective. But an average of 10,231 compounds may be screened to achieve a single commercial pesticide. In 1973, it cost an average of $6,112,963 to bring such a pesticide on the market, with an average time expenditure of about 80 months (Easly, 1977). Much of the expenditure of time and money is for investigation of environmental impact.

## Registration and Labeling of Pesticides

For the first two centuries of American agriculture, few pesticides were available to farmers and they were of very limited value. Concern about pesticides was mainly in relation to quackery and fraud connected with their preparation and sale. By 1903 only six states had passed pesticide laws, principally to protect farmers against fraud. Federal legislation relating to standardization and regulation of pesticides first appeared in the Insecticide Act of 1910, which was revised in 1947 in the Federal Insecticide, Fungicide, and Rodenticide Act (FIFRA). FIFRA and the Pesticide-Residue Amendment (Public Law 518) to the Food, Drug and Cosmetic Act, were at first administered by the U.S. Department of Agriculture (USDA) but began to be administered by the newly established federal Environmental Protection Agency (EPA) in 1970. This agency also issues licenses and registers pesticide labels.

If the pesticide is to be used on food or feed crops, the Food and Drug Administration (FDA) of the Department of Health, Education and Welfare publishes a regulation that specifies the amount of residue—the "tolerance"—that may legally remain on the crops for which the pesticide is intended. The tolerance is one-hundredth of the largest amount that causes no effect on test animals. The legal tolerance is expressed as the quantity of toxicant in a single dose for each animal (usually rats) in milligrams per kilogram of body weight (mg/kg), that would kill 50 percent of the species of animal tested. This quantity is referred to as the median lethal dose ($LD_{50}$ or MLD). Thus the safer the toxicant the larger the $LD_{50}$.

EPA solicits the opinions of other federal agencies, such as the USDA, the Public Health Service, and the Fish and Wildlife Service before deciding on a registration application. If there is any doubt in the mind of the user of pesticides as to the current registration status of the compounds he wishes to use and for the purpose he wishes to use them, he should consult the appropriate state authority, for the registration of pesticides is under constant review by both state and federal agencies. For safe use of pesticides, they should be kept in the original or properly labeled container and kept out of the reach of small children, persons who cannot read the language in which the label is written, or pets. Directions for use, on the label, should be followed carefully.

### Risk vs. Benefit

As with all aspects of existence in a technologically oriented society, benefits of a technology must be balanced against the risks incurred by abandoning it. According to a report prepared by members of the Cooperative Program of Agro-Allied Industries with the Food and Agriculture Organization and other United Nations organizations, and published in five languages, pests destroy up to a third of the world's food crops during growth, harvesting, and storage. In developing countries, losses are even higher. Sometimes we are unaware of the true seriousness of crop losses from insects, plant diseases, and weeds until effective pesticides are used to control them—often, for some initial period, doubling or trebling crop production (UN, 1972). Increased yields of certain food crops as the result of insecticide usage has brought benefits of as much as 4, 18, and 29 dollars per dollar invested, and the benefits of pesticide use in public health programs is even more noteworthy, for who can measure the value of the millions of lives that have been saved?

Nevertheless, the use of pesticides is often beset with serious problems, for these compounds have the following limitations: "(1) selection of resistance in pest populations, (2) destruction of beneficial species, (3) resurgence of treated populations, (4) outbreaks of secondary pests, (5) residues in feeds, foods, and the environment, and (6) hazards to humans and the environment" (Luckmann and Metcalf, 1975). These adverse side effects are called *externalities.* They commonly accompany human interventions in the agroecosystem and in natural systems generally, but should be judged with regard to the benefits obtained in comparison with the risks.

Lately, pesticides, if used at all, are generally used in minimum dosages and in conjunction with all possible bioenvironmental controls and in such a way as to cause minimal environmental contamination and minimal adverse effect on natural enemies of the pest. Such control programs are referred to as "integrated control" or "integrated pest management." The objective may be to merely bring the pest population down to economically tolerable levels (Smith and Allen, 1954; Kilgore and Doutt, 1967; van den Bosch and Messenger, 1973; DeBach, 1974; Giese et al., 1975; Metcalf and Luckmann, 1975; Dethier, 1976). What is new about integrated pest management is the terminology. In our increasing awareness of its ecological significance we sometimes forget that, in both theory and practice, it has evolved over a period of many decades.

## THE SETTLERS' LIVESTOCK

It is easy for those of us whose earliest memories go back only to the horse and mule to forget the important role that the ox played in the early days of American agriculture. Oxen were particularly suitable for work among tree stumps and rocks; the single chain running from yoke to load was an ideally simple tackle. In an age when nearly everyone was poor, the low initial cost of an ox was an important factor, the standard price for a yoke (pair) of oxen being five dollars. The ox was free of a number of diseases and disabilities that affected the horse. After it was no longer useful as a draft animal, the ox was deemed, in those times, as acceptable beef. "The ax and the ox-team together will always symbolize the conquest of the wilderness of eastern America" (van Wagenen, 1953). With improved roads, and more of them, the faster horse came into favor. Mules proved their worth in the South.

Cattle found little pasturage in the forest and uncultivated clearings. No root crops were grown for livestock, as was the custom in Europe. Browsing on twigs and weeds, cattle were thin, scrawny, and sickly. Cows became lost in the woods and might not be milked for two or three days; such cows soon "dried up." Insect pests sapped their vitality. Cows usually gave only a quart or two of milk a day. Many died, possibly from feeding on some poisonous weed. The breeds of livestock the colonists had were very poor.

Least attention was paid to the hogs that ran wild in the woods, even in winter when wolves and bears were not too great a menace. In years when acorns and beechnuts were abundant, hogs might thrive, at least in summer. An early traveler to New York State described the "wood-hog" as "more like a fish called a perch than anything I can describe." But probably better-bred hogs would not have fared so well in the forest. Hedrick (1933) says the "narrow-bodied, long-nosed, arched-back swine could make speed in the woods, and in an encounter with bears and wolves gave a good account of themselves." If livestock were fed, the feed was often the Canada field pea, particularly for swine. This was not the pea now used for human food, but a hard, round, smooth, gray pea, no longer grown (van Wagenen, 1953).

Sheep breeds were also poor, but European countries prohibited the exportation of blooded stock. Poor pasturage in forest country was an added handicap. Sheep proved to be the most difficult of all domestic animals to rear successfully under adverse conditions. The sheep of the early settlers produced only two or three pounds of coarse, short wool per animal. Yet the farmer strove desperately to obtain enough wool for the handloom; wool clothing from England was beyond his means. Many sheep were killed by wolves and dogs. Bounties of 5, 10, or even 20 dollars were offered for wolves' scalps.

## SOME FACTORS RETARDING AGRICULTURAL DEVELOPMENT

During the first century and a half of colonial settlement, the constant conflict with the Indians and the French in Canada tended to drive immigrants to the safer provinces such as Delaware, New Jersey, and Pennsylvania. In the time context of the twentieth century, this seems like an unbelievably long period of almost no agricultural progress, but there was constant fear during the colonial period that the French might at any time conquer and hold not only the northern parts of the New York province but even the Hudson Valley. It was not until the fall of Quebec and the Treaty of Paris in 1763 that the power of France in the New World ended. However, that same year the British Proclamation of 1763 forbade settlement west of the Allegheny Mountains until some time in the future, and English troops ordered western settlers to return to east of the mountains. Restrictions on land, inheritance, and sales of agricultural products were major causes of the American Revolution (Rasmussen, 1975*a*).

Thousands of settlers moved from the old towns of southern New England into New

*Fig. 7. A virgin eastern forest; of the two largest trees, the one in the foreground is a white pine and the other is an oak (courtesy U.S. Forest Service).*

Hampshire, Maine, and Vermont, where practically all the lands were taken up by 1812. Eventually, New England settlers also moved via the Mohawk Valley into the Genesee country of New York State, and Pennsylvania emigrants crossed the Alleghenies into the fertile valleys of Ohio. Important causes for all these emigrations were (1) exhaustion of the soil by the cropping systems then in vogue; (2) increased cost of land in the older areas, in part the result of land speculation; (3) inability of some people to fit into the rigid, puritanical social and ecclesiastical systems; and (4) pursuit of relatives and friends who had gone and a desire to see the new country, live its new life, and seek new experiences and adventures. The deprivations and economic struggle were likely to be more severe for five to ten years on the frontier than in the older settlements, but the pioneer was creating capital goods in the way of land and buildings on a scale otherwise impossible. Eventually the pioneer family might live more comfortably than if it had remained behind.

## The Forest Barrier

The forest was in important ways a boon to the pioneer, providing him with logs for his cabin, fuel for his fireplace, and wild game, fruits, nuts, berries, and herbs for his table. But the dense eastern forest (fig. 7) was also a formidable obstacle to farming in the early centuries of our nation, just as forests have been in the development of farming throughout the world. Therefore, in the East there was little migration over the Appalachian Mountains for farming for almost two centuries.

According to van Wagenen (1953), the best time to cut down trees to clear an area for farming was in early summer, for the leaves[5] clung to their twigs and evaporated the moisture from the branches of the felled trees. Then on a dry, breezy autumn day the cutover area was burned, the fire consuming the leaves, smaller branches, and underbrush, but leaving the fire-blackened stumps, trunks, and larger limbs. Too hot a fire could burn down through the humus layer and destroy the fertility of the soil. The limbs were cut off and piled up for burning later. Logs were cut into lengths that could be dragged by a yoke of oxen and rolled into a heap. Although it was not feasible to burn a single half-seasoned log, a fire once started in a large pile would consume everything. The farmer could either dig the stumps out, a slow and backbreaking job, or leave them in and till the soil between them. If he already had cleared land for crops, the burned-over area might be fenced in and used for pasture; the stumps eventually rotted out. Browsing cattle ate off all young growth and there was no chance for the clearing to revert to forest. Fortunately for the hard-pressed farmer, there was no immediate need to fertilize virgin soil and he had practically no weeds with which to contend; nearly all the troublesome weeds of the East were later introductions from Europe.

The role of women in the grim battle with the forest in New York State has been graphically described by Hedrick (1933):

> They complained that the daytime silence of the world about them added tortures to spirits already tried by the monotony and solitude of the new life. In winter there were no sounds throughout the day except the pistol-like reports of trees cracking with frost in the neighboring forest and the ringing sound of the axe, not even the whisper of leaves, the dropping of fruits, or the flutter of birds. Nature seemed to be in an everlasting sleep and the silence oppressed womenfolk, snowed in and half starved for four or five months. The noises of the night were scarcely less trying than the silence of the day to ears attuned to the subtler sounds in the silence of the forest. Wolves, attracted by the livestock, howled about the cabins; there were the screams of the wild-cat, lynx, and an occasional panther in a neighboring swamp; and the lesser noises of bears, porcupines, coons, and other night prowlers on errands of gastronomy or progeneration. The gales and piercing cold that swept down from Canada and tales of men lost and frozen in the woods added to the horrors of the long winter for surely time must have been slow.

[5]Sometimes the trees were girdled and allowed to become seasoned before the area was burned over, as was the Indians' custom.

Frosts sometimes singed and blackened the season's crops and the settlers then began the winter with scant food supplies. The stinging cold of winter penetrated the cracks and chinks left by inadequately seasoned logs, which cracked from the severe cold with the sound of a rifle shot. Every wintry gale blew in drifts of snow through the cracks to cover the beds in the garret; bed coverings were fringed with congealed breath. Food, water, moisture-soaked clothing, and boots froze solidly. Before bread and meat could be cut and consumed, it had to be thawed at the fireplace. Feet and hands often exposed to the cold became swollen and inflamed and schoolchildren might have to take off shoes and stockings in the schoolroom to attend to chilblained feet.

During the winter when physical vitality was low, death from contagious disease was common and in late summer dysentery was equally disastrous. In western New York malaria[6] was so bad that it became known as "Genesee fever," named after the valley of the Genesee River. Western New York, which came to be known as "Genesee country," became associated with sickness and death in the minds of inhabitants of the Atlantic seaboard. Despite descriptions of the beauty and fertility of the land, Easterners were reluctant to migrate to the area (Hedrick, 1933).

The settlers had much more to contend with than incessant toil and privation; they had to face the possibility of extermination. It is understandable that the Indians should resent the ever-increasing intrusion of the frontier into their hunting grounds, a resentment exacerbated by many instances of greed, brutality, and betrayal by white men. On the other hand, the Indians, with a long tradition of fighting among themselves—warlike nomads preying on tillers of the soil or tribes battling against one another for other reasons or without rationality—had an unquestioned capability for keeping the frontier constantly aflame. Terror and massacre were commonplace, abetted by guns and "fire water" the Indians, ironically, had obtained from the whites. Only stockades, militiamen, and eternal vigilance preserved many of the settlements from destruction (Beard and Beard, 1944).

## THE NEW ENGLAND "FAMILY FARM"

The first settlers in the East were, of course, the English. It was their industry, self-reliance, and indomitable will that anchored a new nation in the inhospitable climate and rocky soils of New England. The first settlers had their land allotments in "commons," not fenced separately, but surrounded by a common fence. Commons provided some advantages to the earliest settlers, who wanted compact settlements for protection against Indian attacks and in order to more easily maintain religious and social intercourse. In any case, the scarcity of labor made separate enclosures impossible (Bidwell and Falconer, 1925). However, the commons system had serious disadvantages, from the standpoint of farm management and agricultural progress, as already discussed in chapter 1 with regard to the commons of medieval England. After it had served its initial purpose in the newly settled land, the commons system was abandoned. One remembers the English settlers of New England principally for having given Americans their initial impetus toward the free, independent, self-reliant farm family that would one day loom so large in the dreams of Thomas Jefferson for our

[6]The protozoan parasite that causes malaria was not discovered until 1881 and the role of anopheline mosquitoes as vectors of the parasite was not known until 1897. The colonists referred to the dreaded disease as "chills and fever" or "ague" and usually correctly associated it with lowlands and stagnant water—places where mosquitoes breed. Even as recently as the 1930s there were six to seven million cases of malaria annually in the continental United States (Russell, 1959).

nation. They established a unique pattern of small farms owned by the families that tilled them.

New England became increasingly self-sufficient and its manufacturing began to vie with its agricultural pursuits. Industry, frugality, thrift, and self-reliance became famed New England traits. Accumulated wealth was reflected in better houses, more cleared land (and stone fences), improved livestock, bridges, roads, new schools and churches, and better and warmer wearing apparel (Carrier, 1923). The English also developed the vital fishing industry of the North Atlantic and built ships that reached all parts of the world and eventually became an important factor in encouraging and sustaining the American presence on the Pacific Coast at a time when the area was eagerly coveted by Spain, Great Britain, and Russia.

Meanwhile in agricultural enterprises the European big-estate system, with modifications, was frequently adopted in the South and even in the valley of the nearby Hudson River. Fortunately, not only the English, but also later-arriving European ethnic groups, continued the New England pattern, particularly in the North and West. It eventually became the model for American public land legislation as embodied in the Land Ordinance of 1785 and the Homestead Act of 1862 and its later modifications for the needs of the arid West. Land continued to be owned in family-sized tracts, with as much of the work as possible done by the owner, who believed that work was dignified and honorable.

Confidence was developed in the ability of the common man to become educable and capable of leadership in a democratic society. Common men from the farms became leaders in farmers' organizations, cooperatives, land-grant colleges, and even in government, always retaining understanding and empathy for the problems of farm people. This is a factor missing in the political life of predominantly agrarian nations today.

Unlike the other colonial powers (France, Spain, and Portugal), England allowed anyone, of any faith or allegiance, to settle in her colonies. This resulted in English colonies surpassing all others in numbers, strength, and prosperity. Ethnic diversity had a profound influence on the development of the colonies and, later, the new nation. Nowhere was this influence more beneficial than in agriculture.

## THE ETHNIC AGRICULTURAL COMMUNITIES

The principal European ethnic groups that became farmers during the first two centuries of the nation's development were the English, Scotch-Irish, and Germans. The Dutch, Swiss, and Swedes were good farmers but were relatively few in numbers. The Scandinavians did not come in large numbers until about 1870 and settled principally in the North Central States, where they became prominent in agriculture. Although the Irish emigrated to the United States in large numbers, particularly beginning with the potato famine of 1846-1848, very few became farmers; they tended to be attracted to the cities. The Italians, excellent farmers in their native land, came to America in substantial numbers too late to reap the benefits of the Homestead Act and therefore had minimal effect on the course of early American agriculture. The Italian influence was felt much later and principally in California. The same may be said of the Japanese and Filipinos.

The Spaniards made the earliest introductions of plant crops and livestock into the Southeast and the Spaniards and Mexicans played a similar role in the Southwest. The English pioneered the settlement of the Atlantic Coast and the French settled and gave the initial impetus to agriculture in Louisiana. But in none of these cases was adaptation to alien culture and language involved. For the immigrants who had to make this adaptation, the ethnic community played an important role. It gave the newcomer time to adjust to the alien environment, overcome his homesickness, learn the English language and New World customs, and establish family roots in the new land.

After the English (and the Dutch in some areas) established their settlements along the Atlantic Coast, the immigrants who followed were likely to settle farther inland. The Scots, Irish, Germans, and French Huguenots were generally poor and were attracted to the cheaper lands far to the interior. Also they were less likely to benefit from the largess of the king of England. The thrifty, hardworking Germans in particular were especially good farmers, following the sound farming traditions of their upper Rhineland (Palatinate) homeland, where their envied harvests had been repeatedly ravaged and looted by foreign armies in the terrible Thirty Years War. Likewise in the American colonies their abundant crops and livestock were always favorite targets of the Indians. Their settlements in the Mohawk and Schoharie valleys of New York, begun about 1713, suffered more from Indian attacks during the critical period of the Revolution than any other frontier area. Those who could gather in a protected stockade could generally save their lives, but had to abandon their crops and livestock. The Indians of the Six Nations in New England were particularly fierce fighters. Armed and encouraged first by the French, during the French and Indian War, and after the beginning of the Revolution by the British, the Indians were very aggressive.

## INDENTURED SERVANTS AND REDEMPTIONERS

Indigent immigrants from the British Isles usually immigrated as "indentured servants," with fixed indentures specifying the conditions of service. Germans unable to pay for their passage usually immigrated as "redemptioners," with no definite understanding of the nature of their service in America. In either case the immigrants agreed to work for some colonist family until the price of their transportation was paid to the shipowner who had advanced it. The period of servitude was usually from three to seven years. They became indentured servants to any colonists who "bought" them. After serving the agreed term, they were released, given a suit of clothes, sometimes given money or land, and thereafter had all the rights of a free citizen.

Many fell prey to "immigrant agents," who were either employed by transportation companies or acted on their own initiative. They were paid a commission for every immigrant they were able to bring in, so they circulated pamphlets in Europe with glowing and exaggerated accounts of a life of wealth and ease in the New World. There was more profit for them in servants than in paid passengers. Many immigrants who had the money to pay a legitimate fare found that costs were greater than they were given to believe they would be. Others were robbed outright, or their baggage might be put "on the wrong ship" and never recovered. All such persons became indentured servants; people of rank fared no better than peasants. The sale of servants was not abolished until 1820 (Faust, 1909).

In two important respects the status of the servant resembled that of the slave: the master could sell the servant's services, and thereby transfer him for the unexpired period of service, and he had the right to chastise his menial. However, servants could usually make complaint of ill treatment before a justice of the peace. Laws were different in the different colonies or states. In Maryland, a whipping by more than ten lashes was considered to be mistreatment. The master might apply for the right to administer a greater number of lashes but no more than thirty-nine (Ballagh, 1895). Referring to the South, Gray (1933) stated that planters were known to be so cruel to a servant during the last few months of his term of service that he would agree to forego his rights of freedom dues in order to gain his freedom a month or so earlier.

The greatest tragedies befell the immigrants during the months at sea in the sailing vessels. Living conditions on these ships were appalling. During the long sea voyage, passengers were packed closely together under filthy, unsanitary conditions. Many died of scurvy, dysentery, smallpox, and other contagious diseases. Death from starvation and thirst was common. One passenger wrote in his diary in 1738 that of 312½ passengers (a child was counted as one-half), 250 died, not including those who died after landing. One authority reported that 160 people died on one ship, 150 on another, and only 13 survived on a third. In 1745 a ship destined for Philadelphia with 400 German passengers arrived with only 50. Few children under seven years of age survived the trip. Likewise few pregnant women survived; they were then cast into the sea with their unborn child. To top it all, shipwrecks were common and there was the ever-present danger of being captured by hostile fleets or pirates (Faust, 1909).

Conditions for transatlantic travel improved, of course, particularly after the advent of the steamship. Yet many living Americans recall the tales of their fathers and mothers who found it necessary to cross the Atlantic under the crowded and dehumanizing conditions of the steerage section of transoceanic passenger ships around the turn of the century.

Despite the many unfortunate abuses, it can still be said of those who profited from the human bondage business, that they enabled people to come to America who would not otherwise have been able to come. The system helped to populate the colonies. To many, the period of bondage was a training period during which they could learn the language and decide what to do when they were released. Many, through industry and thrift, became highly successful and purchased land, sometimes even the estates of their former masters, and became farmers (Faust, 1909).

## AN AGRICULTURAL PRIZE OF THE REVOLUTION

During the Revolution, the Germans were particularly strong in support of the patriot cause, there being very few Tories among them. The intense loyalty of the Germans was to bear fruit during the many bitter campaigns of the Revolution, including the critical battle of Oriskany. When British General Burgoyne began his march from Canada in the middle of June 1777, with the object of cutting off New England from the rest of the colonies, he was to be aided by another British expedition, coming up the Hudson from New York, and another, made up of Indians and Tories, from the west under Colonel St. Leger. The latter was to join Burgoyne at Albany after subduing the entire Mohawk Valley, which had been settled by Germans,[7] and rob it of its rich harvests and fat herds, which were to supply Burgoyne's army with food. Hearing of this plan, farmers of the Mohawk Valley summoned to arms all men between sixteen and sixty years of age and organized four battalions (eight hundred men). Under the com-

[7]The Mohawk and Schoharie valleys of New York and the rich limestone valleys of Pennsylvania and western Virginia, which were settled predominantly by Palatinate German immigrants and their descendants, were an indication of their unerring ability to ferret out the most productive agricultural lands, similar to those they left behind in southwest Germany. The good land and excellent farming methods of the Germans were an important factor in meeting the food requirements of the patriot fighting forces of the colonists during the Revolution.

mand of General Nicholas Herkimer, these farmer battalions marched to the confluence of the Oriska and Mohawk rivers, where Oriskany is now located, to give battle and save their fertile valley from looting and destruction. In this fierce battle Herkimer's hastily gathered volunteers were victorious, although about a fourth of the patriot force was killed or severely wounded. But the valiant fight of the Mohawk Valley farmers, according to George Washington, was the battle in the northern campaign that "first reversed the gloomy scene." Without that victory, the rich harvests of the Mohawk Valley would have been used to feed Burgoyne's army. A further benefit from the battle of Oriskany was that the Indians became discontented with their British and Tory allies and the latter in turn considered their Indian allies a failure.

The German settlements were not again attacked until the fall of 1778, when a surprise attack by a superior force of Indian warriors forced a thousand men, women, and children in German Flats to hastily retreat to their forts, leaving their recently harvested crops to the Indian marauders. No effective help was received all along the flaming New York frontier until after the massacres of settlers in Wyoming Valley (July 3, 1778) and Cherry Valley (December 10, 1778), when a punitive expedition under Sullivan in 1779 devastated the villages of the Six Nations (Faust, 1909).

By 1790 the German population in the Schoharie Valley had risen to only 2,073, but the valley had already gained fame as a wheat-growing center. Then the "Yankees" moved into the region in large numbers and advanced the zone of settlement into the uplands. They were surprised to find German women working in the fields, a custom unknown in New England. The New Englanders helped to increase the population of the Schoharie district about eightfold by 1810 (Ellis, 1946).

## AGRICULTURE IN THE MIDDLE COLONIES

Along with the original English and Dutch settlers, the Swedish, German, and Scotch-Irish settlers developed thousands of acres of rich, gently rolling farmland in the "Middle Colonies," growing wheat and corn and raising livestock for market. Even farther south, rich farmlands followed the many rivers nearly 100 miles inland to the "fall line," as in the piedmont regions of Maryland and Virginia, with tobacco an important crop. Gradually a surplus of farm produce was available and increasing numbers of farmers began to have as their primary goal a crop that could be brought to market, as in the case of the wheat farms of Pennsylvania and the tobacco plantations of the Chesapeake region. In the Middle Colonies, farmers had a relatively easier time than in colder, rockier, hillier New England and New York. Pennsylvania had most of the features of Middle-Colony agriculture and may be used as an example. Although agriculture got an earlier start in seaboard colonies, the colony and state of Pennsylvania became the "breadbasket of the nation." From 1725 until 1840, Pennsylvania was foremost among the colonies and states in food production.

Pennsylvania from its beginning was colonial America's melting pot, unlike some other colonies such as Virginia and Massachusetts, which were originally ethnically quite homogeneous, with one language and one religion. Thus the Pennsylvania pioneers brought with them from Europe different agricultural techniques, and this may have accounted in part for the preeminence of Pennsylvania agriculture in the colonial period. However, the ethnic groups generally settled in distinct areas of the state. Agricultural historian Theodore Saloutos (1976) decries the fact that the constructive role of the immigrant in the agricultural development of this country has been "blacked out from our history," possibly because of the desire to deemphasize the ethnic factor and thereby strengthen the bonds of national unity.

## Indian Farms

The Indians the white settlers first contacted in the middle colonies did considerable farming. However, they rarely stayed in the same place longer than twenty-five years, for by that time, despite their practice of shifting agriculture, the soil in the entire area became exhausted by repeated cropping with corn, and game and firewood were no longer close at hand. The Indians depended mostly on fish and game for their subsistence and the products of wild plants and trees were also important. Fletcher (1950 writes:

> Among the wild plants used by Indians as food, the ground nut (*Apios tuberosa*) was most important. The roots were roasted. A marsh plant called Taw-ho (arum family) also was used freely. Other wild food plants were huckleberries, chokecherries, plums, strawberries, blackberries, raspberries, dewberries, gooseberries, cranberries, elder-berries, mulberries, grapes, persimmons, crab apples, pawpaws, black walnuts, butternuts, chestnuts, chinquapins, hickory nuts, and acorns. Wild peaches were abundant in the Delaware Valley. Chestnuts were dried, ground into flour and made into bread. Many fruits were dried for winter use.

The "wild peaches" were no doubt seedling trees. The Indians were quick to recognize the superiority of European fruits to their wild ones. For example, during the Revolution an expedition of colonists found bearing orchards of apple and peach in a previously unexplored region west of the Genesee River in what is now New York State (van Wagenen, 1953).

From December to March, Indians subsisted mainly on stored corn, beans, nuts, dried fruits, and fresh, dried, or smoked meats. Indians ate dogs, their only domesticated animal, when food was very scarce.

Indian villages were located on river-bottom land and consisted of 10 to 50 families. The villages were surrounded by cropland, commonly 20 to 200 acres, usually with girdled or dead trees standing on it. Farther south even larger acreages were cultivated and at least one Indian village in Virginia had about 3,000 acres of cleared land, much of it in corn (Shannon, 1934). Beans, pumpkins, squash and tobacco were also grown. A family ordinarily cultivated one to one and a half acres of corn. Girdling of forest trees was done with tomahawks, which were wedge-shaped pieces of flint or hornstone inserted into the end of a split stick and bound with rawhide thongs. This crude tool was sharpened by rubbing the edge against other stones, a task requiring several days. Trees might also be felled by setting fire to brush piled against the trunk, then chipping away the crusts of charcoal formed by a succession of such fires. Heavy mattocks or hoes were made of hardwood or by tying the shoulder blade of an elk or bear to a stick. These implements were used to grub out brush in the crop field (Fletcher, 1950).

Indians planted corn in "hills" five to six feet apart and beans, squashes, and pumpkins were often planted in the same field, especially in rich ground, in the manner of the originators of the corn-bean-squash culture in Mexico, as related in the first chapter. Successive plantings of corn in April, May, and June ensured the Indians a succession of roasting ears. Soil was drawn up around the stalks and, since the same hills were used in successive years, eventually mounds of considerable height were formed. Only the soil in an area of twelve to twenty inches from the stalks was stirred; the areas between hills were not broken, but wild growth was pulled up or chopped off (Fletcher, 1950). Commenting on the amazing industry of Indian women, an early observer wrote: ". . .an other work is their planting of corne, wherein they exceed the *English* husband-men, keeping it so cleare with their clamme shell-hooes, as if it were a garden rather than a corne-field, not suffering a choaking weed to advance his audacious head above their infant corne, or an undermining worme to spoile his spurnes [roots]" (Wood, 1634). Thus the modern soil-conserving "no-tillage" system of farming was utilized by the Indians in the pre-Columbian era. The arduous work of the white settlers in preparing soil for planting, and all the subsequent cultivation, was always a source of wonder to the Indians. Only relatively recently has the white man recognized that on many soils cultivation is of value principally to control weeds. Weed control accomplished by the Indian women can now be accomplished by herbicides.

According to Fletcher (1950), the largest and best ears of corn were hung up in wigwams, the remaining ears were dried in the sun or over smouldering fires, usually with husks on, and at night were covered with mats to protect them from dew. The dry ears were husked and shelled, and the kernels were placed in bark or willow baskets. Swedish botanist Peter Kalm, who traveled extensively in eastern North America in 1748 and 1749, said that corn was kept in holes in the ground, seldom more than six feet in depth, on the bottom and sides of which were placed broad pieces of bark, or if no bark was available, a kind of grass the white settlers called "Indian grass" was used (Kalm, 1770).

Corn was a crop not only ideally suited for the primitive agriculture of the Indians, but also for the pioneer farmers. Because it was grown in "hills" spaced a considerable distance apart, corn could be grown among stumps and down timber and even among trees killed by girdling. Only a hoe was required for planting corn and the ground between plants did not have to be broken up. This was an advantage to pioneers who had neither plows nor draft animals. For small grains, land must be more fully cleared. A much smaller percentage of the corn crop had to be kept for seed, and yield was considerably greater than for small grains. Corn could be pounded in a hollow block with a wooden pestle (Schoepf, 1789), while small grains required a more difficult milling process and the aid of windmills or water mills. Another advantage of corn was that cornstalks were superior to wheat straw as forage for livestock. Not enough emphasis has been placed by historians on the vital role that corn played in the economy and even the very survival of the early colonists.

## The Dutch and the Swedes

The first white farmers of the Middle Colonies were Dutch, but most of the early Dutch were more interested in buying furs from the Indians than in farming. Later some Dutch farmers bought land from William Penn and developed dairy farming, especially the farm manufacture of cheese. By 1750 few spoke Dutch and most had merged with the more numerous Swedes, English, and Germans by intermarriage.

Soon after the Dutch, came the Swedes and Finns, who might all be called Swedes, for Sweden and Finland were united in those days. Their first permanent settlements in Pennsylvania were along the Delaware River in the region of what is now the city of Chester. More land was then purchased from the Indians and the region settled by the Swedes was called "New Sweden." New Sweden lasted only from 1638 to 1655, when Peter Stuyvesant, governor of New Netherland, took possession of the area, and ten years later the Swedes came under English rule. There were then only about 650 Swedes on both sides of the Delaware; yet they were the most numerous element of the white population of Pennsylvania until William Penn arrived. They developed a prosperous agriculture, raising rye, wheat, buckwheat, corn, and tobacco with surplus quantities for trade. Likewise they had surplus quantities of cattle, sheep, swine, and poultry. The English, German, and Scotch-Irish settlers could not have become settled so quickly without the surplus of the herds and flocks of the Swedes. One can hardly imagine frontier life in the East without the log cabin. It was introduced by the Swedish settlers.

Peter Kalm in his *Travels in North America* recorded that the oldest Swedish inhabitants in New Jersey remembered when grass was more abundant in woods, fields, hills, and pastures and cows gave about four times more milk. Many years of overgrazing, before seed could be formed, resulted in great deterioration of pastures. Thus overgrazing became a serious problem early in the nation's history. Kalm (1770) wrote, "However, foresighted farmers have procured seeds of perennial grasses from England[8] and other European states, and sowed them in their meadows, where they seem to thrive exceedingly well." The more common solution to overgrazing in those days was to move farther west.

Even before 1700, English grasses were sown in the colonies either by accident or design. During the eighteenth century they were increasingly sown on tilled ground to develop "artificial meadows" with a gratifying increase in hay production. A grass known in England as herd's grass was used as a hay crop near Portsmouth, New Hampshire, in 1720 and was then distributed elsewhere by Timothy Hanson, whence it obtained its name, "timothy." It was the American farmers who first recognized its value and began its cultivation as a forage crop (Piper and Bort, 1915). Red clover was another important forage crop and sometimes clover and timothy seeds were sown together (Strickland, 1801).

The pace of change in colonial agriculture seems unbelievably slow to us now. For over a century the colonists followed the native Indian corn culture practically unchanged. Kalm noted in 1748 that, in New Jersey, corn was usually planted in squares, five and a half feet between each hill in both directions. This is the way the Indians had been growing it in the Delaware Valley when white men first arrived. Also the soil might be hoed only around each plant, as was the custom of the Indians, although Kalm (1770) also recorded that "In some places the ground between the corn is plowed and rye sown on it, so that when the corn is cut the rye remains upon the field." The Swedish settlers carried their preference for rye bread with them to America and continued to grow rye rather than wheat. Rye also had the advantage of not suffering from "blast"[9] (black stem rust), a devastating disease of wheat. the Pennsylvania Germans ate rye bread, but possibly because wheat brought good prices when exported.

## William Penn's Grant

In 1680 Charles II, king of England, granted the Quaker William Penn a charter for a vast province on the west bank of the Delaware River that was named Pennsylvania for Penn's father, to whom Charles II had owed a large debt that was canceled by this grant. Penn was made proprietary of the province. Penn's purpose was to provide a refuge for Quakers and other persecuted people and to erect an ideal commonwealth. A series of treaties based on mutual trust with the Indians preserved Pennsylvania from Indian hostilities until 1755. Credit is due to the Swedes, whose purchase of Indian lands and friendly and honest relations with the Indians had given the latter confidence

[8]Annual and perennial grasses of the New World were far inferior to those of Europe in nutritive value and their ability to withstand trampling and grazing. English grasses sown in scattered areas of the Northwest spread rapidly, replacing indigenous species, but they did not thrive from Virginia southward and as a consequence it was difficult to fatten free-range cattle in the southern colonies (Jones, 1974).

[9]John Winthrop, Jr., in Connecticut, wrote in 1668, "What the cause was, whether natural or a blasting from heaven we know not" (Bidwell and Falconer, 1925).

in their ability to work out satisfactory relations with white men. A constantly increasing flow of immigrants, not only from the British Isles, but also from the upper Rhine country of Germany, was attracted by Pennsylvania's liberal and tolerant principles of government.

### English and Welsh Immigrants

Most of the immigrants immediately following Penn's arrival were English and Welsh Quakers. They obtained the 360-acre site for the town of Philadelphia from three Swedish brothers, giving each of the latter 250 acres of land elsewhere, plus "a yearly rent of one half bushel of Wheat for each one hundred acres" (Fletcher, 1950). By 1685 Penn reported that the English were the majority of the 7,200 population of Pennsylvania province. By 1840 the areas that were predominantly English were along the northern border of what is now the state of Pennsylvania and in the southeast, including Philadelphia. By 1840 the Germans predominated in the south central area of the state and the Scotch-Irish in the southwest. Except for the neglect of livestock by the early settlers, the English were said to be good farmers, quick to take advantage of new practices and implements. They were particularly good at the business aspects of agriculture and with passing years many became merchants and tradesmen and dominated the business and political life of the colony (Fletcher, 1950).

Kalm frequently remarked on the abundance of fruit in the environs of Philadelphia, in sharp contrast to his native Sweden, where hardly any people besides the rich could eat it. This was characteristic of the prodigality and waste in the New World, once an area became well developed. Kalm (1770) wrote: "But here every countryman had an orchard full of peach trees which were covered with such quantities of fruit that we could scarcely walk in the orchard without treading upon the peaches that had fallen off, many of which were left on the ground. Only a part of them was sold in town, and the rest was consumed by the family and strangers, for everyone that passed by was at liberty to go into the orchard and gather as many of them as he wanted. Nay, this fine fruit was frequently given to the swine." The Pennsylvania-New Jersey peach-growing area continues to be important to this day, being exceeded only by California and the South Carolina-Georgia areas.

### The Pennsylvania Germans

The Germans, whose descendants are often erroneously called "Pennsylvania Dutch," eventually became the most numerous of the ethnic groups in Pennsylvania. Most of them came to escape religious persecution but many were victims of the political and economic ills brought on by the Thirty Years War that ended in 1648. Most were farmers, some were skilled craftsmen, but nearly all were poor. The first German immigrants were mostly Mennonites who arrived in Philadelphia in 1683 and founded Germantown six miles from that city. These were people who were much impressed by the religious views expressed by William Penn during his trip to Germany in 1677. Mennonites constituted the majority of German immigrants for the next thirty years, most of them settling in what are now Montgomery and Lancaster counties and the valleys of the Schuylkill, Perkiomen, and Lehigh.

Ethnic groups tend to settle on land that most closely resembles the land from which they came, which, in the case of the Germans and Swiss, was the gently rolling country of the Upper Rhine. This was very fertile land and one of the most productive wheat-growing districts in central Europe. The Rhinelanders found this kind of land in the limestone valleys of Pennsylvania. Traveling through the wilderness in search of land for farming, they judged the land by the kind, number, and size of trees growing on it. They preferred land with hardwoods—oak, maple, hickory, beech, and particularly where black walnut grew to large size. They were not deterred by the backbreaking work of clearing forested land. They cut down all trees and dug out all but the largest stumps. Saplings and underbrush were pulled or grubbed out. Large logs were cut into eleven-foot lengths and split for rails, or into four-foot lengths for firewood. The largest stumps were left to rot for five to ten years before removal. Families cooperated in clearing the land. The advantages of clear-cutting were immediate use of most of the land for cropping, working the soil, and harvesting the crops. Logs not otherwise used could be burned and converted to potash and sold. Ten acres might produce a ton of potash selling for $200.

Besides the Pennsylvania Germans, the New Englanders were the only other eastern settlers who used this clear-cutting method. All others followed more or less the Indian method of girdling and burning. After girdling, dead branches began to fall in a few years and the "deadened" trees eventually blew down. Branches were piled on the fallen trunks and burned. In a log rolling bee the charred butts of the trees, twelve to fifteen feet long, were rolled off the land and burned. Among the disadvantages of this method of land-clearing were falling branches, which fell on the grain, demolished fences, toppled on horses and cattle, killing or maiming them, and occasionally killed men and boys (Fletcher, 1950).

The original density of the forests and large size of the trees in Pennsylvania (Penn's Woods) were the subject of frequent comment by such early observers as Kalm (1770). The transition from virgin forest to agrarian landscape, much of it occurring before the advent of the bulldozer, tractor, and power saw, beggars the imagination.

The Germans built potteries and brickyards very early in the period of their settlement and made bricks, tile, and earthenware of excellent quality. Substantial and comfortable homes were built of brick and stone, and stoves were built of tile. The massive "Switzer" or "Swisser" barns were so solidly built that they, as well as the brick houses, are still in use—a conspicuous feature of the landscape. The spacious barns allowed the Pennsylvania Germans to give their livestock good care, preserving the customs of the old country (Bordley, 1801). The Dutch and Swiss settlers, present in smaller numbers, had similar good agricultural traditions.

These massive structures usually represent the contributions of several generations of farmers, for the father wanted to hand the farm down to his sons in better condition than he found it. The pioneer settlers had neither the time and help to build a massive structure nor the large herds of stock to require it. At first just a small rude structure built of logs from a clearing in the forest had to suffice. Likewise the farmhouse increased in size to accommodate the needs of the steadily increasing size of the family. Examples of all types, from simple log structures to the most elaborate ones built of stone and brick, may be seen to this day. We are indebted to Amos Long (1972), in his *The Pennsylvania German Family Farm,* for an exhaustive treatment of the subject, illustrated with 234 photographs.

In colonial times farmers elsewhere in the colonies were generally actually brutal to their animals. Cattle were given almost no shelter and were poorly fed; losses were great in winter. Hogs foraged in the forest summer and winter (Benson, 1937; Clark, 1945; Fletcher, 1950). Hogs starved in winter developed a disease called "black teeth." In northwestern Pennsylvania, McKnight (1905) observed that a common remedy was to knock out the teeth with a hammer and spike. If the animal died the treatment was considered to be given too late. There were also diseases and sores of horses and cattle resulting from exposure, starvation, and beatings, and for these also there were cruel and painful "treatments." Horses were treated with almost unimaginable brutality.

"They were fed barely enough to keep them on their feet, and were driven with whip and spur. When they collapsed, they were left to die, for a few dollars would buy another unfortunate beast. About the most valuable part of a horse was the hide, which was made into leather."[10] The Pennsylvania German "stood almost alone in colonial times in treating his animals decently" (Clark, 1945).

Artificial meadows, planted to red clover, timothy, and other European grasses, provided more nourishing food for farm animals than native grasses. In the stalls of the great barns, horses, oxen, cattle, and other livestock were protected from the severe weather and fed with hay, cut straw, and rye meal. The Germans developed the Conestoga horses, the finest on the continent during the colonial period, from stock originally brought over by English settlers. They also developed the Conestoga wagon, sturdily built and able to carry two or three thousand pounds of farm produce to market. The wagons were covered with linen[11] cloth. Both horse and wagon were named after a stream in Lancaster County, Pennsylvania (Rush, 1789). There were at least 7,000 Conestoga wagons in the colonies in 1750 (Schlebecker, 1975).

The accumulation of manure in barns led to its use to fertilize fields. After oats were harvested, compost (manure mixed with corn stalks and straw) was spread on oat stubble and plowed under in preparation for wheat. Unlike other colonies, in Pennsylvania wheat became the principal cash crop; corn was second in importance and was mostly utilized on the farms. The black stem rust, so destructive in New England, was much less serious in Pennsylvania. Besides grain-growing, important endeavors were general farming, tobacco-raising, and livestock husbandry, and Pennsylvania became the wealthiest agricultural colony (Carrier, 1923).

"If you steal from the fertility of the soil," said the devout Mennonites, "you steal from God" (Fletcher, 1950). If civilization is to survive, farmers everywhere will have to develop the tradition exemplified by the Pennsylvania Germans and their descendants—to consider it a sacred trust to leave the soil in a better condition than they found it. It has only been in recent decades, with an enormous infusion of money and technical help from federal agencies, such as the Soil Conservation Service, that most American farmers have begun the good soil-conservation practices that were an old tradition of the German and Swiss immigrants from the Upper Rhine region, as well as new practices developed by research and extension agencies.

The Germans began irrigating meadows before 1750 and by the end of the century this became a common practice in Pennsylvania, a few "watered meadows" also being found in New England and New York. Brooks that originally flowed through meadows were diverted to channels built along hillsides; then the water was distributed by lateral ditches as widely as possible over the lowlands. Hay crops were greatly increased and land was valued in proportion to the acreage that could be irrigated (Ellis and Evans, 1883).

It seems that authorities on early Pennsylvania land use were unanimous in their praise of the high state of agriculture of the Pennsylvania Germans. For example, W. F. Dunaway (1935) in his *A History of Pennsylvania* wrote:

> Their chief contribution was to the promotion of agriculture, in which they excelled all other groups. By their industry and skill they made the wilderness a gar-

[10]There are large areas of the world even today where animals are characteristically treated with great brutality, particularly horses. The tourist is shocked to see small, skinny horses pulling heavy loads and being continually beaten—sometimes collapsing in harness.

[11]Unlike other colonists, the Germans grew much flax and were expert linen weavers (Frame, 1692).

den spot. Conservative, religious, industrious, frugal, they were a most substantial body of people; close to the soil, they added to the province an element of strength without which it never would have reached its remarkable prosperity.

Good use of the soil became a tradition handed down from father to son. When further improvements in soil conservation were encouraged by the United States Soil Conservation Service, which was established in 1935, nowhere did a region respond more rapidly and thoroughly than Lancaster County, Pennsylvania. Figure 8 shows two photographs of the same area in that county, indicating how completely the farmers shifted to soil-conserving contour cultivation in ten years (1937 to 1947). During most of that early period the landowner received little or no financial assistance from the federal government such as generally provided in later years.

The Amish (a Mennonite sect) in particular retain their old way of life to this day. Their excellent farming methods and their way of life and types of homes and farm structures may be seen in certain areas where they tend to congregate, as, for example, an area extending about twenty-five miles east of Lancaster, Pennsylvania. Before they came to America the Amish lived for centuries under governmental restrictions that confined them almost exclusively to farming. This resulted in an Amish church regulation prohibiting residence in towns and cities and in occupations not clearly related to farming. The Amish consider their marked success in farming as a form of divine blessing. They seek to make it possible for each son in the family to have a farm of his own. They refuse to send their children to high school and college, believing that higher education will lead them away from the farm. However, they have had great success in training their children to be good farmers. Although an increasing number of young Amish are becoming "worldly," most of them retain the old way of life. They are willing to pay high prices for land in a given area in which they can live as an Amish community, as in the Lancaster colony (Fletcher, 1955).

The Germans eventually passed beyond the forks of the Ohio at Fort Duquesne (Pittsburgh) and also settled the beautiful Valley of Virginia, particularly the more fertile northern half of it, then passed on into the upper reaches of the Clinch and Holston rivers and on to the fabulous Blue Grass region of Kentucky and into Tennessee. The Ohio, Cumberland, Tennessee, and Kentucky rivers pass through or near important limestone basins, and the Germans continued their limestone-seeking tradition in these river valleys. Everywhere they demonstrated the values of limestone soils, manure, and clover in the production of superior crops, as well as the finest cattle, horses, and hogs. Sometimes they were pioneers, but they also followed the paths of earlier Scotch-Irish and English subsistence homesteaders, buying and improving the land. But they did not follow them into the thin-soiled hollows of Appalachia.

To the Germans farming was more a business than a matter of mere subsistence. This was a fortunate circumstance, for the New Englanders were more interested in fishing, lumbering, and trade than in commercial agriculture. In the South, commercial agriculture concentrated on a plantation tobacco culture close to the coast and navigable rivers. Away from the coastal and river valleys, indentured servants, released from plantations, took up homesteads in the hilly uplands, but only for subsistence agriculture supplemented by an occasional sale of handicrafts or an occasional cow or pig. Thus the "Pennsylvania Dutch" had a ready market both north and south for agricultural produce (Higbee, 1957).

Both New England and the Middle Colonies were fortunate in their trade with the West Indies, where the sugar planters' cash crop was so valuable that they could not afford to grow anything else. Thus the West Indies were a good market, accessible by

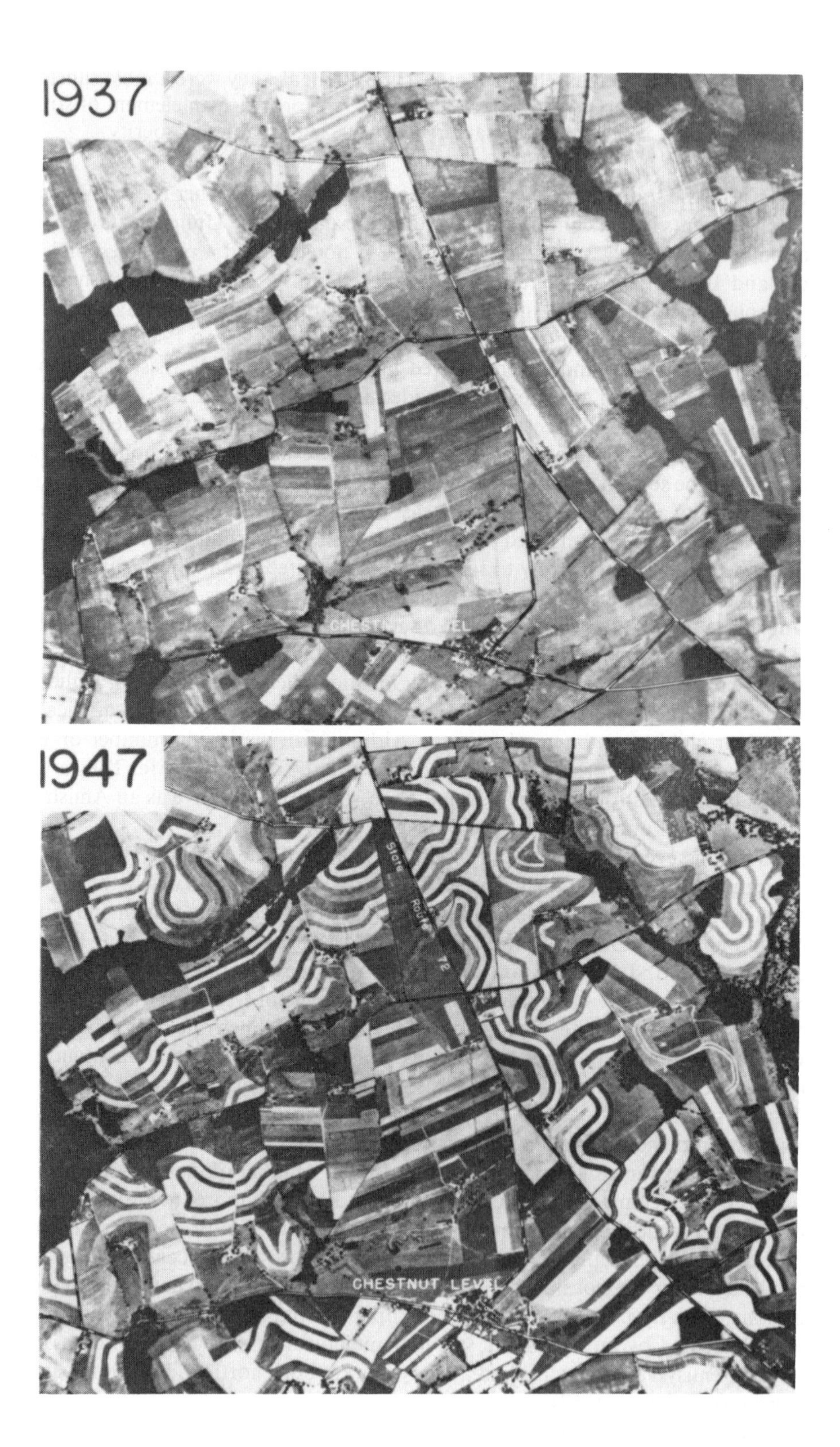


*Fig. 8. Two photographs of the same area in Lancaster County, Pennsylvania, showing how a shift from square to contour farming changed the pattern on the land in a ten-year period, 1937-1947 (courtesy USDA Soil Conservation Service).*

sea, for northern products. Likewise in later years the southern colonies, with their monoculture of cotton and rice, were to become a good market for the diversified northern agricultural products.

## The Germans in New Jersey

The Germans also had a hand in New Jersey's enviable agricultural tradition. In 1707 a group of Germans on an overland trek from Philadelphia to New York decided to go no further when they reached the beautiful valleys of the Musconetcong and Passaic rivers in northern New Jersey. Along with subsequent immigrations from Germany they settled in "German Valley" (Morris County), thence spread to Somerset, Bergen, and Essex counties, developing in these areas the gardenlike farmlands that were to become characteristic of the "Garden State." One of the immigrants, arriving from Germany in 1733, was Johann Peter Rockefeller (Rockenfeller), whose famous descendant John D. Rockefeller, founder of the Standard Oil Company, erected a monument to his ancestor's memory near Flemington, New Jersey, in 1906. Following a long tradition of philanthropy, the Rockefeller Foundation was to eventually become one of the principal private agencies massively supporting research and development in agriculture on a worldwide basis. This effort reached the apex of its accomplishment and influence in the Green Revolution, bringing new hope to a world threatened with famine. It escalated plant breeder Norman Borlaug to center stage among the dramatis personae in the drama of scientific discovery in agriculture and to the 1970 Nobel Peace prize.

## The Scotch-Irish

The Scotch-Irish immigrants were descendants of Scots who had been colonized in Northern Ireland for a few generations, beginning about 1610. They left Ireland for economic and religious reasons. At first they settled in predominantly German areas, but so much friction developed between the two ethnic groups that in 1743 the proprietary refused to sell Scotch-Irish settlers more land there and made liberal offers to induce them to move farther west. Much of the displacement of Scotch-Irish and English families from the better farmland was the result of the wearing out of the lands from lack of manuring and proper care. A group of adjacent German families would buy the land, plow it well, let it remain fallow for three or four years, then place an immigrant German family on the land. By hard work and the use of manure, the land was made productive and the new owner paid for it in installments and lived comfortably. The displaced family could go two or three hundred miles west and buy as much land as it pleased for a dollar or less per acre.

By the middle of the nineteenth century the Scotch-Irish were the predominant population in southwest Pennsylvania, as they still are today. They also settled much of northwest Pennsylvania excepting those areas bordering New York State, which were settled by the English. Unlike the Germans, who preferred limestone valleys, the Scotch-Irish preferred the shale soil of the foothills, which most closely resembled the land to which they had been accustomed in the old country. This land was less fertile than the limestone valleys, but had more springs. In limestone lands, water generally sank below the surface and expensive wells were required (Day, 1843). Also the Scotch-Irish were not accustomed to woodland and avoided the heavily wooded valleys, which required prodigious effort to clear and were more likely to have frost. The Scotch-Irish were also less interested in livestock husbandry and soil conservation than were the Germans. They were highly politically minded, and were foremost in promoting farm organizations. They were said to be excellent frontiersmen, impatient with the restraints of civilization, and often sold their partly developed farms and moved farther into the wilderness.

Among the Scotch-Irish who migrated south into the fertile Valley of Virginia was the McCormick family. In a primitive shop on his family's farm in the shadow of the legendary Blue Ridge Mountains, Cyrus Hall McCormick, at the age of twenty-two, invented the famed McCormick reaper, patented in 1834. Thus he freed thousands from the brutal, backbreaking drudgery of reaping grain with the "cradle." Later he was to become a highly successful and innovative Chicago captain of industry, and he and his descendants became a dominant force in the development of Midwest agriculture.

Another important agricultural legacy from the pioneer Scotch-Irish began with John Wallace, who came from Ireland at the age of eighteen and started farming in the hills above Pittsburgh in 1823. In 1836, the first of a line of three famous Henry Wallaces was born at Spring Mount, the Wallace farm. During Henry's boyhood the wheat belt passed inland along with the great western migration of a population that was swollen by the Irish potato famine of 1846-1848 (caused by the potato blight) and the German Revolution of 1848. Henry moved with the tide and started the Iowa dynasty of Henry Wallaces that was to have a profound influence not only on American agriculture but eventually in the higher echelons of national politics (Lord, 1962).

### The Colonists' Social Fluidity

An important aspect of the life of the colonists was its social fluidity. About 90 percent of the colonists were reared on farms. New land was available to anyone, and the possession of land provided the opportunity for upward mobility unattainable to European ancestors. In the absence of aristocracy and hereditary titles, upward mobility could be attained by anyone of any trade or profession who possessed ability and ambition, whatever his family name. America became known the world over as the land of social as well as economic opportunity, and wealth became the most important prerequisite for social advancement.

Eventually large landowners emerged as a distinct factor in colonial life, particularly along the northern river valleys and the tobacco plantations of the southern colonies. They lived the luxurious life of "country gentlemen." The wealthy landowners maintained a staff of household servants and many field-workers—"hired hands" in the North and Negro slaves and indentured servants in the South. Others who became wealthy were the merchants and shipmasters of New England, who developed a lucrative foreign and coastal trade in the mid-1700s. These were often men who had lived the strenuous and hazardous life of fishermen and seamen in their youth. The American nouveaux riches spent much of their wealth on such items of "conspicuous consumption" as liveried servants, expensive coaches, silverware, linens, fine wines, and jewelry and fashionable gowns from London for the ladies of the family. However, unlike their European counterparts, the American upper classes abandoned "conspicuous leisure" as defined by Thornstein Veblen (1899) as a measure of social status; they generally kept themselves well occupied. Fortunately, the landed gentry sometimes developed importation of plants as a hobby.

## AGRICULTURAL SOCIETIES AND PERIODICALS

Colonial agriculture fell behind that of Europe in developing new inventions and methods. This was recognized by the members of scientific societies that had been established at the time (American Philosophical Society, founded in 1743, and the American Academy of Arts and Sciences, founded in 1780), and which encouraged the investigation of European ideas, experiences, and agricultural experimentation. Soon societies devoted entirely to agriculture were formed. Apparently the first of this kind was established in New Jersey in 1781, but the Philadelphia Society for Promoting Agriculture, organized in 1785, was the first to publish the results of its work. The South Carolina Society for Promoting and Improving Agriculture was formed the same year in Charleston, with George Washington as president, and the Society of Maryland for the Encouragement and Improvement of Agriculture was organized a year later, followed by others in a few years. The early agricultural societies were made up of men of all professions who would seek out and adapt to American conditions new crop varieties, new equipment, and new agricultural methods developed in other countries, and conduct experiments themselves when possible. Prizes were annually awarded for the best solutions of problems of general significance. Four of the men in the Philadelphia society had signed the Declaration of Independence and two were members of the convention which drew up the Constitution of the United States. None depended solely on farming for a living.

Among journals devoted exclusively to agriculture, *The Agricultural Museum,* which was the organ of the Columbia Agricultural Society, was apparently the first, appearing in Georgetown, D.C., on July 4, 1800. The first to achieve prominence and relative permanence was *American Farmer,* established in Baltimore in 1819, followed later that year by *The Plough Boy* in Albany, New York. Agricultural periodicals were influential throughout the nation by the middle of the nineteenth century (Rasmussen, 1960).

## UNITED STATES DEPARTMENT OF AGRICULTURE

Although George Washington recommended the establishment of an agricultural branch of the national government as early as 1796, and many individuals took it upon themselves, officially or otherwise, to introduce new or improved seeds and livestock breeds from abroad, it was not until 1839 that Congress took official action in the form of a $1,000 appropriation of Patent Office fees for collecting agricultural statistics, conducting agricultural investigations, and distributing seeds. Then Henry L. Ellsworth, Commissioner of Patents, established an Agricultural Division in the Patent Office, which by 1854 was employing a chemist, a botanist, and an entomologist (Rasmussen, 1960).

The United States Agricultural Society, established in 1851, presented resolutions and memorials to Congress and these, combined with pledges of the Republican party in 1860 for agrarian reform, led to the establishment of the Department of Agriculture in a bill signed by President Abraham Lincoln on May 15, 1862. The department was headed by a commissioner responsible to the president. In 1889 the Department of Agriculture was raised to cabinet status (Rasmussen, 1960).

The research agencies of the United States Department of Agriculture (USDA) are the Agricultural Research Service, Economic Research Service, Forest Service, Statistical Reporting Service, and Farmers Cooperative Service. Scientists are located in the area of Washington, D.C. and in various places throughout the country, sometimes at land-grant colleges and universities and sharing facilities provided by the host institution. The USDA's Agricultural Research Service shares its employees with the state experiment stations for cooperative studies of mutual benefit.

The USDA's Plant Quarantine Division organizes and enforces the provisions of the National Plant Quarantine Act to prevent the introduction into the United States and the spread therein of injurious plant pests. It cooperates with state and county quarantine officials.

## A NATIONWIDE SYSTEM OF AGRICULTURAL RESEARCH AND EDUCATION

As soon as agricultural societies and farm journals were established they began to point out the need for agricultural education. During the first half of the nineteenth century there were a few agricultural schools and academies, and some private colleges gave instruction in agriculture and in the application of science to agricultural problems. In 1855, Pennsylvania and Michigan obtained state agricultural colleges as a result of acts of the state legislatures, followed by Maryland in 1856 and Iowa in 1858.

In 1857, Justin S. Morrill, representative in Congress from Vermont, introduced a bill for donating public land to the states for colleges of agriculture and the mechanical arts. The bill was passed in 1859, but was vetoed by President Buchanan. It was then reintroduced in 1861 and was signed by President Lincoln on July 2, 1862 (ESCOP, 1962).

The Morrill Land Grant Act was an important step, but there was still need for the development of experiment stations where research could be done to provide basic knowledge upon which college-level courses could be developed. On March 2, 1887, the Hatch Act was approved in Congress and signed by the President. It provided for a yearly grant to each state for the support of an agricultural experiment station. A few stations had already been established, beginning with the Connecticut Experiment Station in 1875, but within a year after the Hatch Act became law, every state had accepted its provisions and within a decade staff members were doing original research. The agricultural experiment stations not only gave direction to the land-grant colleges but were probably the principal factor assuring their continuation (ESCOP, 1962).

## COOPERATIVE EXTENSION

Research findings would be worthless, of course, if not brought to the attention of the farmer. A great part of this task is performed by the vast network of the Agricultural Extension Service. This is a cooperative undertaking of the USDA, the state land-grant institutions, and the farmers, through the provisions of the Smith-Lever Act of May 8, 1914. The cost is borne cooperatively by the federal government, the states, and the counties.

The Smith-Lever Act provided federal aid for agricultural extension work. For a state to obtain its share of federal funds each year it must submit its plans for agricultural extension work under the act to the federal Secretary of Agriculture and secure his approval in advance; it must also provide matching funds. Counties also cooperate by contributing financial support. The responsible person in each county is called county director and farm adviser or home adviser. The county cooperative extension agent can be called upon for information or counsel. County educational programs include direct contacts such as farm calls, meetings, newsletters, and use of mass media. The county extension agent utilizes resources of the county, state, and federal governments but actually is usually an employee of the state agricultural college, conducting off-campus, noncredit teaching programs. In addition, there are extension specialists, such as extension agronomist, entomologist, economist, horticulturist, forester, and the like, to assist in bringing information and research results from the agricultural college of the

university to the county extension agents. The federal extension service also e
technically trained specialists to carry the practical results of the researchers
Department of Agriculture to the research departments of the state agricultu
leges, thereby incorporating them into the extension program of the state.

The fact that farmers are also experimenters and innovators deserves emphasis. They often permit the county agent to develop test plots and conduct experiments on their land to develop local results and information for the benefit of producers in the area. The farmers are termed cooperators, working with the university in developing and carrying out plans, each supplementing the knowledge and experience of the other.

In 1917 federal aid to vocational education was broadened with the passage of the Smith-Hughes Act, which provided about $7.2 million a year for vocational programs in high schools, including agricultural, trade and industrial, and home economics education. The act also established a Federal Board of Vocational Education to administer the program, but in 1933 the responsibilities of this board were transferred to the office of the Commissioner of Education. In 1936 the George-Deen Act authorized an annual appropriation of $13 million for vocational education in agriculture, trade and industry, home economics and, for the first time, distributive occupations.

## MECHANIZATION IN AGRICULTURE

Understandably, the genius of the colonial farmer lay in his ability to utilize his mechanical skills to modify and adapt the technology of the past to new environmental conditions rather than in the invention of new machines or equipment. New inventions of value to agriculture were more likely to be made by blacksmiths. For example, the principal contributors to the development of the plow were blacksmiths: Jethro Wood, John Deere, John Lane, William Purlin, and James Oliver. Blacksmiths in those days could make about anything the farmer asked for—and more. A farmer might invent something new if he was also a blacksmith, as in the case of Cyrus McCormick, one of the inventors of the mechanical reaper (Wik, 1975). An historically interesting exception to the general rule occurred in 1798 when Thomas Jefferson was the first in America to design a moldboard for the plow according to true mathematical principles, removing the construction of plows from the domain of empiricism.

Development of new equipment was made difficult owing to superstition and resistance to change. Plows of the eighteenth century were made of wood covered with odd scraps of iron to prevent rapid wear, but the uneven surfaces caused much friction. A cast-iron plow made by Charles Newbold of New Jersey was patented in 1797. It worked well but was not accepted by farmers, who believed that iron poisoned the soil and made weeds grow. An important disadvantage of Newbold's plow was that all parts were in one casting. When the share was dulled or broken the whole plow had to be replaced. A cast-iron plow made by Jethro Wood of Scipio, New York, patented in 1814 and improved in 1819, had interchangeable parts joined together and fastened by lugs and interlocking pieces. In part through the efforts of agricultural societies and farm journals, farmers gradually lost their fear that iron plows would poison the soil (Bidwell and Falconer, 1925; Rasmussen, 1960).

In the earliest years of the eastern frontier the sickle and scythe were used for harvesting grain. Then the "cradle" was invented in New England and began to be used at about the time of the Revolution. It was soon used in Europe also. The cradle was a scythe with a wooden framework of fingerlike bars, to catch and bunch enough stalks for a sheaf, making them easier to bind by hand. Construction of the cradle demanded extremely skilled craftsmanship, for it had to be perfectly balanced. The all-day use of the cradle called for strength and endurance that is now almost unimaginable. Under favorable conditions a good man could cut four acres of grain in a day (from sun to sun), although two or three acres were probably the most common area cut (van Wagenen, 1953). A cradler demanded, and obtained, two or three times the wages of a common laborer (Bidwell and Falconer, 1925).

Obed Hussey made a reaper in Cincinnati and patented it in 1833. McCormick's reaping machine, made in Virginia, was patented in 1834 and again in 1845, and by that time both machines were coming into use in the East. McCormick's reaper had an advantage in that the grain was raked from the side of the platform, thus avoiding the necessity of removing the grain before the next round of the machine (Miller, 1902). Van Wagenen (1953) believes the reaper was not common in New York State until as late as 1875. The next step forward was the harvester, on which two men could ride and bind the cut grain as it was delivered to a platform. A driver and two binders could cut and bind eight to ten acres in a single working day. Next came the mechanical binder, "that wonderful combination of reels, elevators, and steel fingers." Eventually came the "reaper-combine," or simply "combine," which displaced the reaper, harvester, binder, threshing machine, and the various power engines formerly required to operate each. One reaper-combine displaced 170 unskilled or 145 skilled workers. Likewise with the mechanical corn sheller one man did the work formerly done by a hundred (Hedrick, 1933). The opening of the vast trans-Mississippi virgin lands to agriculture, the rapid expansion of railroads, and the efficiency with which grain crops could be mechanically harvested, resulted in an era of disastrously low prices and "the great agricultural depression" in the last quarter of the nineteenth century (van Wagenen, 1953).

## EASTERN AGRICULTURE BECOMES SPECIALIZED

As the nation developed, the East lost its preeminence in American agriculture. As mills, factories, and shops absorbed capital and men, farming became of secondary importance. Grass grew undisturbed where once there were fields of wheat, corn, buckwheat, oats, rye, and other crops. The stubborn forest, subdued by centuries of back-breaking toil, marched back over the agricultural lands until today there is more forest in New England than there was in the year 1800. It is an eerie feeling to walk out from the perimeter of a New England town through miles of forest, climbing over brush-covered stone walls that once marked the boundaries of farms dating back to early colonial times.

What remained of eastern agriculture became highly specialized. Specialized areas for intensive gardening and fruit-growing developed rapidly between 1820 and 1840 in eastern Massachusetts, Rhode Island, Long Island, and New Jersey because of high prices for such produce in the rapidly growing urban areas and the short haul to city markets. It involved the use of hired labor, liberal use of manure from the farm and brought from the cities, drainage of meadows, irrigation of dry lands, and garden-type rather than field-type tillage. Some vegetables, such as onions and beets, were shipped to the West Indies and the southern states. The rapidly increasing price of land forced less-intensive farming back onto cheaper land. Many immigrants who had learned gardening in Europe leased truck farms and were usually able to purchase their farms in a few years (Bidwell and Falconer, 1925).

As previously related, New Jersey, despite its high population density, became known as the Garden State. Even out-of-season crops such as tomatoes, cucumbers, grapes, roses, chrysanthemums, and carnations, grown in acres of greenhouses, became highly profitable. Certain geographic areas became famous for specialized crops, such as onions and a high grade of tobacco, excellent for cigar-wrapping, in the rich soil of the Connecticut River Valley. In 1976, Connecticut and Massachusetts tobacco amounted to 9,034,000 pounds, only 0.42 percent of the total domestic crop, but sold for much higher prices than tobacco raised in other areas (USDA Agr. Stat., 1977). Cranberries, asparagus, and mushrooms are other examples of local specialties.

### Apples

Although the Puritans brought over known varieties of nursery stock from England, further propagation of fruit varieties was by seed. No two seedlings were alike and were generally inferior to the fruit of the parent tree; yet there was endless opportunity for selection. Distribution of apple seedlings was facilitated in a curious way by John Chapman, an eccentric missionary who became known as "Appleseed Johnny," or "Johnny Appleseed." He was born in Springfield, Massachusetts, in 1775. In his extensive wanderings, preaching to whites and Indians over a period of forty years, he gave away crudely printed little Bibles and planted apple seeds in likely spots, particularly throughout the forests and fields of what was then called "Northwest Territory." Appleseed Johnny's strange odyssey ended in Indiana, where he died and was buried under an apple tree, with one of his little Bibles in his hands.

Eventually from 85 to 90 percent of American fruit varieties were obtained as naturally occurring seedlings. Superior varieties could thenceforth be maintained by vegetative propagation (budding or grafting). The Rhode Island Greening apple, for example, started about 1750 from a seedling raised by a tavern keeper named Green, near Newport, Rhode Island. Visitors who tasted the delicious fruit of this tree obtained scions; the variety was widely distributed in this manner. In the early days fruit was generally eaten only locally. Most of it was made into beverages: apples into cider and applejack, pears into "perry" and brandy, peaches into brandy, grapes into wine and brandy. Much fruit was fed to livestock. Growing fruit for market began about 1840 when transportation began to make it possible to ship fruit to the cities. Some of the early eastern apple varieties are still among the world's best.

By 1850, Ohio began shipping apples to the eastern markets and was soon followed by Michigan. Nevertheless New York, with 372,000 metric tons produced in 1976 (USDA Agr. Stat., 1977), remains an important apple-growing state, and in fact it is second only to Washington in the quantity produced. Likewise apple-growing is important in southeastern New England, where the sea moderates winter temperatures to some extent and summers are cool. Spring comes late; then the danger of frost is reduced and the hilly terrain assures good air and water drainage.

Apple trees are long-lived and shifts in the important commercially grown varieties are very slow. Also the big chain-store buyers deal in large volumes, demand a uniform product, and show little interest in new or different varieties. Delicious, McIntosh, Rome Beauty, Staymen Winesap, Rhode Island Greening, Jonathan, Northern Spy, Ben Davis, and Baldwin—varieties we oldsters knew as small children—are still grown, and some are among the most important modern varieties. Delicious, Golden Delicious, and McIntosh are the first three in importance in the United States.

Sixty-six percent of New York's 67,000 acres of apples are grown in western New York, mostly in counties bordering Lake Ontario (fig. 3, Region L), in the "Lake States fruit, truck and dairy region" that borders the southern Great Lakes and extends from New York to Wisconsin. In eastern New York, the industry is concentrated mainly in the southeast corner.

Mutations, or sports, occur naturally quite often in apple varieties, possibly caused by cosmic rays that pierce the growing tip of a shoot and cause gene or chromosome change. Sports can also be induced artificially by exposing dormant stems to mutagenic chemicals, radioactive isotopes, or X rays. Several promising apple and sweet cherry varieties have been developed in this way. Branches developing from the mutant bud can be used as scions for grafting and starting new strains. The lucky grower whose sharp eyes detect a promising new sport can sell its propagation rights to nurseries, possibly for thousands of dollars. The mutations may result in better fruit color or a spur (dwarf or semi-dwarf) habit of tree growth. The spur type of tree is usually about two-thirds the size of the parent variety. This is a good feature because pruning, harvesting, and other tree care costs less for small trees. The spur types may also have other desirable characteristics, such as increased winter hardiness, thicker twigs, thicker and greener leaves, and greater fruit-production efficiency. (Way, 1973; Hull et al., 1975).

McIntosh, Red Delicious, Rome, Greening, Cortland, and Golden Delicious lead among the 32 standard apple varieties in New York State, comprising 79 percent of the 1,925,000 trees. Among the 33 dwarf and semi-dwarf varieties, the Red Delicious, McIntosh, Idared, Golden Delicious, and Rome are the leading varieties, comprising 61 percent of the 1,629,916 trees (Anonymous, 1975*b*).

The wonders of chemistry are as spectacular in production of apples as with more basic food crops, although they are possibly less well known. Spray thinning with growth-regulating chemicals has resulted in annual flowering and eliminated the alternate light and heavy production of former years. Other growth regulators bring trees to bearing at an earlier age (particularly important for the Delicious variety), and still others are used as sprays to prevent preharvest drop of fruit (particularly important for McIntosh). Other chemicals promote abscission and facilitate mechanical harvest of apples (and cherries) that are to be processed. Dipping harvested fruit in dilute solutions of DPA or Ethoxyquin has eliminated scald, a nonparasitic malady that used to cause apple skin to turn brown in storage, sometimes causing heavy losses. The most important contribution has been made by the newer fungicides, less toxic to the trees than sulfur for scab control. None of the chemicals, so useful in the production and storage of the nation's most important deciduous fruit, has presented problems in environmental pollution or in contamination of the fruit (Hoffman, 1970).

### Peaches

Peaches have been grown in great abundance since early times. As previously related, Swedish botanist Kalm (1770) was amazed by their abundance in the environs of Philadelphia in 1750. According to the annual report of the Patent Office in 1849, a Major Reybold had 1,000 acres of peaches in Maryland and Delaware. Steamboats were constantly transporting the produce of these orchards to New York, Boston, and Philadelphia.

### Grapes

Certain species of grape, the cranberry, and the blueberry are three of the six plant crops of significant commercial importance that are native to the United States, the others being the pecan, sunflower, and Jerusalem artichoke. Most of the commercial table, wine, and raisin varieties of grape in the United States are European, grafted onto phylloxera-resistant American rootstock. Native American grapes are ordinarily of small size and inferior quality, but in 1852 a seedling from a fruit borne on a vine of the fox grape (*Vitis labrusca*) in a garden in Concord, Massachusetts, was recognized as being of good size and excellent quality. Named the "Concord," it has continued to this day to be the principal commercial variety of grape in the Northeast (Schery, 1972). It is the usual slipskin type of American grape, with a tough skin that separates readily from the pulpy flesh. Its pronounced fruity flavor makes it a good dessert grape and it is the only important variety for sweet juice, jelly, and preserves. Much of the Concord crop is also used for wine production. Niagara is the leading American-type white grape, used fresh and for wine. Delaware is a red grape, likewise used fresh and for wine. Catawba is an important constituent of New York champagnes and table wines (Einset et al., 1973).

New York is the principal producer of grapes in the East, averaging about 170,000 tons a year (USDA Agr. Stat., 1977). Of the state's 42,653 acres planted to grapes, 65 percent is of the Concord variety (Anonymous, 1975*b*).

### Cranberries

The cranberry (*Vaccinium macrocarpon*) is a creeping, vinelike plant growing in thick mats, hugging the ground, in "cranberry bogs" with acid soil. It is propagated by cuttings. The round berries become red toward autumn. They are borne on erect branches with small oval leaves. The berries are harvested from both wild and cultivated stands by means of toothed "cranberry scoops." The berries are very tart and are cooked with sugar to form a cranberry sauce or jelly (Schery, 1972). By means of water-control gates and ditches, a cranberry bog can be flooded to a uniform depth or drained when desirable. By controlled inundation, the fruit can be protected against winterkilling. Also, weeds can be suppressed and adequate moisture can be assured during dry seasons.

The principal cranberry areas of the nation are in southeastern Massachusetts, southern New Jersey, west-central Wisconsin, and the coastal regions of Washington and Oregon. Cranberries are the principal export crop of Massachusetts. The state produces an average of about 40,000 metric tons of cranberries per year, only slightly exceeded by Wisconsin with 41,000. The two states produce about 76 percent of the nation's total crop of cranberries (USDA Agr. Stat., 1977).

### Blueberries

Another Eastern indigenous berry is the blueberry (*Vaccinium* spp.), related to the cranberry and adapted to acid, generally infertile barrens, where it spreads by means of rhizomes, which are underground, rootlike stems sending up leafy shoots from the upper surface and possessing roots on the lower surface. The pemmican of the American Indians consisted of wild blueberries combined with venison. Blueberries are still harvested from wild stands, especially in Maine and eastern Canada. Two "lowbush" species, *V. myrtilloides* and *V. angustifolium,* are seldom cultured, but wild stands are biennially burned over in early spring to control weeds and pests and rejuvenate growth. A "highbush" species, *V. corymbosum,* of eastern swamplands has been the source of several notable selections that have been domesticated.

Blueberries have long been handpicked, but harvesting is becoming increasingly mechanized. The wild berries from burned-over stands are raked up with toothed scoops. Foliage and debris are winnowed out and the berries are used chiefly for processing. Most of the berries to be eaten fresh are from cultivated plants. North America is the source of almost all of the approximately 50,000 tons of blueberries produced in the world annually (Schery, 1972). Maine grows 90 percent of the domestic blueberries.

## Potatoes

The potato was taken from its native home in South America to Spain about 1585, then much later returned to the New World in improved form. It was first cultivated in the English colonies by Scotch-Irish settlers of Londonderry, New Hampshire, in 1719. Potatoes became one of the most successful crops of New England and New York. After local markets were supplied, large quantities were annually hauled to seaport towns and sold for the southern market. Cost of production was so low that the tubers were widely used as feed for livestock and for manufacture of starch. The flourishing potato industry in the East was abruptly checked, beginning in 1843, by a new disease called "rot," or "potato disease," now recognized as late blight of the potato. This affliction caused losses ranging from 20 to 30 percent. Decreased yields resulted in increased prices and diminished use of potatoes for stock feed and starch manufacture.

Since the potato is native to the cool highlands of the Andes in South America, and can grow far above timberline in that region, it is not surprising that in the United States it is one of the few vegetable crops that thrive where the growing season is short and cool, as in northern Maine, northeast North Dakota, the Snake River Valley of Idaho, and high altitudes of Colorado. However, potatoes are also grown during cool growing weather in areas such as California, south Texas, and Florida, which produce late- and early-season potatoes while the ground is still frozen in the North. Among the major crops, the small grains, and hay, flax, and peas will mature during a short and cool growing season and in the North and high-altitude West they are generally grown in rotation with potatoes. The potato keeps well if protected from heat, frost, and dampness, as in simple masonry storage rooms on high, dry ground, where the soil serves as excellent insulation against heat and frost.

Maine is exceeded only by Idaho and Washington in the production of potatoes. Northern Maine is important in the production of seed-potatoes because of its cold climate and, therefore, less likelihood of insects transmitting viruses. By planting only disease-free stock, potato fields in the North Central States have been kept remarkably free of virus diseases and bacterial ring rot since about 1950. The potato is a labor-intensive crop and Aroostook County, Maine, where most of them are grown, is so geared to the agricultural community that schools start in mid-August and then recess for three weeks in September and October so that schoolchildren can help in the harvest of the $70 million potato crop. The small farmers could not harvest their crops without them. In 1974 Congress prohibited the use of children in agriculture by passing a rider attached to the minimum wage bill. Although the law has not been stringently enforced, potato growers are bitter. An important social principle is involved in the controversy, well expressed by a former schoolteacher who evidently inherited the old New England belief in work as a "character builder": "These kids are so proud of the work they do and the money they earn. We're pretty fortunate. We come from an area where we have to work and I'll tell you what: we've only had one criminal boy in this town in 100 years. The boys here go pretty straight because they know what it means to have done something constructive and seen the results" (Lamb, 1975).

## Beef Cattle

Colonial agriculture lagged behind European and nowhere was this more evident than in connection with livestock. Not only were the animals generally mistreated, as related earlier, but little effort was made to improve the breed. Early in the nineteenth century humanitarians began to emphasize kindness to animals. Agriculture being the nation's most important business, men of means, usually already established as leaders in other fields of endeavor, began to operate stables, barns, and sheds and import the best animal breeds the world had to offer. Being amateurs, they usually employed experts to do the work. These early benefactors of agriculture, soon followed by professional breeders, were able to greatly improve American livestock. Eventually the federal government became involved and farmers' organizations such as the Grange and 4-H clubs taught and demonstrated the value of pedigreed livestock. But as late as 1841 the records of an abattoir near Boston showed the scrubby nature of livestock: cows weighed an average of only 450 pounds, steers 600 pounds, and oxen 875 pounds. No distinction was made between beef and milk production, because there was no market for milk, and butter and cheese were produced by the farm wife and locally consumed.

During the first four decades of the nineteenth century, beef production as a specialized industry was considerably expanded in the Connecticut Valley, whence most of the cattle were driven to the historic and picturesque Brighton market just outside of Boston, described by Nathaniel Hawthorne (1887) in his *American Note Books.* Here a weekly cattle fair had been held since the days of the Revolution. Some cattle (and sheep) were also driven down from the pastures of Maine. In hilly sections pasture grass and hay were supplemented by potatoes and, where corn could be grown, by cornmeal. For the Philadelphia and New York markets, cattle were fattened in southeastern Pennsylvania, especially in Chester County. Cattle were driven in from Maryland, Virginia, South Carolina, Ohio, and northern and western New York to the famed southeast Pennsylvania pastures for summer feeding. This business was so prosperous that as early as 1819 grazing farms near Philadelphia were selling at from $100 to $300 per acre. After grazing, the cattle were stall-fed throughout the winter on a meal made of corn and oats ground together (Bidwell and Falconer, 1925).

### Dairy Farms

By 1840 railroads made possible the cheap transportation of beef cattle from the Midwest; eastern farmers could not compete and, especially in New England and New York, they turned to dairy farming. Before the opening of the Erie Canal in 1825, Orange and Ulster counties in New York State had long been producing high-grade butter and cheese for market. In *My Boyhood,* John Burroughs (1922), whose father kept thirty Durham cows, describes in great detail how he helped his mother make butter. All dairy chores from milking to making butter and cheese were considered to be "women's work" in those days.

The canals offered cheap transportation of wheat from the Genesee region, and this competition, plus the ravages of the "grain worm" (a midge, *Cecidomyia testies*), encouraged a shift from wheat farming to dairying in the Mohawk Valley, particularly in Herkimer and Oneida counties. The canal was, of course, also a great benefit to the dairymen. The principal export was cheese. In only one small town, Steuben, was butter produced for market, believed to be the result of a special ability for the production of butter by the Welsh immigrants who settled there (Bidwell and Falconer, 1925).

Land in the East was generally relatively poor for most crops, but the cool summers and even distribution of precipitation throughout the year made it ideal for pasturage. The rapid growth of cities created good markets, and railroads enabled dairymen to rush fresh milk to the market. The fresh-milk market was so lucrative that farmers discontinued producing butter and cheese. These products, along with canned milk, were shipped in from Midwest dairy regions such as Wisconsin. Economics compelled farmers to improve their dairy herds and to give them better feed and better care. Famous European breeds such as the Jersey, the Ayrshire, and the Holstein were imported. Today eastern states from Maine to Maryland and those bordering the Great Lakes from New York to Minnesota comprise the nation's principal milk-producing areas.

Although dairying is now intensively developed in the East, the dairy farmers there raise little grain, preferring to purchase it from the Midwest. In the East there is some good soil for growing grain, but it is interspersed with thin, gravelly, or rocky soil, rough wooded terrain, and slopes that are subject to erosion, particularly in New England. But eastern dairymen have some advantages over those of the midwestern Great Lakes dairy region. The average annual rainfall is heavier, the climate is milder, and the frost-free growing season is longer. The properly managed grasslands support a longer grazing season and have higher yields of forage than those of the Great Lakes region. Proximity to the nation's greatest concentration of population results in eastern dairymen selling a larger percentage of milk for direct human consumption, rather than processing it, and this is the milk that brings the highest prices (Clark, 1945).

In 1975, New York State, with 917,000 milk cows, was exceeded only by Wisconsin's 1,812,000. Pennsylvania, with 699,000 milk cows, also ranked high among the dairy states (USDA Agr. Stat., 1977).

### Sheep

Sheep were unimportant in America until 1807, when the Embargo Act forbade American ships to engage in foreign trade and woolen fabrics could no longer be obtained from Europe. About that time a famous Spanish breed, the Merino, was imported and Merino wool sold in the United States for about two dollars a pound compared with forty cents for common wool. Great herds were established in the East. One farmer sold four full-bred rams at $1,000 each; another sold two rams and two ewes for $6,000. A collapse in the woolen industry, beginning in 1816 and resulting in fine wool selling for fifty-seven cents a pound by July, 1821, led to the abandonment of sheep-raising by most farmers and a ruthless sacrifice of their stock, for slaughter, for as little as one dollar a head. In 1830 another period of prosperity in eastern wool-growing began as woolen manufacturing became firmly established and encouraged by tariff protection. By 1840 wool-growing had become particularly well established in Vermont and the Berkshire region of western Massachusetts and Connecticut. By 1850 Ohio had become the center of the sheep industry and, following the Civil War, the industry moved farther west to the Great Plains (Bidwell and Falconer, 1925; Clark, 1945). A few flocks can still be seen in the states of the Northeast.

### Swine

Improvement of the razorback swine of colonial days at first came about mainly through fattening them in pens on corn with admixtures of buckwheat, potatoes, peas, and beans. Previously they had been allowed to run wild in the forest. Swine became increasingly important as agriculture progressed, and by 1800 farmers generally kept a half-dozen or more pigs to supply their families with salt pork, which had become the most important article of diet in the American home. Salt pork was also important for provisioning ships, including the fishing fleets, and in the export trade to the West Indies (Hedrick, 1933).

Swine, coming to maturity early, are easily modified by selective breeding. By the end of the eighteenth century, crossing with improved breeds from England, China, and Spain began to show striking improvements in swine types. Such breeds as the Byfield and Mackay in Massachusetts and the Chester County Whites in Pennsylvania were developed from imported animals (Bidwell and Falconer, 1925). With the importation of the Berkshire hogs about 1830 a speculative fever in swine-breeding developed, but these fine animals were too delicate for the rough treatment received in this country and many farmers who bought them lost heavily. The Duroc breed, and later the Duroc Jersey, were developed in New York. Most new breeds adapted themselves well and by 1840 the old race of "native swine" had almost disappeared. When the butter and cheese industries became important in the middle of the century, cheese, whey, and skimmed milk were abundant to feed pigs and by 1860 it was estimated that in the dairy regions one pig was kept for every four cows. Pigs were also fed on the mash from breweries and distilleries. However, the eastern hog industry eventually succumbed to the competition of the Midwest (Hedrick, 1933).

### Poultry

Wild turkeys, ducks, geese, and other wildfowl were originally abundant in some areas of the East and some hunters made a regular business of supplying city markets. Wild turkeys weighed as much as thirty or forty pounds (Benson, 1937). But as wild game vanished near the big towns, poultry became increasingly important and, unlike beef cattle and sheep, they retained their importance in the East, particularly in Maine and from New Jersey to Maryland. The introduction of incubators and brooders, and later the cold storage of eggs, gave a great impetus to the poultry industry in the East and eventually throughout the country (Hedrick, 1933; Durost, 1969).

Of the breeds of chickens with which the present generation of Americans may be familiar, the Plymouth Rocks originated in New England in the 1860s and the Rhode Island Reds in Rhode Island a few years prior to 1900. But the best commercial breeds of chickens in America today are quite modern. They will be discussed in chapters 4 and 9.

### Humanized Nature

"Humanized nature," as popularized by Dubos (1972, 1973), is so universally attractive that in some areas, particularly in the East, many well-managed farmlands have become valued as much for their aesthetic and social appeal as for their production. Farmlands that are especially noteworthy in this connection are in the counties of Essex in Massachusetts, Fairfield in Connecticut, Westchester in New York, northern New Castle in Delaware, Montgomery in Maryland, Chester and Bucks in Pennsylvania, Loudoun in Virginia, and in the bluegrass regions of Kentucky and Tennessee. In such areas many of the most beautiful rolling farmlands with woodlots have become rural estates that figure prominently in the social life of their proprietors, whose principal business may be in the nearby city.

## RESURGENCE OF FOREST AND WILDLIFE

The East, particularly New England, provides an example of how man could manage nature, not only to provide for his material needs, but also to fulfill his aesthetic desires, provided only that he can curb his own population growth. Forest vegetation is returning to land formerly cleared for farm and pasture. As early as 1850, half of New Hampshire and three-fourths of Massachusetts had been cleared of forest growth, but today the percentage of forest land in those states is 85 and 65 percent, respectively. The trend is the same in other New England states. Maine, where Henry David Thoreau wrote that man seemed determined to "drive the forest out of the country," is still 90 percent covered with forests of pine, spruce, fir, hemlock, and sugar maple—the most heavily forested state in the nation. Although the woods have been cut and renewed as many as five times in the last 350 years, through the efforts of public and private foresters to protect, improve, and rehabilitate forest land, Maine produces almost twice as much wood as it cuts each year.

Between the cities and their surrounding "bedroom" adjuncts lie extensive areas of second-growth forest rising 30 to 40 feet over the old stone walls that marked the boundaries of farms of earlier days, dating back as far as the colonial era. Cluster-housing developments encircled by common lands kept in their natural state are situated throughout the forested areas, but the houses, largely hidden by trees, are barely detectable to the amazed visitor flying over the New England states for the first time (Larson, 1972).

No part of the United States fills me with such hope for beleaguered nature as the rural East. The East has had a heavy date with growth and is healing its wounds. There are, of course, some Easterners who take a dim view of this development, fearing that the "cancer of Appalachia" will spread to their "economically faltering region" (Peirce, 1976). However, the plight of Appalachia came about through gross environmental abuse and overpopulation, which the East has been more successful in avoiding.[12]

Economic decline is particularly evident in New England as the result of the specially high cost of energy there. Yet the well-watered area possesses 229 flood-control dams, many of which can be made to accommodate hydroelectric power installations.[13] Also many small- and medium-sized hydro installations were abandoned in the belief that "bigger is better." Now David Lilienthal (1977), past chairman of the Atomic Energy Commission and the Development and Resources Corporation, and leading advocate of regional resource development, reminds us that "Nobody foresaw what would happen to the cost of oil and gas and coal, or to the cost of transporting these fuels, or to the cost of constructing huge generating plants and mighty transmission lines. No one foresaw the rapid diminution of the world's reserves of oil. We placed major reliance upon nuclear energy, without being fully aware of its costs or its hazards." Lilienthal suggests that the flight of so many industries and the chronic unemployment in New England has been coincident with diminishing reliance on hydropower. Hydropower is in the aggregate an enormous renewable resource that does not pollute the air and need not make a violent impact on the environment, especially where dams have already been built. Small hydro plants provide communities with a chance to show that they can do something to help themselves. Yankee ingenuity, where are you now that we need you?

If a latter-day Rip Van Winkle awakened from the sleep of a century to stroll through a section of eastern farmland, he would find little besides his long beard to indicate the duration of his sleep, assuming, of course, he didn't happen to meet one of the mechanized monsters that now till the soil. The disconcerting transience that characterizes modern life pertains only to our artificial creations, but not those of nature, from which we draw our sustenance. Rip Van Winkle would see the same crops: corn, wheat, oats, barley, timothy, red clover, June grass, and white clover. He would see the same weeds: quack grass, wild mustard, white daisy, yellow dock, and bull thistle. In apple

[12]While the overall aspect is pleasing to the tourist, there is a groundswell of resentment in the upper three New England states over absentee land speculators who bought land cheaply from poor farmers and made fortunes from a tenfold increase in land values in about a decade, meanwhile exacerbating the financial problems of such people as remaining farmers, woodcutters, and fishermen (Faux, 1975).

[13]Considering the nation as a whole, the Federal Power Commission estimates that only about 1,400 of the 50,000 small dams have been developed for power generation. The Commission asserts that the United States has undeveloped hydro potential almost twice the 66,000 megawatts of existing capacity (Lilienthal, 1977).

orchards he would see the same old familiar varieties (or their sports)—Baldwin, Ben Davis, Cortland, Gravenstein, Jonathan, McIntosh, Northern Spy, Rhode Island Greening, Rome Beauty, Winesap, Twenty Ounce, Wealthy, Yellow Newtown—that he had once proudly boasted were the world's best in flavor. He would note that some of the farmlands of his youth had reverted to forest. The same denizens of the forest would be there, but in larger numbers than he could remember. Plunging deeper into the woods, he might cross a rocky ridge where somehow the virgin timber had escaped the ax and saw. Nowhere else on earth has man seen a deciduous forest of such grandeur and beauty.

It is easy to walk through the eastern forest and become acquainted with the different trees (Emerson and Weed, 1936; Gleason and Cronquist, 1964). The maples are large, beautiful trees, often used for parks and shade. The yellow, orange, and red autumn foliage of these trees is perhaps the most strikingly beautiful of all eastern foliage colors. The beech and the maple were the characteristic trees of the deciduous forest encountered by the early settlers who crossed the Alleghenies. They allowed their swine to roam the woods to feed on the nuts that fall from the beech trees after the first frosts. Apiaries situated where basswood trees are abundant produce a specially prized honey. The white ash is easily distinguished by its grayish trunk, furrowed in zigzag fashion, its curious pollen-bearing and seed-bearing flowers, usually on different trees, and the silvery undersides of the leaves that give the tree its name. The white ash can be readily grown from seed.

The eastern deciduous oaks are divided into two groups, those with rounded leaf lobes and bearing acorns that mature in a single season, for example the white oak, and those with pointed leaf lobes and with acorns that require two seasons to mature, for instance the red oak. The butternut hickory or "swamp" hickory is in many respects the most beautiful of the hickories and has a yellow autumn foliage.

The American or white elm that once occurred throughout the East and Midwest along the banks of streams, in river-bottom lands, and on low hills of rich soil, and along many a city street throughout the country, became the victim of the Dutch elm disease. The disease was first discovered in the United States near Cleveland in 1930 and spread rapidly from the East Coast to the Rocky Mountains. The noble chestnut tree that shaded the village blacksmith of Longfellow's immortal poem is now little more than a nostalgic memory. In the 1930s a distinctive fungus disease called chestnut blight was accidently introduced from Europe and soon killed all the larger chestnuts. Some young trees may still be seen, mostly sprouts from roots of old trees whose tops have died, but they too will eventually be killed by the blight.

Although the deciduous trees dominate the remaining eastern forests, evergreens are found throughout the area. A beech-maple-hemlock community extends from New England west to Minnesota and southward along the slopes of the Appalachian Mountains to Alabama and Georgia. The slender twigs and small flattened leaves along each side give the hemlock a specially graceful appearance. The white cedar is a common evergreen, especially in the North in the dense, tangled woods in boggy places. The red cedar is one of the most abundant of the eastern evergreens. Many birds feed on the "cedar berries" of this tree. Particularly in the North, white and red pines; jack pine; pitch pine; tamarack; white, black, and red spruces; Canada balsam; and arbor vitae may be found. The white pine, *Pinus strobus* (fig. 7), was once the principal commercial species of the East, much sought after by lumbermen. Sometimes it grew in large pure stands and sometimes in forests of mixed species. The pure stands developed mainly after a fire or other disturbance, then the pines tended to be replaced by a hemlock—hardwood forest. White pines are now relatively scarce.

The gratifying comeback of the forest is matched by that of wild animal species that were largely or even entirely extirpated in the East. They have returned in large numbers and are now a managed, protected resource with both tangible and aesthetic values. Such species as whitetailed deer, beaver, black bear, bobcat, and porcupine are back in large numbers, to the delight of lovers of wildlife as well as hunters. Deer in particular are found surprisingly close to large cities. More than $2 million worth of meat and hides are harvested each year in New York State, and the recreational business generated by deer has been estimated at $56 million per year (McNeil, 1974).

Not everyone is pleased. Beavers cause plugged culverts and flooded fields. Authorities receive frantic calls from worried mothers when bears pass near school-bus stops. Transplanted city dwellers in their new homes at the edge of the forest, especially if they own a riding horse, have mixed feelings about deer hunters (Larson, 1972). Hundreds of farmers suffer small losses to their trees and a few growers of high-value crops such as grapes, cherries, and nursery shrubs and trees may suffer serious losses (McNeil, 1974).

# 4 The South

*No other human occupation opens so wide a field for the profitable and agreeable combination of labor with cultivated thought, as agriculture.*
—Abraham Lincoln

Long before the English had made any attempts at colonization along the Atlantic Coast, the Spaniards had settlements in Florida, many missions along the Georgia and Carolina coasts, and a fort among the Upper Creek Indians in western Georgia. In 1526 the Spaniards established, and maintained for about a year, a settlement called San Miguel de Guandape not far from the spot where the English later established Jamestown. In the Carolinas the French Huguenots tragically failed in colonization projects from 1562 to 1564 and from 1627 to 1633.

In the spring of 1586 the first crops planted by Englishmen in America were planted on Roanoke Island by potential colonists brought across the Atlantic by Sir Walter Raleigh, an English courtier and soldier to whom Queen Elizabeth had given a patent, practically a deed, to any unoccupied territory he might find in the New World. The crops probably consisted of barley, oats, and peas. But shortly before harvesttime the colonists were taken back to England by Sir Francis Drake. In 1603, members of Pring's expedition planted wheat, barley, oats, peas, and garden vegetables. During the seven weeks of the expedition's stay, these crops seemed to thrive (Smith, 1608).

## THE JAMESTOWN COLONY

In the spring of 1607, the first permanent English settlement was established at Jamestown, along the east shore of the James River estuary near where the historic town of Williamsburg, Virginia, is now situated. This was land granted to the Virginia Company of London. Early in June, the settlers began to sow grain. They also prepared a garden in which they planted various fruit trees and vegetables from the West Indies, as well as cotton. They further planted some crops, such as oranges and pineapples, for which the Virginia climate proved to be too severe. Unfortunately for the fate of the Jamestown colony, few of the settlers were practical farmers. In any case, the primary purpose of the colony was to be the discovery of precious metals and a passage to the "South Sea," and to find numerous products for which England was then dependent on her European rivals. Too much of the energy of the colony was spent in preparation of clapboards, pitch, and pearl ashes, presumably of interest to the English navy, and in experiments with exotic and ill-adapted plants (Bruce, 1895).

Fish and game were abundant in the Jamestown area and provided enough food for the fall months, but during the winter, 67 of the original 105 colonists perished of disease and malnutrition. The following spring 200 more arrived and about fifty men were employed in clearing land and planting crops, but again the yield was insufficient. Once more much energy was expended in the manufacture of pitch, tar, soap ashes, lumber, and glass and in vain search for precious metals. The year 1608 happened to be a poor crop year for the Indians, but after six weeks of trading, Captain John Smith succeeded in obtaining 279 bushels of maize (corn), which helped to tide the colonists through the second winter (Smith, 1608).

The aboriginal inhabitants all along the South Atlantic Coast were quite actively cultivating the most important native American crops: corn, beans, squash, sweet potatoes, and tobacco. Corn was the principal grain of the natives from Canada to Brazil and throughout the West Indies. The natives of Virginia also used the seed of a wild rye (*Elymus*) they called *mattoume,* and buttered it with deer suet to make bread. The Jamestown colonists soon became acquainted with corn by trading with the Indians and began planting it themselves. Early attempts to grow wheat were unsuccessful. In the rich virgin soil wheat had extraordinarily vigorous stalks but produced little grain. The settlers later found that English wheat could be successfully grown on old Indian fields and abandoned tobacco fields (Gill, 1978). Wheat and other small grains were destined to remain relatively unimportant in tidewater areas during the seventeenth and early eighteenth century. The climate and the soil of the coastal plains were less well adapted than the metamorphic soils and cooler temperature of the piedmont areas for growing small grains, especially south of Virginia and Maryland (Gray, 1933).

Under Captain John Smith's vigorous management, in the spring of 1609, 30 to 40 acres of ground were planted to corn under the instruction of two Indian captives. The number of hogs had increased to sixty and poultry to five hundred. But the colony continued to be plagued by misfortune. Half the corn spoiled and the remainder was eaten by rats. A large new contingent of settlers arrived from England, but much of their food was lost in a wreck of the flagship. In October 1609 there were nearly five hundred persons in the colony, but with only ten weeks of food provisions left, besides the livestock. Increasing hostility of the Indians cut off food supplies from that source. A new expedition from England in May 1610 found "only about sixty gaunt and haggard wretches, who were subsisting on roots, herbs, acorns, walnuts, wild fruits, and snakes, with occasionally a little fish" (Gray, 1933). The decision was made to abandon the colony and the dejected colonists were already near the mouth of the James River on a trip back to England when they were met by an expedition headed by T. W. Delaware and brought back to Jamestown. When Delaware departed in March 1611, he left enough provisions to feed the colonists for ten months (Brown, 1890).

As recounted in chapter 1, hunter-gatherers generally obtained an abundant and nutritious diet with minimal effort. The Indians of the Jamestown area supplemented such a diet with corn, but this was such a productive crop that only a small amount of land needed to be planted. According to Smith (1608), the colonists observed that Indians had a way of living without much work. The little work required was done by women, for Indian men disdained any form of menial labor. Some colonists slipped away from the colony and lived with the Indians. This despite the certainty of being executed if they should later be captured and returned to the colony. The fugitives thereby enjoyed a nutritionally superior diet and a life of sport and leisure and avoided the incessant toil, malnutrition, scurvy, and recurrent starvation that was the lot of the wretched, mismanaged colonists (E. S. Morgan, 1975).

Curiously, while civilized peoples considered work to be highly virtuous, they envisioned heaven as a place where this earthly virtue was absent. Thus the white fugitives and the male Indians shared with the numerous gentlemen (nobility and gentry) shareholders of Jamestown the "conspicuous leisure" (Veblen, 1899) with which the latter demonstrated their superior social status and which provided all three groups—fugitives, savages, and gentlemen—with a preview of heaven.

One wonders how the harsh and precarious life in the English colonies continued to lure new recruits. But England was no paradise either for the poor people who were the raw material of colonization. At the time of the founding of the Virginia Colony, the agricultural laborer in England enjoyed little freedom of movement under the State of Apprentices. Political and religious disturbances in England and in all of Europe during the first three quarters of the seventeenth century were potent factors encouraging migration. Also, even in those early times the pressure of population was recognized as one of the principal factors that pauperized a nation. People were willing to take their chances in a new land of limitless space, whatever the initial cost in toil and suffering might be.

## SOCIAL ORGANIZATION IN JAMESTOWN

The colony at Jamestown was at first a commune. Like communes everywhere and throughout history, the Jamestown colony succeeded only so long as its members were subject to strict discipline and unchallenged authority.[1] After vigorous leader Captain John Smith departed for England in the autumn of 1609, the colony barely escaped extinction. According to one of its members, "...we reaped not so much corne, from the labours of 30 men, as three men have done for themselves." Upon the arrival of a new governor, Sir Thomas Dale, "most of the companie were at their daily and usuall works, bowling in the streets." Dale reorganized the colony into an industrial autocracy with forcible control and direction of subordinate laborers. These laborers were all indentured whites; the first Negro slaves were introduced in 1619 (Gray, 1933).

Depending on the functions required, the laborers were divided into squads commanded by "captains." In recognition of the fact that the colony could not survive from cash crops alone, beginning in the year 1614, private gardens of three acres each were allotted to some of the colonists for a yearly rental of 2½ barrels of corn and a month of labor for the company. Later, families that came at their own expense were assigned a 4-room house, 12 acres of fenced land, tools, and livestock, and provisions for 12 months, on condition that only "wheat, roots, maize, and herbs" (no tobacco) be raised on this land and that the family become self-sustaining after the first year. By the year 1616, a third of all the male workers were tenants, each with his obligation to the Virginia Company in labor and a specified quantity of grain. The company had become essentially a plantation proprietor; the southern plantation system was on its way (Gray, 1933).

In the summer of 1610, Delaware constructed Forts Henry and Charles at the mouth of the James River. The following year, under the vigorous administration of Sir

[1]Examples of present-day successful and prosperous communal societies with these qualifications are the more than 200 Hutterite agrarian colonies in the United States and Canada.

Thomas Dale, colonists cleared the planted lands in the neighborhood of the forts, built a fortified town, and erected a "pale" (palisade) two miles across the neck of the peninsula, enclosing a large plat of corn (Brown, 1890).

When Sir George Yardley became governor of Virginia in 1619, he granted each settler who came at his own expense 100 acres of land and as much more as he could pay for at the rate of 12 pounds, 6 shillings for each 100 acres. Others were to be granted 100 acres after they had worked out their period of service to reimburse the company for their transportation. Virginia received 1,216 persons that year, arriving in 11 ships. Martial law and communistic production were abolished (Carrier, 1923).

As the indentures of the first settlers ended, they became sharecropping tenants on the Virginia Company's land. Also, after 1619, each immigrant who could pay his own way over received from the Virginia Company 50 acres for himself and 50 acres for each person he brought to America. Each piece of land was called a *headright.* The system was continued when the Crown took over the colony in 1624 and was adopted by other colonies of the Middle Atlantic and South. The Crown also occasionally gave large tracts of land to various individuals. These proprietors tried to establish a sort of feudal tenure, but this system failed to work satisfactorily in America and the land was eventually sold or disposed of in some other way.

In view of the changed policy, ninety young maidens "of good character" were brought to Virginia to be distributed as wives to the settlers upon payment of 120 pounds of tobacco to cover the charges of transportation. (Tobacco had become legal tender.) This was about all one man could produce in a year if he also raised sufficient corn for his own needs as well as contributing to those members of the colony not engaged in agriculture. Despite the great Indian massacre of 1623 and the abolishment of the London Company in 1624, shiploads of colonists and livestock continued to arrive in Virginia from England. Brush hovels with thatched roofs gave way to frame houses, the material for which was brought from England in some instances.

## Tobacco

Despite the survival of the colony at Jamestown, England was disappointed at what were considered to be meager results in the development of desired trade commodities. Tobacco, not envisioned as a crop of any importance in the early plans for the Virginia Company, soon proved from practical experience to be, from the company's standpoint, the most important gift of the Indians. The Indians found tobacco growing wild in the southern region of the Atlantic seaboard, but elsewhere, where climate permitted, they cultivated a long-leaved variety. They hung mature plants in the sun to dry and then stored the dry tobacco in a hogan or tepee. When they wanted to smoke they pulled off a few leaves and crumpled them into a pouch. Only the men smoked. Their pipes were made of red sandstone or talc with a long reed or hollow branch for a stem. Improved curing methods and the cigar, cigarette, and snuff, were inventions of white men.

The grade of tobacco grown in Virginia by the Indians was much inferior to the varieties raised in the West Indies. In 1612, John Rolfe experimented with the growing of tobacco, using seed he obtained from the Spanish colonies. The success of the new product was soon apparent (Brown, 1890). In 1613, Rolfe had enough tobacco to export—the first cash crop shipped from English America. It was the first American crop that lent itself well to mass production for sale at a high price in Europe. By 1615, 2,000 pounds and by 1619, 20,000 pounds of tobacco were shipped by the Virginia Colony. Thus commercial agriculture in America started earlier and more profitably than many historians have acknowledged.

Early shipments of tobacco to England brought such large returns that there was danger that not enough food crops would be raised. Dale had to promulgate a resolution that each person responsible for his own maintenance must plant two acres of corn for himself and for each manservant, as a condition of raising tobacco. Enough tobacco was sold to pay for many imports from England. In 1620, imports included 200 head of cattle, 400 goats, 20 mares, and 80 asses. In 1621, imports included "bees, pigeons, conies, and mastiffs" (Neill, 1669).

Tobacco prices fell drastically in the 1630s; yet tobacco continued to be planted and corn neglected. A law requiring every man to plant two acres of corn had to be repeatedly reenacted. People continued to be brought into court for delinquency in planting or tending corn. Nevertheless increase in corn production and pasture farming were undoubtedly factors in Virginia's ability to feed her growing population. By 1634 a six-mile palisade had been created near present-day Williamsburg to isolate a large segment of the peninsula between the James and the York rivers. Here and elsewhere, by fighting back the Indians and placing bounties on wolves, the settlers were able to increase their herds. Fencing laws ensured the protection of corn fields. Once their numbers passed a certain point, cattle and swine multiplied much faster than people and the threat of famine disappeared. Virginia could then export cattle to other colonies (E. S. Morgan, 1975). Peaches and apples were free of many of the pests that now infest them and they became so abundant that they were fed to swine (Hubert, 1626). In those days, instead of using manure, the Virginians followed the Indians' example of shifting agriculture, for plenty of land was available (E. S. Morgan, 1975).

Virginia's population grew rapidly as the result of heavy immigration. Those arriving were predominantly indentured servants—young males, mainly vagabonds and those who, because of England's chronic underemployment, would probably have landed in prison. The Virginia Colony considered them to be most suitable for exploitation and England was glad to send them abroad. They were likely to be suffering from scurvy by the time they landed at Jamestown.

The death rate in Virginia continued to be appalling even after the "starving times" of the early years. Summers were called "seasoning time." New arrivals who survived the summer were said to be "seasoned" for future summers, probably having acquired immunity to typhoid fever and other diseases. The high water table in tidewater Virginia, resulting in construction of shallow, easily contaminated wells, would indicate typhoid fever was rampant during seasoning time (E. S. Morgan, 1975).

Virginia tobacco planters did not buy many Negro slaves until about 1700. With about 50 percent chance of a white servant worker dying in five years (before the end of his period of servitude), it was more profitable to use indentured servants, for they cost only about half as much as slaves. With a decreasing supply of servants, because of improved economic conditions in England, and a decreased rate of mortality among workers, economics began to favor the purchase of Negro slaves. Although generally not a land of opportunity for the servant who came with nothing, Virginia offered much to the man with £300 or £400 sterling. Half of this would buy ten slaves and in a good year they might "make" 20,000 pounds of tobacco, worth £100 to £200. The cost of housing and clothing them was very little, and feeding them cost nothing (E. S. Morgan, 1975).

## THE MARYLAND COLONY

The territory bordering Chesapeake Bay and the lower Potomac was explored by Captain John Smith in the summer of 1608. Other expeditions followed for the purpose of trading with the Indians for corn and furs. Later a joint-stock company was formed by Cecilius Calvert (Lord Baltimore) to promote the Maryland Colony as a refuge for Roman Catholics. Profits were to be shared in proportion to stock subscribed. Plantations were not to be established on a communal basis as in the early years of the Virginia Colony, but rather on the basis of individual initiative and freedom of enterprise. In the fall of 1633 about two hundred persons were sent to the Colony, mainly servants indentured for three to five years, but also about twenty "gentlemen." They were provided with livestock (from Virginia) and other necessities in accordance with the instructions of the English king. The colonists arranged with the Indians to use the houses of an Indian village and part of the cleared corn land. They immediately planted staple crops, including corn and many kinds of English garden seeds. The colonists were able to obtain abundant seed corn from the Indians and within a few years were shipping the crop to New England in exchange for salt, fish, and other products. As in Virginia, two acres of corn had to be planted and carefully tended for each laborer engaged in planting tobacco. Thus the Marylanders avoided the mistakes of the early Virginia colonists, who emphasized cash crops and paid too little attention to food staples. Nevertheless, once again tobacco soon became the principal cash crop and medium of exchange. In 1642 an act was passed regulating the use of tobacco as legal tender, and various penalties, assessments, and fees were made payable in tobacco.

## SUSTAINING THE "COLONIAL GOOSE"

The importation of tobacco from the colonies resulted in large revenues for King James and the merchants, and led to the prohibition of the planting of tobacco in Britain, first within a certain distance of London and finally, in 1619, throughout Great Britain and Ireland. In addition, the imports and duties on Spanish tobacco were so much heavier than for tobacco grown in British colonies as to nearly completely cut off the Spanish imports. It is unlikely that the tobacco industry in the colonies would have survived in its early decades if this policy had not been maintained. The king's policy was encouraged by the Virginia Company's agreement to pay a higher duty on tobacco than required by its charter. The policy was so vigorously opposed by British farmers, however, that the military had to be called upon to destroy tobacco plantings. Despite protests by the liberal group in Parliament concerning this infringement of personal liberty in the interest of royal revenues, the law prohibiting the growing of tobacco in Britain was not repealed until 1779, when the Revolutionary War made it desirable to encourage growing it in the mother country.

As a result of the British policy of sustaining the "colonial goose," the tobacco industry was the principal support for commercial agriculture and the plantation system in the South during the first colonial century. Next in importance as export crops were rice, indigo, and "naval stores"—pitch, resin, turpentine, tar, hemp (used in making rope), masts (for ships), clapboards, and barrel staves. Along with tobacco, these products retained their importance for the remainder of the colonial period (Gray, 1933).

## THE CAROLINAS

Many settlers were attracted to North Carolina by cheap land, mild climate, and the rich meadows and savannas, which provided ample range for cattle and horses. Roots in the swamps provided feed for hogs. The extensive pine forests were exploited for the production of naval stores and various kinds of lumber. As in the other colonies, corn became the principal grain for man and livestock, but considerable wheat was grown. Corn was raised in virgin soil for several years until productivity diminished. At first 80-100 bushels per acre could be obtained and 60-70 bushels were common. This is considered good production of hybrid corn even today in the Corn Belt, although now such production is on a sustained basis.

Nevertheless, it is interesting to note that despite the use of hybrid corn and with close planting and heavy fertilization, modern technology has barely equaled the production of inferior strains, with many less plants per acre, and without fertilization, but grown in virgin soil. This adds emphasis to the suggestion made by Emanuel Epstein, Professor of Plant Nutrition at the University of California, Davis, that various disciplines should be brought to bear upon a basic investigation of the very complex below-ground ecosystem. He suggests that the multidiscipline approach be recognized as having an identity of its own—the science of *geobiology* (Epstein, 1977).

After corn, peas, or beans were grown for a year, yielding 30 or 40 bushels per acre, wheat was raised for two or three years. Much of the potato and peach crop was fed to hogs. When the soil became exhausted, new land was cleared (Carrier, 1923). Some rye was grown in the piedmont, made into bread by the Germans and into whiskey by the Irish. Much tobacco was grown until the land was exhausted. It was the principal cash crop.

Most of the necessities of life were raised on the land or made from wood. The women made cloth of cotton, flax, and wool and produced the clothing required by the family. The mild climate enabled shiftless families to eke out a bare livelihood by hunting, fishing, raising a little corn, and keeping a few swine and possibly a cow or two. However, there were also many thrifty, hardworking farmers with good-sized herds of cattle, swine and sheep. They produced surpluses of tobacco, corn, beans, livestock, and timber products for export. By 1715, North Carolina had a population of 11,200, including 3,700 slaves.

In South Carolina, as elsewhere in the southern seaboard colonies, it was difficult to predict what would become the principal trade commodities. Expecting silk, tobacco, indigo, cotton, wheat, and ginger, the proprietor might have to settle for pitch, tar, turpentine, furs, deer hides, barreled beef and pork, corn, and rice. Although the proprietors did not encourage the development of a livestock industry, the colonists recognized the advantage for livestock of the warmer climate, compared with the more northern colonies. Agriculture was always preceded by the development of a livestock industry. The range was almost unlimited, the only obstacles being depredations by wild animals and Indians. The Indians in turn complained that range stock destroyed their cornfields. The practice of allowing cattle, swine, and horses to roam in the woods led to an overabundance in some areas. As early as 1682, some planters had as many as a thousand cattle, and for one individual to have 200 cattle or 300 hogs was common. Wolves were a troublesome predator, particularly against sheep.

New and improved breeds of livestock were eventually developed. Many cattle driven down from the mountains were fattened for market, developing on a small scale the type of economy that was later to reach enormous proportions in the West, when High Plains range cattle were shipped to Midwest feedlots to fatten them for slaughter. For the early mountaineer cattlemen of Kentucky, Tennessee, and North Carolina the terms "roundup" and "cattle drive" meant the same thing as they did to the famed post-Civil War "cattle kings" of the western High Plains. In his *Ranch Life and the Hunting Trail,* Theodore Roosevelt pointed out that the "rough rider" of the plains was the "first cousin" to the backwoodsman of the southern Alleghenies.

An interesting fact emerged regarding the lean, "razorback" hogs that roamed the woods throughout the year in all the colonies from north to south, living almost entirely on mast (wild nuts). They produced pork that commanded premium prices in England, where it was considered to be superior in quality to English pork.

## THE GULF COAST AND MISSISSIPPI VALLEY

The early agricultural experiences of the French explorers and colonizers along the Gulf Coast and the lower Mississippi Valley were similar to those of the English along the South Atlantic. Similar vegetable foods had been used by the Indians. Wild rice (*Zizania aquatica*) was abundant in marshy areas and was said to be harvested by the natives by shaking it into their canoes as they passed along. Wild rice was used by all Indians wherever it grew, from the Gulf of Mexico to the Great Lakes, where it was the main food of some of the tribes (Carrier, 1923; Bakeless, 1950). Most of the tribes grew corn, beans, squashes, and pumpkins. Some French explorers reported apple, pear, and peach trees before the establishment of the first French Biloxi (Mississippi) settlement, again reflecting the agricultural influence of the early Spaniards. The Indians grew peaches in large quantities, dried them, and pressed them into cakes (Gray, 1933).

In their exploring expeditions along the Gulf, beginning with De Soto (1539-1543), the Spaniards took large numbers of livestock with them. Some were dispersed by the Indians and formed the nuclei of large herds of wild cattle and horses and droves of wild hogs. These herds increased rapidly and became widely distributed, forming cheap sources of foodstuffs for many generations of settlers throughout the South and Southwest. The horses provided a new form of greatly increased mobility for the Indians. La Salle reported in 1682 that it would not be necessary for the French to import from Europe horses, oxen, swine, and various fowl, including, of course, turkeys, for they were already present in different parts of the vast area in which the French were interested (Carrier, 1923; Gray, 1933).

Near the close of the seventeenth century news of the awakening of English interest in the seizure of the Mississippi region convinced the French that quick action on their part was necessary. In January 1699, a French expedition found that the Spaniards had built a small fort at Pensacola, so they proceeded westward and established a settlement at Biloxi, west of Mobile Bay. The Indians were able to supply the French settlers with chickens, the original stock of which had been obtained from a ship wrecked on the coast about four years earlier. The settlers soon exhausted the light, sandy soil of the area. Even with occasional shiploads of provisions from France, the colony occasionally subsisted entirely on corn, acorns, and wild game, or the colonists were forced to distribute themselves among the Indians (Gray, 1933).

In 1717, the French *Compagnie d'Occident,* with increased resources and many new settlers, arrived and a settlement was soon established at New Orleans, designed to be the capital of the colony. Miles of levees and drainage ditches were constructed. By opening the levees, the land back of them could be flooded for rice culture. Then the water was drained away and removed at a point lower down the river. Rice was said to be used for feeding slaves and poor people who could not afford flour. It was cut with a sickle, bound in bundles, threshed at convenience, and hulled by beating in a wooden mortar with a heavy wooden pestle ten or twelve feet long (Gray, 1933).

In Louisiana the large "concessions," which were large grants to individuals or companies for planting crops and developing large colonies, were the beginning of the plantation system. The concessions resembled the joint-stock companies of Virginia. They were large plantation organizations composed of indentured servants and slaves and with a few full administration officials. Most concessions were short lived and feeble.

Delay in shipping caused thousands to die from contagious diseases and famine before they could embark for newly formed concessions.

In Louisiana most of the concessions started with little or no cleared land. The terms of the indentured servants, representing heavy capital expenditures, expired before the managers, most of whom were inexperienced, could realize a substantial return. The speculative bubble burst and supplies were cut off, again demonstrating the futility of trying to establish large-scale agricultural projects by enterprisers dwelling across the sea. Nevertheless, the concessions served the purpose of initial colonization; by 1724 there were about 5,000 inhabitants of the Louisiana Colony, including 1,300 Negroes, and large stores of tools, munitions, and livestock. Levees had been built, lands cleared and drained, and considerable agriculture had been developed.

As had been the case of the other nuclei of settlement in the South, for example, Virginia and South Carolina, private plantations were favored by not having to assume the expenses and responsibilities of initial colonization. In some cases private planters were able to acquire at small cost the land, improvements, and equipment of the unsuccessful colonizing agencies. As regular trade developed, the planter was able to obtain servants, slaves, and equipment on credit and was able to obtain market outlets for his products.

Along with the plantations, small holdings were encouraged. About 1755 and 1756 a few Acadians[2] came to Louisiana and near the close of the French regime a thousand more arrived, settling between Baton Rouge and Pointe Coupée. In 1785 3,800 Acadian refugees arrived, settling along the Mississippi above and below New Orleans, in Bayou Lafourche, and in the districts of Attakapas and Opelousas. All the Acadian settlers were poor and established only a small-scale economy. Under the Spanish regime, which began in 1762, an active policy of encouraging smallholders was adopted, along with increased activity in the slave trade. Thus the population more than doubled in the sixteen years beginning in 1769.

In 1778 the Spaniards imported many families from the Canary Islands. They were settled below New Orleans on the Amite and on Bayou Lafourche. The government built a home for each family, gave it some financial aid, and supplied cattle, poultry, farming utensils, and rations for four years. Under similar arrangements a number of families from Malaga were settled at New Iberia. The government also allowed free land and exemption from duties on goods and chattels brought into the country.

Corn was a favored food among all classes in Louisiana. Negroes were said to prefer it to the best bread.[3] Attempts to grow wheat were defeated by rust and by the tendency of wheat, as well as oats, barley, and rye, to grow too rank in the rich alluvial soil of lower Louisiana. Some was shipped in from Illinois and France. The orange, fig, peach, pear, apple, and grape were grown. One early traveler (Bossu) reported orange and peach trees to be so numerous that much fruit was left under the trees to rot. Orange trees were periodically killed by low temperatures.

Louisiana grew much indigo and a tobacco of high quality during the French and later the Spanish occupations, but at the close of the Latin regime both crops were being rapidly displaced by sugarcane and cotton along the lower Mississippi. The first shipment of sugar was sent to France in 1764 or 1765. The industry was not established on a commercial basis, however, until a few years before the transfer of Louisiana to

[2]Acadians came from Acadia, a French colony of the seventeenth and eighteenth century consisting of what is now principally Nova Scotia.

[3]Corn was the first familiar item Negroes found in the New World. They had grown it in West Africa since the Portugese took it there early in the sixteenth century. They used it as a paste called "cooscoosh," formed by pounding the kernels in a mortar (Giles, 1940).

the United States in 1803. Sea island cotton had been tried unsuccessfully as early as 1733 and was followed by "Siam cotton." The plant was a perennial that was cut to the ground every two or three years because the planters thought this increased production. It was used mainly for domestic use, however, and not until mid-century were small quantities exported.

In the early eighteenth century dried and salted buffalo meat was shipped down the Mississippi River to New Orleans in large quantities. There was little in the way of domestic livestock until the *Compagnie d'Occident* took over the colony in 1719. This company made vigorous efforts to obtain cattle but could not obtain an ample supply for about a quarter of a century. By 1746 there were about 10,000 head of cattle and many large flocks of sheep and droves of hogs in Louisiana. As in the English colonies, livestock were allowed to range in the woods throughout the year with little care. Nearly every family kept poultry and some sent large numbers to the New Orleans market.

## MIGRATION FROM NORTH TO SOUTH IN THE ENGLISH COLONIES

Emigration of Europeans to southern colonies was never as great as it was to those in the north. In 1755, when the white population of the southern colonies was 251,000 it was 795,000 in the northern ones. However, during the last four decades of the colonial period and for several decades thereafter, a steady stream of predominantly German and Scotch-Irish immigrants entered the ports of Philadelphia, Newcastle, and Baltimore and, finding the good lands of the middle colonies occupied, spread southwestward along both sides of the Blue Ridge Mountains, occupied the Great Valley of Virginia, and spread out into the thinly settled areas of the western piedmont. Many moved into the piedmont of the Carolinas and eastern Georgia. The tidewater districts had already been occupied by the English. In the tidewater there were many large plantations with Negro slaves, whereas the piedmont was occupied by small farmers. In the back country the immigrants of similar religious beliefs tended to settle in compact communities that afforded some protection against Indian attack.

As in the northern colonies it was principally the Germans and Scotch-Irish who settled the frontier and bore the brunt of Indian hostility. However, credit must be given to the English, who during the seventeenth century and the early decades of the eighteenth carried on an almost continuous struggle with the Spaniards for the disputed territory along the South Atlantic Coast. The English and Spaniards took turns plundering and terrorizing the towns of their rivals. Gradually the Spaniards were forced to withdraw their missions from the Georgia coast and later from Florida.

## GEORGIA

A project for the settlement of Georgia promoted by James Edward Oglethorpe and his associates, beginning in 1732, called for small landholdings, careful safeguards against land monopoly, close settlement in agricultural villages, exclusion of Negro slavery, provision for free transport and equipment for reputable debtors and indigents, a refuge for persecuted Protestants of southern Germany and Austria, prohibition of rum and spirits of high alcoholic content, and the development of intensive tropical and subtropical crop culture so as to avoid competition with the staples of the other southern colonies. These philanthropic ideals were not realized, however, and, despite the utmost exertions of their promoters, it was the plantation system and slave-holding that ultimately triumphed. Oglethorpe's colonists, encouraged to rely unduly on subsidies, put forth little exertion to help themselves. The servants shirked their work and became obstinate and dishonest. Guided by their humanitarian ideals, the colony's trustees proscribed stern discipline. South Carolina planters, eager to develop plantations in Georgia, joined with the Spaniards across the border in Florida in fomenting discord. The trustees lost control and South Carolinians began to invest in Georgia rice plantations and move in with their Negro slaves.

## FLORIDA

The Spaniards in northern Florida, at St. Augustine, never developed a thriving agriculture. They did not become self-sufficient and were always partially dependent on food supplies from Havana. Delay in the arrival of provisions always caused distress and sometimes the governor of St. Augustine had to appeal to his counterpart in South Carolina for food to prevent starvation. However, crops were grown around the missions and presidios and there seemed to be plenty of fruits for domestic use, particularly citrus. When the British occupied Florida in 1763 they found "China and Seville oranges," lemons, limes, citrons, shaddocks and bergamots, as well as figs, peaches, apricots, guavas, pomegranates, and plantains. Oranges were likewise extensively planted in other parts of the peninsula. The Spaniards were also in possession of various kinds of livestock. In addition, herds of livestock were maintained by the various Indian tribes of northern Florida.[4] South Florida remained largely a wilderness.

It seems hard to believe that what is now the rich agricultural state of Florida was almost a complete wilderness as late as 1819, when it was ceded to the United States by Spain. Many of the Spaniards then left the territory. In 1822 Florida's population was estimated to be about 5,000, much of which consisted of Indians, half-breeds, and runaway slaves, some of whom had themselves become slaveholders. The best lands were held by the Seminole Indians, who were not subjugated until the end of the fierce Indian wars that raged from 1835 to 1842. Then immigrants became interested in the prospects for raising citrus fruits, olives, almonds, coffee, cocoa, tea, sugarcane, flax, sea island cotton, Cuban tobacco, various tropical fruits, and drugs and spices. Cotton and tobacco proved to be the most important staples. The herding of cattle and hogs in the vast unoccupied areas was also an important industry (Gray, 1933).

## EIGHTEENTH-CENTURY LAW AND ORDER

In the mid-eighteenth century there were many wild and desperate individuals living on the fringes of colonial settlements who flourished by stealing horses and cattle and changing brands. The depredation became so serious that the entire countryside was terrorized. Citizens complained of the lack of police and judicial protection. The colonies tried to cope with the problem by the imposition of severe penalties. Despite humanitarian advances being made in that period, in this particular case lynch law prevailed and, in addition, legal penalties became increasingly severe.

[4]Also in the Gulf colonies some of the Indian chieftains maintained large herds of cattle. Some possessed a considerable number of slaves. The Indians learned about herding of livestock from the whites.

The colonists' *bête noir* was the horse thief. A North Carolina act of 1786 provided that "...for first offense the culprit should stand in the pillory one hour, be publicly whipped with thirty-nine lashes, nailed to the pillory by the ears, which were afterwards to be cut off, and be branded on the right cheek with the letter 'H' and on the left cheek with the letter 'T.'" Such penalties proved to be insufficient and four years later the penalty was changed to death without benefit of clergy. Some other southern colonies likewise provided the death penalty for first offenses, but in Georgia this was said to be ineffective "because of the tenderness of prosecutors and witnesses" (Gray, 1933).

## PLANT CROPS OF THE COLONIAL AND ANTEBELLUM SOUTH

As abundantly documented to this point, corn was the basic subsistence crop of the early colonists in the South as elsewhere, and continued to play an important role in the economy. Wheat and other grains eventually became important crops in piedmont areas. Tobacco was an unanticipated but important boon to the southern economy. It was the first important cash crop and even served for a time as a medium of exchange. Some semitropical crops with which the early colonists experimented unsuccessfully eventually reappeared for a while and were then abandoned or, in the case of rice, cotton, and sugarcane, became permanent major southern plant crops.

### Rice

As already noted, rice was one of the crops the promoters of the Virginia Colony planned to establish, and experiments in rice cultivation were immediately undertaken. However, rice was not commercially successful until about the end of the seventeenth century, in South Carolina. The rice industry owed its success in part to the fact that in 1694 a ship sailing from Madagascar was forced to touch at Charleston. The captain left a bag of rice with Landgrave Thomas Smith, who distributed the contents among friends. This rice was found to be of superior quality but was probably just one of several strains introduced from different parts of the world during a period of intense experimentation in the two decades from 1695 to 1715.

After the settlement of Georgia had removed the menace of Spaniard and Indian, the rice industry expanded from the Charleston area to the southeastern part of South Carolina. It also expanded into northeastern South Carolina and along the lower Cape Fear River in North Carolina. Later the cession of Florida to Great Britain removed the danger of Spanish invasion and led to rapid progress in rice culture in Georgia.

The rice crop was usually cut with a sickle until about the time of the Civil War, although occasionally scythes or cradles were used. After it was dry, the straw was bound and carried to the stack or barn on the heads of slaves or on rafts in the canals of the rice fields. Even as late as 1853-1854, when Frederick Law Olmsted visited the seaboard states, 95 percent of the rice grown in the United States came from a narrow strip along the seacoast of the Carolinas and Georgia (Olmsted, 1856).

The bundles of rice were stacked to await the threshing season. Threshing was done with flails, with great difficulty, or trodden out by horses or cattle. (It was not until 1829 that a machine was invented that could thresh 200 to 300 bushels per day when driven by horsepower or 480 to 700 when driven by steam. The steam threshers cost from $3,000 to $7,000, however, and were employed only by the more progressive large planters.) Exports of rice from the colonies increased from 330 tons in 1699 to nearly 2,600 tons in 1711. In the decade 1730 to 1739, inclusive, rice exports totaled almost a half million barrels (600 pounds each).

## Indigo

Although a perennial variety of indigo grew wild in South Carolina and could be manufactured into a good quality of dye, the main commercial varieties and the methods of cultivating them came from the West Indies. Credit for initiating the indigo industry in South Carolina is given to Eliza Lucas, daughter of the governor of Antigua. In 1794, after several years of experimentation, she devoted her crop largely to seed. This was distributed among a number of planters. She also made a little indigo dye herself. The industry was favored by bounties offered both by the colony and by the British government. At the same time planting of indigo was declining in the West Indies, where it was being replaced by the more profitable sugar industry. Some indigo was later cultivated in Florida and Louisiana (Gray, 1933).

## Cotton

Cotton has been used since prehistoric times in nearly all parts of the world; it is indigenous to both the Old and New Worlds. Experimentation with native sea island cotton began very early in the various colonies, starting at Jamestown soon after its settlement in 1607 (Gray, 1933). The long tobacco depression beginning in 1702 resulted in increased cotton production in the colonies, for domestic use. During the Revolution there was again increased interest in the crop. There were two types: upland or green-seed varieties (*Gossypium hirsutum*), with the lint adhering closely to the seed and the sea island or black-seed varieties (*Gossypium barbadense*), with smooth seeds. Upland cotton was a short-staple cotton and sea island was one of the long-staple varieties that was apparently brought to the islands off the coast of South Carolina and Georgia from Barbados and/or Jamaica. It was the quality cotton of colonial times. *G. hirsutum* types seem to have originated in the Mexican-Central America center, whereas the longer-lint *G. barbadense* types seem to have originated in South America (Stephens, 1976).

Both short-staple and long-staple cotton are grown throughout the world, the species in a given area depending on local climatic and soil conditions. The smooth-seed (long-staple) varieties were preferred in the American colonies, particularly after "roller gins" were introduced, for it was easier to separate the seed from the lint with these varieties. Roller gins were used before the Revolution from Virginia to Florida and along the lower Mississippi, replacing the earlier custom of having slaves separate the seed by hand as a nightly task. These gins, consisting of two grooved rollers operated by treadles, were greatly improved by the French in western Florida. They attached two of the roller gins to a large flywheel that could be turned by a boy rapidly enough to gin 70 to 80 pounds of clean cotton in a day (Romans, 1775).

Late in the eighteenth century, textile invention had led to an increase in importation of cotton into Great Britain. Exports of sea island cotton from South Carolina increased from 9,840 pounds in 1790 to 8.3 million pounds in 1801. Sea island cotton was narrowly restricted as to geographic area. It was originally grown successfully only on islands off the coast of the Carolinas and Georgia and on the mainland within thirty miles of the coast. Later its range was extended to the islands along the Florida coast. About the middle of the nineteenth century the interior of northern Florida was found to offer special advantages for sea island cotton and by 1858, Florida was producing slightly more of it than South Carolina.

Short-staple cotton, which had been produced in the back country of South Carolina and Georgia, mainly for domestic use, was given new life by Eli Whitney's invention of the cotton gin, the first model of which was completed in the spring of 1793. The tightly clinging seeds of short-staple cotton escaped the rollers of the machine used for sea island (long-staple) cotton and required a different device to separate them from the fibers. As adapted for modern use for short-staple cotton, circular saws project through narrow slits in a steel grating and pull the lint through the grating. The seeds are left behind, for they are too large to pass through. They then fall to a seed pan and are conveyed to the seed house. (The sawteeth are too harsh for extracting the long fibers of long-staple cotton; a roller gin with gentler action is used.) The fiber is removed from the saws or rollers and passes through a lint flue to one or more cleaners. It is then fed into a gin press and tightly packed into bales.

Expansion of the upland cotton industry in the Appalachian piedmont, from upper South Carolina to central Georgia, rapidly changed the character of that region. Tobacco, grain, and cattle, except for what was needed locally, gave way to cotton. River navigation was improved and new roads were built. Plantation mansions began to replace the rough cabins of pioneer farmers, and the social and economic interests of the people began to be identified with those of the older coastal region. In a typical piedmont area of South Carolina the slave population increased from 18 percent of the total in 1790 to 61 percent in 1860. The cotton industry also moved westward rapidly; by 1820 there was considerable growing of cotton in Tennessee and Alabama, and by 1860 Mississippi was the leading cotton-producing state.

While this expansion was taking place, cotton was deteriorating in quality. The staple was becoming shorter and the pods did not open as widely as formerly, making picking difficult. An average of only 30 to 40 pounds of long-staple cotton and 75 to 100 pounds of short-staple could be picked in a day. Then a superior short-staple cotton was introduced from Mexico. Its large, wide-open bolls could at first be picked at the rate of 150 pounds per day, gradually increasing to several hundred pounds (Gray, 1933). The several varieties of short-staple cotton now comprise about 99 percent of the cotton grown in the United States. The last crop of sea island cotton in the United States was harvested on John's Island, South Carolina in 1956 (Stephens, 1976).

## Silk

Attempts to produce silk were made soon after the beginning of settlement of Virginia. Armenian silk experts were imported and about 1654 a Virginian by the name of Edward Digges produced eight pounds of silk. In 1656 an allowance of 4,000 pounds of tobacco was made to an Armenian to remain in the country and several years later he won an award for having successfully produced ten pounds of wound silk. In 1662 Virginia passed an act requiring every landowner to plant 10 mulberry trees, as food for silkworms, for every 100 acres of land owned. In 1730, 300 pounds of raw silk was reported to have been exported from Virginia (Gray, 1933).

From 1742 to 1755, inclusive, South Carolina exported a total of only 651 pounds. In Georgia, Italian families were brought in to wind the silk cocoons brought in by farmers. A nursery of mulberry trees was established at Savannah and a public filature was constructed in that city. Export of silk had reached 438 pounds by 1755 and 1,084 pounds by 1766, then declined to 485 pounds by 1772. The chief blow to the industry was a reduction in the fixed price allowed by the British government. Exportation of cocoons was not feasible and the reeling of silk could not be done in competition with the low-priced labor of Italy and the Orient.

## Sugarcane

In the United States the sugar industry started in Louisiana in the last decade of the eighteenth century when the Spanish planters began planting sugarcane and making syrup and taffia, a spiritous liquor (Gray, 1933). Then in 1795 Etienne Bore began producing sugar, devoting thirty Negroes to the purpose and importing an experienced sugar maker from Santo Domingo. This expert and other refugees from Santo Domingo brought the century-old experience in sugar production from the West Indies. Nevertheless, the art of sugar production was in a relatively primitive state.

Although sugar production had begun in other states as early as 1805 (in Georgia), in the nine years preceding the 1860-1861 season Louisiana produced over 314 million pounds of sugar per year, compared with less than 18 million pounds for all the other southern states combined.

There was much hard manual labor connected with growing sugarcane and producing sugar. Best production was in rich alluvial soils of river bottoms or bayous. Because of heavy rainfall the fields were ditched and cross-ditched, the main ditches about 180 feet apart and the cross-ditches about 100 feet. Profits became so great and expansion so rapid that in 1846 thousands of Irishmen were imported to dig new ditches.

The one-crop system prevailed with sugarcane as with other cash crops in the South until the land was exhausted. Then growers were forced to rotate crops. A third of their land each year was planted to corn and peas. The corn was harvested and the peas were then plowed under to enrich the soil. Growers also plowed in their cane, bagasse (residue of cane after the juice has been extracted), and whatever animal manures were available.

## TRANSMONTANE MIGRATION

Following the French and Indian War, and after the best lands of the Great Valley of Virginia and the North Carolina piedmont were already occupied, pioneer families began to move westward through the Appalachians, following, more or less, the courses of the rivers: New River, Greenbrier, Big Sandy, Holston, and the French Broad, then over the Wilderness Trace to what were to become the states of Kentucky and Tennessee. This migration had been prohibited by a royal proclamation of 1763. The British, as the French before them, wished to retain the vast transmontane wilderness for its lucrative fur trade. Agricultural products were too bulky to be transported from the interior to the seacoast and enter into the seaborne commerce upon which the British empire was based (Smith, 1950). The proclamation of 1763 by no means retarded the westward migration but only caused bitter resentment among the settlers.[5] The full meaning of the king's opposition was to be felt when the pioneers came up against the hostile Indian bands, well armed and incited into aggressive activity by British agents.

The drama of this thrilling episode in American history was well expressed by Professor Lewis Cecil Gray (1933):

> The first thrusts of Southern population across the mountain barriers were achieved largely by the hardy backwoodsmen who had occupied the Great Valley

[5]The colonies made some efforts to enforce the king's proclamation before the Revolution. But after hostilities began on April 19, 1775, the Americans formulated land policy to meet the need for troops and revenues. The Continental Congress, with no national army and no treasury, offered land in lieu of money to induce enlistments. The size of the grants offered by the Congress ranged from 500 acres for a colonel to 100 acres for an enlisted man, provided, of course, that they serve to the end of the war. The states also offered land bounties on similar terms. Most of the land was obtainable only west of the Alleghenies. Through federal acts from 1776 to 1855, the national government alone disposed of over 73 million acres through military bounties (Schlebecker, 1975).

> and the piedmont region of North Carolina. They were fitted by character and experience for the stern task that awaited them. It is one of the marvels of social history how the Scotch-Irish from the compact villages and tiny farms of Ulster and the German redemptioners from the crowded lands of the Rhine were transformed in a generation into hardy frontiersmen, skilled in woodcraft and in the wily cunning of Indian foray and ambuscade. These backwoodsmen were possessed of sufficient hardihood to settle on an exposed frontier, separated by miles from other white settlers; and man and boy could be counted on to fight desperately and courageously against attacking Indians or to bear a part in counterraid or expedition against the inveterate enemy. Not infrequently women were little behind the men in the desperation with which they defended their firesides. Under the most favorable conditions the daily lives of these sturdy pioneers were characterized by hardships that seem almost incredible to the present generation.

Probably never before in history was a people so severely tested, bringing to a New World stage the dimly remembered drama of the valor and sacrifice of ancient heroes in the wild, rocky moors of Scotland and the gloomy depths of Teutoburger Forest. Understandably, the iron will and inexhaustible courage of the transmontane pioneers did not fail to capture the fertile imagination of Theodore Roosevelt and were thrillingly documented in an immortal American epic, *The Winning of the West,* published in 1889. "Nowhere else on the continent has so sharply defined and distinctly American a type been produced as on the frontier," said Roosevelt, "and a single generation has always been more than enough for its production."

The earliest settlements in transmontane Kentucky and Tennessee were developed in regions of mostly gently rolling topography with phosphatic limestone soils well adapted to the production of grain and forage grasses. Most of the area of these two future states was covered with heavy timber, and as previously related, forests presented a formidable problem of land clearing to eighteenth-century settlers, who had no better means of tree removal than their medieval European ancestors. But there were also large patches of open lands, often with nutritious cane that grew to ten or twelve feet in height and was said to be a good cattle feed. Buffalo grass, related to the famed buffalo grass of the Great Plains, was excellent for hay or forage. However, white clover and bluegrass displaced the indigenous grasses with remarkable rapidity (Bidwell and Falconer, 1925).

In the limestone valleys the rich soil could yield 50-60 bushels of corn and 30 bushels of wheat per acre. The bluegrass that became associated with rich soils had been identified by the Swedish botanist Peter Kalm (1750), in a mixture with white clover, as the principal hay crop of the French in Canada. The French probably took these grasses to Indiana and Illinois about 1700, whence they spread to Ohio and Kentucky, where English explorers found them growing luxuriantly in 1750 (Carrier and Bort, 1916). Salt, indispensable to pioneers and their livestock, abounded in "salt licks," congregating points for the abundant game of the region and therefore for Indian and white hunters (Gray, 1933).

Although most of the area was unoccupied by Indians, it was their favorite hunting ground and had long been a battleground between northern and southern tribes; understandably it would not be relinquished to the whites without a struggle. The whites could maintain their foothold only by protracted and most desperate border warfare. The struggle consisted principally of resisting surprise attacks by small Indian raiding parties, directed against isolated homes or small groups of fortified log cabins, or "stations."

To clear a plot of land, build a cabin, and raise a patch of corn was itself an arduous task. But this was accomplished under continual harassment from bands of Indians. A laboriously built cabin could be burned, livestock stolen, the entire family killed and scalped, or, if prisoners were carried off, they stood the chance of being put to a horrible death at the stake, by slow fire. Contrary to popular myth, the settlers suffered greater loss of life than the Indians. They were shot as they worked in their clearings, gathered their corn crops, or ventured outside the walls of the stockades. Hunters were stalked as they still-hunted game or lay in wait for it to frequent salt licks, as they returned with their horses laden with meat, or as they stopped to drink at springs. They might be lured to ambush by imitations of the gobbling of a turkey or the cry of a wild animal.

The Indians were skilled in ambush and forest warfare, a skill gained by generations of intense intertribal fighting. They had been well armed, first by the French and then by the British or their agents, as well as by fur traders of all nationalities. By adopting Indian methods, the backwoodsmen became just as skilled as the Indians at wilderness warfare, but no better. In their punitive raids, whites could defeat Indians if they outnumbered them, and vice versa. It was only the major, well-organized punitive campaigns under experienced Indian fighters such as John Siever, George Rogers Clark, and Benjamin Logan, and which included the destruction of Indian villages and their stores of corn, that brought peace, or at least a temporary respite.

The earliest settlers to migrate beyond the Appalachians were the bold, impetuous, and adventurous Scotch-Irish Presbyterians. The intensity of their struggle for survival is indicated by a description in Roosevelt's *Winning of the West* of a typical church service. Preacher and congregation alike went to church armed, the former placing his rifle beside the pulpit, the latter placing theirs in the pews or next to the seats. "In more than one instance, when such a party was attacked by Indians the servant of the Lord showed himself as skilled in the use of carnal weapons as were any of his warlike parishioners." During the week, preachers worked in the fields with the lay settlers, all with rifles close at hand and with a guard stationed. Here was democracy in its purest form. Poorest and richest met on terms of perfect equality, often sharing shelter and food. Yet certain families rose to acknowledged leadership by dint of their industry, thrift, ability in civil affairs, or the prowess of some of their members in time of war.

In the transmontane migrations, the Scotch-Irish made up the larger percentage of the pioneer population that lived more by hunting than farming and moved continually westward out of sheer restlessness and love of adventure. The Germans were more interested in developing a serious and permanent farming community, and, as related previously, they were attracted particularly to such fertile limestone valleys as those in the Kentucky Bluegrass region. This did not relieve them from their full share in the struggle against the Indians; in fact, their prosperous farm settlements were a favorite target for marauding bands.

The two types of settlers could be identified by the types of cabins they built. The restless, nomadic settler built a crude hut and grew enough corn to supplement the food he obtained by hunting and fishing. He soon moved elsewhere. Thomas Lincoln, father of our nation's sixteenth president, was this type of pioneer. As a boy his life had been saved when his brother Mordecai shot an Indian who had moments before killed the boys' father. Thomas was continually on the move, even after having acquired a wife and family; he lived principally a hunting-gathering type of existence. He moved his family on horseback from Kentucky to a remote spot in the Indiana wilderness. There he had previously constructed a crude "half-faced camp" of poles and brush, open on the south side. By the following year the Lincolns had a rude cabin without door, window, or wood floor. The entrance was covered with deerskins. They had no livestock and at first they lived exclusively on wild game, which was abundant, and nuts, fruits, and wild honey from the forest. Eventually some corn and vegetables were raised. Corn kernels were broken by pounding them in a hollowed-out place on top of a hardwood tree stump (Beveridge, 1928).

The true pioneer farmer was constantly repairing his cabin. After he had cleared sufficient land for his small patches of corn, he cleared more land for flax, hemp, or cotton and the thrifty housewife spun and wove clothing, blankets, and sheets. The farmer tanned leather, from which he made wearing apparel, shoes, chair seats, and harness. His patches of corn and small grain were enlarged and protected by rail fences, and a vegetable garden and fruit trees were added.

## CHEAP LAND IN KENTUCKY AND TENNESSEE

The beginning of extensive small-scale transmontane farming could be said to have begun when Judge Richard Henderson of North Carolina organized the Transylvania Company and attempted to establish title to land simply through purchase from the Indians. Lands between the Kentucky and Cumberland rivers were purchased from the Cherokees on March 17, 1775. One of Henderson's agents was Daniel Boone, who, besides being a guide, hunter, trapper, Indian fighter, and occasional farmer, was also a surveyor. Self-educated, he remained a notoriously poor speller, but was able to obtain the considerable knowledge required for surveying and road-building, and even held a number of political offices. In 1773, Boone had sold his farm in North Carolina and started west with five other families. Indian hostilities caused most of the party to turn back, but Boone and his family reached the Clinch River and settled for awhile where today, two centuries later, the United States is building its first prototype nuclear breeder reactor.

At Henderson's request, Boone set out with thirty men to lay out the Wilderness Road and establish Boonesborough on the Kentucky River. The Wilderness Road started in southwestern Virginia at the Block House, where two feeder trails converged, one down the Great Valley of Virginia, along the Holston River, and the other from North Carolina following the Watauga River. (Both river valleys eventually became rich agricultural regions.) Then the trail crossed two mountain chains, with the Clinch and Powell rivers between, breaking through the second range at Cumberland Gap and proceeding on to the game-filled savannas of Kentucky, a region the Iroquois Indians called *Ken-ta-ke* (meadowland).

Boone brought his wife and children to Kentucky in 1775. In 1778, Boone was seized by Shawnees and saved his life by feigning cooperation with them, becoming a member of the tribe. On June 16, 1778, he was able to escape, and, racing 160 miles in four days, he alerted Boonesborough of a planned attack and aided in the successful defense of the settlement (Elliott, 1976).

The hostile Indian-inhabited forest was a formidable obstacle to westward expansion. As late as 1779-1780, 172 years after the English landed at Jamestown, the area called Kentucky, destined to become Virginia's neighboring state to the west, was penetrated mostly by following buffalo "traces" (trails) through the vast virgin forest. In 1777, the dreadful "year of the three sevens," Indian attacks reached their peak of ferocity and the ability of the transmontane frontier to survive was questioned. Although these attacks continued unabated there was nevertheless a great influx of settlers in the decade following 1780, especially in Kentucky.

The frontier farmers from Virginia and the Carolinas, in particular, streamed over the Wilderness Road in search of cheap land. Under the laws of Virginia, by building a blockhouse and raising one crop, however small, a group of settlers could obtain four hundred acres of land for each family and a preemption of another thousand. This system was to end in 1780, prompting crowds of settlers to move into Kentucky before the deadline date. Predictably, Indian resentment was raised to a new pitch and the fury of border conflict increased, but the tide of immigration could not be held back. There was strength in numbers. Particularly important was the protection afforded for the growing of corn and other vegetables, for the steady winter diet of smoked buffalo and boiled beef and with only enough corn for an occasional hoecake, was not healthful.

### The Bend of the Cumberland

As late as 1774, the rich, beautiful part of western Tennessee, known only as the "Bend of the Cumberland," with fertile prairie soil and abounding in buffalo, elk, deer, bear, and fur-bearing animals, was known to only a few hunters and French Creole trappers from Illinois. The only American "settler" was a hunter by the name of Spencer, who lived in a hollow sycamore tree and planted a field of corn. Then James Robertson, a prosperous leader among the frontiersmen of the Holston settlements of east Tennessee, decided, like so many early Americans, to leave security and financial success behind and plunge several hundred miles farther into the perilous and uncharted wilderness, seeking the strong excitement of danger to add zest to his life. Not to be discounted, however, was the desire of Robertson and others like him to obtain title to new lands and develop new settlements. Although land sharks, grafters, and currency speculators could be a distinct menace in newly settled regions, Roosevelt (1889) believed the speculative spirit was a strong incentive without which western migration would have been greatly delayed.

Robertson's special partner, John Donelson, led a party of probably two to three hundred, including all the women and children and including also Robertson's wife and five children and Donelson's daughter Rachel, the future wife of Andrew Jackson, on an amazing journey to the fabled "Bend of the Cumberland River." This party traveled entirely by water, first down the Tennessee River, then up the Ohio and Cumberland rivers to the new settlements, using thirty flatboats, dugouts and canoes. The party embarked, at what is now the town of Holston Valley in the northeast corner of Tennessee, on February 27, 1780, and reached the fortified city of Nashborough (now Nashville) on April 20. Space does not permit a recounting of the many mishaps, tragedies, and adventures which beset this group of brave and hardy pioneers, but their trials can well be imagined and are interestingly narrated in Teddy Roosevelt's thrilling epic of frontier adventure.

## THE RAPE OF THE CHEROKEE NATION

The Cherokees were one of the Five Civilized Tribes, which included also the Creeks, Seminoles, Chickasaws, and Choctaws. These Indians lived in what are now the southeastern states. Before the arrival of white men they lived in towns with streets and a central square. The Cherokees became remarkably well adjusted to the ways of civilization. They were peaceful, progressive, intelligent, and had a written language and a newspaper. Numbering some 15,000, they had among their possessions 22,000 cattle; 7,600 horses; 46,000 swine; 2,500 sheep; 1,300 slaves; 1,488 spinning wheels; 762 looms; 31 gristmills; 172 wagons; 2,948 plows; 10 sawmills; 8 cotton gins; 62 blacksmith shops; 18 ferries; and 18 schools. According to an Indian Department report, they built many public roads and carried on considerable trade, shipping cotton down the Tennessee and Mississippi to New Orleans. They had many apple and peach orchards, made butter and cheese, and manufactured cotton and woolen cloth. They had adopted Christianity as the religion of the Cherokee nation (Fey and McNickle, 1959; Terrell, 1972).

Having successfully driven out the Creek Indians by 1826, with the connivance of the federal government, Georgia officials directed their attention to the Cherokees. They used as their pretext the fact that the Cherokee nation had drafted a constitution in 1827, and that it was illegal to do so without the consent of the state in which they were located. A series of legal wrangles finally led to a bill in Congress authorizing the president to move any Indian tribe to the West by force if necessary. President Jackson ordered the Cherokees to leave Georgia and they were driven from their lands (Terrell, 1972). Europeans had previously justified the seizing of Indian territory with the assertion that civilized people had a higher claim upon the land because of their superior use of it; in fact, this was a doctrine formalized in Emmerich von Vattel's *Law of Nations* (1758). The Indians were considered to be no more than roaming hunters. In the case of the Cherokee nation this doctrine received its supreme test and was found to be "not so much a doctrine as a subterfuge" (Fey and McNickle, 1959).

The Cherokees appealed to the Supreme Court, which found for them, declaring that they were a distinct community occupying their own territory in which the citizens of Georgia had no right to enter. In 1835 Jackson countered by sending agents among the Cherokees to find a few weak individuals who could be bribed to deed all the tribal lands to the United States for $5.6 million. When after two years most of the Cherokees still refused to leave, the army was sent to drive them out. The Cherokees and many other Indian tribes were forced to cross the Mississippi River, meeting no better fate in the West than they had in the East (Terrell, 1972).

## THE RAPID GROWTH OF TRANSMONTANE AGRICULTURE

In Kentucky by 1790 the barter economy was changing to a money type, although money was still scarce. Here and there rude log churches and schools were built, the latter ambitiously designated as academies or colleges. Although the pioneer settlers were mostly illiterate, they had a high regard for education. In 1787, only ten years after the frightful "year of the three sevens," the town of Lexington, in the rich "Inner Bluegrass" region of Kentucky, had a newspaper. In its December 16, 1787, issue it announced the formation in Lexington of a "Kentucky Society for Promoting Useful Knowledge"! (Gray, 1933).

During the last decade of the eighteenth century the Wilderness Road was improved and continued to be an important artery of transport. Flatboats carried goods from southwestern Virginia down the Holston and Tennessee river waterway. Roads were opened to Nashville and on to Alabama and western Georgia. Some trade began to trickle down the Mississippi to New Orleans, the first shipment of consequence being a boatload of hams, butter, and tobacco from the vicinity of Frankfort, Kentucky, in the spring of 1787. Flour, tobacco, and bacon were shipped to New Orleans in exchange for slaves, horses, and cattle. Spain recognized the free navigation of the Mississippi in 1795, then withdrew the right in 1802, but it was soon restored. Despite high charges for freight and commissions, the New Orleans outlet for agricultural produce was a great boon to trans-Appalachian farmers.

One way of reducing the exorbitant transportation costs was to convert grain to whiskey and float the highly concentrated crop down the Mississippi to New Orleans, where it was sold or bartered for tools and other items urgently needed by the settlers (Kramer, 1973). Later the steamboat greatly facilitated trade. In Kentucky, Louisville became a more important trade center, at the expense of Lexington. As early as 1801

Louisville shipped 41,149 barrels of flour and large quantities of tobacco, pork, bacon, lard, butter, cheese, beef, hemp, and corn (Gray, 1933).

### The Natchez Trace

The agricultural area of middle Tennessee was especially handicapped by remoteness from markets. Nashville became the central market for middle Tennessee, using large flatboats to ship cotton, hemp, pork, lard, bearskins, deerskins, butter, salt, whiskey, hides, beeswax, and cattle. Merchants sailed to New Orleans with their loads of produce and returned by horseback by way of Natchez or on keel ships up the Mississippi, Ohio, and Cumberland, the entire journey requiring four months. The return by horseback took the traveler along the Natchez Trace. Today the 450-mile Natchez Trace Parkway roughly follows the original trace through the states of Mississippi, Alabama, and Tennessee, connecting the cities of Natchez, Jackson, Tupelo, and Nashville. A strip of land averaging a little more than 400 feet from the centerline of the parkway, except where it widens out to include scenic, historic, and recreational sites and features, is administered by the Department of the Interior, National Park Service. Thus one travels through hundreds of miles of native forests, occasionally interspersed with pastures with grazing cattle. One may visit sections of the "Old Trace" at various points along the entire parkway.

Because of the long time required to return from New Orleans, for many years most of middle Tennessee's imports arrived by wagon from Philadelphia and Baltimore to the Ohio, thence up the Cumberland to Nashville. Later steamboats brought imports from New Orleans, requiring little more than a week for the journey.

### The Kentucky Colonel

The people who migrated to Kentucky prior to 1783 were mainly backwoodsmen from Pennsylvania, Maryland, and Virginia. Then came the gentry from the more densely populated parts of the country—some with considerable money to invest. By 1817, Lexington, Kentucky, had three-story brick stores, several good hotels, and residential brick mansions with fine lawns, and in the vicinity were fine country estates. In the vicinity of Louisville a planter might sow as much as five hundred acres of wheat, and keep sixty horses and several hundred Negro slaves. Kentucky had already become noted for fine breeds of saddle and carriage horses, derived from the best Virginia strains. Large herds of cattle were already being driven to Baltimore markets. However, even in the Inner Bluegrass[6] area of Kentucky and the Nashville Basin there were thousands of small, middle-class farmers, many of whom owned no slaves (Gray, 1933). Ownership of slaves carried prestige; as many as a half-dozen entitled the owner to the sobriquet of "Colonel" in a land "Where the corn is full of kernels, and the colonels full of corn."

## SOUTHERN POOR WHITES AND HIGHLANDERS

Westward expansion resulted in a continuous succession of pioneer villages that, in rich limestone valleys and prairie lands, soon developed into thriving communities.

[6]The fertile Bluegrass area is in the approximate center of the state and extends north to the Ohio River, comprising an area of about 8,000 square miles. It attracted the earliest settlers and is now Kentucky's richest agricultural section.

Meanwhile, in extensive areas of Appalachia and the coastal pinelands, isolation, unfavorable natural environment, and soil erosion or exhaustion, resulted in some communities lingering in the pioneer stage for many decades. Subsistence farming also took place in areas handicapped by remoteness from markets or roughness of terrain, or in areas of inferior soil, such as the "wire-grass" region of southeastern Georgia, the sand barrens back of the South Carolina coast, the pine flats of the Gulf States, or in any area where the exploitive and extensive plantation system had exhausted the soil and moved west.

Even in districts that passed into a higher stage of economic development, some families were unable or unwilling to advance beyond the pioneer stage and became known as "poor whites." The poor whites of the South lived for many generations in a manner differing little from that of the earliest pioneer farmers. They lived in crude huts, and on eroded, exhausted soil they cultivated small patches of corn or rice, sweet potatoes, cowpeas, and garden products. They might have a few hogs, an emaciated horse, and a crude, homemade cart. Generally the women and children did much of the work while the men spent their time in hunting, fishing, or idleness. As in pioneer days, clothing was spun and woven by the women. The men were generally inveterate drunkards, often joined by the women, who also dipped snuff and smoked clay pipes. Men, women, and children commonly ate a kind of sweet clay, a habit believed to be an effect of hookworm, and which gave poor whites the name "dirt-eaters." Clay-eating or cachexia Africanus is believed by Piero Mustachi (1971) to have been beriberi (wet type) complicated in some cases by hypochromic anemia and hookworm. A thiamine (vitamin $B_1$) deficiency was indicated, possibly caused in part by the long-curing of pork with salt or pickling it with brine. Hookworm and pellagra were common, giving the poor whites a characteristic gaunt and sallow appearance, with high cheekbones and sunken eyes. They were as uneducated and superstitious as the Negroes of that era, and in fact the latter regarded them as beneath contempt (Gray, 1933).

It is likely that soil erosion or exhaustion leads to malnutrition. In Russell Lord's *Care of the Earth* (1962), he cites the experiment of a Dr. William A. Albrecht, who observed the reaction of cattle to stacks of hay harvested from fully fertilized, partially fertilized, and unfertilized meadows. They fed greedily on a stock of hay from the first, nibbled "disdainfully" at the second, and left the third almost untouched. Deficiencies of calcium, phosphorus, and iron were found in the diets of children in Georgia from areas of great soil erosion. The conclusion was reached that such erosion had definite and serious effect upon human health. A point generally overlooked by "organic" gardeners or farmers is that if a soil is deficient in a certain mineral then a manure from local animals or composts from decaying vegetation of the region cannot provide it, for minerals cannot be created by any living things. Organic gardening can then perpetuate a local soil deficiency rather than correct it (Rosenberg and Feldzamen, 1974).

A distinction could be made between poor whites and highlanders, the latter exhibiting less squalor and degradation. Their idleness and laziness were considered to be the result of the absence of an impelling motive for work, rather than listlessness and inertia. Both men and women worked in the fields and both smoked clay pipes.

Appalachia has more than its share of "poor whites" and highlanders to this day, although increased demand for coal is revitalizing some areas. Still there are families practicing principally subsistence farming, with their humble huts often incongruously embellished with a radio or television set. In such areas, "welfarism" is the principal industry (Fetterman, 1967).

Before the turn of the present century, agriculture in the South suffered from what was considered to be laziness and shiftlessness of Southerners in the low-income classes. C. W. Stiles, a parasitologist for the USDA Bureau of Animal Industry, attributed this condition to hookworm, a theory that obtained little support in those days. Nevertheless, Stiles was able to obtain Rockefeller backing amounting to a million dollars to fight the hookworm; the ridicule that had been heaped on Stiles ended abruptly. One of the methods of hookworm eradication was to give poor Southerners shoes, greatly reducing their contact with infested soil. In 1902 Stiles was hired by the U.S. Public Health Service, but collaborated with the Bureau of Animal Industry and the staff of the Rockefeller Institute at the Hygienic Laboratory. The social and economic levels of millions of people were raised to a large extent through sanitary and other measures suggested by Stiles (Harding, 1947).

## PIONEER FARMERS, SOUTHERN YEOMEN, AND SMALL PLANTERS

The pioneer lowland farmers, "southern yeomen," and the small planters were types intermediate between the highlander and the large planter. They might have a few slaves. Most of the general farmers, however, had few slaves or none. To a large extent, they were the descendants of the German and Scotch-Irish immigrants who came over as redemptioners or indentured servants. They were true pioneer farmers who came into the wilderness to build permanent homes. They were stable, home-loving, thrifty, hardworking, and migrated only to improve conditions for their families. They often grouped themselves according to nationalities and religious denominations. They gradually crowded out the nomadic pioneer types. Many crossed the Appalachians and settled the commercial farming regions of Kentucky, Tennessee, and Missouri, while the mountaineers in the same districts of those states developed the characteristic features of their own economy and social life. Among the true pioneer farmers who settled in the back country of the Carolinas and Georgia, many later became planters, but much of middle and western North Carolina, the northwestern part of South Carolina, and northern Georgia continued to be occupied by small commercial farmers. This was also true of the valleys of northern Alabama, certain districts of upper Mississippi, northern and northwestern Arkansas, northern Louisiana, and eastern Texas. In the lower South, where general farming was practiced it tended to be combined with cotton production.

The above farmer types were characterized by sturdy independence and self-respect. Sociable, democratic, and almost universally hospitable, they delighted in group activities such as quiltings, cornhuskings, candypullings, weddings, and shooting matches, often imbibing freely of the products of their numerous distilleries. Generally with little education, but deeply religious, they sometimes were inclined to "shout" at religious camp meetings. Since they developed a diversified economy rather than clinging to a one-crop system, they enjoyed a more comfortable type of life than the middle-class planter. Staple foods, vegetables, and fruits were generally abundant (Parker, 1834; Drake, 1948).

## BIG AND MIDDLE-CLASS PLANTERS

In many areas, particularly in the deep South, the plantation and slavery tended to supplant other systems of agricultural economy, for small farmers could not compete with slave labor given the plantation arrangement. They either had to become big planters, for which many had neither the ability nor the capital, or to go back to a self-sufficient farm economy, generally in regions least favorable for agriculture. Even in

Georgia, where for a time an economy of small landowning farmers was favored by the trustees, slavery and the plantation system triumphed.

Frederick Law Olmsted, a farmer, architect, journalist, and antislavery Whig, who spent much time traveling in the South, most of it on horseback, staying at plantations or farmers' homes at whatever point he happened to reach at nightfall, made the following observations concerning a "first-rate" cotton plantation in the valley of the Lower Mississippi (Olmsted, 1860): There were between 1,300 and 1,400 acres under cultivation with cotton, corn, and other hoed crops and much swampland with 200 hogs in it running at large. A large and attractive mansion stood on the highest ground, but the owner lived 200 miles away. The intention was to raise enough corn and hogs to feed the slaves and cattle. There were 135 slaves of all ages, of which 67 went to the field regularly, and there was a blacksmith, a carpenter, a wheelwright, 2 seamstresses, 6 cooks, a stable servant, a cattle-tender, a hog-tender, a teamster, a house servant (overseer's cook), and one midwife and nurse. A nursery for the sucklings of the slaves had twenty women in it at a time; they left work four times a day, for half an hour, to nurse their babies.

In the field there were thirty plows in action and from thirty to forty hoers, mainly women. A Negro driver walked among them with a whip "which he often cracked at them, sometimes allowing the lash to fall lightly on their shoulders. He was constantly urging them also with his voice." The slaves ate their breakfast before daylight, then dinner was brought to them in a cart. The hoe gang ate its dinner in the field and the plow gang drove to sheds where the mules required two hours for feeding. Work continued until visibility became poor in the evening. The slaves then prepared their evening meals at their cabins. Thus the hoe gang at that time of year worked sixteen hours, relieved by only one short interval of rest, although during the hottest weather it was customary to extend the dinner time rest period to an hour or two for all hands. (In South Carolina the legal limit of a slave's workday was fifteen hours.)

At this plantation, fieldwork generally ceased at 8 A.M. on Saturday except during harvest. Olmsted presumed that such tasks as washing, patching, wood hauling and chopping, and corn grinding were done on Saturday. The food allowance per slave was a peck (one-fourth bushel) of corn and four pounds of pork a week. However, the Negroes had gardens, raised many vegetables, and also poultry and plenty of eggs. At Christmas the plantation owner sent the overseer from $1,000 to $1,500 worth of molasses, coffee, tobacco, calico, and "Sunday tricks" to give to the slaves. (The general custom in the South was to declare a holiday for slaves, except for house servants, from Christmas to New Year's.) The Negroes, through Saturday and Sunday work, were able to earn money with which they could buy things for themselves, such as tobacco. On Sundays some Negroes went into the swamps with an ax to make posts; one had sold fifty dollars worth the previous year. These were times when the usual wage for a white farm laborer was a hundred dollars per year (Olmsted, 1860).

The opinion was commonly expressed in the South, in justification of black slave labor, that the white man was unsuited for farm labor under the climatic conditions prevailing there. There were many examples to refute this opinion. Olmsted (1856) related the opinion of a Virginia planter that Negroes never worked so hard as to tire themselves and were "lively and ready to go off on a frolic at night." Olmsted's own observation was that they seemed to go through the motions of labor but kept their powers in reserve for their own use. Probably the blacks were less susceptible to malaria than white men; yet the planter hired Irish gangs for draining his swampland, for that work was too dangerous (presumably because of malaria) to risk his slaves, valued at about $1,000 each. The use of Irish gangs was common throughout the South for the disagreeable and dangerous work of draining swamps, as in the extensive cane and rice fields of Louisiana. Olmsted interviewed another Virginia planter who had freed his slaves, urging them to go to Liberia, and who agreed with Thomas Jefferson (himself a slave owner) that slavery was more pernicious to the white race than to the black. The planter had found the Irish to be the best workers. He paid them $120 a year and Irish housemaids he paid $3 to $6 a month. Freed Negroes he considered next in effectiveness as workers and the native white Virginians the least effective, but both classes of laborers were demoralized by the influence of adjacent slave labor. He had found the hiring of free men economically superior to the use of slave labor.

The best workers by far were the European immigrants, particularly those working their own land, but they tended to shun the South. They did not like the aristocratic character of the upper classes, the absence of a large employing class, and the low premium that Southerners placed on human toil. European immigrants also detested slavery. In 1880, foreigners comprised about 8 percent of the population in Texas, 6 percent in Louisiana, 4 percent in Florida and Kentucky, and less than 2 percent in the rest of the southern states (Saloutos, 1960). The few immigrant enclaves, however, proved that foreigners could toil and produce in the southern climate. This was demonstrated, for example, by the striking success of the German immigrants in raising cotton in New Braunfels and the other German colonies in Texas, in sharp contrast with the general situation in the area (Olmsted, 1857), or by the Italians in their various enclaves in Louisiana (Stone, 1905, 1907). Five hundred Italians imported from Italy, many of whom had never been on a farm, working in competition with Negroes as cotton choppers in southeastern Arkansas, produced an average of 2,584 pounds of lint cotton per working hand, compared with an average of 1,174 pounds produced by Negroes. They were equally superior to native whites in their capacity for agricultural labor.

The new feudalism associated with cotton culture would never have developed if the planters had been forced to rely on European immigrants, or their immediate descendants, for their labor. The plantation system was not necessary. Devoid of any moral restraint, it developed solely for the purpose of producing profits for planters. Particularly in the growing social and environmental awareness of modern times, there is ever-increasing criticism of the remnants of this attitude in the American economy.

"The attitude is still present among those who imagine that economic systems are mechanisms for producing entrepreneurial profits rather than that profits are as yet a very imperfect mechanism for driving the economic system in a regime of freedom" (Graham, 1944).

Although the planter class reflected European social distinctions, many were from professional, commercial, and small-farmer classes who had risen from the lower classes through their own merit. The life-style and personal characteristics of the antebellum upper-class planters, as well as the stately mansions, with well-kept grounds, are well known to Americans, having had a prominent place in history and fiction. A bad feature in many cases was the tendency of the plantation owners and their families to be absent from the plantations, sometimes because of residence on another plantation, sometimes to avoid the discomfort of the climate, and often because of travel to Europe. Despite the great attention they attracted, such plantation owners were in the minority. Socially beneath them, but much larger in numbers, were the middle-class and small planters, the "backbone of southern society." According to Gray (1933):

> Generally the middle-class planter and his wife exercised an active personal interest in the details of the slaves' welfare. They rarely worked with their slaves, who were under more rigid discipline than in the farming districts. These planters had little time for luxurious or ostentatious living or systematic self-indulgence. They were compelled to rise early to lay out the work for the day. At best, their

manner of living was comfortable; at worst, it was slovenly, careless, and comfortless. At best, the houses were large, airy, comfortable, with glass windows and ample verandas. Their comforts were supplemented by a dairy, an orchard, and a well-kept garden. At worst, the house was an enlarged log cabin, dirty and full of flies and mosquitoes; the food a dreary monotony of bacon, corn pone, hominy, and coffee.

### The Overseer

Unfortunately, Negro slaves were generally completely at the mercy of white overseers, particularly on the large plantations. The overseer system was conducive to hard driving and poor treatment of slaves. Plantation overseers might be indentured servants whose terms had expired, sons of small planters, or more rarely, of "poor white" families. Gray stated that overseers tended to be men of "dense ignorance and narrow vision," exacerbated by a strong prejudice against "book farming," that they were often unreliable and dishonest, and, in their treatment of slaves, were often "cruel, drunken, and licentious tyrants." Washington and Jefferson despaired of obtaining good overseers on their plantations. Olmsted (1861) pointed out that the overseer's wages were ordinarily $200-$600 per year, but a real driving overseer would often get $1,000 or more if he had demonstrated his ability to "make" a maximum number of bales per hand. Overseers sometimes were able to save enough money to buy slaves and start a plantation of their own. The short-range objective of maximum production per hand was generally attained at the expense of both land and human resources. The greater the number of slaves, the greater the need for rigid and impersonal discipline. Olmsted considered the treatment of slaves by middle-class and small planters to be generally much better than the treatment they received on large plantations.

### Exploitive Plantation Agriculture

Along with the growth of new plantation communities and westward expansion of agriculture, there was a tendency of the plantation system to decline in the older areas. In the South, land had been abundant in relation to population and the system was based on the policy of exhausting the soil and moving on to newer land—an extensive and exploitive method of agriculture compared with the intensive and soil-conserving methods of areas of high population density such as New England and even more so in Europe.

On the southern plantation, soil exhaustion forced continuous expansion, new capital, and a rural economy characterized by extravagance both in production and consumption. A large proportion of the income was concentrated in the hands of a relatively few people. Many whites were forced into isolated areas to pursue a self-sufficing economy involving much work, unrelieved by labor-saving devices. There were Southerners who condemned the extravagance and wastefulness of the prevailing plantation system and suggested soil conservation, greater plantation self-sufficiency, and the discontinuance of the dependence on credit obtained at interest ranging from 10 to 20 percent. Such advice was not only generally ignored, but actively denounced by leading planters.

The objective of the plantation was to obtain the greatest production per laborer rather than the greatest production per acre. It was most profitable to abandon old lands and apply labor and capital to new lands. The results of this policy were graphically described by Gray (1933):

"Over the upland soils from Virginia to Texas the wave of migration passed like a devastating scourge. Especially in the rolling piedmont lands the planting of corn and cotton in hill and drill hastened erosion, leaving the hillside gullied and bare. Even in Southwestern areas of comparatively recent settlement the work of devastation was manifest on every hand before the Civil War.''

## MODERN REAPPRAISAL OF SLAVERY

Quite aside from moral considerations, the rather somber view of southern slave agriculture presented to this point could be amplified further with the opinions of many other contemporaneous observers, including such influential economists as H. R. Helper (1857) and J. E. Cairnes (1863) and the majority of economists and historians up to modern times. This ''traditional'' view, however, was vigorously challenged by R. W. Fogel and S. L. Engerman, who presented a rather benign interpretation, based on a reappraisal of slavery's economic aspects, in *Time on the Cross* (1974). Discussion of the arguments of these two modern economists is justified at this point because their opinions became widely known by being featured in *Time, Newsweek,* and many other popular journals and in newspapers. In an article by T. Fleming (1975), *Time on the Cross* was said to contain ''new findings about slavery discovered by a new generation of unbiased historians'' using ''computers, massive research and sophisticated mathematical techniques.''

Most historians, including even the noted black scholar W. E. B. Du Bois (1866), would agree that the *material* living conditions of slaves in the South compared well with those of the lower echelons of laborers elsewhere and were not as bad as they were sometimes made out to be by contemporary abolitionist crusaders or by modern ''neo-abolitionists.'' Discontinuance of slave importation and high and rising prices of slaves resulted in the maintenance of minimum levels of material care and living arrangements compatible with the growth of the slave population by natural increase. The slaves' diet was monotonous but adequate to allow them to keep up their body weight and general health, although there was some indication of occasional nutritional imbalance. In these respects, at least, southern slaves were relatively better off than those of the Caribbean and South America.

But Fogel and Engerman went a step farther and depicted slavemasters as shrewd, profit-maximizing captains of agriculture who were forced in a highly competitive industry to act benevolently and even compassionately toward their human chattel. The slave field gangs were said to be ''highly disciplined, interdependent teams capable of maintaining a steady and intense rhythm of work.'' Slaves learned to respond to the economic incentives created for them through cooperation and identification with the economic interests of their masters. Said to be managed principally by black overseers, requiring only a minimal show of force, and motivated primarily by ''a wide-ranging system of rewards,'' slaves became devoted, diligent, industrious, responsible, and self-improving—veritable Horatio Algers in black skins—who were harder working and more efficient than white agricultural workers. In this startling new cliometric reappraisal of the slave economy, it emerges as a dynamic, capitalistic, agricultural system, which in 1860 was far more efficient than that based on the labor of free men, North or South. It was said to provide a high rate of economic growth and a high standard of living for all Southerners, the enslaved as well as the free, Fogel and Engerman wondered ''how so many previous scholars could have been so badly misled.''

Among the ''misled'' were a group of historians and economists (P. A. David, H. G. Gutman, R. Sutch, P. Temin, and G. Wright) who collaborated in a devastating point-by-point exposure of what they found to be ''a misleading interpretation of antebellum southern society and its relationships to the economy of the region" (David et al., 1976). In a book titled *Reckoning with Slavery,* published in 1976, these critics condemn what they consider to be a paucity of documentation and references, insufficient data or gross misuse of fragmentary data, and widespread misinterpretations in *Time on the Cross.* Here we shall consider only their criticism of Fogel and Engerman's appraisal of the efficiency of slave agriculture.

David et al. claim that, in the calculations presented in *Time on the Cross,* "virtually all of the South's apparent advantage was created by questionable measurement procedures which understate the amount of labor and the amount of land inputs used in that region relative to the amounts used in the North." More important, it was not the physical productiveness of northern and southern agriculture that was compared, they say, but rather the "overall revenue efficiency." Fogel and Engerman's comparisons were made for the year 1860, a "banner year" for cotton and a "low-water mark" for food crops such as grown in the North. The demand for cotton had increased rapidly in the 1850s right to the end of the decade. It was a period of enormous profit for slave-holding cotton planters, much of it realized in the form of capital gains—increase in the value of slaves. Fogel and Engerman believed that if it had not been for the Civil War, southern prosperity would have continued. However, while demand increased approximately 5 percent per year from 1830 to 1860, it increased less than 1½ percent per year between 1866 and 1895. Although demand increased more rapidly in the late 1890s, it never again returned to the growth rates of the antebellum decade.

In the prewar South, small farmers favored raising food crops, considering them to be less risky than cotton and providing at least subsistence for the family. They grew relatively less cotton than the big planters and did not share as much in the cotton boom of the 1850s.

While the southern economy grew in the antebellum era, it did not *develop* as did the northern. Alternate ways of generating income, so as to cope with later decrease in the growth rate of demand for cotton, were not developed. The South failed to use its earnings to develop the institutions and acquire the skills needed for sustained growth in the modern era.

Exhausting land and moving ever westward to virgin soil could not have continued indefinitely. But while the westward movement lasted, forced separation from family and friends and uprooting from familiar environments and understood social contexts resulted in Afro-Americans reliving the shock and horror suffered by their forebears who had left their ancient homeland in chains. No amount of sophisticated cliometrics can refute the fact that the slave economy, efficient or not, was hated and resisted in every practicable way by its victims.

## POSTWAR CHANGES IN SOUTHERN AGRICULTURE

The Civil War freed the South of the abomination of slavery, allowing it to slowly and painfully make its way into the mainstream of American history, but under exceedingly difficult circumstances. Much of what little manufacturing the South had developed had been destroyed. The cotton mills of Atlanta and Jackson were in ashes. The investment in slaves, amounting to between two and four billion dollars, was lost. The loss in livestock was very heavy. Farm buildings still standing showed the ravages of four years of neglect. Fences had been burned as firewood or had rotted away. Not until 1879 was cotton yield as high as in 1860. The sugarcane and rice industries were nearly completely wiped out. Farmland that sold for fifty dollars an acre before the war brought from three to five dollars, banks were engulfed in Confederate paper, Confederate bonds were worthless, and insurance companies became bankrupt. Guerrillas and deserters from both armies ravaged the land. The scanty resources remaining in the South that could be interpreted as being property of the Confederate government were subject to confiscation by the North. "Carpetbaggers" representing themselves as government agents, engaged in lawless plunder and were believed to have seized some three million bales of cotton, as well as livestock, tobacco, rice, sugar, or anything of value. Negroes, formerly cared for by their masters, were left to their own resources, with neither money, property, nor friends. Their hope for "forty acres and a mule" did not materialize. Negro colonies sprang up in the cities under deplorable sanitary conditions. Estimates of Negro mortality in such areas, from starvation, disease, and violence, ran as high as one-third (Ezell, 1963; Saloutos, 1964).

Agricultural areas normally recover from war more easily than industrial ones, for the land remains. After the Civil War, high prices encouraged the resumption of cotton- and tobacco-growing, but the end of slavery necessitated revolutionary changes in the method of production. The required changes were made difficult by lack of capital, heavy taxes, and uncertain labor supply. Many planters were forced to sell their land cheaply to Northerners, yeomen and poor whites of the hill sections, and the few Negroes who had the money to buy land. Those who did not sell their land, or those who bought large acreages of cheap land, divided it into "one-horse" or "two-horse" farms for cultivation by tenants or sharecroppers, including most of the Negroes who went into farming. The sharecropper who contributed no more than his labor retained a fourth of the crop. He might retain as much as half of the crop, depending on how much of the essential equipment and provisions he was able to provide. The sharecropper, generally illiterate, could not keep his own accounts and generally felt he was being cheated, which no doubt often was the case. On the other hand, the sharecropper might be lazy, incompetent, or lack the managerial ability that successful farming demands. In a year of bad crops he might simply leave the area (Saloutos, 1964).

By 1880 a third, and by 1920 two-thirds, of all farmers in the lower South were tenants. The tenancy system proved to be self-perpetuating and the ideal of landowning, prosperous yeomen failed to materialize (Ezell, 1963). Southern farmers hoped for industrial growth that could form new markets for southern products and use some of the mass of workers not yet assimilated in established economic pursuits. Some degree of industrial growth did indeed take place, particularly in spinning and weaving and in lumber and food manufacturing, but much of the wealth created left the South. The domination of single-crop agriculture on the region's economy had not been broken even as late as 1930 and prewar "colonialism" had not been ended. The products of southern mines, farms, and forests continued to find their way to the North or abroad in the form of raw or crudely processed materials.

## The Country Store

There were few banks in the South after the Civil War; many counties had no banking facilities whatever. Most of the meager capital attracted went for urban and industrial enterprises, not for agriculture, which supported the bulk of the population. The impoverished farmers were compelled to seek credit at abnormally high rates of interest (Coman, 1918). The landowner, seldom with sufficient cash, had to depend on the local merchant for credit, with the planter's share of the crop pledged for security. To protect themselves against loss, the country storekeepers took crop liens and, if possible, chattel (property other than real estate) and land mortgages also. After signing the lien note, the landowner (or tenant) could purchase goods at an agreed monthly rate, chargeable against his account. Goods bought on credit were much more expensive than those bought for cash, for the storekeeper charged a substantial percentage, known only to himself, to make up for bad debts, absconding tenants, and interest. This percentage, known as the "carrying charge," amounted to from 25 to 75 percent. Tenants who had hoped to become landowners sank increasingly into dept and landowners found the crop-lien system the path to tenancy (Ezell, 1963). Yet one can hardly imagine life in the post-Civil War South without the country store, with its typical, homely, single square front—a familiar landmark in every town.

## THE POSTWAR NEGRO

By 1900 three-fourths of the Negro farmers were still sharecroppers or tenants. Although by 1925 they comprised a third of the southern population, they owned only one-seventh of the farms and the land they owned was less desirable. The Negro's lack of experience in farm management, the westward shift of the Cotton Belt, the ravages of the boll weevil, which migrated from Mexico to Texas prior to 1892 and moved northward and eastward at the rate of 40 to 50 miles a year, and the droughts of 1916 and 1917, all militated against the success of the Negro in agriculture. Frequent failure in farming, along with the difficulty Negroes encountered in finding employment in many industries and the increasing acceptance by whites of jobs formerly monopolized by Negroes in the South, combined to induce the blacks to move north, particularly during and after World War I. By 1930 nearly two million had moved to industrial centers in the North and Midwest. The desire for economic improvement was their chief motivation; nevertheless, the majority were compelled to solve their problems without leaving the South.

One of the most progressive ideas to come out of the Reconstruction period after the Civil War was industrial training for Negroes. Generally credited with the idea was General S. C. Armstrong, an agent of the Freedmen's Bureau. Through his efforts the Hampton Normal and Industrial Institute was established, opening in 1868 "with the basic aims of teaching respect for labor, building up character, and providing graduates with an economically useful trade" (Ezell, 1963).

Booker T. Washington was the best example of a Hampton graduate. Born into slavery in 1857 or 1858, he worked in a coal mine after emancipation. He was allowed to work his way through Hampton Institute, then pledged to give other Negroes of the lower South the same chance given him. Following the Hampton model, in 1881 he established Tuskegee Institute in Tuskegee, Alabama, with a borrowed log shanty, one teacher, and thirty students. He opposed migration to the North and taught that friction between blacks and whites would "disappear in proportion as the black man, by reason of his intelligence and skill, can create something that the white man wants or respects." He taught that the Negro should concentrate on education and economic opportunity and remain aloof from politics. This was contrary to the philosophy of W. E. B. Du Bois, a professor of economics and history at Atlanta University and a guiding spirit of the National Association for the Advancement of Colored People (NAACP). Du Bois bitterly attacked Washington's strategy and argued that salvation for the Negro would come only through aggressive political action.

It was Booker T. Washington who invited George Washington Carver, a son of slaves and a graduate of the state agricultural college at Ames, Iowa, to head a newly organized department of agriculture at Tuskegee Institute in 1896. Carver formulated more than 300 by-products of peanuts and sweet potatoes, which were said to be "a potent factor in changing the economy of the South." He made the plans for a "school on wheels," to give demonstrations about soil fertilization to those unable to travel to

Tuskegee. Carver helped the poor blacks in the area by demonstrating the value of substitutes like acorns for feeding hogs and swamp muck for enriching croplands. This was of great value to farmers who could not afford commercial feed and fertilizer.

Thomas Edison was among the many persons offering Carver employment at a high salary, but he would not leave Tuskegee and he would not accept a raise above the $1,500 received when he first came to the institute. Following Carver's death on January 5, 1943, his entire estate was added to the Carver Research Foundation's endowment, to which he had already contributed most of his savings (Elliott, 1966). Henry Ford declared that "Professor Carver has taken Thomas Edison's place as the world's greatest living scientist." Harry S. Truman said that "the scientific discoveries and experiments of Dr. Carver have done more to alleviate the one-crop agricultural system of the South than any other thing that has been done in the history of the United States." In 1943 Carver's birthplace became a national monument—an honor previously granted only Washington and Lincoln. In 1977, Carver was enshrined in the Hall of Fame for Great Americans (Mackintosh, 1977).

Carver's actual accomplishments, remarkable for a man born as a slave and bucking the barriers of discrimination, were in themselves sufficiently remarkable to gain for him universal admiration. But both blacks and whites found it to be expedient—but for different reasons—to build a legend around the man. Carver the legend far exceeded Carver the scientist in accomplishments. "The progress of the Carver myth," says Barry Mackintosh (1977) "may be traced in the writings of journalists, popular biographers, publicists, politicians, and professional historians from the early 1920's to the present."

Carver's teaching at Tuskegee was disappointing to students and fellow teachers, as well as to Booker Washington. The bulletins, circulars, and leaflets that were published under his name contained little that was new. By the time his peanut bulletin, *How to Grow the Peanut,* appeared in 1916, peanut production in the South had already reached 40 million bushels, up from 19.5 million in 1909 and 3.5 million in 1889. The greatest rate of increase preceded Carver's identification with the peanut. Peanuts continued to go almost entirely into confections, baked goods, peanut butter, and oils, with no significant contribution from the 287 peanut commodities attributed to Carver by the Carver Museum. A *New York Times* article erroneously gave Carver credit for originating dehydrated foods.

Although credited with "formulas in agricultural chemistry that enriched the entire Southland, indeed the whole of America and the world" (Hughes, 1954), in actual fact Carver left no formulas and kept no laboratory records. He declared that he depended on divine revelation for his product ideas and methods and never consulted books for this purpose. Some Negro scholars have compared Carver's scientific work unfavorably with that of such unpublicized black contemporaries as biologist E. E. Just and an authority on insect behavior, C. H. Turner. "What he did," says Mackintosh, "was less important than what he was and the larger purposes his existence served for blacks and whites alike."

The zeal of Negroes for education was shown by the fact that over 150,000 were in some kind of school in 1866. Three years later 9,593 teachers were working with Freedmen alone, nearly all the white teachers coming from the North. The North also financed these schools by means of the Freedmen's Bureau and private organizations. There was no other way Negroes could have been educated in a region where traditional antagonism toward publicly supported education was combined with postwar financial inability to supply it.

Booker T. Washington argued for vocational education for Negroes and Du Bois favored equal and similar instruction for Negroes and whites. With certain notable exceptions, the vocationalists had their way, which was probably the more realistic of the two possible courses in view of the situation that confronted the blacks in the post-Civil War South.

By the end of the nineteenth century thirty-four institutions for Negroes gave college training in the South and northern schools were beginning to accept them. After 1900, most growth took place in public-supported colleges. By 1890 there were more than 90 southern colleges for Negroes, with about 12,000 students (Ezell, 1963).

In 1890 the Second Morrill Act authorized the establishment of separate land-grant colleges for Negroes. Seventeen states in the South decided to do so, although none offered college-level courses until 1916. By 1970 Negro land-grant colleges enrolled approximately 50,000, about 20 percent of all black students in college at that time. Five percent of the students enrolled were white (Payne, 1970).

All southern states now have excellent state universities with strong research and teaching programs in agriculture. Integration has made them available to blacks as well as whites. But Booker T. Washington probably would have been most pleased with the philosophy of Berea College, in the small town of Berea in eastern Kentucky. Berea College is bringing modern technology to the youth of Appalachia hills and hollows—young people from communities like Plain Talk, Gravel Switch, Meathouse Fork, Meadows of Dan, Mouth of Wilson, Coolridge, Odd, Quicksand, Hazard, and Stinking Creek. It has been a leader in interracial education since its beginning, and at the present time approximately 13 percent of the students are black.

The agriculture curriculum comprises twenty-nine courses. It trains the student for career opportunities in agriculture service occupations, farming, conservation, veterinary medicine, landscaping, agriculture economics, soil technology, and research. The head of the agricultural experiment station at Quicksand is a graduate of Berea and has many Berea graduates on his staff, upgrading the agriculture and thereby the entire economy of depressed areas (Nelson, 1975*b*).

## FARMER ORGANIZATIONS

Two farmer organizations, the Grange and the Alliance, deserve great credit for their efforts to provide economic, social, and cultural amelioration by striving for agricultural education, improving the comfort and attractions of farm life, dispensing with the surplus of middlemen, improving transportation, opposing monopolies, promoting nonpartisan and clean politics, ending sectionalism, and advancing the position of women. The Grange, opposed by the powerful landlord and merchant classes, declined in the mid-1870s, but much of its work was continued by the Alliance, headquartered in Washington, D.C. The Alliance was more politically motivated than the Grange and was responsible for some legislation of benefit to farmers. Growing out of the Alliance was the Populist political movement, which rallied the many farmers and farm leaders who believed agriculture was not sharing in the national prosperity. The Populist Party reached its highest point in 1896, but failed to elect its candidate for President. Thereafter the party declined rapidly, but not without achieving some of its goals, such as the free delivery of rural mails. The next two decades also witnessed federally sponsored rural credit facilities and federal regulation of futures trading. Farming achieved a period of comparative equilibrium with the rest of the economy that lasted through World War I (Saloutos and Hicks, 1951; Saloutos, 1964; Rasmussen, 1975*a*, p. 1166).

The National Grange is now the least active of the voluntary farmers' associations. More politically oriented are the huge American Farm Bureau Federation, the National Farmers' Union, and the National Farmers' Organization. In this era of agricultural specialization, specific agricultural commodity organizations are of increasing importance, most of them maintaining lobbyists in Washington (Shover, 1976).

## THE NEW DEAL

Although for most of the nation the Great Depression began in 1929, farmers had been in economic distress since 1920 and this was particularly true of the southern farmer. A worldwide decline in demand for cotton goods came at a time of maximum cotton acreage in the South. Growers had increased output, to offset low prices, to a total of 15 million bales in 1929, competing on the world market with a foreign production of 11.5 million bales. The price dropped from 20 to 12 cents a pound; yet the South grew 17 billion bales in 1931, whereupon the price dropped to 5 cents. Whereas southern farmers had received $1,245 million for the 1929 crop, they received only $484 million in 1931 and $374 million in 1932. Forced farm sales in the United States came to 46 per 1,000, but they were 68.2 in North Carolina and 99.9 in Mississippi. Many farmers abandoned their land and fled to the city, but with little improvement in their situation. Only the tobacco industry was stable. Whites began to compete with blacks for "Negro jobs" they formerly disdained.

The South had about 97 percent of the nation's cotton and 87 percent of the tobacco, but only 16 percent of the manufacturers were southern in origin. About 27 percent of the nation's property value was in the South, but only 14 percent of the bank deposits—and 32 percent of the state government debts. Half of the South's inhabitants had substandard housing, 60 to 88 percent of the city dwellers were ill-fed, and sickness and death rates were the highest in the nation (Ezell, 1963).

The New Deal of the Franklin D. Roosevelt administration began its attack on the South's specific problems in 1933. Among its devices to rehabilitate farmers were government credit, debt adjustment, crop controls, soil conservation, crop loans, and aid to tenants to purchase homes. These were measures aimed to controvert tenancy, monoculture, and the poor financial arrangements in the South. The Agricultural Adjustment Acts of 1933 and 1938 regulated farm production and maintained purchasing power at pre-World War I levels, but crop restrictions caused thousands of tenants and sharecroppers to leave the farms and go on relief. Efforts to force landlords into a more equitable sharing generally failed; in many cases planters abandoned sharecropping and rehired their tenants as wage hands without rights to benefits (Ezell, 1966; Saloutos, 1974).

Reducing acreages did not solve the problem. Modern technology increased production per acre so rapidly that land could not be taken out of production rapidly enough to affect increased productivity. The big growers, who could afford increased mechanization and heavy fertilization of the soil, were the ones who gained most from the New Deal programs (Saloutos, 1974).

In 1933 Congress created the Federal Farm Mortgage Corporation for low-interest loans on agricultural lands. The Farm Credit Act provided loans for production and marketing of farm produce, and the government thereupon loaned $300 million to tenants and sharecroppers to buy livestock and equipment. The Federal Security Administration helped more than 300,000 low-income farm families get started in about 16,000 small cooperatives. However, the larger and more successful farmers believed that there was little hope for the depressed elements of the farm population and that it would be to their best interests and those of agriculture if they would seek their livelihood in urban areas. The efforts of the USDA, colleges of agriculture, and agricultural experiment stations were seldom directed specifically toward the problems of the submarginal farmer, for they considered their responsibility to be the establishment of prosperous farm families through applied research. They shared with the American Farm Bureau Federation, the National Grange, and the agricultural establishment in general the view that agriculture suffered from a surplus of farmers (Saloutos, 1974).

The Public Works Administration constructed new courthouses, post offices, libraries, bridges, hospitals, college and university dormitories, as well as sidewalks, drainage ditches, sewer systems, and roads. The Civilian Conservation Corps helped combat erosion by reforestation. The Social Security system gave thousands of low-income Southerners financial help, as in the form of old-age and unemployment insurance and old-age assistance programs. The South obtained about half of the Federal Housing Administration's slum clearance funds, particularly helpful to the blacks. Inexpensive electrical power was made available in remote rural areas—bypassed by private utilities—in the Rural Electrification Administration program. This made the rural areas more attractive and made possible the diversification of farming and the establishment of small industries there.

There was frequent abuse of New Deal programs. Rise in cotton prices brought complaints about crop restrictions. Some landlords cheated on their acreage allotments and increased fertilizer to such an extent that in 1937 average production was fifty pounds per acre in excess of former records. The record-breaking 19-million-bale production of 1937 resulted in the price of cotton dropping from an 11- to 14-cent range down to 8 cents a pound. Agencies such as the WPA and CCC were criticized as havens for loafers and for interfering with the labor supply by offering shorter hours, lighter work, and year-round employment at public expense. Relief payments were criticized as being more than a sharecropper or many tenants could make when employed. Nevertheless, the New Deal is usually considered to have been beneficial to the South. It enabled the region to survive until new technologies and laborsaving mechanization in agriculture, coupled with increased industrialization, allowed the area to get back on its feet (Ezell, 1963; Saloutos, 1974). How these new trends affected the major crops will be told in the following pages.

## COTTON

Following the Civil War, the inadequate transportation facilities militated against alternate cash crops that might have obviated dependence on a single crop, usually cotton or tobacco. Farmers were restricted to nonperishable products that could stand rough handling and the crudest kind of processing for marketing and storing. Storekeepers with advances from cotton buyers could most conveniently pay off their debts with cotton. Thus agriculture in the South was carried on largely by the small farmer raising a cash staple with hand tools and a mule, directed by his landlord and his supplier of credit (Clark, 1945).

The old process of exhausting land and then moving on to new fields could continue only briefly, chiefly in the new fields of Texas and Oklahoma. Thus the timely contribution of a well-educated Georgia farmer, F. C. Furman, is worthy of note at this point. Furman had the cotton plant analyzed and got the idea of putting back into the soil the elements the plant extracted, in about the same proportion. His land was particularly deficient in humus, so Furman made a compost of stable manure, acid phosphate, kainite (a mineral containing potassium and magnesium), and cottonseed. He covered it with three to six inches of good topsoil and allowed it to stand six weeks. The kainite attracted moisture from the air and the moisture was then absorbed by the humus. When the mixture was drilled into the furrows of his soil it not only provided the essential nutrients but kept the soil moist when plants in other fields were suffering from drought. Furman's formula was cheap and practical. Starting in 1879 he increased his cotton production about sixfold with a ton of compost per acre and the following year twelvefold with two tons. His formula was then sold commercially. Furman preached the gospel of soil renewal and conservation and called for a revitalized "New South" through diversified and intensified agriculture. In 1883 an attack of malaria resulted in Furman's untimely death at the age of thirty-eight, but by that time he had left incontrovertible proof that southern agriculture could be revived (L. D. Stephens, 1976).

The cotton-growing area of the South became stabilized around 1930 and by that time was producing close to 14 million bales of cotton, about 60 percent of the world's cotton supply. King Cotton dominated the life-style and attitudes of an enormous area extending 1,600 miles from Texas and Oklahoma to eastern North Carolina. Minimal attention was given to food crops; the farmer's diet was largely limited to the easily raised pork, corn, and molasses. There was great need for field labor—both sexes, including children—encouraging a high birth rate. The least prosperous states were the ones most completely devoted to the growing of cotton (Ezell, 1963).

### Mechanization in Cotton Production

Even after machines were developed for planting and cultivating cotton, there was little incentive to mechanize these operations because much labor was required anyway for picking—30 to 40 man-hours per acre. (This was as much time as required to grow and harvest three acres of corn in Iowa.) The mechanical cotton picker began to be commercially produced in 1941. This uncanny machine draws the lint from open bolls and leaves the unopened ones for a later picking. It delivers cotton grades about the same as if the cotton were picked by hand. Cotton pickers were first used in California and Arizona, the larger fields and higher production favoring mechanization. By 1952 only 18 percent but by 1967 nearly 95 percent of the cotton crop in the United States was mechanically harvested. Using mule power and "half-row" equipment, for cultivation and with all weeding, thinning, and picking done by hand, the production of a bale of cotton required about 155 man hours of labor (Street, 1957). Complete mechanization reduced this to about 10 or 12 man-hours. Most of the reduction resulted from mechanization of harvesting, in which one machine (fig. 9) might do the work of forty men. Only through mechanization have cotton growers of the United States been able to compete successfully against foreign competition and synthetic fibers (Gittens, 1950). As with all crops, the efficiency of machine-picking of cotton is enhanced by the work of plant breeders, who develop varieties that lend themselves more readily to machine-picking.

### Pest Control

Few plant crops have been so severely plagued by pests as has cotton. Contrary to common opinion, only about 5 percent of the total crop acreage of the United States is treated with insecticides (12 percent if pastures are excluded), but nearly half of the insecticides are applied to cotton (Eichers et al., 1970; Pimentel, 1973).

The experience of cotton growers in the control of the formidable array of cotton pests is commonly used as a classic example of the "pesticide-syndrome." Bottrell and Adkisson (1977) reviewed the history of this interesting post-World War II phenomenon for cotton pest control in the South. Insect and mite pests had prevented maximum yields of cotton until shortly after the war when DDT and other organochlorine insecticides such as benzene hexachloride, toxaphene, and chlordane, became available. These insecticides were spectacularly effective during the late 1940s and early 1950s. Pesticide combinations were applied regularly, usually once a week, from the time the vulnerable cotton squares were formed until the green bolls had hardened and become immune to boll weevils (*Anthonomus grandis*) and bollworms (*Heliothis zea*).

*Fig. 9. Cotton picker in a southern cotton field (courtesy Deere and Company).*

The initial confidence in the new organic pesticides began to falter after repeated destruction of the beneficial insect parasites and predators and increasing resistance of pests to the new compounds resulted in serious outbreaks of not only major pests but also some that had previously been considered to be of minor consequence. Then for a while a new class of pesticides, the organophosphorus and carbamate compounds, were effective, and at very low dosages, but were ecologically even more disruptive than the organochlorine compounds. Some growers went bankrupt or discontinued growing cotton.

Chemical pesticides still play an important role in the control of cotton pests. They are generally applied in dilute aqueous solutions, three to ten gallons per acre, by means of airplanes (fig. 10). However, in the current integrated control strategy, as explained in the preceding chapter, pesticides are used only to suppress pest outbreaks that have attained damaging levels. The pesticide used, dosage, and the time of treat-

*Fig. 10. Aerial application of a dilute pesticide solution by use of a Pawnee Brave 300 Ag airplane (courtesy Piper Aircraft Corporation).*

ment are carefully coordinated to minimize resurgence of the principal pests and outbreaks of formerly secondary pests, and to conserve natural enemies.

Entomologists are continually seeking pesticides that cause minimal ecological disruption. Synthesized insect pheromones, which have long been used successfully in pest survey and detection, show promise of successful use in suppressing insect pest populations, as has been demonstrated in field tests with gossyplure (synthesized female sex pheromone) as a male confusant to suppress mating of pink bollworms (Shorey et al., 1974; Gaston et al., 1977). Gossyplure obtained the approval of the federal Environmental Protection Agency (EPA) in May 1978. The compound is packed inside hollow plastic fibers, 200 microns (0.2 mm) in diameter and 1.5 cm long, which are distributed over cotton fields by aircraft or tractor. The pheromone gradually evaporates from the tubes in precise, uniform quantities. Since the sex attractant is distributed throughout the field, the males cannot identify females and mating seldom takes place. In tests in California in 1977, cotton plants treated with the new pheromone method suffered only 10 percent damage from the pink bollworm compared with 80 percent damage to untreated plants. Beneficial insects are unharmed.

Modern integrated pest management depends heavily on the utilization of natural manipulative factors worked out by some of the early cotton insect entomologists, beginning in the late nineteenth century. For example, insecticidal control of prehibernating boll weevils and early mandatory stalk destruction on an area-wide basis is a part of the current program. In addition, new short-season, but highly productive, cotton varieties, new production techniques, and new varieties resistant to certain pests, should lead to further improvements in pest management (Bottrell and Adkisson, 1977).

By utilizing integrated pest management strategies, farmers can produce crops more profitably while lessening the previously prevailing environmental and health hazards of cotton pest control. To ensure that pest-control measures are applied at the right time and the right way so as to have minimal disruptive effect on the pest-natural enemy balance, it is desirable to employ specialists skilled in pest-management strategies to aid in the decision-making process (Reynolds et al., 1975).

### The Present Status of Cotton

In 1976, the United States produced 18.4 percent of the world cotton crop of 57,584,000 bales. (A bale of cotton weighs 480 pounds.) The leading states in cotton production were Texas with 3,314,000 bales, California with 2,482,000, and Mississippi with 1,151,000 (USDA Agr. Stat., 1977). The average yield is about a bale per acre, but in some irrigated areas, as in Arizona and California, the average is two bales per acre. In years of good production, in certain restricted areas, as in the Imperial Valley of California, production might average 3½ bales per acre and in some fields more than 5 bales. Excellent yields are also possible in the deep South with good agricultural practice. Figure 11 shows a contoured, skip-row planting in a field with a 2 to 8 percent slope, 12 miles east of Yazoo City, Mississippi, (Region O, fig. 3). The field yielded an average of 2¾ bales of cotton per year over a 5-year period.

Cottonseed, once considered largely a waste product, is now one of the main assets of the South. A special ginning machine removes the short fuzzy fibers (linters) that remain after the first ginning process and these are made into plastics and hundreds of synthetic products. For every pound of lint there are about two pounds of seed. The seeds are crushed and processed into cottonseed oil and livestock feed (Lent, 1968).

## TOBACCO

Of great importance as a plant crop staple in the South is tobacco, of which there are six types, each used for a specific purpose: for cigarettes, cigars, pipe tobacco, chewing tobacco, cigar filler, cigar binder, and cigar wrapper. The type depends on the kind of soil in which it is grown and the method of curing (Gage, 1942; Garner, 1946). Economically, the most important is the flue-cured (bright-leaf) type, produced in the Carolinas, Georgia, and Florida, and used mostly in the manufacture of cigarettes. The demand for cigarettes, in the first half of the twentieth century grew from 32 to 2,607 per capita per year. This was an increase of from 2 to 75 percent of the total national tobacco consumption (Ezell, 1963).

Tobacco is a notably labor-intensive crop. Many families obtain their cash income from three to six acres. Often all members of a family are employed full time in growing a crop of tobacco. In January or February the seedbed is prepared and seeds are planted; then in May the plants are transplanted by hand to a field that had been prepared in April. Thinning and replanting, as well as hoeing and cultivation to keep the weeds out, are additional tasks. Inferior leaves and suckers are removed to improve the quality of those remaining. There are many diseases and insect pests, generally controlled by seed treatment, treatment of seedbeds, use of resistant varieties, suitable rotations with other crops, and good sanitary practices. The most widely distributed and

*Fig. 11. Contoured skip-row planting of cotton near Yazoo City, Mississippi (courtesy USDA Soil Conservation Service).*

most destructive pests are the large hornworms, which can devour entire leaves and are often handpicked to avoid the use of insecticides that might discolor the leaves or affect taste. Beginning in late July and continuing into the middle of September the leaves are picked as they reach maximum size. After they are dried they are sorted, tied into bundles weighing from 30 to 60 pounds, and taken to the market.

In 1976 North Carolina was the leading state in the production of tobacco, having produced 903 million pounds, followed by Kentucky with 499 million, Virginia with 154 million, South Carolina with 153 million, Tennessee with 139 million, and Georgia with 124 million. The nation's total for 1976 was 2,134 million pounds, 17.5 percent of the world total (USDA Agr. Stat., 1977).

Smoking had long been known to be something less than an innocent pastime to be enjoyed in the company of attractive and sophisticated companions in outdoor settings of breathtaking beauty, as depicted in glossy magazine advertisements (and in television commercials until they were banned in 1972). But it was not until 1964 that the Public Health Service of the federal government declared that cigarette smoking was an important health hazard—a major cause of lung cancer and bronchitis and figuring prominently in heart disease, prenatal damage, vascular diseases, and other health problems (Fergurson, 1975). The Department of Health, Education and Welfare (HEW) spends $14 million to research lung cancer and $10 million on health education. Yet the USDA pays out $70 million annually to run a tobacco price-support program and has lent tobacco farmers about $3 billion over the last 20 years (Pearce, 1978).

## RICE

After the Civil War, rice production, which had been heavily dependent on slave labor in the coastlands of the Carolinas and Georgia (fig. 3, Region T), rapidly declined from the 6,732,627 bushels of 1860 to 2,648,742 in 1870 and to barely 1,000,000 in 1900. Rice cultivation required much more capital than cotton and it was hard to find tenants who would subject themselves to the arduous and disagreeable task of raising the crop. Except in areas where there is rain practically every day, rice must be grown in fields kept flooded until just before the rice is harvested.

In the 1880s farmers from northwestern wheat fields began to utilize wheat-farming techniques for rice production along the coast of Louisiana and Texas, the western extremity of Region T (fig. 3). Artesian wells that provided a level, steady flow when needed, and lacked the drawbacks of the previously utilized system based on tidal flow, transformed a strip of coastal prairie into the world's best rice-producing land. The level fields had a subsoil of clay that held water but dried quickly when drained and therefore supported heavy machinery. Mechanization revitalized the rice industry (Ezell, 1963).

In 1976, four states produced 94.3 percent of the rice grown in the United States: Arkansas, 1,831,000 metric tons; Texas, 1,108,000; California, 1,061,000; and Louisiana, 1,007,000. (USDA Agr. Stat., 1977).

### Rice Processing

Rice is an example of how important crops, including wheat, lose much of their nutritive value through processing. In many villages in Asia rice husks are still removed by hand pounding when this cereal is to be used in the home or sold locally. This type of rice is also available throughout the world as "brown rice." Brown rice retains the "bran" (seed coat or pericaps and aleurone layer) and embryo largely intact. The aleurone layer contains vitamin $B_1$ (thiamine) and other valuable substances. In areas where this kind of rice is consumed, beriberi, a disease caused by vitamin $B_1$ deficiency, is rare. With another method of processing, the unhusked rice is parboiled, gelatinizing the outer layers of starch in the endosperm. These outer layers are then able to absorb part of the vitamins of the aleurone layer so that they are retained after the grains have been milled and polished. Commercial milling of rice leaves the aleurone layer more or less intact, when removing the husk and seed coat, but the rice is then scoured and polished, removing the last bits of bran and embryo and thereby removing most of the vitamins and minerals (Kik and Williams, 1945). The only positive thing to be said for white polished rice is that it keeps longer than unpolished rice in storage, for it possesses less nutrient value for fungi, mites, and insects.

## SUGARCANE

Sugarcane accounts for two-thirds of the world production of sugar, the remainder being derived from sugar beets. On his second voyage in 1493, Columbus brought sugarcane to Hispaniola and it was soon carried to Cuba and Puerto Rico, but this was the thin-stalked species. The superior thick-stalked or "noble" cane (*Saccharum officinarum*) was transported by Captain Bligh from Tahiti to Jamaica in 1791 (Ochse et al., 1961). Sugarcane was introduced into Louisiana early in the eighteenth century, but almost a century of experimentation and trials was required before sugar was successfully produced (Klose, 1950). The thick-stalked varieties soon replaced the previously grown varieties wherever they were introduced and even to this day they give enormous yields on newly cleared land.

In nearly every country where sugarcane is grown the industry has been saved from extinction by new, resistant strains developed by plant breeders. New, higher yielding strains, especially adapted by selection for specific soil and climatic conditions, are sometimes the reason for replacing previously established varieties. A relatively new development is the spraying of chemical ripeners (herbicides used at low doses) on sugarcane a few weeks before harvest. Vegetative growth is slowed, carbohydrates (sucrose) accumulate, and increase in sugar production approximates two tons per hectare per year. In Hawaii drip or trickle irrigation systems are being installed at the rate of 5,000 hectares per year. All new plantings now have them. Cost of installation is about $1,500 per hectare, but this is amortized in a year as the result of savings in labor (Wittwer, 1976). Probably in connection with no other crop have continued inputs of advanced technology been a greater factor in its commercial success. Greater attention to various cultural details would increase sugar production in most areas of the world where sugarcane is produced.

There has been an astounding increase in the amount of sugar consumed in affluent countries in the past century. In the United States and Britain people are consuming an average of two pounds or more of sugar per week, compared to about that many ounces a century ago. In Norway, average sugar intake per person per year was 2¼ pounds in 1835, 11 pounds in 1875, 67 pounds in 1937, and 90 pounds in 1965—a forty-fold increase in 130 years. It is difficult to accept these facts, because less sugar is used in the household than formerly. But a greatly increasing amount of "industrial sugar" goes to the factory and comes to the consumer in the form of such products as candy, ice cream, soft drinks, cakes, cookies, and packaged "convenience foods" (Yudkin, 1972).

In 1972 Louisiana harvested 312,000 acres of sugarcane, Florida 307,000 acres, and Hawaii 104,000 acres. Hawaii had the highest productivity—85.0 tons of cane per acre compared with 37.7 for Texas, 35.0 for Florida, and 25.0 for Louisiana (USDA Agr. Stat., 1977).

## PEANUTS

The peanut (*Arachis hypogaea*), known in most of the world as "groundnut," was probably first domesticated in the foothills of the Bolivian Andes. It is an unusual plant in that the stalk of the flower elongates after fertilization and pushes the developing pod underground (Heiser, 1973). Peanuts became a crop of considerable importance in the South in colonial times, but mainly as hog feed rather than as a cash crop. They were commonly grown in cornfields and hogs were turned in to fatten on the nuts (Gray, 1933). Following the Civil War there was considerable diversification of agriculture in the South, including the growing of peaches in the piedmont areas of Georgia and the Carolinas and apples in the mountain valleys of Virginia, but of particular importance among the new cash crops were peanuts, of which 740 million pounds were grown annually by 1930. In the area of Enterprise, Alabama, peanuts alone were bringing in as much money as cotton had formerly. The town built a monument to the boll weevil, which was an important factor in dethroning King Cotton, with the inscription: "In Profound Appreciation of the Boll Weevil and What It Has Done as the Herald of Prosperity."

Peanuts are very nutritious and are a particularly important part of the diet in countries where the majority of the people cannot afford animal fats and proteins. The pods may be roasted or boiled and the nuts may be eaten out of the shell or they may be purchased as roasted and salted shelled nuts. Also whole or chopped nuts may be used in candies, cakes, cookies, and other confections. Peanut butter is a popular item of diet, particularly in the United States, where the greatest part of the crop is used (Heiser, 1973).

Peanuts contain from 25 to 32 percent protein, much more than the cereal grains, and about 40 to 50 percent fat, in addition to cystine, thiamine, riboflavin, and niacin. However, the protein is deficient in three amino acids: methionine, lysine, and threonine, and should therefore be supplemented with other, higher quality protein (PSAC, 1967). A pound of peanuts contains more protein, minerals, and vitamins than beef, but in the United States a negative factor in marketability is the fact that peanuts also contain more fat than heavy cream and more calories than sugar. USDA researchers are experimenting with a process by which oil is squeezed out of peanuts, thereby eliminating what, to Americans, is their principal dietary disadvantage. Much research is now directed toward new peanut by-products such as flour, soup, and peanut butter and jelly sandwiches, the latter to be shipped to school cafeterias (Harvey, 1975). According to present law, the Department of Agriculture must purchase peanuts not sold in regular trade and sell them at a loss on the glutted world market or donate them to foreign aid and domestic school-lunch programs, at a total cost to the American taxpayer in 1975 of between $125 million and $150 million a year.

Although a large quantity of peanuts is consumed directly by man, about 65 percent of the world's production is crushed for oil, forming about 20 percent of the world's edible oils. After crushing and solvent extraction, a meal remains that contains about 50 percent protein. This may be treated further to remove carbohydrates, leaving a concentrate with more than 70 percent protein. Peanut-oil press cake is an excellent animal food supplement and is also eaten by human beings in some tropical countries after a processing that involves the use of fungi. The cakes are cut into oblong strips and fried or used in soups (Ochse et al., 1961; PSAC, 1967).

In the United States, commercial peanut-growing is almost completely confined to the South. The leading states in 1976 were Georgia, with 529,000 harvested acres, Texas 310,000, Alabama 216,000, North Carolina 168,000, Oklahoma 124,000, Virginia 104,000, and Florida 63,000. These seven states had 97.8 percent of the nation's 1,521,500 acres harvested for these nuts, for a production of 1,702,000 metric tons. (There are an additional 300,000 acres of peanut vines harvested for hay.) Countries with greater peanut acreage are India, with 18.5 million acres annually harvested; China 5 million, Senegal 2.9 million, Nigeria 2.4 million, and Sudan 2.1 million (USDA Agr. Stat., 1977).

Despite the fact that millions of people throughout the world suffer from malnutrition, peanuts are a glut on the world market. In the United States a 32-year-old law allots growers 1.6 million acres of peanut farmland. When the law was passed, the yield was 500 pounds per acre, but through improved fertilization, pest control, and other cultural measures, yield has been increased to 2,500 pounds or more per acre.

At a question-and-answer session with Department of Agriculture employees, President (and peanut grower) Jimmy Carter pointed out that in the early 1950s, the average production of peanuts in Georgia was about 800 to 1,000 pounds per acre. Now the production is 2,500 to 3,000 pounds per acre. Carter added, "It is almost directly attributable to basic research that discovered that the more you plow peanuts, the lower the production is. So, when we quit cultivating our crops, we not only saved a tremendous amount of energy and expense but we also derived tremendous financial benefit, and so did the rest of the world in getting cheaper food."

## SWEET POTATOES

The sweet potato (*Ipomoea batatas*) originated in some unknown way from a widespread American plant genus. It is sometimes erroneously termed "yam," but the true yam is of another genus (*Dioscorea*) and is a tropical lowland root crop. A ten-month growing season is needed for yams to mature, so they are not a suitable food crop outside the tropics.

The sweet potato is in the morning-glory family (*Convolvulaceae*) while the Irish potato is in the nightshade family (*Solanaceae*), so the two are not closely related despite both being called "potatoes." The most important part of the plant is the swollen, tuberous root, although the leafy tops may be used in silage or for animal browse. The root is also fed to stock in some countries.

The sweet potato is no longer found in the wild and its ancestry is uncertain. It was introduced into Spain from Middle America by the early Spanish explorers, probably Columbus. Navigators of the sixteenth and seventeenth centuries carried it from Europe to all parts of the world. The possibility has been suggested that its appearance in Polynesia in pre-Columbian times may have been the result of its water-resistant seedpods. The sweet potato is now a very important food crop in West Africa, Indonesia, the Pacific Islands, India, China, and Japan. In Japan its importance is second only to that of rice, and it is considered to be the most productive crop per unit of upland area (Schery, 1972).

In the southern United States, the sweet potato has assumed some importance, principally as a source of carbohydrate; protein content is only 2 percent, about the same as the Irish potato. However, the sweet potato is unusually rich in vitamins, iron, and calcium. Throughout the world, sweet potato varieties usually have a white or a pale yellowish flesh, but in the United States the trend has been toward dark yellow or reddish-orange types rich in carotene. In the North the "dry" types with a mealy, starchy flesh have gained favor while in the South the soft, more gelatinous, sugary types, often called "yams," are preferred. In the United States sweet potatoes are usually marketed as they are dug from the ground and are baked or boiled for eating.

In the tropics the vines grow continuously and propagation is by stem cuttings. In this country propagation is by shoots taken from "seed" roots, kept over from the previous growing season. Sweet potatoes yield heavily, as much as 20 tons per acre under the best conditions, but their culture is labor-intensive. Mechanization has, however, been initiated. In 1976, domestic production was 621,000 metric tons, with North Carolina, Louisiana, and California the largest producers (USDA Agr. Stat., 1977).

## COWPEAS

The cowpea or black-eyed pea (*Viga sinensis*) is probably native to central Africa. Despite its common name, it is more closely related to the bean than to the pea. Early in the eighteenth century it was introduced into the West Indies and from there into the Carolinas. It is grown principally in the South and is also more popular as a food there than in other parts of the country. The seeds consist of approximately 23 percent protein, 1 percent oil, and 57 percent carbohydrate. With modern agricultural technique, yields may be more than a ton per acre.

Like the true pea, the cowpea is sensitive to cold or frost, but grows well on almost any soil. Cowpeas are generally rotated with cotton or corn in the United States, or they may be planted with sorghum for forage. In North America cowpeas are perhaps grown more as a green manure and forage plant than for human consumption (Schery, 1972).

## CITRUS FRUITS

Citrus fruits are native to the Orient, probably southeastern Asia, whence they were slowly distributed westward—first the sour orange and lemon and later the sweet orange, which probably did not reach Europe until about A.D. 1400. The seeds brought over by Columbus on his second voyage in 1493 were probably those of sour orange, citron, and lemon. The first orange seeds introduced into Florida by Spanish explorers were also those of sour oranges. Indians planted them in inland areas, and later did the same with seeds of sweet oranges when the earliest settlers established plantings in coastal areas. Some travelers finding orange trees growing wild in Florida's forests believed they were part of the native flora (Camp, 1947).

By 1579 there were citrus plantings in the vicinity of St. Augustine. After Florida was annexed by the United States in 1821, settlers steadily expanded the groves found along the coast at such points as St. Augustine and Tampa Bay, and along the St. John's River south of Jacksonville. Oranges were packed in barrels padded with Spanish moss and shipped by sailing schooner to the snowbound North for the Christmas holiday season, selling for fabulous prices when other fruits were largely unavailable. This was before the days when vitamins and minerals loomed large in dietary consideration. What sold the orange was not only its delicious, refreshing flavor, but also the aura of romance and glamour of distant lands that was associated with this beautiful, exotic fruit.

The freezes of 1894 and 1895 resulted in the citrus industry moving to the south of Florida, but today most of the industry is in the area called the "South-Central Florida Ridge," running north and south through the middle of the upper half of Region U in figure 3. When sweet orange is grafted to rough lemon stock, it thrives on the high, warm, sandy ridges in that area. Nevertheless, frosts sometimes occur two or three nights in succession. Many orchards have wind machines—giant propellers on high columns. These are used to prevent cold air from stratifying in "frost pockets." They may suffice to protect the orchard from damaging frosts during some winters, but sometimes the grower may have to add oil-burning heaters, as many as seventy to an acre. He may fire them up on cold nights to prevent damage to trees and fruit, using wind machines and heaters at the same time. Sometimes the heaters need to be added only in some low spots where cold air has settled. Under extremely cold conditions, the wind machines are not as effective as heaters.

The thrifty Valencia orange trees (shown in fig. 12, laden with fruit) are growing on sandy, well-drained, rolling lands of the "ridge." On the ground are the remains of weeds that have been sprayed with herbicides and will be disked into the soil to supply humus.

After the great freezes of earlier years, and as new acres were developed, the planting of grapefruit began. Florida is now the greatest citrus-growing area of the world. Its production of 7,076,600 metric tons of oranges and tangerines in the 1974-1975 season was 22.6 percent of the world total. The production of oranges and tangerines, in metric tons, for the other countries or states in the top ten was as follows: Japan, 4,234,000; Brazil, 4,068,000; Spain, 2,693,000; Italy, 1,931,000; California, 1,827,000; Argentina, 1,131,000; Israel, 983,000; Egypt, 953,000; and Mexico, 905,000. Florida's production of grapefruit in the 1974-1975 season was 1,720,000 metric tons, 48.6 percent of the world total. Florida's production of oranges and tangerines in the 1975-1976

*Fig. 12. An orange orchard in the sandy ridge citrus area of Florida (courtesy Florida Department of Citrus).*

season (7,445,000 metric tons), was 2.57 times larger than that of the entire nation's production of apples (2,900,000 metric tons) (USDA Agr. Stat., 1977).

By 1976 Florida's production of oranges had increased to 186,621,000 boxes (7,226,000 metric tons) of which 92.5 percent was processed. Of the processed fruit, 82.1 percent was made into frozen concentrate, 3.9 percent was canned single strength, and 14.0 percent chilled. Of the 49,083,700 boxes of grapefruit, 58.5 percent was processed: 31.7 percent was made into frozen concentrate, 50.8 percent canned single strength, and 17.5 percent chilled (Florida Department of Citrus, correspondence).

The principal Florida varieties of orange are the following: in early season, Parson Brown and Hamlin; in midseason, Pineapple, Homosassa, Jaffa, and Temple, the latter probably a natural hybrid between the sweet orange and mandarin, and of outstanding flavor; and among late varieties, Valencia and Lue Gim Cong, the latter named for a Chinese who developed the variety. The grapefruit varieties are the Marsh or Marsh's Seedless, Thompson or Pink Marsh, with pink flesh instead of the characteristic yellow, and the Foster, also with pink flesh. The principal mandarins are the Dancy Tangerine and the Satsuma. Modern growing and merchandising methods favor standardization on as few varieties as possible, preferably seedless or "sparsely seeded." For this reason, Parson Brown, Homosassa, Jaffa, Thompson, Foster, and Satsuma are rapidly disappearing.

In the light, sandy soils where most Florida citrus production takes place, not only was a suitable rootstock required, but minor elements, such as magnesium, manganese, copper, and zinc had to be supplied along with the usual nitrogen, phosphorus, and potash. Some minor elements such as iron, zinc, and manganese are most effectively applied as a foliar spray. Fertilization, cultivation, and the application of large amounts of sulfur for rust mite control causes the soil to become very acid. Dolomite is added to reduce acidity and to add magnesium to the soil. Yet the pH must not rise above 6.0, so as to avoid tying up zinc, manganese, and copper in the soil. The introduction of many beneficial insects, greater sophistication in pest-management programs, and decreased need for superior surface appearance of fruit, now that 92.5 percent of the oranges and 58.5 percent of the grapefruit are being processed into concentrates and juices, have all combined to greatly reduce the volume of pesticide use in Florida. Nowhere have continuous technology inputs had a greater influence in the development of an agricultural industry than in the case of the great citrus industry of Florida.

### Humanized Nature

Florida's central highlands, a rolling terrain of orange groves interspersed with many lakes, could be used as a prime example of "humanized nature"—an area of man-made beauty exceeding that of nature. It is important that the area maintain that character, soaking up rainfall and recharging the Floridian aquifer, in which respect it is said to be even more important than the "Green Swamp" farther south (Carter, 1974).

One might at first expect a state with 50 to 60 inches of annual rainfall to have a surfeit of water. Yet, except for the panhandle area in the northwest, Florida has no rivers bringing water to its land from distant regions. The peninsula must depend for water on aquifers supplied by its own rainfall, an environmental problem of great complexity. The southern part of the peninsula is the most vulnerable. There the development of agricultural land by draining wetlands has caused soil subsidence, salt-water intrusion, muck fires, and loss of wildlife and natural ecosystem. Agriculture has sometimes made exorbitant demands on water in areas where the supply is limited and where pollution by agricultural runoff is especially undesirable. Yet even in heavily settled areas of Dade County, where tourism is the principal industry, the county's Metro Commission advises that some farmlands be devoted to specialty crops such as avocados, limes, mangoes, and winter vegetables to serve as "green belts" to "set off one community from another and help keep overall population densities within desired levels" (Carter, 1974).

## CATTLE

As previously related, the early settlers in the South had found the region very favorable for raising cattle, and cattlemen always preceded farmers all the way from Virginia to Texas. Yet during the depth of the depression in 1932, there were only 8,200,000 head of cattle in the South, a number that increased to 18,260,000 head within three decades (Clark and Kirwan, 1967). In the eleven states entirely within the South, as indicated in figure 2, thus excluding eastern Texas and southeastern Oklahoma, there were 21,437,000 cattle and calves in 1976, or 18.5 percent of the nation's total of 122,896,000 head (USDA Agr. Stat., 1977).

More important than increased numbers, cattle quality had been greatly improved owing to improved breeding. Effective control measures were worked out for Texas fever ticks and other cattle pests, including the screwworm, which was controlled by the release of sterile male flies according to a procedure worked out by entomologists of the USDA.[7]

Particularly in the piedmont and upper coastal plains of the South (fig. 3, P), soil that had been eroded and exhausted from a century of cotton and corn cropping was brought back into production chiefly as grass and legume meadows on which beef and dairy cattle could feed. The abundant rainfall and yearlong grazing season of this region made an ideal pasture country. Increase in number of beef cows in the Southeast took place at a rate that was twice the national average. The best meadows there could carry a beef cow and her calf through a year on from one to three acres, compared with twenty acres in the sand hills of Nebraska and from about 100 to 150 acres in the drier sections of the intermontane West (Higbee, 1958).

USDA and state experiment stations devoted much attention to the growing of grasses and proved that sufficient grass could be grown during the hottest summer months in the South, contrary to the previous prevailing opinion (Clark and Kirwan, 1967). Nevertheless, many pastures in the South, improved with considerable expenditure of capital, are now irrigated. The farmer finds that it costs less to irrigate these pastures than to add new acreage that would be dependent on natural precipitation. This is especially true where low-cost electrical energy, resulting from development of power plants in the area, can be used to operate the irrigation pumps in community projects.

Along with irrigation, heavy fertilization and rotational grazing make the pastures ideal for intensive livestock husbandry, especially in view of the yearlong pasturage. When pastures are amply supplied with fertilizer, the protein, phosphorus, lime, and vitamin content of the forage is increased (Higbee, 1958).

## THE BROILER INDUSTRY

There are two kinds of chickens: one kind specialized for egg-laying and the other for rapid meat production. The latter are called "broilers." They are young chickens up to 3 pounds, a weight they may reach in 8 weeks while consuming as little as 1.8 pounds of feed per pound of gain. They are by far the most numerous type of chicken—over 3 billion were raised in the United States in 1972, compared to 302 million hens and pullets intended for egg-laying. The broiler industry is concentrated in the South, the leading states being Arkansas, Georgia, Alabama, North Carolina, and Mississippi. The Delmarva Peninsula of Delaware, Maryland, and Virginia, where the broiler industry originated, is still important in production, and if the figures for the three states are combined, the sum is exceeded only by Arkansas, Georgia, and Alabama (USDA Agr. Stat., 1977).

### Breeding

The young chicken may be "the most researched animal in this much-researched world" (Bird, 1968). From the standpoint of the consumer, it may also have the most

[7]Screwworms are reared on cheap meat and sterilized by exposure to gamma radiation. Released by aircraft over an entire infested region, the overwhelming numbers of sterile males compete with the relatively few wild males of the region in mating with wild females. The mated females do not mate again, and they lay infertile eggs. Repeated releases of sterile males result in eradication of the pest. (Knipling, 1960*a*, *b*).

to show for this intensive research. In 1938 there was only one major broiler-producing area, the Delmarva Peninsula. At that time 82 million broilers were produced in the United States and there was grave concern about "overproduction" and "low prices." Live broilers were then selling for 17 cents a pound. But in 1966, 2.5 billion broilers were sold at 15 cents a pound (Bird, 1968). Consumption of poultry, which had been around 16 pounds per person annually from 1910 to 1940, rose to 50 per person in the early 1970s (Shover, 1976).

During World War II in particular, there was a big increase in consumption of broilers, encouraging the leaders in the industry to believe that there were still greater opportunities for expansion if better-quality birds could be produced. A series of contests was held, supported by the Great Atlantic & Pacific Tea Company, in which breeders submitted samples of eggs. These were hatched at a central point and the chicks were reared to broiler weight and graded as to various criteria of marketability. The New Hampshire breed provided everything except "conformation" (shape or build). The goal was a chicken with broad breast and thick drumsticks. These were features for which the Cornish was noted. But the Cornish grew slowly and inefficiently, feathered slowly, laid poorly, and had low fertility and hatchability. Some breeders believed the best approach to an ideal combination of growth rate and conformation could be obtained by improving the White Plymouth Rock. But the first national contest, in Delaware, was won by Charles Vantress of Live Oak, California, with his Cornish-New Hampshire crosses. They again triumphed in the second contest, and by 1940 the Vantress Poultry Breeding Farm became one of the largest international poultry-breeding operations.

Broilers are now produced by crossing two strains having different characteristics. The female parent must lay well and brood well to produce broiler chicks efficiently. Growth rate and conformation are also important. For the male parent, growth rate and conformation are the primary consideration, with egg production and hatchability secondary. The female lines are derived mostly from White Plymouth Rocks and the male from Cornish and New Hampshire.

## Diet

Diet was found to be another important consideration. Research on vitamin D at the University of Wisconsin in the 1920s made production of poultry independent of sunshine. Later, research at various land-grant colleges and the USDA led to the use of synthetic vitamins to reduce the levels of wheat by-products and increase the corn, thus raising the energy content of the feed and putting some fat and yellow color into the skin. Research also revealed how soybean meal could be supplemented with vitamins and minerals to make it suitable as a source of protein. Research in the USDA laboratories at Beltsville, Maryland, showed that if an unidentified vitamin from cow manure was added to soybean meal, the combination was as effective as animal proteins. In 1948 the laboratories of Merck and Company isolated vitamin $B_{12}$ and also proved it to be identical with the "cow manure vitamin." They proceeded to produce it in large quantities.

Thomas H. Jukes and his coworkers at American Cyanamid's Lederle Laboratories discovered that the residues of chlortetracycline (Aureomycin®) fermentation contained vitamin $B_{12}$. Whereas vitamin $B_{12}$ itself increased the growth of chicks by 19 percent in a 25-day period, chicks fed the antibiotic residue containing vitamin $B_{12}$ grew 56 percent more rapidly than the controls. The fermented residue that contained vitamin $B_{12}$ increased chick growth 37 percent more rapidly than the vitamin alone. Field trials with the antibiotic on chickens, turkeys, pigs, calves, and sheep were even more spectacular. Researchers also found that the more the animals were exposed to stress, adverse climate, and disease, the greater the improvement in growth, livability, and feed conversion induced by the antibiotics in the diet (White-Stevens, 1975).

Antibiotics suppress deleterious microorganisms. Some of these microorganisms can readily invade massed animal groups. Whereas formerly it was not possible to raise livestock crowded together in large numbers without the risk of great losses from various diseases, antibiotics have made it possible to do this. Flocks of broilers could be increased from around 3,000 to 5,000 to as many as 60,000, and cattle in feedlots from a few hundred to tens of thousands (White-Stevens, 1975).

Low levels of antibiotics in feeds were found to improve growth and conversion of feed to meat, saving millions of tons of broiler feed; yet the most sensitive methods detected no antibiotic in the meat (Bird, 1968).

### The Broiler Factory

In the United States the farmer is generally not an independent producer of broilers. A financial sponsor, generally a feed supplier or operator of a slaughtering plant, provides the chicks, materials for the enterprise, and technical advice, while the farmer provides the housing and labor. More than any other agricultural enterprise, the production of broilers has become a vertically integrated,[8] assembly-belt, factory-type industry. Poultry are moved through the factory in batches and are kept in a strictly controlled and crowded environment.

## FISH PONDS

In 1927 a group of Auburn University faculty members became interested in the development of farm ponds in which fish could be raised to provide recreation and a welcome addition to the family income. Emerging as a leader in this group was an entomologist, H. S. Swingle. Largely through his enthusiasm, vision, and dedication to this project, fish farms were put on a scientific basis. Organic and/or inorganic fertilizers were added to increase the amount of algae in the water. Much research was done on feed mixes. Swingle described procedures for growing channel catfish for food, based on his research, that were to become the basis for much of the present-day channel catfish industry.

Auburn became a widely recognized center of research on farm-pond fish culture. Fish-farming spread rapidly through Alabama and the nation. Students and research workers came to Auburn from all over the world to study water-farming methods. Swingle became project director of a group sponsored by the U.S. Agency for International Development. He and his staff visited more than twenty countries to train scientists in producing food fish. USAID and Auburn University established an International Center for Aquaculture at Auburn in 1970, with Swingle as its first director (Shell, 1975).

[8]Vertical integration is "the combination of two or more successive stages of production and/or distribution and the ownership and/or control of one firm." It can be accomplished through contracts ("contract farming") or by ownership, usually by profit-type firms and cooperatives which own two or more successive stages ("ownership integration") (Roy, 1963).

## TENNESSEE VALLEY AUTHORITY (Wengert, 1952)

The Tennessee Valley Authority (TVA) is a federal agency created in 1933 to develop the Tennessee River and its tributaries, a river basin of 40,000 square miles covering parts of 7 states. TVA demonstrated the feasibility of a multipurpose approach (flood control, navigation, and hydroelectric power) to stream development in the nation's fourth-largest river in streamflow. After congressional bills providing for governmental operation of the project were vetoed by President Calvin Coolidge in 1928 and Herbert Hoover in 1931, Franklin D. Roosevelt not only endorsed the plan but recommended an even broader plan to Congress that included reforestation and improvement of the economic and social well-being of the people of the vast river basin. The last of nine multipurpose dams was completed in 1944. A great inland waterway with a 9-foot channel 650 miles in length was created by a system of high-lift locks connecting the lakes.

TVA helped reforest more than a million acres of eroded land and valuable watershed, and promoted better farming methods and conservation techniques in a region of formerly exhausted soil and depressed economy. It developed a successful mosquito-control program and cooperated with other agencies in promoting better health services.

Before TVA, the Tennessee Valley contained a mixture of low-level subsistence farms and highly organized commercial family-type enterprises rivaling the best that might be found anywhere in the country. The small farm might consist of an acre or two of tobacco and a few acres of corn, several hogs, and perhaps a cow. Income might be supplemented by seasonal employment in the forests or in the industries of the area. It was largely the subsistence farms in overpopulated rural areas that were the source of the many workers who migrated to northern industrial centers seeking employment in 1933 when TVA began its work.

The farm systems were often not adjusted to the soils and other factors in the environment. The row crop culture encouraged erosion and made a reversal of the declining economy difficult. TVA's objective was to convert the less efficient cotton farms to various types of diversified farming. Often hay, legumes, and small grains were substituted in a suitable rotation plan. The need for terracing and strip cropping had to be determined, as well as fertilizer and lime requirements. TVA provided the economic stimulus in the region as well as the fertilizers to make the desired agricultural conversion possible.

The Tennessee Valley happens to be rich in phosphates. When munitions requirements ceased, TVA was the major producer of concentrated phosphate in the country and one of the largest producers of ammonium nitrate for fertilizer purposes. But mineral fertilizers alone could not solve the problem. Humus, organic matter, and the physical condition of the soil are as important to fertility as the application of chemical plant nutrients. There was need for sound farm systems that would include crop rotations, cover crops, and similar conservation practices—systems that contrasted sharply with the traditional cash-and-row-crop farming of the South. The amount of readjustment on the part of the farmers and the challenges presented to the extension specialist can well be imagined. It might take five or six years to substantially eliminate row crops, substitute small grains, legumes, and grasses and learn about lespedeza and alfalfa, poultry and cattle, and about tractors, seeders, threshers, and hay driers. The success of the immense project is obvious to anyone visiting the area today.[9]

[9]On the debit side, the reservoirs took much good bottomland out of agricultural production. One southern congressman quipped that TVA had indeed solved the flood-control problem: "all the land" was flooded, obviating the need for control.

## SOIL EROSION

Soil erosion is perhaps the most important of all environmental degradations. The earliest and most spectacular manifestation of its potential for irreversible destruction became evident in the South. In 1928, H. H. Bennett and W. R. Chapline (1928) had this to say on the subject:

> A much lighter rain than formerly now turns the Tennessee River red with wash from the red lands of its drainage basin. Added to the severe impoverishment of a tremendous area of land throughout this great valley, and its extensions southward into Georgia and Alabama and northward into Virginia, are the gullied areas, which are severely impaired and completely ruined by erosional ravines that finger out through numerous hill slopes and even many undulating valley areas. Field after field has been abandoned to brush, and the destruction continues.
>
> Much erosion of the same type has taken place over the smoother uplands of south-central Kentucky; that is, in the rolling parts of the highland rim country; over much of the Piedmont region, and through many parts of the Appalachian Plateau. Land destruction of even worse types is to be seen in the great region of loessial soils that cover the uplands bordering the Mississippi and Missouri Rivers and many of their tributaries, from Baton Rouge, Louisiana, northward. Numerous areas, small and large, have been severely impoverished and even ruined in the famous black lands of Texas.

The above authorities referred to an experiment at the Missouri Agricultural Experiment Station where, over a 6-year period, an average of 41.2 tons of soil were annually washed from an acre of land plowed 4 inches deep on a 3.68 percent slope.[10] From a grass-covered area of the same slope and soil type, less than 0.3 ton of solid matter was removed each year. On the latter, 88.4 percent of the 35.9-inch rainfall was retained compared with 68.7 percent of the rainfall on the cultivated slope.

At an erosion station in the piedmont area of North Carolina, during a year of 35.6 inches of rainfall, an uncultivated but bare plot lost 24.9 tons of solid matter per acre per year, whereas on the same slope the erosion from grassland amounted to only 0.06 ton to the acre. Grass held back 415 times more surface soil than was retained on untilled bare ground and 215 times more soil than on cotton plots on the same slope. The uncultivated plot retained 64.5 percent of the rainfall, the cotton plot 74.4 percent, and the grassland 98.5 percent.

Nationwide, topsoil is being lost at an estimated average annual rate of 12 tons per acre but is produced by natural processes at the rate of only 3 tons per 2.5 acres of cultivated land per year. About 4 billion tons of sediments are delivered to waterways in the 48 coterminous states each year (NRC, 1974; Pimentel et al., 1975, 1976; Schwab et al., 1966). About three-fourths of it is from agricultural lands (USDA, 1968*a;* Wadleigh, 1968; Beasley, 1972). Erosion caused by the wind is also important, estimated to cause a loss of about a billion tons of soil per year. Of the soil eroded by water, about a billion tons end up in the ocean and the remaining 3 billion tons settle in reservoirs, rivers, and lakes (NRC, 1974), requiring the dredging of about 450 million cubic yards of sediment from U.S. rivers and harbors annually and materially reducing the useful life of reservoirs (USDA, 1968*a*).

[10]An acre-inch of soil weighs about 150 tons. Topsoil on cropland averages about 7-8 inches in depth (Bennett, 1939). Soil may be formed at the rate of 1 inch (2.54 cm) in about 30 years under conditions of ideal soil management, but perhaps 100 years under average agricultural conditions (Pimentel et al., 1976). Soil is formed at the rate of 1 inch in 300-1,000 years under natural conditions (Bennett, 1939; Gustafson, 1937; Olivers, 1971).

*Fig. 13. Stubble mulching, leaving the ground loose and protected by residue that can catch and retain moisture (courtesy Deere and Company).*

Particularly in hilly land, such as much of the farmland of the South, continual planting of crops like corn, cotton, and tobacco, known as "row crops," led to erosion and soil exhaustion to such an alarming extent that some farmers, under the guidance of the Soil Conservation Service and agricultural extension specialists, have gone back to integrating the rearing of livestock with plant cropping. Sloping land is left in pasture, and manure from grazing livestock is used to enrich the level land, the latter often only a small percentage of the farm. Even gently sloping land can be conserved and improved by contour plowing, strip cropping, stubble mulching, and grassed waterways.

In contour farming, plowing, seedbed preparation, planting, and cultivation follow the contour of the land, rather than up and down the slope. This in itself reduces erosion and gullying. Contour farming is most effective in reducing erosion when it is combined with strip cropping. Strips of row crops such as corn, soybeans, peanuts, or cotton might be alternated with strips of close-growing perennial grasses and legumes. The latter serve to retard runoff. Figures 8 and 17 provide good aerial views of contour farming and strip cropping. Stubble mulching or mulch tillage is a soil-management practice that leaves a large percentage of crop residue on or near the surface of the ground (fig. 13). The mulch breaks the fall of raindrops, impedes surface flow of water, and helps to maintain an open soil structure to allow water to penetrate rather than run off. Mulches of leaves, straw, hay, sawdust and other organic materials can be hauled into the field. Mulch tillage on contoured corn land has been known to reduce soil loss over 90 percent.

Other soil-conserving measures are crop rotation, the use of cover crops on otherwise bare soil surfaces, fertilization and liming, and maintenance of an adequate content of organic matter. Organic matter is an important soil ingredient. It provides the structure or tilth that is required for greater fertility. Among organic constituents of the soil is the black colloidal material known as humus, which is highly hydrophylic and is responsible for the crumb structure of soil so necessary for fertility. Thus, organic matter is one of the factors that facilitate maximum penetration of water into soil and minimize runoff and erosion. Organic matter in the soil also increases the availability of soil phosphate to plants (Dalton et al., 1952).

Some authorities believe that the most promising approach to soil conservation may be Section 208 of the Clean Water Act of 1972. That section requires each state to submit to the Environmental Protection Agency (EPA) by November 1, 1978, an enforceable plan for abating pollution from all identifiable sources. Adequate soil-conservation practices could be made a condition for participation in USDA financial assistance programs such as farm loans, crop insurance, disaster relief, and possibly price supports (Carter, 1977*a*).

## No-Tillage Systems

One of the most encouraging developments in the last two decades in the effort to reduce soil erosion has been reduced tillage and even no tillage. The no-tillage system started with the American Indians. When they planted corn, for example, they merely made a hole in the ground with a pointed stick, put in a few seeds, and covered them with soil. Indian women were expected to keep the field completely free of weeds. Indians could not understand the white man's penchant for churning up the soil in a variety of ways, particularly in view of the fact that his crops were no better than theirs. We now know that tilling the soil often accomplishes no more than controlling weeds, but has the disadvantage of increasing erosion hazard.

Tillage increased in intensity until by the 1920s or 1930s a corn grower in the Midwest might plow, disk three or four times to break the clods, then drag or harrow to prepare a fine, firm seedbed. At the time of crop emergence he might rotary hoe or lightly till again. After crop emergence, cultivations continued until the stalks grew too high for most equipment. Then special equipment, the Georgia Stock, could still be used for cultivation, even after the corn was shoulder high. Weeds still remaining were then hoed or pulled out by hand (Triplett, 1976).

The discovery of selective organic herbicides (weed killers) was the first step in breaking the farmers' addiction to excessive cultivation. Herbicides did not initially eliminate moldboard plowing before planting, and only moderately reduced the number of cultivations believed to be required. Yet as early as 1952, herbicides were used to completely eliminate tillage in row crop production (Davidson and Barrows, 1959), albeit with several applications, including several complementary types of herbicide. By 1959 a single herbicide, atrazine, was found to be versatile enough to control all weed species. Soon no tillage whatever was reported successful by researchers in New York, Virginia, and Ohio. The "no-tillage" system, also called "zero-tillage" and "sod-planting," has continued to be tested and has been extended to other crops, including soybeans, small grains, cotton, peanuts, potatoes, and forage crops. The conclusion has been that some soils should be tilled under certain conditions while others should not (Triplett, 1976).

No-tillage not only reduces time and labor demands, increasing labor productivity, but reduces erosion hazard on sloping sites as much as a hundredfold (Triplett, 1976; Carriere, 1976). The no-tillage system also decreases evaporation, resulting in a greater water reserve and helping to carry a crop through periods of short-term drought (Blevins et al., 1971). It facilitates double-cropping, thereby increasing production per acre

(Young, 1973; Lewis, 1976). In 1975, nearly 6.5 million acres were planted with no-tillage and nearly 48 million acres were minimum-tilled in the United States (Lessiter, 1975). A prediction was made in a USDA technology assessment that by the year 2000, 80 percent of the nation's crop acreage will be planted with reduced tillage and 65 percent of feed grains, wheat, rye, and soybeans will be produced by no-tillage.[11]

In no-tillage as practiced in the United States, usually a narrow slot or trench is opened just wide and deep enough to cover the seed properly. This may be done in soil with previous crop residues such as those of corn, sorghum, soybeans, or peanuts, or in soil with a rye cover crop or small grain stubble or in perennial grass sods. The "planter" used for this purpose may be equipped to place fertilizer in the row to the side of the seed, although fertilizer can also be broadcast on the soil surface prior to planting.

The tradition of thoroughly incorporating fertilizers and lime into the soil has been challenged by the no-tillage system. The greater amount of water in the soil under no-tillage, particularly in the upper 3 inches results in a high concentration of plant roots near the surface and more efficient uptake of surface-applied nutrients (Lewis, 1976).

No-tillage or reduced-tillage methods of crop production have sometimes increased insects and other pests, particularly armyworms, cutworms, stalk borer, European corn borer, corn rootworms, white grubs, slugs, and mice, but pest-management methods, coupled with regular field observations, are available to cope with these problems (Gregory and Musick, 1976). On the other hand, a program of no-tillage and no insecticide treatment resulted in less damage to corn seedlings from the lesser cornstalk borer (*Elasmopalpus lignosellus*) than any of the insecticide treatments plus conventional tillage. This was believed to be due to increased soil moisture, a factor tending to reduce infestation (All and Gallaher, 1977). Observations made in the Southeast have seldom revealed any difference in crop yield caused by differences in insect infestation resulting from the various tillage systems (R. N. Gallaher, correspondence).

## Hardpan

"Hardpan" or "plowpan" is a layer of almost brick-hard soil that is several inches thick and 6 to 14 inches below the soil surface. Some soils are much more susceptible than others. The coastal plains of the South all the way from Louisiana to Delaware are especially susceptible to damage from this source. As with many environmental obstacles, hardpan is the result of modern technology (use of heavy agricultural equipment), the solution for which depends not on abandoning technology but by directing it in an environmentally sound direction.

Hardpan reduces the growth and productivity of crops by confining their root systems to the limited area above the impermeable layer. In an effort to solve this problem, representatives of the USDA, the Extension Service, a herbicide manufacturer and a manufacturer of equipment have cooperated to develop what they have called the "no-tillage-plus" system, the "plus" being in-row subsoiling.

The way the "no-tillage-plus" system works has been impressively demonstrated, for example, on a farm in southern Alabama (Beeler, 1977). Here a crop of winter rye is planted in the fall. It is grazed all winter by dairy cows. In the spring, the rye and any weeds that may be present are killed with an application of a "tank mix" of paraquat

[11]No-tillage also happens to be one of the features of the worldwide Appropriate Technology ("A.T.") movement. In Nigeria a low-growing legume provides a constant mulch in fields in which maize is planted with a jab planter. As many as seven crops a year have been produced without the erosion that follows plowing (Ellis, 1977).

and atrazine, the latter for prolonged residual herbicidal effect. Corn is seeded into a slit made under the mulch by means of specially designed equipment called a "Super-Seeder." This is located in front of the planter units. A notched colter (cutting tool) cuts through the cover crop or crop residue (mulch). Next is an in-row subsoiler that can rip as low as 18 inches to open up the hardpan. This in turn is followed by a "spider" wheel that fluffs up the sides of the subsoil slot, filling it to just the right level for seed placement. An attached sprayer applies the herbicide. The next equipment to enter the field is the harvester. On the same farm, soybeans are planted in wheat stubble and millet in rye according to the no-tillage plan.

The advantage of the "no-tillage-plus" system is that it saves on labor and fuel, eliminates some equipment, and precludes wind and water erosion. Both the mulch and the subsoiling are important features. Subsoiling allows the crop roots to go through the slit and spread out in deep moist soil below. This is clearly shown photographically in some USDA soil profiles. The no-tillage-plus system is said to have increased corn production from 50 to 100 bushels in the South and from 100 to 150 bushels on many Corn Belt farms. In the severe drought of 1977, when 30-bushel corn was common in the South, 100-bushel corn was grown in the same kind of soil with the no-tillage-plus system. According to Beeler (1977), the system "is likely to revolutionize the South much like irrigation opened up the deserts of the West."

## UNITED STATES SOIL CONSERVATION SERVICE

At the Cotton States International Exposition at Atlanta, Georgia, in 1895, there were three large models of the same farm property designed to graphically show the destructive effects of erosion and how to go about renovating the land. Contour cultivation and strip cropping were clearly illustrated. These models had been constructed under the direction of Bernard E. Fernow, a German immigrant who became chief of the Division of Forestry, USDA. Like his fellow countryman Interior Secretary Carl Schurz, he was influential in promoting in America the ecological concepts that were then, as they are today, so well demonstrated in the charming rural landscape of his native land. The same year, Fernow (1896) reproduced photographs of the models in an article titled "The Relation of Forests to Farms" and published in the 1895 *Year Book of the United States Department of Agriculture,* nowadays known simply as *Yearbook of Agriculture.* The models, viewed by thousands, demonstrated (1) the disastrous effects of misuse of a section of deforested agricultural land, (2) the land being ecologically reconstituted, and (3) the land fully restored to permanently productive use.

Fernow's models dramatically showed that land misuse was not due to inadequate knowledge of the basic rules of land conservation. Today adjustments in practice have to be made, of course, to be compatible with modern multirow equipment and monocropping. No agency of government has been so instrumental in helping farmers put basic rules into practice as the Soil Conservation Service (SCS). The SCS is a USDA bureau established by the Soil Conservation Act of 1935 to prevent soil erosion, establish a permanent and balanced agriculture, and reduce the hazards of floods, droughts, and siltation. The bureau was headed by Hugh H. Bennett, one of the world's leading conservationists. Every part of the nation was benefited by this valuable public service, but nowhere, with the possible exception of the Dust Bowl of the Great Plains, was the benefit greater than in the South.

In 1973 there were nearly 3,000 soil-conservation districts, covering about 93 percent of the land in farms and ranches, with more districts being organized every year. Nearly 1.5 million farmers and ranchers had SCS men assigned to soil- and water-conservation districts, helping them make conservation plans. Soil-conservation districts furnish help only to those who request it and are interested in carrying on a conservation program. As the soil conservationist and the farmer go over the farm, the former will have with him a soil-survey map showing the different soils on the farm, each classified according to its capability for use and need for treatment. The farmer receives the federal share of the cost (50-75 percent) as he completes each conservation step. Needs vary in different states and counties. In a particular county there might be a high priority for permanent conservation practices such as terraces, grassed waterways, and stream diversion. Cost-sharing for such projects might be at the 75 percent level, whereas for other practices, such as contour strip cropping, pasture management, subsurface drains, farm ponds, and crop residue management it might be 50 percent. The SCS will furnish assistance to landowners through a referral from the Agricultural Stabilization and Conservation Service (ASCS). This agency takes care of the payment for a particular practice while SCS furnishes the technical layout and is responsible for certifying that the practice meets the required standards and specifications (W. M. Archibald, correspondence).

In addition to helping the farmer regarding the *use* of each field the SCS conservationist helps him to decide on how he will *treat* each field to get the desired results. On a bottomland field the farmer may decide to install tile drains, smooth the surface, lime and fertilize, and follow an intensive cropping system. If a sloping field is to be used as cropland, the farmer may decide to make grassed waterways, build terraces to establish contour strip-cropping, follow a conservation cropping system involving grass-legume meadow and row crops, and lime and fertilize for good crop yields. Some fields might be planted to various forage plants that could be used as much of the year as possible for grazing livestock. Such fields might be limed and fertilized to the extent appropriate for pastures. Stock-water ponds or springs might be developed and brush and weeds controlled. If woodland property is involved, conservation treatment might include protection from fire and livestock damage, weeding, thinning, and intermediate harvest cutting.

The soil conservation district supervisor arranges to give the farmer the technical help needed to put the plan into operation and can help him get equipment, planting stock, or other materials that may be needed. The SCS conservationist records the plan on a map and in narrative form and the farmer is given a copy. Changes in markets, prices, or other circumstances might make it desirable for the farmer to change the conservation plan and, in that case, technical help from the SCS conservationist is available. SCS plans have not only resulted in higher sustained incomes to farmers but have served to protect and improve the land resource base of the nation (USDA, 1973*a*).

The Soil Conservation and Domestic Allotment Act of 1936 attempted to secure reduced production of surplus crops by payments for improved land use and conservation practices. The Agricultural Adjustment Act of 1938 stressed balanced abundance and provided for loans, acreage allotments, and marketing quotas for "basic" crops, and aimed for "parity" prices and incomes for farmers. Here again, soil conservation was a major objective. Although frequently modified, this act remained the nation's basic agricultural price-support and adjustment law into the 1970s (Rasmussen, 1975*a*, p. 1995).

### Who Should Pay for Conservation?

The watersheds that sustain municipal and industrial uses, and are the source of such insidious problems as siltation and such calamitous tragedies as flooding, are mostly in the nearly three-fifths of the nation's area that is privately owned rural land, a high percentage being in crops, pasture, range, or other nonforest agriculture. Conservation work that is being done is paid for predominantly by the owners and operators of farm- and ranchland. Of an annual conservation investment of possibly $1 billion in 1970, about a third was paid by the federal government; most of the remainder was paid by owners of the land. But even twenty years ago the USDA estimated the need for an investment of $2.4 billion a year on agricultural lands alone to do the required amount of conservation work. With farmers' net incomes only about three-fourths of the level of American nonfarmers, even in years of relative farm prosperity, it is not realistic to expect farmers to shoulder a larger share of the conservation burden. The required funds will have to come from somewhere "...whether they originate as a part of the food dollar or as a part of a Federal or State or local budget, or a combination of these. The larger society should not expect to be spared a concern and a financial responsibility for its own survival and well being. Nor will it be so spared" (Looper, 1970).

## ECOLOGICAL RESURGENCE IN THE SOUTH

Many of the types of soil under pine woods were soon eroded and exhausted by row-crop farming. Eventually much of this land was again taken over by the forest, and under good silvicultural practice pine timber has now become a profitable crop. The South is now often referred to as "the nation's wood basket."

Ten of the thirteen species of pines that are indigenous to eastern North America are called "southern pines." They provide about a fourth of the timber supply of the thirty-seven states east of the Rocky Mountains. The four major species—loblolly, shortleaf, longleaf, and slash—comprise 90 percent of the total inventory of 64 billion cubic feet. Most of the reforestation is being done with loblolly and slash pines (White, 1976). Southern pines are the source of an expanding multibillion-dollar forest products industry (Sternitzke and Nelson, 1970).

Within the well-managed forest, thinning of dense growth increases sunlight, forage, browse, and wildlife. In the South, wildlife was also favored by the end of cotton monoculture and subsequent diversification of plant crops, for the new plant species provided the grasses, seeds, and fruits that lured back the many species of wild animal life that originally had been so abundant in the South. Also the interspersed forest, pasture, cropland type of country is favorable to wildlife, for it provides the "edge effect" favored by many species.

In the South, as elsewhere in the nation, there remain many areas that could benefit ecologically and economically from reforestation. With such large numbers of teenagers unemployed, programs for educating, training, and employing them in the skills of forest work such as tree-planting, fire control, and trail-building would appear to be a legitimate concern of various levels of government. Some of the commercial and recreational timberlands of today were recovered from erosion in the Civilian Conservation Corps (CCC) era under President Franklin D. Roosevelt.

In March 1977, President Carter proposed to Congress a resurrection of the CCC as part of a $1.8 billion program to put youths to work. Called the National Young Adult Conservation Corps, the program would employ jobless Americans aged sixteen to twenty-four in maintaining and improving forests, parks, and recreation areas (Armstrong, 1977).

It is in the temperate regions of the earth with adequate rainfall that nature has the best chance of restoring itself when given the opportunity. In many areas of the South, confidence has been restored in the harmony that can exist between man and land. The great current migration of people from the North's congested cities to the South is a measure of the latter's ecological, as well as economic, resurgence.

The great influx of Northerners will be a mixed blessing unless sound policies for growth and land use are established. A preview of the problem and possible solutions were provided in an early southern testing area, Florida's Dade County, where Miami is located—an area already overpopulated. The Metropolitan Dade County Planning Department seeks to restrain population growth, setting population limits based on the county's "carrying capacity" and fostering activities expected to employ people already living in the county rather than activities likely to further stimulate immigration. The newcomers who added so much to the severity of the growth problem became interested in preserving and enhancing the special quality of life that attracted them to the subtropical paradise. A political majority emerged to foster concepts of land use and growth control that took into consideration humanistic, esthetic, and ecological considerations as well as economic interests. Dade County's Committee for Sane Growth, which has had opportunities to demonstrate its political muscle, warns that "the more people who come to share our good life, the less good it becomes" (Carter, 1974).

Perhaps there is some consolation to be gained from the observation that even politicians, the last holdouts for the old ways, are beginning to voice the "paradise lost theme." Luther Carter in his *The Florida Experience* (1974) quotes one who reflected that Florida's once paradisiacal Dade County "was a pleasant place to live in the 1930s. even though we were all broke. Now we are affluent and nobody wants to live here." Senator Gaylord Nelson of Wisconsin, when asked if people should be denied water if the choice had to be between people and the Everglades National Park, made what would appear to be the obvious reply: if people know they cannot have the water needed by the park, towns and industries that depend on that water will not be developed.

# 5 The Midwest

*In every great crisis a peculiar dependence has had to be placed upon the farming population. . . . We cannot afford to lose that preeminently typical American, the farmer who owns his own land.*

—Theodore Roosevelt

The difficulties experienced by the early settlers in migrating across the Appalachians via the Ohio River and the Wilderness Trace during the last three decades of the eighteenth century were recounted in the preceding chapter. Nevertheless, by 1840 the populations of the settled states of what was then called the West had already reached the following numbers: Ohio, 1,520,000; Indiana, 686,000; Illinois, 476,000; Missouri, 384,000; Michigan, 212,000; Iowa, 43,100 (mainly along the Mississippi River); and Wisconsin, 31,000. Thus more than 19.5 percent of the nation's 17,069,000 people were living in the mentioned states. Yet farming west of the Alleghenies was mostly of the pioneer type—with abundant land but limited markets and capital. Only a small area around Cincinnati and in the Bluegrass region of Kentucky had a well-developed agriculture.

The settlers had followed the rivers and settled in the wooded regions, avoiding the larger, open prairies. By 1840 many of the small prairies of Ohio and Indiana adjacent to rivers and timber had already been settled, as well as the "oak openings"—sparsely timbered country that had probably been thinned out by fires set by Indians. Wet prairies, or "marshes," frequently occupied the low, level spots, producing a rank growth of wild grasses and enabling the settlers to support their stock from the beginning with little labor or expense.

As the settlers proceeded west, they found increasingly larger open areas. By the time they reached the northern and east-central portion of Illinois, they found land to be largely prairie, with timber usually along the rivers and small streams or scattered through the prairie in groves (Gerhard, 1857).

Settlers from the wooded regions of the East and South and from foreign countries at first chose wooded areas in the West. Time was required for the settlers to adapt to prairie land far removed from woods. The reason was in part psychological; wooded areas resembled those from which they came. Prairie land was most likely to be used along the edge of woodland, the latter providing lumber for buildings, fencing and fuel, mast for hogs, and protection from the fierce, cold winds of winter. In the forest, water was usually abundant, whereas on the prairies deep wells might have to be constructed, with pick and shovel. Frequently prairies were avoided because they were believed to be conducive to fever and ague.

A combination of woodland and prairie provided the advantages of both. The latter provided an abundant range for cattle without the slow and backbreaking task of felling trees. A supply of winter forage needed only to be gathered and stored away. The prairie provided tillable land, although the thick sod was very difficult to plow until the advent of the steel plow, which did not become widely used in such regions until about 1850. The steel plow and well-drilling machinery were of great importance in the settling of the prairies.

Land along forest margins was soon occupied and settlers were forced to move into wide-open prairies. The prices of corn, wheat, beef cattle, and hogs began to rise about mid-century. The reaper became available, allowing the harvesting of larger fields and favoring the large, open, level areas. The construction of railroads across the terrain, during the decade 1850-1860, opened it to profitable agricultural production by providing better access to markets. Fuel and building materials could then be obtained by rail from timbered regions or from Chicago. Most pioneer settlers of the prairie were too far from a railhead and in any case could not afford to import much building material for the construction of their homes. All the way across the continent the pioneer farmer utilized the resources of his environment for this purpose. For the regions east of the great prairies, the forest provided logs and lumber. For some areas stone or adobe could be utilized. On the prairies, sod was a common building material. In the spring the sod could be plowed to a depth of a couple of inches, then rolled up and hauled to the building site. Walls were made of sod, with framed openings for windows and doors. Lumber was used for the roof, then sod was put on it to shed water. Some huts had a wood floor, but for others the bare earth served as a floor. A number of temporary homes were merely caves dug into hillsides, with a chimney rising to the surface. Such was the early Nebraska prairie home of the Bohemian immigrant family around which Willa Cather's great novel, *My Ántonia,* was centered (Cather, 1918).

The land west of the Mississippi was an almost treeless prairie with deep rich topsoil —an expanse that must have seemed limitless in those early years. Here were countless thousands of rich acres "to tickle with a hoe, that they may laugh a harvest" (Rasmussen, 1975*a,* p. 554). With the continuous migration from the East and the great tide of excellent European immigrant farmers, the stage was set for the greatest and most rapid agricultural expansion the world has ever witnessed.

## THE DEVELOPMENT OF FEDERAL LAND POLICY

In 1836 the Treasury Department instructed local land-office officials to accept only gold and silver in payment for public lands, except for actual settlers buying not more than 320 acres. This was believed to be necessary to stop wildcat bank inflation in the West, valuable only to speculators. However, by this time inflation and speculation had already made the severe panic of 1837 inevitable. Sale of public lands fell off sharply.

The previously mentioned "squatting" on public lands in colonial times continued unabated after the Revolution, justified in part by the slowness with which resources were opened for utilization and in part by their apparent inexhaustibility. In the West, survey and sales proceeded too slowly and lands had to be occupied illegally if they were to be occupied at all. On the frontier, the squatters were not considered to be lawbreakers. In 1828 the view was expressed in Congress that squatting was inevitable—even desirable—and that squatters should be permitted to buy without competition the land they had settled, as a reward for their contribution to the development of the country. "Claim associations" with their own rules and regulations arbitrated disputes between legitimate claimants and intruders (claim jumpers) and between rural claimants. If moral suasion failed, claim jumpers often received harsh treatment (Dana, 1956).

Free land was always favored by the West and at first generally opposed in the East, particularly in the South. The matter became a hot political issue in the twenty years preceding the Civil War. The northeastern states finally sided with the West, recognizing rapid development there as a means of expanding the market for their manufactured goods. In the political campaign of 1860 many believed that without the Republican party's pledge to support the homestead bill Abraham Lincoln would not have been elected to the presidency. The South remained in adamant opposition, believing that the plantation system, dependent on slave labor, could not compete with an extensive system of small farms operated by their owners.

### The Homestead Act

With most of its members from the southern states absent, the new Congress overwhelmingly passed the Homestead Act, and it was signed by President Lincoln on May 20, 1862. The act provided that any person who was a citizen, or had declared his intention of becoming a citizen, and who was the head of a family or over twenty-one years of age, might acquire title or patent to not more than 160 acres of nonmineral land after he had resided on it for 5 years and had made improvements on it, or had merely resided on it for 6 months, made suitable improvements, and paid $1.25 per acre. The land could be purchased at its regular price any time after six months from the date of filing. The Homestead Act speeded up the development of the West, opening its agricultural resources to poor and rich alike as long as the supply lasted.

Paul W. Gates, Professor of History, Emeritus, at Cornell University, recognizes the extent to which fraudulent entries were made under the Homestead Act for cattlemen, lumbermen, mineowners, land speculators, and land companies to gain illegally and at little cost land and resources they could not otherwise obtain. (This problem will be discussed further in the following chapter.) Yet he points out that the legislation that provided for free grants of land to 1,622,000 citizens and immigrants moving toward naturalization "constituted an act of extraordinary benevolence." "Few laws," says Gates (1977), "have so shaped American development, so accelerated its growth, so intimately touched the lives of millions of people."

Even after the Homestead Act there remained over 125 million acres of railroad lands, 140 million acres of state lands, 100 million acres of federal lands, and 100 million acres of Indian lands in small or large blocks, leaving much opportunity for speculating and land monopolization. Speculators were generally able to acquire the most desirable land. Settlers had the choice of either buying it at the speculators' prices or of exercising their homestead privilege farther afield, often suffering logistic, social, and economic disadvantage (Gates, 1975*a*).

### The Dawes Severalty Act of 1887

It would be unconscionable to ignore the fact that the land with which the government was so generous was once the communal property of the American Indians. Of the approximately 2.25 billion acres of land of the coterminous United States and Alaska that the Indians once "owned," they have retained slightly less than 100 million acres, and some 40 million acres of this was only recently acquired through the Alaska land settlement. Federal land policy originally envisioned the gradual removal of the eastern tribes to the Great Plains, then known as the "Great American Desert" and considered to be uninhabitable by white men (Kickingbird and Ducheneaux, 1975). Even there the Indian was driven to enclaves of steadily decreasing size as white men found the plains to be great cattle country.

Based on the usual assumption among Euroamericans that private property, like Christianity, had always been the basis for civilization, provisions for "severalty" (allotment of private land) for Indians were incorporated into law beginning in the 1850s. This policy was formalized in the Dawes Severalty Act of 1887. A corollary development was expected to be that the breaking up of Indian reservations would lead to the sale of surplus land to white settlers after the Indians had received their allotment. In addition, the Indians, lacking the white man's zeal for private ownership of land, tended to sell it and use the money improvidently (Hagan, 1961).

Aggravating the problem were the low salaries and usual incompetence of Indian agents and the graft and corruption accompanying land allotments. For example, among 1,735 Chippewas granted allotments of land about 1871, within seven years five-sixths of them had sold their lands or been defrauded of them by unscrupulous whites. In one way or another, Indians were separated from 86 million of a total of 138 million acres of their land between 1887 and 1934. The remaining land was mostly desert or semidesert land worthless to the white population (Hagan, 1961).

Quite aside from the problem of fraud, experience proved that abrupt changes from cultural heritage, whether respecting religion or use of land, were bound to fail and lead to demoralization and deterioration among Indian tribes (Hagan, 1961).

In January 1934, less than a half-century after the Dawes Act, various associations and councils interested in Indian rights and welfare, including the General Federation of Women's Clubs, called for repudiation of the severalty policy and the vigorous promotion of community ownership and control. In June 1934, President Franklin D. Roosevelt signed the Wheeler-Howard Act giving legislative sanction to the new policy. By 1940 nearly three million acres were given back to the Indians. The new policy also condoned or even encouraged native religions, ceremonials, and crafts. Land conservation, irrigation, tree-planting, road construction, and improvement of the quality of Indian herds (and reduction of livestock to the number the range could properly support) resulted, within a 20-year period, in a 23-fold increase in beef production and a fourfold increase in other farm production (Hagan, 1961). Ironically, as their financial condition improved, some Indians developed the self-interest that detracted from clan and family unity. Others, such as some Oklahoma Indians who had already integrated successfully into white society, even opposed the Wheeler-Howard Act, believing it would perpetuate an inferior status.

## Railroad Lands

Federal land grants to railroads began in 1851, when Congress granted more than 2.5 million acres of land to the Illinois Central, which effectively and profitably carried on colonization work. By 1864 railroads were receiving grants of land extending to 20 miles on either side of the right-of-way, and up to 40 miles in the territories. This land was sold to settlers for an average of about $4 an acre, compared with federal land that sold for $1.25 an acre. However, land reasonably close to the railroads was worth much more to a farmer than land farther away. The railroad companies encouraged movements of settlers to the prairies and also conducted agricultural experiments and disseminated information among farmers, thus speeding up the settlement of these vast areas. They welcomed colonies of settlers, often led by ministers. For example, the German-Russian Mennonites bought 60,000 acres from the Santa Fe in Kansas in 1874. Many of the first cultivated tracts were "bonanza" farms that failed in the depression of 1876 and were taken over by smaller operators (Schlebecker, 1975).

It soon became apparent that farming in the semiarid western section of the Great Plains would require methods far different from those with which most American farmers had become acquainted. The Department of Agriculture established the Bureau of Dry Land Agriculture in 1906 largely as the result of urging by the railroads. Dry-land farming allows half the land on a farm to lie fallow each year to accumulate moisture; one year's crop is grown on two year's water. This was merely the rediscovery of a practice dating back to antiquity. The Enlarged Homestead Act of 1909 allowed the homesteader to acquire 320 acres in areas where dry-farming was necessary, recognizing that he could cultivate only one-half his land in any given year. The homesteader needed only to cultivate 80 acres successfully for five years in order to secure title (Schlebecker, 1975).

## EARLY MIDWEST CROPS

Immigrants from Europe and from the East and South had come from regions representing a wide variety of crops and agricultural experience. The most profitable crops for a given area could be determined only by trial and error. For example, trials showed that tobacco could be profitably grown in southeastern Ohio. In 1840, six million pounds were grown. In Missouri, tobacco and mules had become the leading source of income (Bidwell and Falconer, 1925).

### Corn

As in the East and South, the first major crop of the Midwest was corn. Not only was climate favorable, but the fertile, easily worked, river bottomlands produced large yields, sometimes for thirty years in succession. The Ohio, Mississippi, and Missouri rivers and their many tributaries, such as the Scioto, the Wabash, and the Illinois, furnished not only rich bottomland, but also a means by which farm products could be transported to the rapidly developing market in the South. Likewise, from such river valleys as the Scioto, cattle could be driven to the markets of the East. Beef cattle could be raised on corn and the grasses of the open prairie, with the addition of a little clover and timothy. Swine required only corn, clover, and a small amount of oats and peas. Most of the corn was used for the partial fattening of cattle to be driven to the East, but much of the surplus corn was used for raising hogs and in the manufacture of whiskey, a convenient product for export.

The river bottomlands were so fertile that often wheat could not be grown on them profitably for several years after the first plowing, because the rank growth of the stalks caused lodging (falling over) and made the crops especially susceptible to rust.

### Wheat

By 1839 Ohio had become the leading state in the growing of wheat, with 19.5 percent of the nation's crop, followed by Pennsylvania with 15.6 percent, New York with 14.5 percent, and Virginia with 11.9 percent. Wheat production had also obtained a good start in Indiana, Illinois, Missouri, and Michigan. By 1860 Illinois, Indiana, and Wisconsin, in that order, led in wheat production. The center of wheat production moved continuously westward until eventually it reached the Great Plains.

As early as 1839 a writer reported to the *Western Farmer* that in southern Michigan he noted areas of 8,000 to 12,000 acres solidly planted to wheat and that with all men, women, and boys engaged in cradling, raking, binding, and shocking, there still was not enough help available to harvest all the crop—a chronic problem throughout the country.

### Transportation

Late in the eighteenth century it cost less to ship freight from Pittsburgh to Philadelphia via the Ohio and Mississippi rivers, and then by sea, than to send it overland. Nevertheless, about $4.50 of the $8.00 obtained in New York for a barrel of flour shipped from Cincinnati went for transportation. But by the 1840s Ohio (and Indiana) had built canals linking the Ohio River with the Great Lakes and the Erie Canal and savings in transportation amounted to nearly $3.00 a barrel. Buffalo then handled more commerce than the port of New Orleans. Steamboats dominated traffic on inland waters by 1860. But all water transportation systems remained slow, most of them freezing in winter (Schlebecker, 1975).

Railroads in the United States increased from 23 miles in 1830 to 30,626 in 1860. They had been extended to eastern Iowa, reaching the western frontier of agricultural production. According to the U.S. agricultural census, railroads moved two-thirds of the freight to and from the West by 1862. Although rail transportation was more expensive, it was faster and took a more direct route than the canal and river systems. Producers of perishable fruits and vegetables and dairy products were particularly benefited and this type of agriculture began to flourish in farm areas remote from the big urban centers.

The completion of the locks at Sault Ste. Marie in 1855, removing the last obstacle to navigation from the western end of the Great Lakes to the Atlantic seaboard, plus the rapid development of east-west railway trunk lines at about the same period, resulted in the dominance of the lake ports, particularly Chicago, in industry and commerce. Fully automated mills resulted in processing of grain becoming the nation's most important industry. Processing became concentrated in the cities, providing additional work for city dwellers and wealth for purchase of various farm products. Nevertheless, exports still provided farmers with the margin of success (Bidwell and Falconer, 1925; Schlebecker, 1975).

By 1914 the United States, with 257,292 miles of railroads, had more mileage than all of Europe and a third of all the track in the world. Farming had become nationwide and there was a sufficient rail network to serve it. Nevertheless, after the beginning of mass production of the Model T Ford in 1919, ordinary road-building began in earnest. In the brief period 1910-1914, automobiles on farms alone increased from 50,000 to 343,000, trucks from 2,000 to 15,000, and tractors from 1,000 to 17,000. The tractor increased the acreage that could be handled per farm, and the truck expanded the area in which commercial agriculture could be practiced. The eventual dominance of trucks in the transportation of agricultural produce depended on roads, in the building of which the federal government became heavily involved, particularly during the two world wars. By 1948, 99 percent of tobacco, 97 percent of milk, 96 percent of cotton, poultry, and eggs, 91 percent of grain, 88 percent of livestock, 83 percent of fruits, and 82 percent of vegetables were transported by truck (Schlebecker, 1975).

Rivers still play an important role in the transport of farm produce. Of the more than

$24 billion worth of farm products the American farmers sold abroad in 1977, more than $10 billion worth moved down the Mississippi to freighters at Gulf of Mexico ports.

Food processing and preservation, such as canning and the addition of preservatives; improved packaging; refrigeration in railroad cars, trucks, and warehouses; and quick-freezing (after World War I) developed chiefly as a result of and as an adjunct to transport. It made possible the expansion of farming and ranching on a scale and at a profitability that would have been otherwise impossible.

## MACHINERY FOR HARVESTING

As previously related, it was difficult to find farm laborers for hire. An English traveler in Illinois in 1818 noted that a man accustomed to work, if he was also "sober and industrious," could earn enough in two years to enter a quarter section (160 acres) of land, buy a home, a plow, and tools, and begin farming. Even men with plenty of capital could not start large-scale farming in the Midwest, because of the absence of a class of farm laborers. If laborers were imported from England, they were soon able to acquire farms of their own. The Englishman concluded that Illinois (and presumably the entire Midwest) "was a good location only for the small farmer who was willing to work his land without hired labor" (Bidwell and Falconer, 1925).

Laborsaving machinery had not yet relieved the long hours of backbreaking drudgery that characterized frontier farming. There had been no improvement in the quality of equipment or of farming methods from the days of the English manor. The daily routine of the medieval bondsman or serf, as described in the first chapter, would apply equally well to that of the American frontier farmer. Scot immigrant John Muir was put to work on his family's thin-soiled Wisconsin pioneer farm at the age of twelve in the mid-nineteenth century. During harvest season his invariable seventeen-hour work-day routine consisted in first grinding scythes, feeding the animals, chopping stove wood, and carrying water up the hill from the spring. Then, after a quick breakfast, there followed a full day of harvesting, supper at dark, more chores, worship, and on to bed. At other seasons Muir plowed, split rails, chopped saplings, and dug stumps. In only eight years the soil, with no fertilizer, wore out and the excruciating toil of felling trees and breaking sod on nearby virgin land was repeated. "We were all made slaves through the vice of industry," Muir complained in later years (Melham and Grehan, 1976). Such was the boyhood of the best-known American naturalist and conservationist, increasingly admired and appreciated with the passing years.

### The Reaper

The limiting factor in the magnitude of agricultural operations was harvesting. There was no point in planting more wheat than a farmer could harvest. Thus the mechanical reaper was a vital factor in the expansion of Midwest agriculture. As related in Chapter 3, two reaper models were invented in the 1830s. By 1849 Cyrus McCormick, one of the two inventors, had built a factory in Chicago, where he was one of America's pioneers in developing assembly-line mass production and innovative advertising, distribution, and sales procedure. In his *The Century of the Reaper* (1931), McCormick's grandson quotes Secretary of War Edwin M. Stanton as saying, "The reaper is to the North what slavery is to the South." Stanton believed the North could not have won the Civil War without the reaper, which took the place of "regiments of young men in the western harvest fields." Not only was the nation fed with a greatly reduced agricultural labor force, but more than 200 million bushels of American grain were sold to Europe in the 4 years of the Civil War, saving the nation from bankruptcy, and keeping its credit good among foreign nations at a critical period of its history (Casson, 1909).

Cyrus McCormick won gold medals and high honors in exhibitions and contests in England, France, Germany, and Austria. In 1878 the French Academy of Science elected him a member for having done "more for the cause of agriculture than any other living man" (Casson, 1909).

The next step after the reaper was the harvester, on which two men could ride and bind the cut grass as it was delivered to a platform. A driver and two men acting as binders could cut and bind eight to ten acres of grain in a single working day. A mechanical binder was soon developed.

Deering and other former rivals consolidated with McCormick in 1902 to form the International Harvester Company (IHC), the name reflecting the worldwide business in American harvesters. Manufacturing was then expanded to all kinds of agricultural equipment from tractors to the twine[1] to bind sheaves of grain. IHC moved into coal-mining and steel manufacture, and built one of the world's largest iron foundries. Yet even now agricultural equipment represents about 40 percent of the company's total sales; in annual volume of agricultural equipment manufactured, IHC is rivaled only by Deere and Company.

### The Combine

A patent for the combine was issued to Hiram Moore in 1836. No innovation had a greater impact on small-grain production. This machine heads, threshes, and cleans grain as it moves over the field. Projections on the cutter bar divide the crop into bunches. A reciprocating knife cuts the crop and feeds it into a threshing cylinder. With the grain removed, the bound straw can then be poured into sacks, a tank, or a truck. With many combines the straw is blown out on the stubble. The combine was also modified for harvesting corn and other crops.

## THE INFLUENCE OF GERMAN IMMIGRANTS

The famed agricultural traditions of the early German settlers in Pennsylvania and the Valley of Virginia were perpetuated by their descendants who streamed westward through Allegheny passes and by the many new immigrant farmers from Germany. Germans formed by far the largest of the foreign element in agriculture. German farmers had already developed a thriving enterprise in southeastern Pennsylvania by buying emaciated livestock along the drovers' routes to Philadelphia and feeding them back to flesh with clover and corn from their thriving pastures and bulging granaries, then selling them for high profits. As ever larger agricultural areas developed west of the Appalachians, the annual husbandry practices of the early Germans in Pennsylvania were continued by their descendants and by new immigrants farther west, where good productive lands were found near market centers and shipping points (Higbee, 1958).

The translocated farmers again not only raised their own livestock, but also purchased lean hogs and cattle, this time from settlers in the plateau uplands of eastern Ohio, Kentucky, and Tennessee, and fattened them on their lush bluegrass pastures

[1]The first twine, made of flax, was fed upon by grasshoppers. It cost the new company about a million dollars to develop a satisfactory substitute from hard sisal fiber from Yucatan and manila fiber from the Philippines.

and with corn, oats, barley, and hay. Location of farmland near the navigable rivers flowing through the fertile limestone basins of Kentucky and Tennessee and on to the Mississippi gave agricultural communities trade access to the plantations of the South. Cincinnati, on the Ohio River, was nicknamed "Porkopolis" and was the new center for pork-packing, drawing its supplies from the valleys of the Ohio and its tributaries (Higbee, 1958). According to the Annual Report of the U.S. Patent Office for 1845, Cincinnati had also become the center of considerable egg production, having in 1844 shipped 10,700 barrels (90 dozen eggs per barrel), mostly to New Orleans, but some also to New York and Baltimore.

Later, when the peak of the great wave of mid-century German immigrants[2] reached Missouri, the best lands were already occupied by native white Americans of the roving, squatter type, living by hunting and subsistence farming. The latter were willing to sell their log cabin, shed, and scant acres of land for a few dollars and then move on. Thus the Germans again acquired the best lands[3] and took particular pride in keeping them in their possession generation after generation (Faust, 1909).

When the military expeditions against the Indians, incident to the Black Hawk War of 1832, opened up the state of Wisconsin for settlement and the members of those expeditions reported on the fertility of the soil, the good reports spread eastward and across the ocean. Predictably the German immigrant tide flowed into Wisconsin, so that by the year 1900 these settlers and their immediate descendants made up 34.3 percent of the entire population of the state. They concentrated particularly in eastern and north central Wisconsin, and were again able to do some real pioneering, manifesting their predilection for settling in forested country. Their well-kept farms, neat houses, commonly of light-colored brick, and generous barns, reminded travelers of Lancaster County, Pennsylvania (Faust, 1909).

It was not until about the middle of the nineteenth century that German immigration began to exceed that from any other nation, including Ireland. The Irish continued to be attracted principally to the cities and played a small role in agriculture. While the Germans became numerous in some large Midwest cities, such as Cincinnati, St. Louis, and Milwaukee, most of them continued their traditional interest in seeking the best lands and developing small agricultural enclaves. Those in cities often were there only to earn enough money to buy land in frontier regions, where it was both fertile and cheap (Faust, 1909).

German immigrants of the second half of the nineteenth century profited from the encouragement and wise counsel of C. L. Fleischmann, a graduate of one of the many German agricultural schools. After Fleischmann emigrated to the United States in 1832, he made great efforts through his position in the Patent Office to convince Congress of the backwardness of American agriculture as compared with European, and the need for government action to improve it. He emphasized particularly the destructive effect on the soil of most American farming. Much of his effort was later directed toward distributing to prospective immigrants a series of treatises—which he wrote in German—concerning agriculture, industry, commerce, government, and law in the United States. These treatises were written in a restrained and sober tone, unlike much of the propaganda to which those with foreign backgrounds were subjected in those days. While pointing out that America afforded many favorable opportunities, Fleisch-

[2]According to the U.S. Census Bureau, by 1910 German immigrant farmers numbered 221,800, 33.5 percent of the total. The Austrians, with 33,338, had a similar agricultural influence. The Scandinavians were second to the Germans in number, with 155,570 immigrants (Swedes, 67,453; Norwegians, 59,742; Danes, 28,375) (Brunner, 1929).

[3]Including the valuable estate of Daniel Boone in the beautiful valley of the Femme-Osage in St. Charles County, Missouri.

mann also emphasized that only through hard work and the exercise of good judgment could the newcomer expect to succeed. He warned against expectations of early fortunes (Gates, 1961).

The fact that over five million Germans immigrated into the United States between 1830 and 1900, mostly into the Midwest, is a measure of their impact on the agricultural development of that region. Beginning in the last decade of the nineteenth century, Slavic and Italian immigration exceeded all others, but by then the best cheap land had been taken in the Midwest.

Throughout most of the nation—the Middle Atlantic area, the Midwest, the Great Plains, and California—Germans had the greatest impact on agriculture of all immigrant influences, through sheer numbers and through the quality of their agricultural heritage.

## THE INFLUENCE OF SCANDINAVIAN IMMIGRANTS

Following upon the heels of the Germans came another great wave of immigrants, the Scandinavians. From 1850, they began to have a great influence on the course of agricultural development in the upper Midwest. From 1850 to 1938, Sweden alone, with a population of six million, contributed over a million immigrants, with Minnesota receiving by far the greatest number. The history of the different elements of Scandinavian immigration—the Swedes, Norwegians, and Danes—is so sufficiently similar that the experience of the Swedes, the most numerous, can be used to illustrate the Nordic influence.

Estates in Sweden usually passed to the eldest son, so the younger sons and the daughters of landholders had to leave the home farm. Thus the overwhelming majority of Swedish immigrants were already farmers or persons familiar with rural life. At the flood tide of Swedish immigration arable land along the Atlantic Coast and in the Ohio River valley was already in the possession of farmers or speculators, but Minnesota's lands were being opened for settlement (Englund, 1938).

The popular Swedish novelist, Fredrika Bremer, had fired the imagination of her countrymen with her *The Homes of the New World,* published in the 1850s. Her discerning comments on the opportunities in the upper Mississippi Valley for a "glorious new Scandinavia" were a compelling influence. In 1890 about 73 percent of the Scandinavians in the United States lived in what was then known as the Northwest.

The forests and numerous lakes of Minnesota appealed to the Swedes. As did the Germans in Pennsylvania and Wisconsin, the Swedes in Minnesota selected wooded sites for their farms, believing these were the only kinds of virgin lands worth developing. Converting forest to cropland was slow, hard work, but for the frugal, patient, hardworking people inured to the hardship and poverty of farming the tiny plots of thin, rocky or marshy soil of Sweden, the Northwest in those days was a utopia. (This was before reforestation, drainage, dairy farming, and modern technology had transformed Sweden's lands into their present productivity.) Eventually the Swedes found that the open prairies to the west were even richer than forest land and needed only the plow to convert them to highly productive farmland.

Despite the great potential of the new land, the early years of every immigrant farm family consisted of ceaseless toil, struggle, and hardship. The merciless northern blizzards often took their tragic tolls before the one-room log cabin or sod hut could be replaced by more substantial and adequately heated homes. Life on the open prairie was especially hard. Running through the usual story of the conquest of the northern prairie is the epic note that American readers came to expect. The traditional novel has been replete with heroic action, romance, and eventual triumph—and a happy ending. It remained for a talented Norwegian immigrant, free from this cultural obsession, to write the consummate classic of the northern prairie with honest realism. Ole Edvart Rølvaag, in his *Giants in the Earth* (1927), wrote a tale of the tragic fate of a tiny settlement of Norwegian immigrants in the lonely, harsh prairie of mid-nineteenth-century Dakota Territory. Without avoiding the undercurrent of tragedy that blighted the lives of his lovable characters, and in the end destroyed them, Rølvaag wrote a gripping story of surpassing beauty—a monumental literary triumph.

There were gentle souls on the frontier whom fate had not roughhewn for pioneering. Particularly on the wives of the restless frontiersmen did the desolate prairie lay a ruthless hand. Throughout America's pioneer period, women proved on the average to be better able than men to withstand physical stresses such as prolonged exposure to cold and starvation, but were less able to psychologically adjust to the unfamiliar, bleak environment of the plains and deserts and the absence of even some of the simplest amenities of civilized life. Leaving behind home, family, and a familiar social and cultural milieu—even their native language—they often found isolation, desolation, loneliness, homesickness, insecurity, poverty, and cruel bludgeoning by impersonal forces of nature—heat, cold, blizzard, flood, drought, locust plague, and disease —too much to bear. Many were driven to despondency and some to stark madness. But many of the strongest men and women were also broken and even destroyed, perhaps by watching helplessly while their loved ones were slowly consumed by hopelessness and despair. Or perhaps they were suddenly and violently destroyed, as when caught in a blinding blizzard and freezing to death before finding shelter. Persons lost on the prairie were almost certain to meet with death in a blizzard—a blinding windblown snowstorm that neither man nor beast could face. They might become lost in trying to go from the house to the barn. Some plainsmen fastened cords to themselves and to the house so that if the barn should be missed the house would be found again (Webb, 1931). How often has the human cost of empire-building been measured?

Probably most American readers will find greater satisfaction in Willa Cather's *O Pioneers!* (1913). In this beautiful, poignant novel, centered about the trials and tribulations of a Swedish immigrant family that settled on a bleak, unbroken Nebraska prairie, Cather recounted the ruthless savagery of the northern plains. But she also recognized the historical fact that some men and women with unshakable faith in the potentially rich land clung to it stubbornly and tenaciously and had the character and fortitude to conquer it and force it to yield its surprising bounty—and a happy ending.

Many of the Swedish immigrants had to work at railway construction, in lumbering, in factories, or as farmhands before they had enough capital to buy land and the equipment needed for farming. Many young Swedish peasant girls became employed in domestic service in America, to contribute their share toward this end. Fortunately wages were higher in the United States than in Sweden. Land could be purchased for $1.25 per acre or, after the Homestead Act of 1862, it could be obtained free. And the land obtained was not the tiny farms of the native homeland, but a fabulous domain of 160 acres.

Even after farms were purchased, the men often had to work somewhere else for a while to earn the necessary cash. This they could do because, like the Germans, the Swedish women could care for the livestock and work in the fields, as well as spinning, knitting, weaving, and making the clothes for the family, until the farm produced cash income. The Swedes, particularly the first generation, tended to produce many of the necessities of life on their farms. They used their skills in carpentering, blacksmithing, and masonry to good advantage, having brought their chests of tools with them from the homeland. Thus the Swedish communities developed considerable economic security. The Swedes had great faith in producer and consumer cooperatives and tended to be agrarian liberals, active in political movements for solutions of agricultural problems (Englund, 1938).

The Swedes, Norwegians, and Danes usually developed separate settlements in Minnesota. Among Norwegians, linguistic loyalty remained strong longer than with other ethnic Europeans. Norwegian enclaves could be identified in large numbers of Wisconsin, Minnesota, and North Dakota counties well into the twentieth century. As late as 1940, half of all Norwegians in the Midwest lived on farms or in villages of less than 2,500—a greater percentage than for any other ethnic group (Shover, 1976).

Scandinavians showed great interest in education and local self-government. Their devotion to education resulted in such colleges as Augsburg, St. Olaf, Gustavius Adolphus, and Concordia and their political proclivities soon involved them in state politics, an involvement clearly evident to this day. It was the landless number-two sons who came to Minnesota, angry at the old order and transferring some of their rage to America. Abetted by a second wave from the old country, an intelligentsia opposed to the Scandinavian establishment, the stage was set for considerable sentiment toward socialism, social reform, and agrarian populist movements. But with the passing years the sentiment became attenuated (Salisbury, 1976).

Likewise weakened was the strong Scandinavian work ethic. According to Salisbury (1976), "The fathers enjoyed working the land, but schoolteachers taught the sons that physical labor was bad, that they should go on to higher things. That meant that three-quarters of the students stayed dissatisfied for the rest of their lives, despising physical labor. Minnesota has lost eighty-six thousand farms since 1930." And if young people do not inherit or "marry into" a farm, they have little chance of owning one. A study made at the University of Minnesota in 1976 indicated that going into commercial agriculture in Minnesota in 1975 required capital of $250,000, not counting "start-up" expenses. A half-section of land (320 acres) is required for bare subsistence, 600 acres to make a living, and 900 acres to become prosperous.

The Danes, although smallest in numbers among Scandinavians, were pioneers in dairy cooperatives. On the strength of what they had observed of the successful dairy industry in Denmark, they started a cooperative in Clark's Grove in Freeborn County in 1889 that was to have a pervasive influence throughout the state because of its great success. Members brought their milk to the dairy and profits were shared on the basis of what each member contributed. By 1898, Minnesota had 560 cooperative creameries and twenty years later there were 1,400 in the United States, 49 percent of them in Minnesota. As the business grew in volume, cooperative marketing continued to be based on its original principles, first through the Minnesota Cooperative Dairies Association (1911) and a decade later by Land O' Lakes, Incorporated, the latter becoming the largest butter-marketing concern in the world. The idea of cooperatives spread to other areas of agriculture: grain elevators, livestock, potatoes, and other products. In 1923 a Minnesota law was passed giving cooperatives legal sanction (Blegen, 1963).

The steam-driven flour mill should be given its just share of credit. Along with the steel plow and the reaper, it conquered the Midwest. Then these three mechanical innovations, joined by barbed wire and windmills, conquered the Great Plains. Minneapolis and St. Paul grew up around the mills. Wheat flour and cornmeal were shipped down the Mississippi and later on railroads to all parts of the world.

The steam-driven mills produced mostly wheat flour. They also milled some cornmeal, but cornmeal was consumed mostly in rural districts where the old-fashioned, water-driven gristmills were still at work. The mills slowly ground the yellow cornmeal that was favored by rural people. They produced less heat and retained the germ, thereby ensuring better flavor and nutrition. But because of its greater nutritive value, the meal was more liable to insect and fungus attack (Giles, 1940).

## YANKEE INFLUENCES

The inability of the stony, exhausted lands of New England to compete in productivity with those of the virgin and naturally superior land of the Midwest led to transmontane migrations of Yankees. These migrations to the Midwest were not as massive as those of European immigrants, but their influence was nevertheless great, not only in agriculture but also in industry that supports agriculture, in invention, and particularly in the nature of the areas's cultural, social, and political development. The story was interestingly told by Stewart H. Holbrook in *The Yankee Exodus* (1950), while protesting that his 398-page book was "little more than a footnote" and hoping that some "funded professor or wealthy institution" would one day tell the story in "ten thumping big volumes."

Forces within New England attempted to stem the exodus with stories of frontier hardships, including the gory consequences of Indian attack, but reports of fabulous crops of corn and wheat in "stoneless" soil resulted in "Ohio fever" being no more resistible than the "Genesee fever" that led to the settlement of western New York State. The first Midwest Yankee settlement and, in fact, the first permanent settlement of any kind in Northwest Territory was at Marietta, Ohio. Under the leadership of Rufus Putnam, twenty-one men left Ipswich Hamlet (now Hamilton), Massachusetts, on December 1, 1787, and in eight weeks they had arrived at the upper reaches of the Ohio River in Pennsylvania, where they were joined by a second contingent from Hartford, Connecticut. The two groups built a "flagship," a flatboat, and several canoes. On April 2, 1788, the combined party, now numbering forty-eight, along with oxen and horses, floated down the Ohio, and five days later the flotilla was tied up on the Ohio bank just beyond the mouth of the Muskingum. Cabins were built and a strong fort was constructed, the latter with heavy logs and with blockhouses at the four corners, dwellings making up the walls between. A formidable log palisade surrounded the whole. This stronghold was destined to shelter hundreds of refugees from Indian raids.

Cabins and fort were hardly finished when the first families arrived, with the women and children in time to pick wild fruits and berries and harvest the first vegetables the men had planted. Within a few months school was being held, presently to become Muskingum Academy and later to grow into Marietta College.

Stimulated by stories about corn that grew fourteen feet high at Marietta and a seven-acre farm that had produced seven hundred bushels of corn, in 1795 Brigadier General Moses Cleaveland helped to organize the Connecticut Land Company, and a year later led fifty-two other Yankees into the new lands of the Western Reserve along the shore of Lake Erie. This was land held by Connecticut after all the other states except Virginia had yielded theirs to the federal government. With the first *a* deleted, Cleaveland's name was destined to become immortalized in the name of a town that had fifty-seven inhabitants in 1810, but eventually outstripped all others in the Western Reserve. Soon the Ohio wilderness was to feel one of those previously mentioned strong Yankee influences, namely, their extraordinary zeal for education. New Englanders were uncommonly preoccupied in combating the devil's designs, among which was keeping people from reading lest they thus be able to decipher God's word in the Bible. Yankees originally considered education to be an adjunct to religion and for more than two centuries the majority of Yankee schoolmasters were also preachers. As the frontier moved westward, Yankee educators were everywhere a dominant influence. Californians, for example, may recall that the first two presidents of the University of California, Henry Durant and Daniel C. Gilman, were New Englanders.

To get back to our story, two "obscure Congregational preachers and fanatics," P. P. Stewart and J. J. Shipherd, went into the wilderness along the shore of Lake Erie to discover that the backwoods was truly a benighted region of "ignorant and degraded" inhabitants, which they were called upon to save from the clutches of the devil. Like Martin Luther, Shipherd had met the devil personally, but had the good fortune to be mounted at the time and he "triumphantly rode down the Evil One." Working with fanatic zeal to obtain the funds, Stewart and Shipherd were able, in 1833, to found Oberlin Collegiate Institute, the first coeducational place of higher learning in the United States. It not only was open to Negroes, but encouraged and helped them to come to Oberlin. Between 1850 and 1860 they numbered 5 percent of the student body. Oberlin College now has approximately 2,750 students divided equally between men and women. Blacks constitute about 11 percent of the enrollment.

Oberlin became a center for the Women's Rights movement and for antislavery activism. Students were encouraged to work to help pay their way through college. The town of Oberlin soon had a sawmill and a flour mill run by a genuine steam engine. As might be expected in a Yankee-inspired institution, in the town as well as in the college, liquor and tobacco were forbidden (Holbrook, 1950).

The founders and graduates of Oberlin started other colleges with similar objectives and ideals, such as Olivet and Hillsdale in Michigan, Berea in Kentucky (see chapter 4), Ripon in Wisconsin, Northfield (now Carleton) in Minnesota, and Grinnell and Tabor in Iowa.

The most talked about Ohio town developed by New Englanders in those early days was Kirtland. Here on a January day in 1831, a tall, handsome stranger from Vermont stepped out of his sleigh to startle the inhabitants with the news that he was the Prophet Joseph. This was Joseph Smith, the nucleus of a gathering of Mormons who eventually erected in Kirtland the most impressive temple in Ohio.

The Kirtland bubble burst in the panic of 1837, but elsewhere the Mormons were demonstrating their remarkable success in farming, business, and community development. Joseph Smith, unlike most of the fervent mystics of that era, did not mind mixing religion and business. In fact, his frequent revelations from God commonly contained detailed advice on specific business projects, always in happy agreement with the prophet's own plans for himself and for the Mormon community. This practical blend of the divine and the mundane was in keeping with time-honored Yankee tradition, and the majority of the Mormon leaders were Yankees, particularly in the early days (Brodie, 1945). No doubt the rapid and amazing success of the Mormons was one of the factors responsible for the antagonism they encountered among the "Gentiles" of western Missouri and Nauvoo, Illinois, and which soon led to their historic exodus to seclusion and peace in the wilderness of the Great Basin and to a highly significant chapter in the history of western agriculture, as we shall see later.

Yankees found a fertile field for the exercise of their famed business acumen in a little upstart of a town named for a wild onion the Indians called *chicagou.* As early as 1835, Vermonter G. G. Hubbard built an immense warehouse in the town and started to kill pigs and salt and smoke pork in excess of the rapidly growing town's needs. A few years later another Yankee (via New York) by the name of Philip D. Armour started packing meat in Chicago, and after him came Gustavius Swift, from Cape Cod, to pack pork and experiment successfully with a refrigerator car invented by Vermonter W. W. Chandler.

New Englanders were prominent directors and stockholders of the Illinois Central

Railroad Company, which was chartered in 1851. In exchange for 2,295,000 acres of Illinois land, the railroad agreed to build 700 miles of line in 6 years and to pay the state 7 percent of its gross income in perpetuity. The Illinois Central launched a vigorous advertising campaign to lure farmers from the East and South to its newly acquired lands. As a result, many a New England town began a population decline in those days and many new towns sprang up in Illinois with a distinct Yankee flavor, complete with "the necessities of life, namely, a Congregational church, a seminary, and a Temperance society" (Holbrook, 1950).

With the opening of the Erie Canal, the trickle of immigrants to Michigan became a flood. In the month of July 1830 four thousand farmer families passed through the canal, attracted by Michigan's fame in growing wheat and the good climate and soil along the margins of the Great Lakes favorable to growing fruit. Yankee loggers and lumbermen were drawn to Michigan by the magnificent white pine forests.

In Wisconsin the names of New Englanders were linked with nearly every important industry, but were conspicuously absent in one—brewing. Yankee Temperance crusades caused great resentment in the predominantly German and Scandinavian population. Finally all ethnic groups found a common cause in Abolition. Also, the Yankee came to be admired for his enterprise and energy in industry, creating business and providing employment. About the middle of the nineteenth century at least two-thirds of the native born Americans in Wisconsin were believed to be Yankees or born of Yankee parents, compared with a mere 4.4 percent in Iowa, for example, where about 60 percent of the population was from the southern states.

By 1876 there were only approximately 29,000 native Yankees in a total population of 450,000 in Minnesota; yet their prominence in all lines of endeavor, intellectual and otherwise, resulted in the "whole map of Minnesota" being "covered with their names" (Holbrook, 1950). Most of the Yankees who came to Minnesota in pioneer days sought farmlands and timber, but a few hundred shrewd, aggressive, and ruthless industrialists made their fortunes and moved to the larger cities to build big mansions and purchase gold-headed canes, plug hats, and elegant gold-plated cuspidors. Many others, however, bequeathed vast sums of money for academies, colleges, libraries, churches, hospitals, and parks.

## THE SIOUX WAR

Minnesota was only three years old when the beginning of the Civil War was heralded by the bombing of Fort Sumter. The struggle of the immigrant farmers to establish themselves in a new and strange land was intensified by the absence of husbands and sons who volunteered or were recruited to serve in the northern armies, many never to return. A year later the terrible Sioux War caught Minnesotans by surprise. The tide of immigration from abroad and the East, stimulated by the prodigal attitude of the federal government toward settlers, which reached its climactic expression in the Homestead Act of 1862, had compressed the Indian population into narrow reservations and destroyed their hunting grounds. There was great need for Indian agents of greater honesty, reform in the treaty system, and control of the dangerous liquor traffic. If there were any plans for such changes in federal policy, they proved too little and too late. Ironically, messengers carrying gold for long-delayed annuity payments got only as far as Fort Ridgely, along the lower Minnesota River, on the first day of the Sioux uprising on August 16, 1862. The gold never reached its destination, the Redwood Indian Agency. An earlier delivery might have averted the war (Blegen, 1963).

Hundreds of men, women, and children were killed in their homes or while fleeing from their plundered and burning villages. "Shorthairs"—Indians who had accepted civilization—as well as whites were targets of the attack. The greatest destruction was in Brown and Nicollet counties where nearly a half-hundred German farmers and many others were killed. Through their expertise at stealth and concealment, in which the Indians took great pride, they secured initial successes in their campaign, but their inability to subdue the town of New Ulm and particularly the garrison at Fort Ridgely, the military guardian of the entire valley of the Minnesota River, doomed the Sioux campaign to failure. Nevertheless, many whites fled not only from their homes and farms but from the entire state. Many who returned found that their homes and barns had been burned to the ground, their crops destroyed, and their livestock gone. Even after the war ended, occasional raiding parties of Sioux attacked isolated farms in search of horses, food, and even scalps. However, as the older chiefs had warned the young "hotheads," who started the war with such great confidence, the eventual result was that in 1863 Congress abrogated earlier treaties with the Sioux, discontinued annuities, opened the Winnebago Reservation to settlers, and drove or transported the tribe even farther westward (Blegen, 1963).

In the same year, some five million acres on the Minnesota and Dakota sides of the Red River were ceded to the Chippewa tribes, traditional enemies of the Sioux, along with cash annuities, food, and goods. But the Sioux were gone, and the long and agonizing ordeal of the settlers of the picturesque Minnesota River valley was over.

## KING CORN

In 1839, 46 percent of the corn crop of the nation was produced in the South, with Tennessee and Kentucky the two leading corn-growing states. Virginia was third. In what is now the Midwest, southern Ohio and the valleys of the Wabash, the Illinois, and the Missouri rivers were still the centers of corn production. During the forties, corn-growing expanded northward into upper Illinois and Indiana. With the exception of Michigan and Wisconsin, where wheat predominated, corn was the great staple. It could be grown successfully on newly broken prairie sod where wheat was rarely successful. By 1860 the leading corn-producing states were Illinois, Ohio, and Missouri, in that order, and the center of corn production continued to move westward. The cost of growing corn in the Midwest was only a fourth the cost in such areas as Maryland, Delaware, New Jersey, and eastern Pennsylvania, where the use of guano, lime, and superphosphate on the exhausted soils was rapidly increasing. Despite the expansion of corn-growing into what is now known as the Corn Belt, relatively little corn as such was marketed, for it was used mostly as feed for cattle and hogs. In the cultivated areas of Ohio, Kentucky, Indiana, Illinois, Iowa, and Missouri, production of corn exceeded that of wheat from five to eighteen times; yet it was the marketing of wheat that attracted the most attention (Bidwell and Falconer, 1925).

As cotton was king in the "Cotton South," corn was to become king in the "Corn Belt" (fig. 3, Region M). The richness of the black soil of the Corn Belt, resulting from the beneficial physical and biological action of thousands of years of tallgrass prairie vegetation, and a fortunate combination of soil conditions, temperature, sunshine, and rainfall—about 40 inches—resulted in the Corn Belt being ideally suited for corn. To the north, limits to intensive corn-raising were set by the cool summers, to the west by inadequate rainfall, and to the east and south by rough terrain. This does not mean that corn cannot be grown well outside the Corn Belt, and in fact it is now grown in every state of the Union. Until recently the record yield was raised by a young Mississippi 4-H Club member who in 1955 produced 304 bushels of corn on an acre, nearly 50 bushels above the previous record. In 1975 the record went back to the Corn Belt when a central Illinois farmer grew 338 bushels of corn on a one-acre test plot in an area where the average yield was slightly less than 100 bushels per acre. (A bushel of shelled corn weighs 56 pounds.) In any case it was logical that what is now known as the Corn Belt should be devoted principally to the crop that yielded the greatest food or feed per acre.

The United States now annually produces about 5.5 billion bushels of corn, about half that grown in the world. The leading states in corn production in 1976 were Illinois with 31,328,000 metric tons, Iowa with 29,156,000, Indiana with 17,604,000, and Nebraska with 13,074,000 (USDA Agr. Stat., 1977).

Although about 90 percent of the corn grown in the United States is fed to cattle and hogs, the remainder, which is sold for direct human consumption and for industrial purposes, still is enormous. The chief commercial product is cornstarch, widely used industrially as well as in foods. Plant breeders have developed a strain that yields an especially large quantity of high-amyglose starch, particularly well adapted to paper sizing and binding. Starch may be hydrolized or fermented into syrups, adhesive polymers, and other products. When corn is used for industrial purposes, the 10 percent protein in the grain is reserved as a valuable constituent in animal feeds. Corn steep liquor is useful in the production of antibiotics, while the embryo is valuable as the source of corn oil. Usually the remaining residues and hulls are incorporated into cattle feeds (Schery, 1972). In this day of increasing emphasis on recycling, it should be pointed out that probably nowhere is the process more complete than in agriculture.

The prominent position of corn in American agriculture justifies some detail regarding its origin, dispersal, improvement, current status, and future possibilities.

## Origin and Dispersal

Corn (*Zea mays*) is a member of the grass family, which probably has more economically valuable species than any other family of plants. In fact, the grasses supply over 75 percent of the world's food. In one subfamily are corn and the crops most closely related to it—the sorghums, sugarcane, and millet; in another subfamily are the major cereals, such as wheat, rice, oats, barley, and rye, as well as the forage crops and the bamboos. Caribbean tribes introduced corn to Columbus as "mahiz," from which the word *maize* was derived. On November 5, 1492, one of Columbus's scouting parties saw cornfields, in what is now Cuba, that extended for a distance of about 20 miles. All the common types in use today—popcorn, sweet corn, and flour, flint, and dent corn—were developed and improved by the Indians and would be unknown to us today if they had at any time during a period of thousands of years failed to preserve their supplies of this precious food crop. There is no recorded instance of corn as we now know it having ever survived in the wild (Crabb, 1974). A fascinating account of corn's long and romantic history, and an eloquent eulogy to its well-deserved fame, may be found in Dorothy Giles's *Singing Valleys—The Story of Corn* (1940).

Much information has been obtained from prehistoric vegetal material found abundantly in caves thanks to lax housekeeping among their prehistoric occupants. In accumulated trash, garbage, and excrement in Bat Cave, New Mexico, tiny cobs about one inch in length were found at the bottom of a pile of refuse some 6½ feet deep. Radiocarbon determinations of associated charcoal indicated that the cobs were about 5,600 years old. The kernels were of a popcorn type and were enclosed by floral bracts (glumes). A distinct evolutionary sequence was found from the lower to the higher level of refuse in Bat Cave and in many other caves later discovered in Mexico. Teosinte, the closest relative of corn, was found in one cave.

According to R. S. MacNeish (1964), cobs found in a cave in Mexico, and shown to be about 7,000 years old, indicated that "corn's wild progenitors probably consisted of a single seed spike on the stalk, with a few pod-covered ovules arranged on the spike and a pollen-bearing tassel attached to the spike's end." The spike bore both male and female inflorescences. The most primitive cobs unearthed in Tehuacán Valley had "the stump of the tassel at the end, each had borne kernels of the pod-popcorn type and each had been covered with only a light husk consisting of two leaves." Seeds could be readily disseminated by the wind. The reconstructed progenitor of corn, according to this theory, is shown in figure 14, left. The two other grains shown in the figure, teosinte (*Euchlaena*) and gama grass (*Tripsacum*), readily hybridized with wild and cultivated corn, with modern corn as the end product.

*Fig. 14. Three New World grasses believed to be involved in the history of domesticated corn: wild corn (reconstruction at left), teosinte* (Euchlaena) *(center), and gama grass* (Tripsacum) *(right). (From R. S. MacNeish, 1964. Copyright 1964 by Scientific American, Inc. All rights reserved.)*

The earliest prehistoric corn in Tehuacán caves was considered by some authorities to be wild. It was ¾-1 inch long, with 8 rows of kernels (occasionally 4), and with 36 to 73 kernels per cob. There has been no substantial change in 7,000 years in the corn plant's basic botanical characteristics, even though there is more foodstuff in a single grain of some varieties than there was in an entire ear of the Tehuacán wild corn (Manglelsdorf et al., 1964). According to Galinat (1977) only two alternatives remain as viable options: "1a) Present day teosinte is the wild ancestor of corn; 1b) A primitive teosinte is the common wild ancestor of both corn and Mexican teosinte; 2) an extinct form of pod corn was the ancestor of corn with teosinte being a mutant form of this pod corn." Galinat believes that there may be South American *Tripsacum* germplasm present in the northern flint parent of the Corn Belt dent corn.

Most modern maize specialists consider some race or races of teosinte to be the most likely progenitor(s) of corn (Galinat, 1971; Beadle, 1972). Teosinte was probably formerly much more variable than it is now and there may have been races more maizelike than they are now (Harlan, 1975). The center for dispersal of the modern race of maize was probably southern Mexico. Thousands of years ago maize may have been distributed to South America (see fig. 1), and strains evolving there could have spread north, eventually reaching eastern North America. Wherever the early white explorers set foot on the shores of the Western Hemisphere, from New England to Peru, they found the Indians cultivating this important crop. It has since been carried to all parts of the world, improved and adapted to local climatic and soil conditions and, from a world standpoint, is exceeded only by wheat and rice as a food or feed grain.

The corn that white men found in the possession of the American Indian was already remarkably developed. This was in large measure the result of hybrid vigor. The southern dent "gourd seed" corn of the southeastern United States and the New England flint corn of the Northeast were destined to meet in the American Midwest to form the famous Corn Belt dent, benefiting from the "explosive evolutionary effects" of blending diverse germ plasms. Finally plant breeders applied complex genetic theory to develop the hybrid and double cross hybrid. This development, along with improved farming techniques, doubled corn yield per acre in a thirty-five-year period. In the same period the yield per acre of oats, a self-pollinated grass, increased only 29 percent in response to improved farming techniques (Galinat, 1965).

Sweet or sugar corn, for human consumption, is also grown in large quantities in this country and is sold as fresh corn on the cob and for canning. It is sweeter than other varieties, a recessive character that can be introduced into any variety by hybridization. Sweet corn (*Zea mays* var. *rugosa*) originated on the highlands of South America, where the Indians used it in making maize beer or *chicha.* In pre-Columbian times it spread northward through Central America and western Mexico, mixing with indigenous races of maize enough to enable it to adapt to new climatic conditions. The Plains Indians had a sweet corn that was almost like our Golden Bantam before hybridization, except that it still had some reddish kernels, as does the *maiz dulce* or sweet corn of Mexico. The early colonists obtained their sweet corn varieties from the Indians of the Six Nations of New York (Anderson, 1952).

W. Atlee Burpee put the first Golden Bantam on the market in 1902. Before that time only white varieties were used as roasting ears. These differed from the dents and flints in their low starch and high sugar content. Later, inbred strains of Golden Bantam were hybridized (Giles, 1940). There are now many varieties of sweet corn. These are consumed boiled or roasted in the green stage—"corn on the cob." In 1976 about 2.9 million tons of sweet corn were grown in the United States, 76 percent of which was canned or frozen and the rest sold for the fresh market (USDA Agr. Stat., 1977).

Although known the world over, popcorn (*Zea mays* var. *everta*) has become an important crop only in the United States. The ripe kernels are small and have a hard, horny endosperm. When heated, the pressure of the steam formed within the kernel causes it to "pop" to turn the endosperm inside out. The two main types, the "rice" varieties with pointed kernels and the "pearl" varieties with rounded ones, each have many forms. Hybrid varieties are developed the same as for dent corn. In fact, popcorn growers accepted hybrids even more rapidly than did growers of field or sweet corn. Hybrids had greater yields and higher ratio of popping expansion than the former open-pollinated popcorn varieties (Walden, 1966).

Although popcorn is now considered a delicacy, among primitive peoples it probably possessed a practical advantage in that it merely had to be heated to become edible; grinding was not required. French explorers found Iroquois Indians popping corn in pottery vessels in the early seventeenth century (Walden, 1966). Mangelsdorf (1950) believed that primitive man may have discovered corn as a food when small kernels accidentally exposed to heat were transformed into nutritious morsels.

## Corn As Grown by Prairie Indians

Corn was the basic crop of the *agricultural* Indian tribes of the prairies. Flourishing agricultural economies were developed in the valleys of the Missouri, Platte, Arkansas, and Washita rivers, where agricultural Indians traded with the nomadic buffalo hunters of the plains (Jablow, 1950). Even on the Great Plains, there were tribes that depended principally on agriculture, although some left their villages for the annual summer buffalo hunt. Some tribes abandoned agriculture completely when the horse increased their efficiency in buffalo-hunting.

Corn plants gradually diminished in height with increasing distance up the Missouri River Valley, ranging from Pawnee corn six to ten feet high in the South to corn never more than two feet high possessed by the Assiniboins of Montana and Canada. The period required to mature the crop decreased correspondingly. As in all areas where corn was grown in North America, it might be red, black, blue, yellow, purple, or white; sometimes all colors were seen on a single ear. The corn varieties were hardy and dependable, well adapted to each region. Apparently the only crop failure occurred during those years when grasshoppers descended in large clouds and stripped the corn patches bare in a few hours. Drought often greatly reduced the crop, but no system of irrigation was developed to avoid this particular source of crop loss (Will and Hyde, 1917).

Besides corn, tobacco and many varieties of beans, squashes, pumpkins, melons, and gourds were grown. Sunflowers, grown for their very nutritious seeds, were planted around the borders of these gardens, which usually ranged from one to three acres per family. The gardens were enclosed by a fence formed of bushes and branches of trees skillfully woven together (Dunbar, 1880). Potatoes were introduced to the Mandan Indians about 1832 and became a favorite crop among all the upper Missouri tribes (Will and Hyde, 1917).

One to three acres might seem to be a very small "family farm" by modern standards, but it probably loomed large in the life of the Indian woman who, possibly with the help of a daughter, did all the work with a digging stick, a wood or antler rake, and a hoe made of the shoulder bone of a buffalo or elk and tied to a stick. Arising at sunrise (3 or 4 A.M.), the women went immediately to the gardens. They seemed to enjoy gardening, singing special field songs as they worked. As temperatures rose, they left the gardens for their regular daily tasks, which never ended (Will and Hyde, 1917).

Growing corn in the Upper Missouri Valley appears to have been much more laborious than along the Atlantic seaboard, where, as related in chapter 3, the Indians merely planted and weeded, finding the tilling of the ground, as practiced by the whites, to be rather curious. Before the corn-growing Plains tribes left for the summer buffalo hunt, the women had already given the cornfield two hoeings and had weeded it thoroughly by hand (Will and Hyde, 1917).

Much green corn roasted in the husk was consumed during the harvest season. Green corn to be stored was removed from the roasted ears by means of the sharpened edge of a clamshell. The kernels were then spread out on skins or blankets to dry, then stored for winter use. According to one white observer, green corn roasted and stored retained its good flavor. Later, mature corn was dried without roasting, then shelled. The best ears were set aside for seed. Pumpkins and squashes were cut in long, thin strips, hung upon poles until dry, then stored away with the corn.

The caches of the Omaha Indians were described (Fletcher and La Flesche, 1970) as follows:

> Near each dwelling, generally to the left of the entrance, the cache was built. This consisted of a hole in the ground about 8 feet deep, rounded at the bottom and sides, provided with a neck just large enough to admit the body of a person. The whole was lined with split posts, to which was tied an inner lining of bunches of dried grass. The opening was protected by grass, over which sod was placed. In these caches the winter supply of food was stored; the shelled corn was put into skin bags, long strings of corn on the cob were made by braiding the outer husks, while the jerked meat was packed in parfleche cases. Pelts, regalia, and extra clothing were generally kept in the cache; but these were laid in ornamental parfleche cases, never used but for this purpose.

### Hybrid Corn

There is probably no other modern food crop that has been improved, from the standpoint of man, as greatly as corn. But hybridization was a relatively late development in American agriculture. Before hybrid corn, the seed for the next year's planting would be selected out in the cornfield, generally in early October. Choice ears, thick, even, and filled to the top, were selected. After the ears were thoroughly dried, a kernel or two was selected from each and tested for swelling or sprouting ability. The best achievers then identified the most promising ears to be stripped to supply the seed corn for the spring planting.

When "the oak leaves were as big as squirrels' ears," the time had arrived for planting. Planter wire was unrolled from one end of the field to the other. The wire had big knots at intervals of forty inches. As the wire ran through a slotted device on the side of the planter, the knots would move the "plates" one notch and drop two or three kernels in a hill. The planter wire was then moved again and the process was repeated, resulting in the hills of corn being forty inches apart in each direction. The cornfield could be cultivated in both directions with the same cultivator. This was before the days of herbicides and the main objective of the many cultivations was weed control. The details of corn selection, planting, cultivation, and all the many other day-to-day tasks of farming in the Corn Belt in the first decades of this century, are charmingly recounted in the popular *In No Time at All* by Carl Hamilton (1974).

The principal basis for improvement of corn was hybridization, although improved cultural measures must also be given credit. It is noteworthy that the development of modern hybrid corn was based on the theories of two great European scientists, Gregor Mendel and Charles Darwin. It was in a study of plant hybridization that Mendel developed the basic laws of inheritance of specific characteristics. Although he published his results in 1866, the importance of his work was not recognized until about 1900. Darwin published a book on cross- and self-fertilization of plants in 1876. He pointed out a phenomenon that was to loom large in the work of the American scientists who developed hybrid corn, namely, that inbreeding usually reduced plant vigor and cross-breeding restored it (Rasmussen, 1960).

Formerly the breeding and selection of corn consisted of isolating superior strains obtained through open pollinations. Because the distribution of pollen was not controlled, a limit was eventually reached beyond which repeated crossings gave no improvement. Then the increasing basic knowledge about genetics, particularly the genetics of the corn plant, led to the development of inbred strains. When varieties of corn with desirable qualities were repeatedly inbred for many generations, uniform "lines" could be obtained. These were low in vigor and yield. When the inbreds were crossed, however, the resulting hybrids turned out to be greatly superior to the original open-pollinated varieties, in accordance with Darwin's generalization. Eventually resistance to disease and insect pests was bred into strains of corn by selection and hybridization.

The story of the ingenuity and dogged determination of the small group of dedicated young men who made agricultural, economic, and even world political history by manipulating the genes of *Zea mays* was brought to light in an intensive study by agricultural journalist Richard Crabb. Crabb's investigation was made in the early 1940s when only one of the men prominently associated with the earliest hybridization had died, and he contacted each of the living persons who had had a major role. He published his findings in a fascinating and easily readable book titled *The Hybrid Corn Makers: Prophets of Plenty* (Crabb, 1947). Thus geneticists and plant breeders such as E. M. East, H. H. Love, G. H. Shull, H. K. Hayes, and D. F. Jones attained their rightful place in the amazing story of hybrid corn. They were all from poor families in rural communities and worked their way through college, usually as assistants in some line of university research in which they had great interest. They all worked with great intensity and made contributions of outstanding scientific significance early in their careers. One must admire the courage and tenacity of these young investigators, for both "experts" and "practical men" ridiculed their ideas. In fact, Love had to surreptitiously remove some kernels from several ears of the inbreds he and East had developed at the University of Illinois at the turn of the century when the director of the study of soils and crops for the Agricultural Experiment Station had ordered the project to be discontinued. The work was continued by East, beginning in 1905, at the Connecticut Agricultural Experiment Station, where the famous Burr-Leaming double cross, the first successful strain of hybrid corn, was developed in 1910. It produced about twice the yield of open-pollinated corn in the experimental plots.

Hybrid corn is one of the great boons bestowed upon mankind by the combination of a prodigal nature and the imaginative, probing mind of man. Even before hybridization, the corn plant was one of the marvels of the food crops. Male and female elements are produced on the same plant—the male cells in the tassel at the top and the female cells on the ear shoots at the nodes of the stalk, usually about midway between the base of the plant and the tassel. Pollen grains from the tassel land on the tender, sticky silk emerging from the young ear-shoot. A tiny pollen tube emerges from the pollen grain and grows down through the corn silk. The male sperm cell advances as the pollen tube elongates and eventually it reaches the female cell. A kernel of corn immediately starts to develop. Corn produced in this way is termed "open-pollinated." On an ordinary field-corn plant there are about 800 to 1,000 silks, and eventually approximately that many kernels of corn, on an average ear. This abundant production meant the breeder was assured ample seed once he had proved the value of a particular hybrid strain.

The development of a hybrid involved first the selection of outstanding individual plants among the old open-pollinated varieties. These served as parent stock and were "purified" by self-fertilization, also called "selfing." Pollen from other plants was not allowed to settle on the silks of the superior plant, and the corn breeder could therefore separate the highly desirable types from the inferior. Repetition of this process for several generations resulted in the strain becoming "stable" or "true breeding." The self-pollinated plants were called "inbreds" and were used exclusively as breeding stock in producing superior hybrids.

An enormous amount of work is involved, for plant breeders discard more than 99 percent of the plants they develop in order to arrive at outstanding new hybrids. The first crop of seed from a selfed plant produces a wide variety of offspring, smaller and less vigorous than the original open-pollinated corn because of the absence of hybrid stimulation. The inferior strains are eliminated, and after several generations of this procedure the inbred lines that have survived are relatively uniform. However, only a small percentage of these strains will have real value. The superior ones can be maintained year after year by ensuring their self-fertilization; they continue to produce progeny of known capacities.

Obviously the large volume of single-cross seed required each year cannot be obtained by hand-pollinated controlled crosses. The corn breeder chooses a well-isolated field and plants two rows of one inbred designed to bear the seed, alternated with one row of inbred male parent. Tassels are removed from those plants expected to bear hybrid seed and the only pollen in the field is produced by the inbred male plants. The latter are also fertilized themselves and produce another generation of inbred seed. Two carefully selected and previously proved single-crop hybrids can be combined to produce a double-cross type; likewise a three-way hybrid can be produced. Strains of corn can be developed that are not only superior producers but are well adapted to particular conditions of weather, soil, disease, and insects—custom-built for particular needs. Some hybrids can produce much greater yields during short growing seasons than were possible with open-pollinated corn, and this has resulted in corn being successfully grown in areas farther north than was possible before the advent of the hybrid strains.

In agricultural research there is always a gap between experimental results and practical farm utilization. Hybrids had to be developed that were adaptable to various soils and climates. Companies of trained professionals had to be formed to produce the seeds for new hybrid strains each year because farmers no longer merely set aside the best-looking ears of corn from which to grow the next crop. Like the mule, hybrid corn was superior in every respect except that it could not reproduce itself.

The first company ever to be formed exclusively for the purpose of developing and selling hybrid corn seed was organized by Henry A. Wallace and some friends. It was called the Pioneer Hi-bred Corn Company. A building designed by Wallace and an engineer named Casady was the first modern hybrid seed-corn drying and processing plant. Wallace became Secretary of Agriculture early in 1933 and later was vice-president under Franklin D. Roosevelt.

Many excellent hybrid strains were developed by seed companies in cooperative experiments with the Department of Agriculture. Hybrids were developed that were not only high yielding but also resistant to chinch bugs, rootworm, European corn borer, Japanese beetle, and weevils. Hybridization of corn also resulted in greater uniformity of plants and ears. This was eventually to make possible rapid mechanical harvesting and the completion of harvest operations throughout the Corn Belt before the beginning of cold, wet, and windy fall weather. The inbred lines from which hybrids were developed were selected for their resistance to lodging, so the hybrids were also superior in this respect (Ochse et al., 1961).

Observations of great differences in the tolerance of different inbreds to drought and cold likewise led to drought-resistant and cold-resistant varieties. At the University of Wisconsin, beginning in 1923, many cold-resistant hybrids were developed, under the

**Table 1**
**Protein and Food Energy Yield and Fossil Energy and Labor Input per Hectare for Certain Crops in the United States (adapted from Pimentel et al., 1975)**

| Crop | Crop Yield in Protein (kg) | Crop Yield in Food Energy ($10^6$ kcal) | Fossil Energy Input for Production ($10^6$ kcal) | Labor (man-hours) |
|---|---|---|---|---|
| Alfalfa | 710 | 11.4 | 2.7 | 9 |
| Soybeans | 640 | 7.6 | 5.3[a] | 15 |
| Brussel sprouts | 604 | 5.5 | 8.5 | 60 |
| Potatoes | 524 | 20.2 | 8.9 | 60 |
| Corn | 457 | 17.9 | 6.6 | 22 |
| Corn silage | 393 | 24.1 | 5.5 | 25 |
| Rice | 388 | 21.0 | 15.5 | 30 |
| Dry beans | 325 | 5.0 | 4.5 | 15 |
| Oats | 276 | 7.4 | 3.0 | 6 |
| Wheat | 274 | 7.5 | 3.8 | 7 |

[a]Includes 1.1 x $10^6$ kcal for processing the beans to make them edible for livestock.

direction of a Canadian, R. A. Brink, who took his graduate work at Harvard under E. M. East, mentioned earlier. By the time of World War II, these new hybrids grew on nearly ten million acres of land in the northern United States and Canada, formerly too cold for corn. Tens of thousands of farmers who never grew corn before could then grow the precious crop and, through animal-feeding, convert it to high-energy food like cheese, pork, milk, butter, and eggs, in time for these products to be sent to the war fronts and home fronts of the allied nations. This was a crucial factor in the critical balance of the scales of victory in the fateful conflict.

Most crops can utilize more light than is commonly available. Among these is corn, which is the most efficient of the staple crops in utilizing solar energy in photosynthesis. Table 1 shows that corn is also the most efficient as to yield of food energy per unit of fossil energy input among the crops consumed by man. Much has been said about the high energy input in American agriculture. In relation to energy inputs in the agriculture of developing countries it is indeed high but, in relation to energy inputs in the total American energy budget, it is surprisingly low when one considers that agriculture is the sine qua non of existence. Of the total domestic energy budget of 18,900 x $10^{12}$ kcal/yr, agriculture uses only 498 x $10^{12}$ kcal/yr or 2.6 percent of the total (Heichel, 1976). This includes energy expended both on and off the farm to produce crops. The manufacture of fertilizers requires the most energy (28 percent) and the production of pesticides the least (0.6 percent). On the farm, petroleum products account for 51 percent of the energy used. At least five times more energy is expended in processing, distribution, and preparation of food than in producing it (Pimentel et al., 1973, 1975). Note the relatively small quantity of protein and food energy produced per hectare by oats and wheat when compared with corn (table 1). But oats and wheat can be grown on land where rainfall would not be sufficient for corn and the land is therefore less expensive.

Corn is also one of the more efficient crops in relation to labor input per hectare (2.47 acres). For any of the crops in table 1, labor could be exchanged for energy. For example, corn has been grown to produce 175 kilograms of protein and 6.8 x $10^6$ kcal of total food energy per hectare with 1,144 man-hours of labor and only 0.053 x $10^6$ kcal of energy—5.2 times more labor but only 0.8 percent of the energy indicated for corn in the table (Pimentel et al., 1975).

Corn plants of the future in the United States will probably be cone-shaped and have small tassels. The uppermost leaves will probably be more vertically oriented and lower leaves more horizontal, so as to capture the maximum sunlight. The new plants will probably be single-cross, short-stalked, prolific hybrids with a high grain-to-stalk ratio, adapted to equidistant planting and plant densities of 30-40 thousand per acre, rather than the current 20-27 thousand. This will result not only in greater use of sunlight but also will increase the efficiency of water use and reduce weed growth. Whereas in 1970 the average yield of corn in the United States was 80 bushels per acre, yields of over 300 have been attained, so the potential is already evident within the framework of current technology. With improvements in technology, based on current knowledge, yields of 400 or more bushels per acre should be possible (Wittwer, 1970).

Hybridization opened the way for changing the nature or percentage of starch, protein, or oil content of corn. A mutant high-lysine gene, called Opaque-2, was introduced into many of the well-established corn hybrids. Lysine is an essential amino acid that is required for health and optimum development of humans and most animals other than ruminants. High-lysine corn contains 69-100 percent more lysine and 60 percent more tryptophan, another valuable amino acid, than other hybrids.

Pigs fed high-lysine corn gained weight over three times faster than comparable animals fed ordinary corn when the latter was not supplemented with some other source of protein. The improved corn is believed to be able to replace about 50 percent of the soybean meal that is ordinarily added to commercial corn to feed pigs. Hybrids with high-quality protein should also be an important dietary factor in areas of the world where corn is a basic food and protein-deficiency is a problem.

High-lysine corn has not lived up to early expectations because yield has not reached that of other hybrids. However, lysine content is continually increasing and has reached 0.43 percent, compared to 0.25 percent in normal corn. Breeders believe "superhigh" lysine hybrids will soon be available with lysine yields as high as 0.7 percent. At such levels, soybean meal could be dropped from swine corn-soy rations as long as the diet was fortified with vitamins and minerals (Anonymous, 1977*a*). Attempts are now being made to improve protein quality in wheat and rice varieties (Hughes and Metcalfe, 1972).

## Efficiency in Corn Harvest

The combine, as modified for harvesting corn (fig. 15), was an important factor in providing the high degree of efficiency for which American corn culture is universally admired. Some farmers store the mechanically harvested ears in an outdoor corncrib with wooden slats or mesh wire to keep them well ventilated. The dried stalks remaining in the cornfield are disked back into the soil, enriching it for whatever crop is to follow. Others store shelled corn in an elevator as with wheat and other grains. If corn is to be fed to livestock, it is generally harvested with a tractor-drawn chopper that picks up green cornstalks from windrows in the field and cuts stalks and ears into about half-inch pieces. The chopped-up material is blown into a trailer attached to the chopper, then hauled in and power-unloaded into feed bunkers or large silos (Lent, 1968; Hilliard, 1972).

*Fig. 15. Combine for harvesting corn: on this model an automatic leveling system keeps the combine level while the corn head or grain platform adjusts to ground contours on slopes of up to 18 percent (courtesy Deere and Company).*

In the airtight silo the cut-up material is compressed and undergoes an acid fermentation that retards spoiling. It is then called "silage" or "ensilage." About 10 percent of the corn crop is grown for silage, mainly in the northern dairy regions. Closely planted corn may be harvested for silage before the ears are formed. Figure 16 shows a forage harvester in operation. This is a relatively small harvester designed for fast one-man harvesting of small acreages with 45-85 horsepower tractors. The harvester blows chopped-up stalks into a trailer. The chopped forage will later be blown into a silo to produce silage.

The silo has played an important role in American agriculture. Typically it is a tall cylindrical structure that is sealed when full so as to exclude air. It makes possible prolonged storage of such products as corn, grass, legumes, and other plants without spoilage and with minimal labor. Biochemical reactions inside the silo increase the nutritional value and palatability of the feed. The danger of bloating of animals fed on alfalfa or clover is greatly reduced.

## Corn and Soil Erosion

Greater yield per acre allowed the retirement from production of the land least suited to corn and other row crops. Much of this was used for forage crops, building up soil fertility, and reducing erosion. Although hybrid corn varieties take more nutrient

*Fig. 16. A corn-forage harvester in operation: the harvester blows chopped stalks into a trailer, and the chopped forage will later be blown into a silo to produce silage for livestock feed (courtesy Deere and Company).*

material from the soil than open-pollinated ones, increased yields have actually stimulated soil-conservation and soil-building progress. Unfortunately, rapid increase in world population, together with increased purchasing power of certain nations, has resulted in much of the rescued land being put back into row crops to meet the ever-increasing worldwide demand for food, the United States and Canada remaining as the only important exporters.

It is difficult to prevent erosion where row crops like corn and soybeans are grown. Only soil that is covered with forest or permanent meadow is not endangered by erosion. Even 20 years ago losses of topsoil in much of the Corn Belt, since cultivation began, were estimated at 25-75 percent! (Higbee, 1958). Corn production in Iowa, without a full measure of soil-conservation technology, results in an annual loss of about 15 tons of topsoil per acre (Schrader et al., 1963). Yet topsoil is produced by natural processes at the rate of an average of only about 1.2 tons per acre of cultivated land per year (Pimentel et al., 1975). Our country's future is being washed into the sea.

Referring to Illinois, Baker (1977) points out that the erosion problem is partly masked by the great natural productivity of the soil and the near-perfect climate for crops. Yet 43 percent of the state's 24.4 million acres of cropland is sloping (2 percent slope or greater) and is subject to excessive erosion. In 1974, when heavy rains doubled erosion to about 16 tons per acre from unprotected sloping cropland, this amounted to the loss of about one-tenth inch of topsoil in a year. Fertility loss for the 272 pounds of nitrogen and 66 pounds of phosphate in the removed soil represented $72 per acre. The cost of fertilizer to replace the plant food washed away and of clearing away sediment deposited in places where it could not be tolerated has been estimated to be more than $600 million per year in Illinois alone. Of the 10.7 million acres subject to erosion there, only 1.6 million now have adequate protective treatment. Yet soil management pays in higher income to the farmer. More important, it is an obligation to future generations.

Devastation comparable to that of the southern Appalachian piedmont has been averted only because of the general levelness of the Corn Belt plains. This fortunate circumstance, as well as the original great depth of the soil, should not lessen farmers' vigilance in protecting it as much as possible from erosion, as by crop rotation, contour plowing, strip cropping (fig. 17), and heavy fertilization to increase crop-plant populations and thereby to increase organic matter. The latter increases the water-absorbing and water-holding capacity of the soil, thus reducing erosion, and in various ways increases soil productivity.

Corn Belt land is generally naturally fertile and, when free of serious erosion, corn has been grown continuously through relatively long periods. Eventually this practice was discontinued because of reduced fertility and because of insect damage, especially by corn-root insects. Later, as commercial fertilizers became more generally available, it was found that heavily fertilized and thickly planted corn could be grown continuously. When it is planted primarily for its grain, the seeds are "drilled" mechanically in rows 40 inches or less apart, to obtain a plant density of 20,000-27,000 per acre in humid regions. A planting machine drawn by a 133-horsepower tractor can plant 12 rows of corn on each trip across the field and can also apply fertilizer, herbicides, and insecticides to protect the young plants. Corn can also be planted with a grain drill to obtain 100,000-400,000 plants per acre, grown for forage or green manure.

High plant density per unit area, made possible by high rates of fertilization, reduces the impact of raindrops and slows down the rate of water runoff. The roots tend to hold the soil in place and make it porous and better able to absorb water. The quantity of organic matter from the cornstalks and roots in heavily fertilized soil was found to be greater than that from the average sod crop, another factor favoring water penetration and reduced erosion. Mulch tillage, which leaves corn (and soybean) residue in the surface and reduces soil erosion hazard, is being increasingly practiced.

### Some Cultural Measures

The quantity and kind of fertilizer required can be based on analyses of soil and leaf tissue. Often just before the seed is sown, a complete fertilizer containing nitrogen, phosphorus, and potassium is worked into the soil, followed later by "top-dressings" of either the complete or nitrogenous fertilizers. Soil insects and weeds are then

*Fig. 17. Contour strip cropping near Westby, Vernon County, Wisconsin (courtesy USDA Soil Conservation Service).*

controlled with appropriate chemicals. It is now common practice to plant corn continuously on relatively level land and forage crops on the rolling areas more susceptible to erosion. In some Iowa experimental plots, land that had been planted to corn continuously for fifty years produced an average of 98 bushels per acre when the proper amount of nitrogen was added. In southern Iowa, in soil low in organic matter, such matter was as much increased under highly fertilized continuous corn as it was under a corn-oats-meadow rotation (Hughes and Metcalfe, 1972).

Chemical ripeners, as discussed for sugarcane in the preceding chapter, can also be used successfully to increase corn production. Applied along with a wetting agent as a foliar spray when the unemerged tassels are about a half-inch in length, vegetative growth is slightly interrupted and development of reproductive parts is accentuated. Pollination occurs earlier, there are more ears per plant, the ears are larger, and grain yield is increased 5-10 percent (Wittwer, 1976).

Corn may be included in many types of rotation. Soybeans may be grown between two corn crops, increasing the yield of the second because of the ability of the legumes to add nitrogen to the soil. Corn and soybeans can also be grown at the same time, as in alternating pairs of rows, with increased yield of corn. Winter wheat can be successfully planted between rows of corn spaced 72 inches apart. The wheat has little or no effect on yield of corn, yet allows the growing of full-season corn hybrids. When corn and wheat crops are alternated, lower-producing early-maturing corn hybrids have to be grown in order to be harvested before the normal time for fall wheat seeding (Hughes and Metcalfe, 1972).

### Insect Pests

Seventy-nine species of insects and mites are listed by Dicke (1977) as pests of major or minor importance to corn in the Western Hemisphere. Corn is subject to attack from the time it is planted until it is utilized as food or feed. Among the principal pests are soil insects (rootworms, cutworms, and wireworms), insects attacking the leaf stalk, and ear (corn earworm, European corn borer, corn leaf aphid, fall armyworm, corn borers (*Diatraea*), armyworms, chinch bugs, grasshoppers, corn flea beetle, and Japanese beetle. There are also many beetles and moths that attack corn in storage. Much of the control effort is directed toward adjustments in plowing and seedbed preparation, planting time, crop rotation, and sanitation. Much emphasis is placed on natural enemies in the pest management program.

### Diseases

Corn is a relatively healthy crop, but the widespread epidemic of southern corn leaf blight in 1970 indicated the potential for disease under certain conditions. Bacteria, fungi, and viruses are the common pathogens. Seedling diseases reduce the number of plants in a stand, but the remaining plants tend to compensate by taking advantage of the greater amounts of nutrients and soil moisture. Leaf diseases reduce functional photosynthetic area. Stalk rots and root rots cause lodging, and lodged corn is difficult to harvest by machine. Ear rots cause reduced quality of the grain and one of them, gibberella ear rot, is toxic to swine when 3 to 5 percent of the corn is infected. Corn plants can be weakened or killed by an accidentally introduced parasitic seed plant called "witchweed," first noticed in North Carolina in 1956. Through use of cultural practices and quarantine, the parasite appears to have been contained within infected areas (Ullstrup, 1977).

## SOYBEANS

The soybean (*Glycine max*) is in the pea family, which also includes peas, beans, peanuts, alfalfa, clover, and vetch among the important agricultural crops. Although soybeans have been cultivated for many centuries and on a very large scale in Asia, particularly in China, Manchuria, and Japan, and form a substantial part of the diet of the people, until recently they were a relatively minor crop in other areas of the world. Then when World War II restricted the import of coconut and palm oils from the Orient, substitutes were sought by vegetable-oil producers. The soybean industry developed rapidly in the United States, and about half of the world production was grown in this country by 1955 (fig. 18). In 1976, this country produced 34.4 million metric tons of soybeans, about half of world production and was the only soybean-exporting nation, sending abroad about a third of its output. The soil and climatic conditions best for corn are also those ideal for soybeans, and about 65 percent of the crop is grown in the Corn Belt. The leading states, in descending order of production in 1976, were Illinois, Iowa, Indiana, Ohio, and Missouri—all in the Midwest (USDA Agr. Stat., 1977).

Because of the great geographic range of the soybean there are many varieties. These are adapted not only to the many worldwide climatic and soil conditions but also the great difference in length of growing season. When compared with corn, increase in

*Fig. 18. A field of soybeans in Illinois (courtesy USDA Soil Conservation Service).*

yield per acre of soybeans in the United States has been relatively modest—a 60 percent increase compared with a 300 percent increase for corn in the 30 years preceding 1965 (Hughes and Metcalfe, 1972). However, drilled, noncultivated soybeans with 6-8-inch spacing yielded up to 50 percent higher than the same varieties grown in wide, cultivated rows. This is made possible with the better herbicide combinations now available, and the closed canopy normally formed within 35-40 days after planting helps to control late-season weeds (Anonymous, 1977*b*).

A number of experiment stations have cooperated with the USDA in attempting to combine the more desirable characteristics of several varieties into one. First a higher oil content of the seed was sought; then with increased prices for soybean meal for livestock and poultry, more emphasis was placed on protein content. Other characteristics desired were resistance to lodging, shattering, and disease; proper maturity; better seed quality; and pods set high enough aboveground to minimize losses when harvested (Hughes and Metcalfe, 1972).

There are varieties of soybeans that contain 45 percent protein on a dry-weight basis. When the oil is extracted and seed coats removed, the protein in the remaining oil meal or oil cake is increased to about 60 percent and additional processing may increase the concentration of protein still further to 70 percent, higher than that of meat on a dry-weight basis. The daily requirements of an adult for protein, vitamins, and minerals can be supplied by a cup (ca. 170 g) of soybean concentrate. The soybean is the most efficient food crop in yield of protein and in relation to total fossil energy input per acre of cropland (table 1). Soybean meal is now being used in meat extenders or meat substitutes and is finding its way into many human foods and high-protein beverages. Soybean oil is used in industry, but over 90 percent of it is used in edible products such as margarine, mayonnaise, shortening, and salad and cooking oils.

The soybean has met the fate of all leading crops in that repeated selections have reduced its genetic diversity, that is, it shares a limited number of ancestors. For example, the six varieties that were grown on two-thirds of the soybean acreage in the North Central States in 1972 all had the variety Mandarin as an ancestor. Therefore the collection of many primitive and wild soybean strains for the germ plasm collection at the U.S. Regional Soybean Laboratory at Urbana, Illinois, made by R. L. Bernard in Japan and South Korea in 1974, was an event of particular importance. The collection was increased to more than 2,500 soybean strains, a veritable goldmine that plant breeders can explore for valuable traits such as disease resistance, better oil quality, or higher photosynthetic efficiency (Steyn, 1974, 1975).

The usual rise in production from increased amounts of fertilizer added to the soil, as obtained with most crops, is not obtainable with soybeans. Legumes such as soybeans, especially when inoculated with nitrogen-fixing bacteria, obtain much of the nutrient from the organisms. Nitrogenous fertilizers merely reduce the activity of the bacteria and do not increase yield appreciably (PSAC, 1967). Nevertheless, in this time of fertilizer shortage it is noteworthy that soybeans produce a valuable crop while adding nitrogen to the soil (Sears, 1939). Whereas crops of corn, wheat, and oats remove 50, 26, and 36 pounds of nitrogen per acre, respectively, a crop of soybeans may add 16 pounds. Corn following soybeans yielded an average of 8.9 more bushels per acre than corn following corn.

With foliar sprays of nutrients, yields of soybeans have been increased 10-20 bushels per acre in fields already highly productive. Two or three foliar sprays containing nitrogen, phosphorus, potassium, and sulfur are applied 10-14 days apart. Foliar application of major nutrients as well as minor elements is an efficient means of boosting productivity of many crops above that which can be obtained by relying on the soil alone. With nonrenewable fertilizers rapidly increasing in cost, foliar-spray technology should receive an added stimulus for further development (Wittwer, 1976).

## OTHER BEANS

Broad beans and soybeans and a few other types were well known in the Old World before Columbus's discovery of America, but the choice food beans of the world, all in the genus *Phaseolus,* were found only in the New World. Archeological discoveries reveal that *Phaseolus vulgaris,* the kidney, field, garden, or haricot bean of English-speaking peoples or the *frijol* of Latin America, was under at least incipient cultivation in central Mexico from 5,000 B.C. (Schery, 1972).

By the time Columbus landed, the various forms of the kidney bean had been distributed from Peru to New England. It is well known in the "bush" and "pole" forms, yielding the familiar "green," "string," "snap," or "wax" beans when the entire immature pod is eaten. The bulk of production is for the dry seeds shelled from the pod. The bean types in the United States in 1976 were, in descending order of importance, pinto, navy, great northern, red kidney, pink, blackeye, large lima, small red, baby lima, small white, and a few others. Combined, they amounted to 781,000 metric tons (USDA Agr. Stat., 1977).

The Midwest leads in the production of dry kidney beans, but principally because of the enormous area planted to that crop in Michigan. Like all other plant crops, beans have responded to improved agricultural technology. Yield and quality have been increased through release of new and improved varieties and with improved fertilization practices. Mutation induced by radiation played an important part in the development of several disease-resistant, early-variety bush beans. The mutant was used in crossing programs that led first to the Sanilac navy bean (1956), then Seaway (1960), Gratiot (1962), and Seafarer (1967). The new varieties have virtually eliminated losses from weather damage, white mold, anthracnose, and common bean mosaic. In Michigan the production of navy beans doubled within a few years (Anonymous, 1971).

In 1976 Michigan produced 221,098 metric tons of dry kidney beans—28 percent of the nation's total. Florida is the leading state in the production of green beans, with 58,258 metric tons—43.5 percent of the nation's total and including the only winter crop. Wisconsin was the leading Midwest state in the production of green beans for processing, its 170,250 metric tons exceeding the 136,350 tons produced by Oregon (USDA Agr. Stat., 1977).

### Nutritional Quality of Legumes

Second only to the cereals as a source of animal and human nutriment, the legumes rank first from a qualitative standpoint. Legumes have two or three times more protein than the cereals and they supply lysine, the essential amino acid that is limited in cereals. Legumes effectively complement the cereals to establish a balance of essential amino acids in the human diet and are particularly helpful as a proteinaceous complement to cassava and plantains in the primitive tropical areas where the protein-deficiency disease, kwashiorkor, is common.

Although they are a significant protein source for many of the developing countries, beans are deficient in methionine, an amino acid vital to human nutrition. Without methionine, the body can live but it cannot grow. Plant breeders are attempting to select plants for use in breeding new bean varieties with a higher methionine content. Such improved hybrids would be a dietary boon to such areas as India and Latin America, where large portions of the population depend on beans as a major source of protein.

## PEAS

The pea, *Pisum sativum,* was probably domesticated in central Asia. There appears to be no indication that peas were eaten "green" until the late Middle Ages, but by the sixteenth century the more sugary green or garden peas had become popular, particularly in France, and their popularity has persisted in the Western world. They are consumed on a larger scale than dried peas, the "split peas" much used for soup.

Peas are adapted to a moderately cool growing season and moderate rainfall. A threshing machine was invented that can remove green peas from vines and pods, permitting development of the pea processing industry. As with most crops, the plant breeders have made their important contribution, having developed varieties that produce mature pods all about the same time. Pea plants may be used as hay, in combination with grasses, and sometimes they are used as manure (Schery, 1972).

The Midwest is the largest producer of green peas, mainly because of the great acreage of this crop in Wisconsin and Minnesota, which in the three years 1974 to 1976 produced an average of 151,153 and 80,767 metric tons, respectively, of shelled green peas. Dried-pea production is concentrated mainly in Washington and Idaho (USDA Agr. Stat., 1977).

## SUNFLOWERS

Sunflowers are composites (family Compositae) and are one of the few native cultivated crops of the United States. Indigenous to the Great Plains, the common sunflower spread into waste places in many other parts of the country. The oily seeds, with their brilliant purple dye, were among the favorite prizes of the seed-gathering Indians. They parched the seeds and then ground them into a meal. The common camp follower sunflower, which resulted from hybridization resulting from man's activities, fitted into the strange new niches developed by man. Eventually a freak unbranched form developed from which the domesticated single-headed sunflowers have descended (Anderson, 1952).

In 1976 the Soviet Union produced over half of the approximately 3.6 million metric tons of the world's production of cultivated sunflower (*Helianthus annuus*) seed oil. Sunflower-seed oil was exceeded in tonnage only by soybean oil (10.2 million) and peanut oil (3.7 million) among the edible vegetable oils. However, it exceeded in tonnage such important oils as rapeseed (2.8 million), cottonseed (2.7 million), and olive (1.7 million) (USDA Agr. Stat., 1977). The high-oil varieties of the sunflower were developed in eastern Europe, but they are becoming increasingly important as a food crop in the United States. This country now has 2 million acres of sunflowers under cultivation and is the world's foremost exporter of whole sunflower seed and a major sunflower oil exporter.

The seeds of the better varieties of cultivated sunflowers may contain as much as 50 percent oil. Some varieties may grow to a height of 15 feet and have heads 20 inches in diameter. In contrast, there are dwarf types with the highest oil content and the most easily harvested mechanically.

Sunflower oil, utilized as a shortening and in the manufacture of margarine, is said to be much like safflower and more stable than soybean. It is very high in unsaturates. After the oil is extracted, residues are 35-50 percent protein, suitable for stock feed but not as valuable as soybean meal because of lower lysine content of the protein. Unprocessed seed may also be used directly as a feed for poultry, caged birds, and livestock. The plant may be used for silage or as a green manure (Schery, 1972). Also, the fiber of the plant and head makes an excellent wallboard.

Kansas has been called the "Sunflower State" solely on the basis of its indigenous wild varieties. The leading producer of commercial sunflower seed is North Dakota, which has 400,000 acres of commercial sunflowers, and produces 60 percent of the nation's crop, followed by Minnesota with 200,000 acres and South Dakota with 60,000 acres. In the Dakotas the principal production areas are in the eastern part of the states, well within the Midwest as defined in figure 2. In 1973 the value of the crop in the three states was $36 million. Currently two-thirds of the seeds are crushed into sunflower oil, one-sixth are packaged as birdseed, and one-sixth earmarked for human consumption.

As with other food crops, continued progress is being made by plant breeders in increasing sunflower seed production. The largest heads on some of the plants have as many as 2,500 seeds. An acre has about 20,000 plants and some growers are averaging 1,400 pounds of seed per acre (Hillinger, 1974). The relatively short growing season of about 100 days results in sunflowers having potential for producing good yields following irrigated wheat in a double-cropping system (Anonymous, 1975).

Since 1965, the USDA has conducted an active sunflower breeding program to develop hybrid varieties with high yield, disease resistance, and uniformity for economical harvesting. Also, in a Public Law 480 project, a multidiscipline research team of Yugoslav scientists is developing hybrid sunflower varieties that are superior to presently grown, open-pollinated types. They have obtained breeding lines and germ plasm from all over the world, primarily from Rumania and the Soviet Union. Much of this genetic material is now available to USDA plant breeders (Guilford, 1976).

## DECIDUOUS FRUITS

A thriving deciduous fruit industry has developed along the shores of the southern Great Lakes (fig. 3, Region L). Spring temperatures remain cold, retarding bud development until danger of frost is reduced, but midsummer temperatures are cooler than they are inland, ensuring highest quality. The well-drained soil of the lacustrine deposits and glacial till that characterize the Great Lakes region is highly regarded by horticulturists. All these factors, combined with proximity to large urban markets, have favored the growing of deciduous fruits.

In the Midwest, Michigan is the principal fruit-growing state. In 1976, its 217,785 tons of apples were rivaled by Colorado and exceeded by New York, but all states were dwarfed by Washington's whopping 1,020,871 tons.

### Cherries

Cherries are the only fruit that the Midwest produces in greater quantity than any other region in the United States, principally because of Michigan's predominance among all states of the nation in the production of this crop. Although known mainly for its tart cherries, Michigan also produces a considerable tonnage of the sweet varieties, being exceeded only by the Pacific Coast states. Sweet cherries are well adapted to western lower Michigan, particularly the northwest corner. About 70 percent of the sweet cherry crop is sold to briners and processed as maraschino or glacé cherries. These cherries are picked before fully ripe and are less likely to crack when it rains. About 20 percent of the crop is canned and 10 percent, principally dark varieties, is sold on the fresh market. All Michigan sweet cherry varieties require cross-pollination with certain others, so two and preferably three types should be planted together. The Montmorency is the only tart cherry variety recommended for commercial planting in Michigan (Hull et al., 1975).

Among tart varieties, Michigan, with 103,000 tons, produced 78 percent of the nation's crop in 1974 (USDA Agr. Stat., 1977). In 1964 only 3 percent of the state's red tart cherry crop was harvested mechanically, increasing to 80 percent by 1969.[4] The cherries are then cooled in water-filled cooling tanks with a capacity of about 25 cubic feet and holding 1,188 pounds of cherries. Holding the cherries at temperatures of 60°F or below for 4-8 hours helps prevent scald, facilitates pitting, and increases yield of the finished product. As soon as a tank is filled with cherries it is taken to a concrete slab and filled with cold water. The water is introduced below and as it circulates through the load of fruit it carries with it dirt, stems, leaves, and heat and overflows at the top (Cargill et al., 1970).

## ALFALFA

Forage crops—grasses and legumes grown as feed for livestock—are mostly consumed on the farm where they are grown, so are underrated in the statistics of agricul-

[4]Mechanical harvesting systems utilize reciprocating limb or trunk shakers with various types of catching devices (Bolen et al., 1970).

ture. However, they take up about five times more land than all grain crops. Forage crops and the animals that feed on them are the backbone of the national agricultural economy. As the demand for grain increases, less is fed to livestock. For example, the percent of forage in the feed of beef cattle increased from 73 percent in 1972 to 82 percent in 1974 as foreign purchases of grain increased (Hodgson, 1976).

The present-day emphasis on preserving and improving our environment has led to special interest in legume crops such as alfalfa (*Medicago sativa*), which help to maintain and improve the soil while providing animal and human sustenance. The United States, with over 26 million acres, is the world's largest producer of alfalfa. It is this country's most important forage crop, with an annual value (1975) of about $4 billion. Argentina, with 18.5 million acres, is likewise a major producer of alfalfa. The USSR, Italy, France, Canada, Australia, and Rumania also have vast acreages. In the United States over 60 percent of the alfalfa acreage is in the north-central region, Wisconsin leading with about 3 million acres.

Alfalfa, also known as lucerne, has been called the "Queen of the Forages." It is one of nature's greatest boons to mankind, being a crop of high quality that yields heavily and is able to adapt itself to a wide variety of climates and soils. It can survive prolonged drought because of its deep taproot, which can reach soil moisture at depths of 20 feet or more. It can become dormant in periods of drought and cold and then again resume growth. Because of a symbiotic relationship with certain bacteria, *Rhizobium meliloti,* which fix free nitrogen from the atmosphere in nodules on the plant's roots, alfalfa increases soil nitrogen. Thus it effectively utilizes photosynthesis and biological nitrogen fixation, the two most important energy-producing processes on earth. The present-day emphasis on conserving and improving our environment has led to renewed interest in crops that help to maintain and improve the soil while providing animal and human sustenance.

Alfalfa also could be an important aid in prevention of erosion. Alarmed at increasing erosion in an area of extensive study in south-central Wisconsin, Brink and his associates (1977) recommend an increase in the ratio of alfalfa or other meadow crops to row crops such as corn. Instead, a decrease is now taking place in response to current market forces. Brink also calls for more forage research funds (now notoriously meager) in erosion-threatened areas.

Probably indigenous to Iran, alfalfa is the only forage crop known to have been cultivated before the era of recorded history. Persian legions under Darius brought it to Greece and sowed it there to provide feed for their chariot horses and cattle. The succulent tops were boiled and used as greens by the soldiers. Alfalfa eventually was carried to Rome and then throughout Europe, South America, and Mexico. In the United States, it did not thrive when planted along the Atlantic seaboard by the early colonists, but was a great success when brought to California during the 1849 gold rush. This "Chilean clover" proved to be well suited to the sunny and dry climate and irrigated soils of the Southwest and it was also soon planted by the Mormon settlers in Utah. It did not have the necessary winter hardiness for the northern United States and Canada, but a German immigrant by the name of Wendelin Grimm brought 15-20 pounds of seed of a hybrid alfalfa from his native land to Minnesota in 1857. With constant selection for a number of years, a very hardy strain was eventually developed, the winter-hardy Grimm alfalfa. The latter is an excellent example of the great contribution to agriculture that can be made by plant introduction coupled with a vigorous program of selection. Plant breeders developed varieties resistant to diseases and insect pests, and improved cultural practices resulted in increased production of forage and seed. Selective herbicides have been an important factor for increased production, particularly when used in spring seedings.

Much alfalfa hay is now cubed instead of baled. "Cubes" measure about 1 x 1¼ inches. Livestock will consume from 12 percent to 20 percent more alfalfa if it is cubed than if it is baled (Hall et al., 1975).

*Rhizobium* nodulates other species of *Medicago* besides alfalfa, such as sweet clover, fenugreek, bur clover, button clover, and barrel medic, and therefore these plants are all very good cover crops. The primary product of nitrogen fixation in the nodule is ammonia, but it is quickly converted to alpha amino compounds and assimilated by the plant. Nitrogen-fixing bacteria are often applied to leguminous seeds or to the soil, usually the seeds. Inoculation is needed in most soils throughout the world, particularly new ones, and has increased production from 15 to 100 percent. It is not considered necessary on the high-calcium soils of most of the western United States, but is strongly recommended on the more acid types of the Pacific Northwest (Weber and Leggett, 1966).

Because alfalfa is not dependent on nitrogen from the soil, it is a dependable and economical protein source for grazing animals. The high quality of its protein makes it especially valuable for nonruminants on pasture, such as swine, poultry, and horses. It is also an excellent source of calcium, magnesium, phosphorus, and vitamins A and D, a fact that has not been overlooked by health-food stores, in such products as alfalfa-juice tablets. In fact, those who advocate less reliance on use of meat as human food, because of inefficient energy conversion in its production, point to alfalfa as a source of vegetable protein for humans because of its high yield of protein per acre.

For thousands of years the seeds of cereal plants have supplied the bulk of the protein for human consumption; yet these seeds contain only 7-12 percent protein. Contrary to popular belief, green leaves, whose biochemical activities directly or indirectly sustain all life, have a far greater potential than grains as a source of protein. Depending on the species and stage of maturity, green leaves contain 20-30 percent protein on a dry-matter basis. Furthermore, leafy forage plants can provide this vital substance of our diet while safeguarding the land on which they are grown from erosion.

One can hardly imagine a plant more suitable than alfalfa for producing leaf protein. It fixes the nitrogen required for its own growth, requires reseeding only every 4-10 years, thrives over an enormous range of soil and climate, and, depending on locality, may be cut 3-11 times a year. Table 1 shows that alfalfa is the most efficient of all crops in yield of protein and food energy in relation to fossil-fuel energy and man-hours of labor input per hectare. However, it is of interest chiefly as an animal feed.

In an effort to improve the present method of dehydrating alfalfa meal for livestock and poultry, USDA scientists have developed a wet process called Pro-Xan II. Fresh chopped alfalfa is run through rolls to squeeze out a large volume of juice. From this a green carotene-xanthophyll concentrate for poultry feed and a white protein concentrate for human food are obtained. The latter concentrate has a good balance of amino acids and can serve as an ingredient for formulating high-protein flours and for fortifying many other types of food.

Much research with alfalfa has been devoted to pest control. More than a hundred species of insects have been recorded as injurious to the plant. Although in any one year only a few species cause serious economic damage, insect loss to alfalfa forage and seed is estimated to be about $260 million annually. Some insects are held in check by natural enemies, others by cultural practices or by growing resistant varieties, but some can be controlled only by insecticides, particularly during outbreaks. It is important to use the right insecticide at the right time and according to directions printed on the registered label (App and Manglitz, 1972).

## HOGS

As might be expected from its fabulous corn yield, the Midwest is also dominant in hog production. (A hog is an adult or growing swine of over 120 pounds. A pig is a young swine, not usually mature.) Iowa with 14.2 million hogs and pigs and Illinois with 6.4 million had between them 37.4 percent of the nation's total of 55 million head in 1976 (USDA Agr. Stat., 1977). Many hogs are also raised in the South, but surprisingly few in the West. All the eleven western states combined had only 1.6 million head, less than 2 percent of the country's total, and 43 percent of these were in Colorado and Montana, the principal agricultural regions of which lie in what is considered the Great Plains region (in fig. 2). Thus the western states, an excess beef- and sheep-producing area, do not raise enough hogs to meet local needs in many areas, including California. Hogs and cured pork products are shipped west from the Corn Belt.

Swine are known to have been domesticated in Asia Minor as long as 8,500 years ago, and have been an important source of meat and fat the world over since ancient times. However, the slow-moving swine did not lend themselves well to the needs of nomadic peoples, and the meat could not be kept long without risk of food poisoning. Possibly this explains the unpopularity of swine among the ancient nomadic Muslims and Hebrews, who banned them from their diet. Among other peoples swine were an important part of the diet. They were so plentiful in ancient Rome, for example, that they came to be regarded as a food fit only for common people. Among the upper classes the pig was used only as a casing to stuff with fowl, vegetables, and fruits. In many countries this custom survives in the form of a boiled or roasted suckling pig with an apple in its mouth (Raskin, 1971).

Probably the first swine to be brought to the New World arrived with Columbus, to provide food for him and his crew; however, some escaped. They and their descendants were able to survive in the New World wilderness, eating rodents, snakes, acorns, nuts, fruits, and bulbs that they could root out of the ground with their tough snouts. Swine were also brought over by the early European immigrants, but the development of breeds was largely left to chance.

Contrary to popular opinion, swine are not by nature dirty animals. They lack sweat glands and are apt to suffer from overheating if they cannot find water in which to bathe and relieve the heat. They wallow in mud only when clean water is not available.

Great ingenuity has been demonstrated in the recipes that have evolved in making almost any part of the swine palatable. Besides the familiar ham, bacon, and various cuts of fresh pork, there are salt pork, pork sausage, barbecued spareribs, pickled pigs' feet, and souse, a cooked and soaked combination of pigs' feet, ears, snouts, and tongues. There is also fried pigskin or "cracklings," and even the small intestine is fried to make "chitlings," one of the "soul foods" of Afro-Americans. From the bones of swine is made bone meal, a food supplement for cattle, sheep, and other domesticated animals (Raskin, 1971).

The trade recognizes two types of hogs: those with just enough "finish" to produce carcasses of desired quality are "lean- or meat-type" hogs, while those with more finish than necessary are "fat-type." In accordance with the current trend to include less fat in the human diet, hogs tend to be leaner than they used to be. In years of experimentation in Beltsville, Maryland, USDA scientists have developed hogs that produce less lard and more meat. The low-fat-type hogs are more efficient in converting high-protein rations into lean meat. Hogs with carcasses weighing 200-220 pounds and having a back-fat thickness of approximately 1.3 inches are the most desirable in the trade. Those weighing more than about 250 pounds have an excessive amount of lard, and the large cuts are considered to be less suitable to the average consumer than those from lighter-weight hogs. Mature swine of any breed may occasionally reach a weight of 1,000 pounds or more, but the raiser generally aims for his animals to attain a market weight of 200-220 pounds at 5 months or more of age.

## Breeds of Swine

There are a surprising number of old swine breeds that have remained commercially important. Most of them were obtained from Great Britain. The British in turn had imported Chinese swine to mate with British varieties. The Chinese swine were noted for great fertility and rapid growth. One of the oldest breeds we have in this country is the Berkshire, which originated in the county of the same name in England. It is black with white points on the feet and usually a splash of white on the face. It is a solid hog, usually free from surplus outside fat. The Chester White originated in Chester County, Pennsylvania. The Duroc was bred from strains of red hogs developed in New York and New Jersey, and is known for hardiness and prolificacy. The Hampshire, which originated in Hampshire County, England, has become one of the popular breeds. A white belt encircles the fore part of the black body and includes both forelegs. The Hereford hog is red with white markings much like those of Hereford cattle. The Poland China hog originated in Butler and Warren counties, Ohio, and was derived from the crossing of several breeds. This strain is usually black, but many are spotted with white. The Spot hog resembles the Poland China in body type, but has many white spots. The Tamworth, one of the oldest of all breeds, is named after the town of Tamworth, England. It is light to dark red and has erect ears. The Yorkshire breed came from Yorkshire County, England. It is white, with occasionally black spots (USDA, 1968*b*).

Prominent among the newer breeds is the American Landrace, descendant of the Danish Landrace hogs that were imported by the USDA in 1934. Some of the foundation animals carried a trace of Poland China. The American Landrace is white, has a good body length (sixteen or seventeen pairs of ribs), and is prolific. More recently the Pietram was introduced from Belgium to conform to the latest trend in swine, for it is small and has a high percentage of lean meat.

Records of all the recognized breeds are kept by various associations with names such as American Berkshire Association, Chester White Swine Record Association, and United Duroc Swine Registry. These swine-breed associations, as well as county agricultural agents and state or local swine-growers groups, are sources of information on raising market hogs. Many commercial producers raise crossbreed hogs of completely different ancestry to obtain hybrid vigor. In these crossbreeding programs rotation of three or four breeds has been found to be helpful in maintaining hybrid vigor. Inbred lines, registered by the Inbred Livestock Registry Association, may be purchased for the purpose.

## Hog Cholera

At one time as many as six million hogs a year died in the United States from hog cholera. As in so many serious problems that beset the livestock industry, the USDA Bureau of Animal Industry came to the rescue. A bureau bacteriologist, Marion Dorset, discovered that the disease, which had been believed to be caused by a bacterium, was actually caused by a filterable virus. Dorset and his co-workers then perfected a protective serum that began to be produced around 1907. In 1935 Dorset developed a crystal-violet vaccine for hog cholera that could be used to better advantage under some conditions (Harding, 1947).

The major tools for regulatory agencies had been provided. A massive federal-state-farmer-industry cooperative effort has eradicated hog cholera from this country. It was a remarkable scientific and technical achievement that had an impact on animal health comparable to the elimination of polio in humans (Memolo, 1978).

### Meat Inspection Act of 1906

In the nineteenth century, diseased livestock were often slaughtered and processed for food. Foreign buyers had all but discontinued importations of our beef and pork because of its poor quality. Finally an effective law was passed in 1891 to provide inspection, but it was used mainly to regain export outlets. The situation remained bad in the United States and was exposed in shocking detail in Upton Sinclair's *The Jungle.* President Theodore Roosevelt became aroused and Congress passed the Meat Inspection Act on June 30, 1906, the same date that the Food and Drug Act was passed. The Secretary of Agriculture was empowered to provide the inspection service. Inspected meats are marked by means of a harmless and indelible marking fluid (Harding, 1947).

## BEEF CATTLE

There are seven genera in the cattle family (Bovidae), including the bison, yak, and the Indian and African buffalo, as well as what we generally recognize as cattle. Nothing is known of the initial phases of the domestication of cattle, but they probably originated in Central Asia and spread to Europe, China, and Africa. Cattle are the most important of all livestock animals, supplying about 50 percent of the world's meat, 95 percent of its milk, and 80 percent of the hides used for leather.

Trenkle and Willham (1977) remind us that historically cattle have "served as power; refuse scavengers; a means of transportation of the grain after consumption; producers of fertilizer; a highly flexible food reserve; sources of fiber, leather, and biochemicals; harvesters of forage from adjacent nontillable land areas; as well as sources of high quality protein that nutritionally complements basic grain diets." Regarding biochemicals, they point out that "there is no efficiency ratio that can be used to include the value of insulin extracted from the pancreas."

The amazingly rapid development of railroads[5] and the introduction of the steel plow and mechanical reaper, opening the tallgrass prairie of the Midwest to agricultural exploitation, resulted in a rapid westward expansion of cattle raising. The English and Scots not only supplied most of the capital that financed cattle-raising on the western ranges but also contributed most of the cattle breeds. The industry grew by leaps and bounds until by 1970 in twenty-two of our fifty states beef cattle were the most important agricultural product, with respect to cash receipts from farm marketings, as shown in table 2. In thirteen states, dairy products were the most important; in four states, broilers; and in one, hogs. Thus in forty states, livestock and livestock products brought the greatest incomes to farmers. For the entire country, livestock and livestock products brought farmers a gross income of $29.6 billion compared with $19.6 billion for plant crops. Nationwide, the relative amount of cash receipts of the five most important crops was, in descending order, cattle, dairy products, hogs, corn, soybeans (table 2). This would not reflect the true importance of corn, however, for without corn livestock could not attain its present importance.

[5]The number of railroads serving Chicago increased from one to thirteen in the brief period between 1852 and 1856.

Texas raises far more cattle than any other state. In 1976 Texas had 15.6 million cattle and calves compared with Iowa's 7.5 million. States with between 6 and 7 million head were Missouri, Nebraska, Kansas, and Oklahoma. California had 5 million head (USDA Agr. Stat., 1977).

Beef accounted for about two-thirds of all red meat consumed in the United States. Per capita consumption of beef rose from 85 pounds in 1960 to 114 pounds in 1970 and was projected to reach 128 pounds in 1980 (Van Arsdall and Skold, 1973).

### Feedlots

The number of beef cattle has been rapidly rising in the United States, but the number of feeder cattle, that is, the number finished in feedlots, has been rising at an even greater rate. As their incomes have risen, consumers have shown a preference for beef from grain-fed animals, as opposed to meat from "vealers" and grass-finished stock. While the number of calves dropped from 25 percent of the total of all cattle slaughtered in 1960 to 11 percent in 1970, grain-fed cattle rose from 51 percent to 69 percent of the total during the same decade. Also the percentage of cattle marketed from feedlots of greater than 1,000-head capacity has been steadily rising, reflecting, as do all other branches of agriculture, the greater efficiency and profitability of larger agricultural operations (Gustafson and Van Arsdall, 1970).

It has not been very long since nearly all beef-cattle feeding was done in the Corn Belt, where surplus grain was fed to cattle from the range areas. Then feeding began in other areas, particularly in the Plains States, California, and Arizona. There is no way of determining the numbers of cattle in the Midwest and Great Plains as these are delineated in fig. 2, for records are kept by states and the boundary separating the two regions runs through an entire tier of states from north to south. The Midwest appears to be surpassed by the Great Plains and the South in numbers raised. However, in numbers of cattle fed in feedlots, the Midwest appears to surpass all other regions. Enormous numbers are also fed in feedlots on the Great Plains, and the two areas combined handled about three-fourths of the nearly 19 million feeder cattle marketed in 1965. California also fed many feeder cattle—about 2.3 million. The South, which raised about a fourth of the beef cattle in the United States, had relatively few feedlots. The percentage of cattle finished in feedlots is increasing continuously, but by far the greatest rate of increase is taking place in the southern Great Plains. Only the Midwest and the Great Plains have a surplus of feed grains over the demand for their feedlots. All other regions must import such grains, the net deficit being particularly great in the South (Gustafson and Van Arsdall, 1970).

Farmer feeders—defined as operators who can feed no more than a thousand cattle at a time—comprised 99 percent of the total number possessing feedlots in the country in 1968. There are farmer feeders in nearly all regions, but most are in the Midwest and northern Great Plains. Their feedlots usually supplement their farming or ranching enterprises, which might entail developing field crops and hogs or other livestock. Although the number of feeders has been decreasing, most of the decline has come among those with feedlots marketing less than a hundred head annually. Feeders with larger enterprises have increased in numbers and so has their production of fed cattle. The farmer feeder has the advantage of being able to utilize off-season labor, nonsalable roughage feed, and other low-cost inputs, when compared with specialized, single-enterprise feedlots.

"Commercial cattle feedlots" are those with a capacity of a thousand or more cattle. In 1968 these represented only 1 percent of all cattle feeders, but produced nearly half of all fed cattle. They are particularly important in the newer cattle-feeding areas. In

**Table 2**
**Principal Agricultural Products of the Fifty States in Order of Cash Receipts in 1970 (from USDA Misc. Publ. 1063)**

| State | Principal Products in Order of Cash Receipts |
|---|---|
| Alabama | Broilers, cattle, eggs, hogs |
| Alaska | Dairy products, potatoes, eggs, cattle |
| Arizona | Cattle, cotton lint, dairy products, lettuce |
| Arkansas | Soybeans, broilers, cattle, cotton lint |
| California | Cattle, dairy products, grapes, eggs |
| Colorado | Cattle, wheat, dairy products, sheep, lambs |
| Connecticut | Dairy products, eggs, tobacco, greenhouse nursery |
| Delaware | Broilers, corn, soybeans, dairy products |
| Florida | Oranges, cattle, dairy products, grapefruit |
| Georgia | Broilers, eggs, peanuts, cattle |
| Hawaii | Sugarcane, pineapple, cattle, dairy products |
| Idaho | Cattle, potatoes, dairy products, wheat |
| Illinois | Corn, hogs, soybeans, cattle |
| Indiana | Hogs, corn, soybeans, cattle |
| Iowa | Cattle, hogs, corn, soybeans |
| Kansas | Cattle, wheat, hogs, sorghum grain |
| Kentucky | Tobacco, cattle, dairy products, hogs |
| Louisiana | Cattle, soybeans, rice, dairy products |
| Maine | Potatoes, eggs, broilers, dairy products |
| Maryland | Broilers, dairy products, corn, cattle |
| Massachusetts | Dairy products, greenhouse nursery, eggs, cranberries |
| Michigan | Dairy products, cattle, corn, hogs |
| Minnesota | Cattle, dairy products, hogs, corn |
| Mississippi | Cattle, cotton lint, soybeans, broilers |
| Missouri | Cattle, hogs, soybeans, dairy products |
| Montana | Cattle, wheat, barley, dairy products |
| Nebraska | Cattle, corn, hogs, wheat |
| Nevada | Cattle, dairy products, hay, sheep, lambs |
| New Hampshire | Dairy products, eggs, cattle, greenhouse nursery |
| New Jersey | Dairy products, greenhouse nursery, eggs, tomatoes |
| New Mexico | Cattle, dairy products, cotton lint, hay |
| New York | Dairy products, cattle, eggs, greenhouse nursery |
| North Carolina | Tobacco, broilers, eggs, hogs |
| North Dakota | Wheat, cattle, dairy products, barley |
| Ohio | Dairy products, cattle, hogs, soybeans |
| Oklahoma | Cattle, wheat, dairy products, hogs |
| Oregon | Cattle, dairy products, wheat, greenhouse nursery |
| Pennsylvania | Dairy products, cattle, eggs, mushrooms |
| Rhode Island | Dairy products, greenhouse nursery, potatoes, eggs |
| South Carolina | Tobacco, soybeans, eggs, cattle |
| South Dakota | Cattle, hogs, dairy products, wheat |
| Tennessee | Cattle, dairy products, tobacco, soybeans |
| Texas | Cattle, sorghum grain, cotton lint, dairy products |

Table 2 *(continued)*

| State | Principal Products in Order of Cash Receipts |
|---|---|
| Utah | Cattle, dairy products, turkeys, sheep, lambs |
| Vermont | Dairy products, cattle, eggs, forest products |
| Virginia | Diary products, cattle, tobacco, peanuts |
| Washington | Wheat, dairy products, cattle, apples |
| West Virginia | Cattle, dairy products, apples, eggs |
| Wisconsin | Dairy products, cattle, hogs, corn |
| Wyoming | Cattle, sheep, lambs, sugar beets, dairy products |
| United States | Cattle, dairy products, hogs, corn, soybeans |

Arizona, California, and the southern Great Plains, more than 90 percent of the fed cattle are marketed in the large, commercial feedlots. The farmer feeder dominates only in the Midwest and Pennsylvania.

Individual commercial lots have a capacity of up to over a hundred thousand head. Most of them purchase all or most of their feeder cattle and feeds and utilize up-to-date technology, modern equipment, and trained management. A large lot might have feed-milling equipment similar to that of many feed companies, and might have its own veterinarians, nutritionists, buyers, and other trained specialists. Cattle and feed are both weighed so that gains, costs, and efficiency can be determined. Detailed records are maintained, often with electronic accounting systems. Sometimes cattle in feedlots are fed by custom cattle feeders who charge a fee for their service without taking ownership of the cattle. Such feeders commonly have cattle of their own. Some obtain capital by custom-feeding for others, then build and operate a feedlot that is large enough to perform economically (Gustafson and Van Arsdall, 1970).

## Cattle Breeds

As in the case of swine, most cattle breeds also came to North America from the British Isles. Differences in pre- and postweaning weight gain are relatively small among the three British breeds—Angus, Hereford, and Shorthorn—and the polled types of the latter two. Also differences in meat palatability and tenderness are not great. Some of the breeds on this continent are brought from Europe or are new breeds based on Brahman-European crossbred foundations. These, and crosses between them and British breeds, grow faster than the British alone, and some produce carcasses with less external fat and higher yields of trimmed preferred retail cuts. However, they ordinarily do not have as much marbling and do not grade as high as USDA quality grade standards (Putnam and Warwick, 1975).

The following brief account of American beef-cattle breeds is based on a paper by USDA scientists P. A. Putnam and E. J. Warwick (1975).

The earliest of the well-known British breeds of cattle to be brought to America was the Devon, "the old red cow" that supplied the early English settlers with milk, beef, oxteams, and leather. It is characterized by its rich red coat of "somewhat mossy" hair, yellow skin, and white, black-tipped, medium-sized horns. The Devon originated in the grassy hills of Devonshire in southwestern England, famous for the high quality of its beef, and was brought over by the colonists as early as 1623. It is now most commonly found in Maryland, South Carolina, Mississippi, Louisiana, Texas, and Oregon.

The Hereford, with red body, white face and front, white legs below hocks and knees, and medium-sized horns is a popular breed that came from Hereford County, England. The permanent establishment of the breed in the United States began with importations made in 1840. Herefords are docile and easily handled, yet tend to produce more calves under rigorous conditions than many other breeds. That may be why they became particularly popular in the West.

The polled Hereford originated as a distinct breed in 1901, when naturally occurring polled animals (ten females and four bulls) from all parts of the country were purchased by an Iowa cattleman. In subsequent breeding, a chance mutation resulted in the gene carrying the polled trait becoming dominant. Polled Herefords were then crossed with the horned breed and the polled descendants were used to improve the type.

The Angus has also been in America a long time. In 1883 a retired London silk merchant imported four Angus bulls into the United States and crossed them with Texas Longhorn cattle. Soon more Angus cattle were introduced from Scotland. The Angus breed is polled and has a black, smooth-hair coat. It is now found in every state and in Canada. Since 1963, it has led all other breeds in total annual registrations, if Herefords and polled Herefords are considered as separate breeds. Angus cattle mature early and produce high-quality, well-marbled meat. A few red cattle have occurred in the Angus breed since early times. When red individuals are crossed with red the offspring are always red.

The Shorthorns originated in northeastern England in the late 1700s and in early times were called "Durham." Mating red to red Shorthorns results in all red offspring; mating white to white gives all white offspring; and mating red to white gives roan offspring. Intermating roans gives red, roan, and white calves in an average proportion of 1:2:1. Thus the Shorthorns are red, white, or roan; have short, incurving horns; possess a good temperament; and are easily handled. Shorthorns were imported into the United States as early as 1783, but the breed was established on a permanent basis from animals imported between 1820 and 1850. They were originally bred for both meat and milk (dual-purpose cattle), but now there are distinct beef and milking types. There are also polled Shorthorns; they have the same characteristics as the horned breed.

The Charolais is a French breed that came to the United States via Mexico in 1936. It has a white or very light straw-colored coat. A mature, purebred bull weighs 2,000 to more than 2,500 pounds, and a mature cow may reach 1,250 to 1,600 pounds or more. This breed has a high rate and efficiency of growth and a large percentage of lean meat with a minimum of excess fat at a young age.

Other less familiar breeds that originated in Europe are the Chianina, Galloway, Gelbvieh, Limousin, Maine-Anjou, Murray Grey, Scotch Highland, Simmental, and South Devon.

A considerable number of good breeds of beef cattle have been developed in the United States—all in the twentieth century. The Brahman breed was developed in the southern United States in the early part of this century from humped cattle of India, the Zebu (*Bos indicus*). Zebus were initially imported in 1849, but the development of the Brahman breed began with importations made by the Pierce Ranch in Pierce, Texas, in 1906 and later. Several Indian breeds or strains were combined and upgraded by crossing with British-breed females. All the while, selection was made for beef conformation and early maturity. The Brahman is the easiest breed to identify. It has a conspicuous hump over its shoulders, loose skin (dewlap) under its throat, and has large, drooping ears. Its prevailing color is light to medium gray, but can also range from light gray or red to almost black. Brahman cattle are known for environmental adaptation, longevity, and mothering ability.

The Brahman has been used to develop most of the new breeds. For example, it was combined with Hereford and Shorthorn to produce the Beefmaster, with Hereford to produce Braford, with Angus to produce Brangus, with Charolais to produce Charbray, and with Shorthorn to produce Santa Gertrudis.

In addition to the animals developed as beef cattle, there are other breeds that have reasonably good beef conformation and are also able to produce milk and butterfat in reasonable quantities. The principal breeds of this type are Milking Shorthorn, Red Poll, Brown Swiss, and Holstein-Friesian.

In addition to the extensive breeding done with the Brahmans, the mating of animals from miscellaneous strains—called "crossbreeding"—is commonly practiced with many types of beef cattle. It results in steers that grow faster and yield as much high-quality meat as others. Another advantage is that crossbred cows not only produce more calves, but also have more milk for their calves than other beef cows.

The farmer wishing to start a beef-cattle herd should obtain advice from established owners in his area or from the county agricultural agent as to the breed best adapted to the region. As in the case of swine, crossbreeding results in hybrid vigor. When selecting heifers for breeding from his own herd, the farmer looks for those that are heavily muscled, with heavy weaning weights and good rates of gain. When buying sires or females from other establishments, the farmer selects animals with the best records from outstanding herds. Calves of horned breeds are generally dehorned before they are three weeks old. Dehorned cattle and breeds naturally without horns (polled breeds) feed together more quietly and need less space than horned cattle. Hornless breeding cows are considered to be easier to handle under most farming and many range conditions. Furthermore, dehorned or polled cattle are less likely to be injured during shipment to market. To raise steers that produce beef that meets American market requirements, male calves must be castrated, usually before they are three to four months old (Putnam and Warwick, 1974).

## Cattle Feed

Development of the enormous American beef-cattle industry was possible only because of our large acreages of range and pasture and the abundance of by-product feeds, hay, silage, and grain used to finish the animals in feedlots. The feed combinations used in feedlots vary with the region. In the Midwest, corn and corn silage are the main feeds. Some hay and hay silage are used. The major source of protein is soybean meal, but an increasingly greater quantity of urea is being incorporated into the feed (Gustafson and Van Arsdall, 1970). An example of the highly fortified protein supplements that are being used today is the so-called Iowa Supplement. To prepare 1,000 pounds of this the following quantities (in pounds) of materials are used: soybean meal, 415; cane molasses, 230; dehydrated alfalfa, 225; urea, 50; dicalcium phosphate, 30; and dried torula yeast, 50. Other supplements might contain such substances as bone meal, cobaltized salt, vitamin A and D concentrate, and various trace minerals (Putnam and Warwick, 1975).

The enormous quantity of grain and soybean meal used in the United States for the feeding of livestock has been the subject of much criticism in recent years by people who believe a greater proportion of such concentrated protein should be consumed directly by the many millions of people throughout the world who are obtaining less than adequate quantities. From the standpoint of efficiency, protein conversion from the plant source as meat, and particularly beef, is wasteful. The low yield per hectare of

**Table 3**
**Animal Protein Yield and Input of Feed Protein, Feed Energy, Fossil Energy, and Labor, per Hectare for Certain Animal Products in the United States (adapted from Pimentel et al., 1975)**

| Animal product | Animal protein yield (kg) | Feed protein input (kg) | Feed energy input ($10^6$ kcal) | Fossil energy input for the production of feed and animal ($10^6$ kcal) | Labor (man-hours) |
|---|---|---|---|---|---|
| Milk | 59 | 188 | 7.0 | 8.6 | 23 |
| Eggs | 182 | 672 | 14.4 | 9.6 | 174 |
| Broilers | 116 | 651 | 8.9 | 10.2 | 38 |
| Catfish | 51 | 484 | 5.0 | 7.1 | 55 |
| Pork | 65 | 689 | 17.0 | 9.2 | 28 |
| Beef (feedlot) | 51 | 786 | 25.0 | 15.8 | 31 |
| Beef (rangeland) | 2.2 | 33 | 1.42 | 0.89 | 1 |
| Lamb (rangeland) | 0.2 | 3 | 0.13 | 0.11 | 0.2 |

cropland when plants are fed to animals and protein is obtained indirectly via the animal (table 3) may be compared with the relatively high yield when the plant protein is consumed directly (table 1). To produce the more than 5.3 million metric tons of animal protein that humans consume annually in the United States, 91 percent of the estimated 27.1 million metric tons of cereal, legume, and vegetable protein suitable for human use is fed to livestock. The conversion is therefore not very efficient. The average feed protein input is 9.75 times greater than the yield of animal protein. The quality of the latter is higher than that of plant protein because of better amino acid balance but, as mentioned, adequate balance can also be obtained by combining plant proteins, as by combining rice or wheat protein with that from legumes.

Feed protein is converted to animal most efficiently in milk (table 3). In the United States, 130 pounds of milk protein may be produced from 413 pounds of the feed grade; 31 percent of the feed protein is therefore converted. But of the 413 pounds of feed protein, almost 200 pounds were derived from forages unsuitable for human consumption; so measured against only the grain protein consumable by humans, the conversion efficiency was 60 percent (Pimentel et al., 1975).

People particularly criticize the feeding of grain (mostly field corn, grain sorghum, and barley) and soybean meal to cattle in feedlots, to produce the "marbling" that results in the meat being more tender and flavorful. In this connection, revisions in the government's beef-grading standards, which went into effect February 13, 1976, will allow slightly leaner beef to qualify for U.S. Prime and U.S. Choice, the two top grades of beef, without sacrificing its eating quality. This is possible because of feeding improvements over the past few years, as well as new genetic developments and crosses of cattle breeds. More important, "yield grading," formerly optional, will now encourage the production of a maximum percentage of meat, compared to fat and bone. This will tend to discourage overfeeding and will reduce excess trimmable fat. These recent changes will result in a reduction in feeding time by two or three weeks and save 200 pounds or more of grain per animal fed—totaling more than 5 billion pounds saved per year.

There are on earth about 2.5 acres of permanent pasture for each acre of arable land suited to production of cultivated crops given the present state of agricultural technology. Nearly all cropland in the United States suitable for growing grain and other food crops is already in use, and most forage land had best be kept for grazing. When the land is not overgrazed, the soil can be kept intact as a precious heritage for generations as yet unborn. Improved forage and range-management techniques might increase the potential for human food production of that 60 percent of the world's land that is non-arable and even lead to grazing in some areas now considered wasteland. Even some land now considered arable and used for growing cereal grains might be more efficiently used for the production of certain high-yielding forages such as alfalfa or Sudan grass, for use as livestock feeds (PSAC, 1967). The problem then is to convert forage plants to human food in the most efficient way and thereby reduce the large quantity of feed grains and forages grown on productive, arable lands and now being fed to ruminants (cattle and sheep). Improved forages, and waste products to replace grain, could reduce the amount of grain fed to livestock by 50 percent and still allow the production of about the same amount of meat (Hodgson, 1974).

In the long run, the solution to the world's food crisis must be sought within the food-deficient nations. Suppose that the American people could miraculously convert to vegetarianism, put all stockmen and their suppliers on welfare, ban leather goods, and send abroad all the grain saved thereby. The food released by these changes in one year would be about the amount of food required to sustain that year's increase in world population—74 million!

### The Ruminant Stomach

Ruminants have a complex four-compartment stomach. Grass and similar roughage pass down the esophagus to the first compartment (rumen), return to the mouth for rechewing (rumination), and then pass into the second compartment (reticulum). While it is being digested, food is moved successively to the third compartment (omasum) and fourth compartment (abomasum) and then into the small intestine. The ruminant is thereby especially well adapted to the utilization of forage-crop sources of food. Two veterinary scientists, A. F. Schalk and R. S. Amadon, cut a "window" (fistular opening) in a goat's stomach in order to easily observe and accurately measure digestive movements in the rumen. Through the fistula they were able to remove any portion of the contents and to insert scientific instruments. Later many scientists used the ruminal fistula to study rumen digestion, synthesis, and diseases (Jensen, 1975).

The rumen can be thought of as a fermentation vat through which food passes on its way to the rest of the stomach; there masticated grass, hay, and roughage are fermented by bacteria and protozoa that synthesize essential amino acids. It is not necessary to feed balanced proteins to ruminants, for they can utilize nonprotein sources of nitrogen such as urea or biuret. One-third of the protein in the diet of ruminants can come from urea, without reduction of such performance factors as growth, milk production, and reproduction, when compared with other sources of protein. When the usual rate of 10 pounds of urea per ton of silage is used, along with 70 pounds of shelled corn, the protein and total digestible nutrients (TDN) added to the silage is the same as when 70 pounds of soybean meal is added, but at a 40-75 percent saving in cost of feed supplement (Hillman, 1964).

Whether grain is fed to livestock depends on supply and demand. When it is not available at a low price it is not used for the purpose. Livestock can thrive on other feedstuffs that generally cannot be used directly as food for human beings. Included in this category are the wastes of food crops and food products, such as by-products from preparing flour, starch, and distillery products, and the rendered wastes from the meat-packing industry. Food grains (wheat, oats, rice, corn, and sorghum) have large quantities of plant residue and mill waste that are a valuable form of energy and protein when fed to livestock. The residue of sugarcane or sugar beets after the juice has been extracted, molasses, spent brewers' grains, oilseed by-products, and fruit residues are other examples. A recent study showed that cattlemen can winter their steers on diets containing cottonseed hulls as the sole roughage source, supplemented only with non-protein nitrogen (NPN), vitamins, and minerals (Nicholas, 1977).

Microbes in the rumen use such sources of nitrogen, plus minerals and the energy derived from other foods, to build proteins needed in animal nutrition. The types of bacteria and protozoa present in the rumen determine the efficiency of fiber digestion. Man may eventually be able to control the nature of the substrate utilized and the quality of the protein synthesized by the billions of microbes present in the rumen (Virtanen, 1966).

While the activity of rumen microorganisms results in ruminants being able to make efficient use of poor quality, highly fibrous diets, they are nevertheless inefficient in their ability to utilize protein in high concentrations in the diet, as when they are fed protein concentrates. Proteins are broken down in the rumen by fermentation and converted into ammonia. The amino acids have to be resynthesized. However, proteins can be modified by treatment with formaldehyde so that they will escape digestion in the rumen and be utilized in the abomasum. This technology will be developed on a worldwide basis by Dow Chemical Company, under a license agreement with the Commonwealth Scientific and Research Organization of Australia (Tollett, 1975).

## INCREASING LIVESTOCK EFFICIENCY

We can expect continued research on ways of increasing the yield of animal products from permanent pasture. For example, the productivity and nutritive value of pastures might be increased through nitrogen fertilization, particularly if nitrogen becomes more available and less expensive. Increases in livestock production have been as spectacular as those in plant-crop production. Progress in breeding, including crossbreeding for hybrid vigor, and in animal nutrition and disease control, have all contributed greatly in bringing us animal products at costs relatively lower than at any time in history. Supplementation of animal feed with amino acids such as synthetic methionine and fermentation-produced lysine, and with minor elements, has been an important factor in conversion of feed to meat. As already related, an equally important factor has been the synthesis of amino acids by bacteria in the stomachs of ruminants from nonprotein sources of nitrogen such as urea or biuret.

Further advances are likely. Many authorities believe breeding yearling heifers will increase efficiency on many beef cattle ranches without increasing numbers of breeding animals. The size and quality of the calf from the heifer when she calves the second time (at three years of age) may thereby be increased. This program requires adequate nutrition and proper care and management (Albaugh and Strong, 1972).

Utilizing new advances in hormone therapy, two calves might in the future be produced each year by each beef cow, instead of an average of less than one. Hogs may be able to produce 18 to 20 baby pigs, double the present number. These might be put on an automatic pig nurser to be reared better than a sow can rear them. Disease problems common when baby pigs are left with their mothers can be eliminated. New breeds of sheep (e.g., Finnish Landrace) produce at least three or four lambs at each lambing, whereas sheep we now have in this country average only one lamb per ewe. With regard to livestock in general, early weaning will make it possible to rebreed females earlier and thereby increase production per female per year. Losses from exposure to disease will be greatly reduced by milk replacers, automatic nursers, and a system of total confinement of the animals (Kottman, R. M., 1973).

While liming their pastures with dust[6] from a cement kiln, three Georgia ranchers dumped some of the dust into their cattle feed. They observed that the cattle ate the feed greedily and gained more weight than expected. They reported their observation to USDA scientists, who found that cattle fed the dust diet for 112 days gained 28 percent more weight, while consuming 21 percent less feed, than animals fed a control diet. The extra weight was all meat, and the meat was also of a higher quality than that of the controls. The scientists had similar results with another group of steers, as well as with lambs (Maugh, 1978).

The Hei-Gro device is a vaginal insert which is not strictly a contraceptive, but suppresses estrus and minimizes sexual activity, resulting in heifers growing faster and using less feed. Tested on more than 250,000 cattle, the device was found to produce a 13-22 percent faster growth rate among foraging heifers and a 5-10 percent faster rate in feedlots. According to the manufacturer, Agrophysics, Inc., the net effect is a $15 to $20 reduction in the cost of feeding each heifer (Maugh, 1978).

## DAIRY FARMING

In the Great Lakes states the soil is generally somewhat inferior and the summers cooler than in the Corn Belt. These factors militate against Corn Belt type of farming. Farmers can derive better income from dairying than from crops, cattle, and hogs. Economic considerations, along with the seven-day drudgery associated with dairying, are tending to discourage the small farmer. The trend is now toward larger, more highly mechanized and better-managed farms. A survey indicated that by 1985 the number of dairymen in the country will decline by about 50 percent and number of cows by 17 percent. Milk produced per cow is expected to reach an average of 12,800 pounds per year, and in that case milk production in the United States will be about 120 billion pounds annually, but per-capita consumption will decline from 540 to 505 pounds (table 4). The majority of farms are expected to continue being family owned and operated (Hoglund, 1975). A prospective dairyman must invest $110,000 to $150,000 in land, cows, buildings, and machinery. For a family to survive on the income from a dairy farm, there should be at least 35-40 cows milking. If he is to produce most of his own feed, the farmer needs to control, through ownership or renting, at least 130-150 crop acres, depending on the land's capabilities (USDA, 1976).

### The Influence of Technological Innovation

The dairy industry developed particularly rapidly in the period 1870-1890, when the number of cows doubled. The silo, first constructed in 1873, permitted better year-round feeding of dairy stock. About 1875 refrigerator cars were first used for transporting perishable farm products. The centrifugal separator for separating cream from milk, invented in Sweden, was brought to the United States in 1882.

An important development in the growth of the dairy industry was pasteurization of milk, which was first recommended by a noted American health authority in 1875, only

[6]The dust is a complex, calcium-rich mixture of minerals carried by hot air pulled out of the cement kiln, but it does not contain the alkalis and hardness required to set cement (Maugh, 1978).

**Table 4**
**Trends in the Dairy Industry for 1954, 1964, 1974 and Projected for 1984 (from Hoglund, 1975)**

| | Year | | | |
|---|---|---|---|---|
| | 1954 | 1964 | 1974 | 1984 |
| Population (thousands) | 161,884 | 191,141 | 212,800 | 238,000 |
| Milk production (million lb) | 122,094 | 126,598 | 114,600 | 120,000 |
| Per-capita consumption (lb) | 699 | 631 | 540 | 505 |
| Milk cows (thousands) | 21,581 | 16,061 | 11,280 | 9,400 |
| Annual milk produced per cow (lb) | 5,678 | 8,099 | 10,100 | 12,800 |
| Cows per farm | 10.0 | 14.2 | 37.6 | 62.7 |

a decade after Pasteur found that heating prevented the souring and abnormal fermentation of wine and beer. At first milk was heated secretly to preserve it and prevent losses to dealers; this was regarded as unethical. Objections were overcome, however, as milk bacteriology and chemistry advanced. It was once feared that pasteurization of milk would destroy the organisms in the milk that form lactic acid and thereby promote the multiplication of putrefactive organisms. In actual fact, however, pasteurization merely tended to reproduce the characteristics of clean, fresh milk. Contrary to previous views, germs grew at the same rate in fresh milk whether pasteurized or raw (Harding, 1947).

Milk is pasteurized either by heating it to 143°F (61.7°C) and holding it at that temperature for at least 30 minutes, or heating it to 161°F (71.1°C), but for only 16 seconds. The milk is then transferred to coolers to prevent growth of microorganisms not affected by the heat.

Some customers insist on having "raw milk," which can be obtained from certain certified herds. Most milk is pasteurized and homogenized. Homogenization is accomplished by passing the milk through a high-pressure pump to reduce the size of the fat globules and thereby produce a more stable emulsion. Milk may also be fortified with vitamin D, usually added as an oil emulsion.

The Public Health Service of the Department of Health, Education, and Welfare publishes a recommended Grade "A" pasteurized milk ordinance. Conditions of producing, processing, and distributing Grade "A" milk and fluid milk products are covered by the ordinance. Adoption and enforcement of this ordinance rests with local municipalities, counties, and states. The U.S. Department of Agriculture, the Federal Food and Drug Administration, and individual states establish specifications and regulations pertaining to conditions of manufacture and grading of manufactured dairy products in the United States. Responsibility for the enforcement of dairy law rests with either the local health department or state health or agriculture departments.

An important factor in the success of the milk industry was a simple test devised by Stephen M. Babcock, of the Wisconsin Agricultural Experiment Station, for estimating the butterfat content of milk. After thorough testing, Babcock announced his invention in 1890. The test facilitated study of individual cow performance and thereby led to the improvement of dairy breeds. It made possible a rational basis for selling milk and cream. The test was not patented, but given to the public. Babcock also developed many new methods for analyzing milk and curing cheese. He also demonstrated the inadequacy of synthetic diets, which in turn led to the discovery of vitamin A by E. V. McCullum (Harding, 1947).

Mechanization and high energy input is as characteristic of the dairy industry as it is of modern farming in general. Nearly all milk-producing dairy cows in the United States are milked twice daily, using electrically operated milking machines. The milk is immediately cooled, again using electricity. The milking parlor and milk room are cleaned with hot water. Fuels and electricity are also used to handle the animals' daily waste and to aid in the control of odors, insects, and diseases.

## CENTERS OF MILK PRODUCTION

In 1972, Wisconsin, with over 19 billion pounds per year, far surpassed any other state in milk production. California and New York each produced 9.9 billion pounds, Minnesota 9.2 billion, Pennsylvania 6.7 billion, Michigan 4.7 billion, Ohio 4.4 billion, and Iowa 4.2 billion. Thus among the eight states producing more than 4 billion pounds of milk, all but California and Iowa bordered one or more of the Great Lakes. Impressive though the milk-production figures for the eight leading dairy states may be, they represent only about 57 percent of the milk production of the United States, indicating that dairies tend to become distributed to accommodate as much as practicable the distribution of population. Areas of high concentration of the dairy industry, such as the Great Lakes region, tend to concentrate more than any other areas on easily refrigerated and shipped dairy products such as butter, cheese, condensed or dry milk, and other processed items.

According to USDA agricultural statistics for 1975, by far the greatest quantity of milk per cow was produced in the Far West: California, 13,566 pounds; Arizona, 12,537; and Washington, 12,829; compared with 10,430 in Wisconsin; 10,120 in Minnesota; 10,706 in Michigan, and 10,215 in Pennsylvania. The eleven states entirely within the boundaries of the region designated "South" in figure 2 averaged 8,803 pounds of milk per cow (USDA Agr. Stat., 1977).

## DAIRY CATTLE BREEDS

Several breeds of cattle are raised principally for their dairy products. The Holstein-Friesian breed is popular throughout the temperate regions of the world. It is commonly called simply Holstein in North America and Friesian elsewhere. These cattle were brought over by the earliest Dutch settlers but the main importations were made in the latter half of the nineteenth century. Holsteins are easily recognized by their black and white markings and their large size. The mature cows often weigh 1,500 pounds and some as much as 2,000 pounds or more. Bulls weigh from 2,000 to 2,400 pounds. A fully grown Holstein cow gives as much as 14,000 pounds of milk a year or 6,800 quarts, with an average butterfat content of 3.7 percent, somewhat lower than that of other dairy breeds. Nevertheless their average butterfat production is 520 pounds a year because of the large amount of milk they produce. The cows, when they no longer produce sufficient milk, have a high salvage value for beef and the calves are often slaughtered for veal.

Jersey cattle, originally from the English Channel island of Jersey, are raised worldwide in temperate regions. They were introduced into America after they had been used for milk on sailing vessels, but most importations were made during the latter part of the nineteenth century. They are famous for the richness of their milk, which has an average butterfat content of 5.2 percent. They now produce an average of 8,600 pounds of milk a year.

Jersey cows vary in color from light gray to dark brown or fawn. They often have extensive black and sometimes white markings. Mature cows can weigh about 1,000 pounds and bulls about 1,500 pounds.

The island of Guernsey, just 22 miles north of Jersey, is the home of the Guernsey dairy cattle. (The ancestors of both Jersey and Guernsey cattle came to the two islands from France.) Some Guernsey cattle were brought to America on sailing vessels in early years, but the importations that formed the foundation of the breed in the United States today were made in 1830 and 1831 and again in 1914.

Guernsey cattle are fawn-colored and white. They rank between the Holstein and Jersey in size and in milk production. Average annual milk production is about 9,400 pounds and the percentage of butterfat is 4.8. Guernsey milk has a yellow color that is associated with the occurrence of a yellow pigment of the skin of this breed of dairy cattle.

Ayrshire dairy cattle came from the county of Ayr, Scotland. Before 1840, only about seventeen of these animals had been imported, but during the next twenty years another two hundred or more were brought over. Ayrshires are red, mahogany, or brown, with many well-defined areas of white. They are medium sized, the cow averaging about 1,150 pounds in weight and the bulls 1,800 pounds. The upturned horns were once considered to be attractive for show purposes, but on dairy farms they can cause injury to the cattle and are now removed at an early age, as with other breeds. In fact, some naturally hornless Ayrshire breeds have been developed since the late 1940s. Mature Ayrshires can produce an average of 10,800 pounds of milk a year and the percentage of butterfat is 4.1.

Brown Swiss dairy cattle originated in Switzerland, where most of the early improvements of the breed took place in the nineteenth century. The breed spread rapidly throughout the United States from the fifty-five head known to have been imported from 1867 to 1906.

Brown Swiss vary in color from light brown with a silvery cast to extremely dark brown and usually with a distinctly lighter color on parts of their heads and in a stripe down their backs. Mature cows average about 1,400 pounds and bulls about 1,900 pounds. Cows produce an average of 12,000 pounds or 5,800 quarts of milk a year, with a butterfat content averaging about 4.1 percent.

Brown Swiss are also considered to be useful for veal and beef production and in earlier times in Switzerland and the United States they were often used as draft animals on the farm (Bayley, 1975).

About 48 percent of the dairy cows of the United States are artificially inseminated. In this way the inheritance of an unusually superior bull can be passed on to many more calves—several thousand per year—instead of the usual twenty to fifty calves produced by natural mating. However, no substantial benefits have been obtained by crossbreeding (mating of animals from different breeds) unlike the experience with beef cattle (Bayley, 1975).

Great economic losses can result from the many diseases to which beef and dairy cattle are susceptible. Practically all cloven-footed animals, especially cattle, hogs, and sheep, are susceptible to foot-and-mouth disease, one of the most contagious and devastating livestock diseases. The disease causes blisters in the mouth and on the feet and the animals may become lame and sometimes even die. The filterable virus that causes the disease is so infective that it will produce it even when diluted a million to one. In the United States there were nine outbreaks of the disease between 1870 and 1929, but they were all stamped out by killing ailing animals and burying them in quicklime and disinfecting the premises with 1 or 2 percent solution of caustic soda. Owners of the destroyed animals were paid specific appraisal rates for them. Since 1929 this country has been free from foot-and-mouth disease through a rigid federal-state quarantine (Harding, 1947).

Pleuropneumonia is another virus disease that was once serious in this country, but was eradicated in a five-year campaign. Great vigilance is exercised in preventing the entry of this disease from other parts of the world where it exists. It should not be confused with pneumonia, which is combated with antibiotics. Tuberculosis, once common in cattle in the United States, was reduced to 0.08 percent by 1965. Humans are slightly susceptible. Other diseases include mastitic calf scours, brucellosis, milk fever, ketosis, pinkeye, anaplasmosis, bloat, and tick fever.

## LACTASE DEFICIENCY

In view of the considerable dietary value of milk, it is unfortunate that many people, particularly among the black population, are not able to satisfactorily digest it. In such people, milk causes diarrhea and gastroenteric pain. Apparently they are deficient in lactase, the enzyme that changes lactose (milk sugar) into the absorbable glucose and galactose. The lactose is then not digested until it reaches the colon, where it induces the rapid proliferation of lactobacillus. Under the anaeorbic conditions of the colon these bacteria break down lactose and produce large amounts of lactic acid which, in turn, causes much water to be secreted into the colon, resulting in diarrhea. Apparently for the same reason powdered milk added to cereals to increase the lysine content has resulted in these food products being rejected in Nigeria, Colombia, Guatemala, and Peru.

Nearly all human infants are born with the lactase enzyme, necessary for the digestion of mother's milk, but may later lose it. American black children generally reject milk because of the diarrhea and pain. Adult lactase production may have evolved as a dominant genetic condition with the origin of dairying in Neolithic times among peoples using cows, sheep, goats, and reindeer. Thus early American immigrants from western Europe might be expected to have a low frequency of adult lactase deficiency. There are some dairying populations, such as the Greeks and Cypriots, that have a high percentage of lactase-deficient adults, but they have developed a cultural adaptation to overcome the disability, namely, the production of cheese or yogurt. In the cheese- or yogurt-making process, lactobacillus bacteria are utilized to digest the lactose, thus avoiding this chemical change within the intestines before reaching the colon (McCracken, 1971; Katz, 1973).

## LIVESTOCK AND ENVIRONMENT

Increasing the efficiency of livestock encourages farmers to use sloping land for grazing and thereby provides environmental benefits. Erosion has taken a dreadful toll in rolling agricultural lands. In cooperation with the federal Soil Conservation Service, many farmers have been practicing sound soil-saving cultural methods by raising food crops, cotton, and tobacco on the richer and more level lands, allowing the more sloping and thinner soils to revert to pasture for livestock, the latter in numbers that can be maintained by the pastures on a sustained basis. Pasture grasses now hold the soil in place, allowing it to regain its former physical characteristics and fertility. In areas of mixed cash-crop and livestock farming, animal manure is used to add organic material to the cropland soil to improve its tilth and water-absorbing ability and to add valuable plant nutrients. Much formerly eroded land has been restored to agricultural productivity.

In early May of 1976, I had the opportunity to travel extensively in the rolling lands of the Ozarks of south-central Missouri (fig. 3, Region N) and was pleasantly surprised at the extent of the forests and meadows in areas that the oldest inhabitants say was once predominantly planted to row crops and extensively damaged by erosion. The land has been restored to productivity through controlled grazing and at the same time has been transformed to a region of great pastoral beauty. "He is the greatest patriot who stops the most gullies," said Patrick Henry.

In the Ozarks of southern Missouri many city workers have bought fifty acres or more of woods and pasture for the satisfaction of residing in a rural environment and having a part-time farm enterprise. Some farmers have sold their overvaluated and overtaxed properties in the Corn Belt and have come south to retire on modest ranches amid the gently rounded, forest-covered Ozark hills, interspersed with bright green meadows of fescue with a sprinkling of ladino clover, lespedeza, cheatgrass, and orchard grass. The meadows are speckled with just enough of the red, white, and black of Hereford, Charolais, and Angus to provide the serene bucolic charm that a refugee from the Corn Belt can appreciate.

# 6 The Great Plains

*For agriculture, as no other industry, develops strong individualism, independent character, initiative and resource. Farm life is free from a certain artificiality of urban life, because it is in close contact with nature, and is less subject to the insidious forces of moral degeneration which are such corroding influences in the life of our great cities.*

—Herbert Hoover

The Great Plains include the region east of the Rocky Mountains and extending to approximately where the shortgrass prairie meets the tallgrass prairie. Thus the eastern boundary is ill defined and might move east or west from an arbitrarily set line, depending on the season. As indicated in figure 2, the Great Plains region extends to 100° longitude in the Dakotas, Nebraska, and Kansas and 98° longitude in Oklahoma and Texas. It is the region that people from the East and Midwest referred to as "The Great American Desert" in the early decades of the nineteenth century—believed to be unfit for human habitation.

The climate of the Great Plains is severe and unpredictable. Cold and dry air masses sweep down from Canada; warm to cold, dry to moist, air masses cross the Rocky Mountains from the West; and warm, moisture-laden air masses move in from the Gulf of Mexico. When all three air masses collide, the atmosphere is thrown into violent turmoil, resulting in heavy rain or severe blizzards such as those of 1886 and 1945 (Kraenzel, 1955). Winter temperatures may drop fifty degrees in a few hours and summer temperatures may be higher than those at the equator. Weeks of drought may be followed by hail or cloudburst.

Spring and summer rainfall in the Great Plains amounts to 12-22 inches per year. Usually the moisture does not penetrate more than 2 feet in depth. After a comparatively short period of growth following rains, the native grasses pass into a drought rest period, short in the north and relatively long in the south. In most of the Great Plains region there is enough annual precipitation for cultivation of wheat, and the dry, warm summers favor the growing and harvesting of this important crop. The fine-textured, dark brown soil is highly productive when it receives adequate rainfall, or, in the drier areas, irrigation water.

The rivers coming down from the Rockies do not have much water because the air masses from the Pacific Ocean carry very little moisture after passing the Continental Divide, and those from the Gulf carry little moisture that far west. The volumes of the Missouri, Canadian, Arkansas, Red, and Brazos rivers increase markedly east of the 100th meridian. The Great Plains region was not a land to delight the hearts of the early eastern and midwestern farmers, who thought of it only as a wretched and perilous area to pass through on the way to Oregon and California.

Nevertheless, the Great Plains includes the shortgrass prairie, the grasses becoming shorter and scantier as one moves west and rainfall decreases. It became evident to perceptive cattlemen that these grasses—blue grama, buffalo grass, and side oats grama—must be highly nutritious, for immense herds of buffalo, antelope, and other herbivores thrived on them. The awesome numbers of buffalo continually astounded early witnesses. At times an observer could stand on a hill and see what looked like a compact mass of buffalo in all directions to the horizon (Gard, 1960). Based on the average carrying capacity of the 1.25 million square miles of the Great Plains grassland, and considering also the impact of other big game animals such as elk, deer, and antelope, biologist Tom McHugh (1972) calculated a buffalo population of 32 million, with another 2 million that probably lived in wooded areas bordering the plains. Roe (1951) estimated the existence of 40 million buffalo on the plains as late as 1830, and there were other estimates much higher, ranging up to 100 million. In earlier times there had also been millions of buffalo east of the Mississippi, ranging as far east as Pennsylvania.

## THE PLAINS INDIANS

The nomadic and nonagricultural Indians of the Great Plains depended for their existence on the buffalo. The buffalo occurred in large numbers and was the only plains animal that could be approached without great difficulty. Nevertheless the livelihood of the Plains Indians was relatively meager until they gained the use of the numerous progeny of the horses that escaped or were stolen from the expeditions of Spanish explorers, beginning as early as the expeditions of Coronado and de Soto in the 1540s. The first to obtain horses and learn to use them effectively were Comanches and Apaches, the two southernmost Plains tribes. Horses then made their way farther northward and were eventually used by all Plains Indians, who were quick to recognize their value. Mobility was greatly increased as horses replaced dogs for carrying packs and dragging travois. The Plains Indians became excellent horsemen and were able to ride down the buffalo and kill them with arrows or spears instead of waiting in ambush for them. Eventually equipment obtained from the whites, such as copper kettles, steel knives for skinning buffalo, and metal scrapers to speed up the fleshing of their hides, facilitated the mobility required to follow the buffalo herds as a way of life. The horse, an unintentional gift from the Spaniards, ushered in the "golden century of the Plains Indians" (Webb, 1931) and, along with the aforementioned utensils, was the only benefit any American Indians ever received from Europeans.

As might be expected, the fighting ability of the Plains Indians, already quite formidable, was vastly improved by their expert utilization of the horse in warfare; in fact, their warriors usually became invincible. The Spaniards never conquered the Great Plains. The universal acquisition of the horse by the Comanches and Apaches in the seventeenth century ended the Spaniards' chance of doing so. The Plains Indians became a scourge and a terror to the sedentary tribes in peripheral areas as well as to the Spaniards who ventured northward. They ruined presidios, pueblos, and missions. In fact, they regularly pillaged to within 500 miles of Mexico City. The period of Spain's greatest effort and final failure on the northern frontier was to come in the eighteenth century (Webb, 1931).

After gaining independence in 1821, Mexico fared even worse than Spain against the Plains Indians. And for a considerable period the Anglos fared no better. The Great Plains were left unoccupied by white men and practically unexplored. Unlike his experience in forested country, on the plains the Anglo faced a mounted enemy, whether Mexican or Indian. The Anglo's traditional weapon, the long rifle, was obviously unsuited for battle on horseback and, in any case, required about a minute to reload. He might carry two pistols, bulky and unwieldy, and with the same disadvantages of loading as the rifle had. The Indian (Comanche) carried a rawhide shield hung on his left arm, a 14-foot spear, a plain or sinew-backed bow, and a quiver of arrows tipped with flint or steel, which he could release with remarkable accuracy while riding at full speed. In most respects the Indian had the advantage. For the minute that the Anglo horseman took to reload his weapon, the Indian could ride 300 yards and discharge 20 arrows (Webb, 1931). Certain types of unpleasantry to which captives were submitted, and for which Plains Indians were particularly notorious, generally encouraged the white man to reserve the last shot for himself in case capture seemed likely (Dodge, 1877; Fehrenbach, 1974).

In its brief history as an independent republic between 1836 and 1845, Texas lost many more of its citizens to a few thousand mounted Indians than to the Mexican army (Fehrenbach, 1974). During this fateful period in plains history, a young Connecticut Yankee by the name of Samuel Colt invented a six-chambered "revolver" of .34 caliber, which a company in Paterson, New Jersey, began to manufacture in 1838. This was a pistol with a rotating cylinder containing six chambers, each of which discharged through the single barrel. In the following year it had found its way into the hands of the Texas Rangers, a body of mounted men legally organized during the Texas revolution (1835-1836) and stationed in the frontier town of San Antonio. To hold their own against the Indians and Mexicans, the Rangers needed equally good horses and needed to become equally good horsemen. For an advantage in battle they needed the Colt revolver, which they named the "six-shooter." The first decisive tests of the Colt revolver came in 1844 in the battle of the Pedernales and later in the battle of Nueces Canyon. Although greatly superior in numbers, the Comanches were caught completely by surprise, having never seen or heard of the six-shooter. In both battles they were routed, with heavy losses. Years after the Nueces battle a Comanche chief said he lost half of his warriors, who died for a hundred miles along the trail toward Devil's River (Webb, 1931).

Thereafter those who left the timbered East and entered the Great Plains went on horseback and with six-shooters in their belts. The Colt revolver was the first of a triumvirate of innovations that conquered the Great Plains, the others being barbed wire and the American-type windmill. One might argue for the railroads, but without the means of making the plains suitable for settlement, the railroads would have remained direct and unbranched roads to the more congenial Pacific Coast, their original goal.

## THE NEAR EXTERMINATION OF BUFFALO

Wayne Gard, in *The Great Buffalo Hunt* (1960), estimated the killing of around 40 million buffalo by hunters in less than a dozen years. As late as 1846, when Francis Parkman, Jr., set out from St. Louis with a friend and a guide for an amazing five-month trip to the Rocky Mountains to live with Dakota and Ogalala Indians[1] in their villages and to learn their language and study "the manners and character of Indians in their primitive state," the herds of buffalo still blackened the prairies. In the preface to the third edition of his great classic, *The Oregon Trail* (1857), Parkman remarked that within six years after his trip the buffalo had vanished throughout the vast area drained by the Platte and Arkansas rivers, where he had made his historic sojourn.

There were various motives for the unprecedented slaughter of buffalo: to starve out the Plains Indians; to save forages for livestock; to supply buffalo tongues, hides, and bones (for fertilizer); to provide freight for railroads transporting the hides and bones; and for trophies and sport. The continual westward extension of the railroads was, of course, a big factor facilitating the slaughter. After the hides and tongues were taken, the succulent meat, superior to that of the Longhorn cattle of those days, was left to the wolves, coyotes, and vultures. By pure chance, some remnant herds in Yellowstone and Canada escaped the thorough and systematic slaughter and their descendants remain on a few western wildlife refuges and in some privately owned herds, including some on certain Indian reservations (Webb, 1931; Fehrenbach, 1974).

### Modern Buffalo Herds

Starting with less than a thousand buffalo in the year 1900, and with hunting prohibited, the buffalo population of the United States increased to more than 40,000, with the largest herds now being raised privately for sale as meat. As a surplus of buffalo developed on western federal and state buffalo ranges, excess stock began to be sold. The National Buffalo Association has more than 400 buffalo breeders. About 60 percent of the buffalo in the United States are confined to South Dakota and Wyoming, but there are also herds in Colorado, Kansas, Oklahoma, Oregon, Arizona, and Montana.

Buffalo like to roam about more than cattle and require a larger range. Also, much stronger fences are necessary on a buffalo ranch. After the initial investment, it costs no more to raise buffalo than cattle. An advantage over cattle is that buffalo do not require supplemental feeding on the range and buffalo cows do not need help when they calve.

Ordinarily buffalo meat sells for about 50 percent more than beef, but in a depressed livestock market it has sold for almost double the price of beef—$1.30 a pound carcass weight. A buffalo bull can reach a height of more than six feet and weigh more than a ton. Chain stores, restaurants, lodges, and individual consumers are customers for buffalo steak or a "buffalo burger." The meat from the younger animals is said to have no wild, gamy taste. It is higher in protein than beef, and lower in fat, thus appealing to many people. The hides and skulls are also sold, the latter put outside for weeks to bleach. Hooves are sometimes turned into ashtrays or lamps (L.A. *Times,* 1976).

## LONGHORN CATTLE

Cattlemen were the first of a hardy and stubborn breed of men who tried to conquer the Great Plains. The fact that they had some early successes is a tribute primarily to the type of cattle available in those days, particularly adapted to the severe climate of the

[1]The Dakotas and Ogalalas happened to be on friendly terms with whites at the time. Parkman also visited villages of the Pawnees and Comanches and at a very opportune time, for he related that soon after he finished his trip these two tribes became hostile along the Arkansas Trail, attacked "without exception" every party, large or small, that passed through the area during the next six months, "killing men and driving off horses."

region. The appearance of cattle throughout the vast ranges of the plains was due initially simply to an overflow from the enormous numbers driven up the famous Chisholm Trail to Kansas, for shipment east, after railheads had been established.

## Origin of the Longhorn

Longhorn cattle shared with hardy cattlemen and settlers the ability to survive and succeed under circumstances in which survival and success depended entirely on certain uncommon innate qualities. Longhorns were ferocious-looking, half-wild cattle, with sharp, spreading horns. The ranches of the Spanish conquerors in the West Indies and Mexico were stocked with these cattle and other livestock brought over the Atlantic, beginning with Columbus's second voyage in 1493. They arrived in Hispaniola (Santo Domingo) on November 3 of that year to be put on ranches the Spaniards had established there. Soon breeding stock and calves were taken to Cuba and later to New Spain (Mexico). Hernando Cortez stocked his ranch in Mexico, a ranch with the now-familiar name of Cuernavaca (Cow Horn). That was in 1521, only two years after he had begun his conquest of Mexico.

The early Spanish explorers of our southern states, such as Ponce de León, de Soto, and Coronado, brought with them large herds of Longhorn cattle, sheep, goats, and swine. However, these were meant not for propagation but for food for the large expeditions. Some livestock escaped and offspring of the hardy Longhorns ran wild from Florida to California. By 1715 a French trader reported that the descendants of the cattle and horses the Spaniards left had increased by the thousands. He believed the Indians feared to kill them. If so, they had lost such fears by the 1770s, when they were said to have driven off 22,000 head in a single raid (Gard, 1954).

Longhorn cattle were well adapted to semiarid lands, first in North Africa, then in Spain, before they were brought to the New World. They found another familiar environment in the Mesa Central of Mexico. The Mexicans had inherited the traditions of cattle ranching in semiarid regions. They awarded about 4,000 acres of grassland to every settler who planned to derive his income exclusively from ranching. This liberal land policy attracted many Anglo-Americans who were frustrated by the restrictions of the Homestead Act, which allowed each settler only 160 acres of land. Later, when Texas became a republic, 4,605 acres were awarded to every citizen who was the head of a family and who had been a resident at the time of Texas's declaration of independence. Texas continued its liberal policy even after joining the Union (Higbee, 1958).

In the 1780s increasing numbers of settlers were migrating to Texas from the United States. These were mostly farmers, and many brought cattle that were mainly from Britain. Despite the blending of Spanish and British breeds, Longhorn cattle continued to appear and behave much like the wilder Spanish type. Many cattle were already being branded in those days; those too wild for human control were called *cimarrones.*

Mexico became independent in 1821 and the Texas revolution occurred in 1836. Soon the English-speaking Texans were rivaling the Mexican *vaqueros* in capturing and taming the wild cattle. These cattle ranged from yellow to black, and had lanky bodies and long legs, providing great speed and stamina. Their huge, horizontally arranged horns, usually from three and one-half to six feet from tip to tip, served well for attack or defense. An angry bull was considered the most dangerous animal on the plains. Cowboys pursuing Longhorns on the range had to be well mounted and well armed.

## The Cimarron and the Brush Hand

The "brush country" of Texas lies more or less between the San Antonio River and the Rio Grande, and between the Gulf Coast and an irregular line running west from San Antonio to the Rio Grande. Many of the thicket species are thorny. Outlaw cattle (*cimarrones*) showed great cunning in finding places of refuge in these thickets. Driving cattle from the thickets was excruciating work done by "brush hands" who, along with their "brush horses," were a particularly hardy lot that developed great skill for this dangerous work.

On long drives, a brush hand typically carried a blanket, a quilt, a frying pan, a jug, and his food ("grub") on a packhorse. The grub consisted of some meal and salt. The jug was for tallow he obtained by killing a fat cow while he hunted meat. Meat was generally cooked on the open fire, but steaks could be put in the frying pan, along with some tallow. Little *chilipiquines* (red peppers) that grew wild in many places were used for seasoning. Sometimes the men were gone from camp two or three days, subsisting on corn bread and dried beef carried in saddle pockets (Dobie, 1941).

The Texas cowboy learned cattle-herding from the Mexican vaquero, who learned it from the Spaniards, who in turn learned it from the Moors. The cowboy also derived much of the terminology of the trade from the Mexicans. Buckaroo derived from vaquero, chaps from *chaparejos,* lariat from *la reata,* coosie from *cocinero,* and ranch from *rancho.* Other words were adopted without change: sombrero, mesquite, latigo, topadero, bandanna, corral, rodeo, and remuda (Michener, 1974).

## The Chisholm Trail

In Texas during the Civil War, prices for cattle had tumbled, ordinary cattle bringing only three or four dollars a head. Yet in the cities to the north and east meat was much in demand. During the war a few of the hardy Longhorn had been driven north great distances to cities or railheads and steadily increasing interest was manifested in developing cattle drives on a much larger scale. Thus began one of the great episodes in the development of the West, the driving of gigantic herds of Longhorn cattle up the Chisholm Trail, named after a Scotch-Indian trader who spoke many Indian languages and whose teams were the first to pass over and mark out the route. The story of the Chisholm Trail is told in a fascinating and meticulously documented book, *The Chisholm Trail,* by Wayne Gard (1954).

Texas was by no means the first state to organize cattle drives but, as might be expected, it developed by far the largest ones. The Chisholm Trail cattle drives were not the first from Texas; large herds had already been "trailed" to Louisiana, Missouri, and Ohio. The acquisition of the Pacific Southwest by the United States in 1848 and the discovery of gold in California led to the trailing of cattle to that state also, despite the rough trail, heat, drought, alkaline lakes, poison weeds, and Apache Indian raiders. Cattle bought for $5 to $15 in Texas might sell for $25 to $150 in California.

Railroads were moving west from St. Louis and Kansas City, reducing the distance cattle would have to be trailed to reach a railhead, and the idea of driving them there began to germinate. In the early spring of 1867 a young man named Joseph G. McCoy, who thoroughly understood every phase of livestock feeding and shipping, had the imagination and initiative to make the initial plunge. It was he who was responsible for the expression "the real McCoy." After a number of rebukes from various railway companies and towns along their routes, he chose the tiny frontier town of Abilene, Kansas, on the Union Pacific Railroad (later the Kansas Pacific) as a potential railhead for cattle to be trailed north from Texas. He was perhaps unaware of a Kansas statute establishing a line about sixty miles east of Abilene, beyond which it would be illegal for Texans to trail their cattle. Texas Longhorns carried a tick-borne fever to which they were immune but from which northern cattle suffered severe losses. Cattlemen had to wait another seventeen years before the cause of Texas fever could be determined and a cure developed and superior breeds of cattle could be safely introduced.[2]

In 1867 the country around Abilene was so sparsely settled that no one seemed interested in enforcing the letter of the new law. McCoy bought 250 acres of land northeast of Abilene and shipped in lumber for the construction of cattle yards, and a pair of ten-ton Fairbanks scales. He also built an office, a bank, and a three-story frame hotel[3] which "boasted such frontier luxuries as plaster walls and Venetian blinds" (Gard, 1954). McCoy induced the Union Pacific to build a hundred-car switch and to build transfer and feed yards at Leavenworth.

The main stem of the Chisholm Trail began at the Rio Grande, near Brownsville, where drovers bought or stole herds of cattle from below the border in Mexico, and swam them across the Rio Grande. The trail extended northward through Austin (with a western route through San Antonio), Waco, Fort Worth or Dallas, and finally through Indian Territory (Oklahoma), reaching the Kansas border at Caldwell and proceeding through Wichita to Abilene. There were many branches leading to the trail from various parts of the enormous Texas cattle country. By standing on a hill, one could see the winding columns of Longhorns extending for miles to the horizon. For most Texas cattle the distance traveled was five or six hundred miles. Wild game was plentiful along the Chisholm Trail, including buffalo, antelope, and turkeys, as well as ducks and geese in the migration seasons. Buffalo were so numerous that they sometimes held back progress. There was danger from buffalo cutting into the herd, or they might be passing in a column ahead of the cattle in such great numbers that it might take as long as two hours for them to pass. There was always the danger that buffalo might stampede the cattle. Sometimes cattle, and even horses, joined the buffalo herds.

Fish were abundant in most of the streams along the trail. Blackberries and wild plums could be gathered in the early summer and pecans, from the bottomlands, in the fall. Nevertheless, the main meal, served in the evening, generally included some form of beef. Sometimes a shoulder of beef could be traded to settlers for fresh fruits, vegetables, and eggs. The cattle found grass upon which to graze over the entire trail and lost little or no weight en route to Abilene, weighing 1,000-1,200 pounds upon arrival. Some of the cattle were trailed on to Colorado, Wyoming, or Montana, where they were in great demand for stocking new ranches.

Despite the brief period remaining in the fall of 1867 after McCoy's facilities at Abilene were completed, nearly a thousand cars of cattle were shipped from Abilene to St. Louis and Chicago, bringing McCoy a large bonus from the railway company.

In 1868 business thrived until late summer, when the market continued strong for cows and calves, but declined for beefs. McCoy decided on a scheme to increase interest in Illinois, where stockmen bought many cattle for further feeding. He had some buffalo, elk, and wild horses roped, including a huge, fierce buffalo bull weighing 2,200 pounds. He sent these to St. Louis and Chicago in a reinforced stock car, along

[2]The establishment of the Bureau of Animal Industry in 1884 set in motion the investigations of a young medical student, Theobald Smith, and his assistants, F. L. Kilbourne and C. Curtice, both veterinarians, on the cattle-tick fever. Smith observed that the red-blood corpuscles of diseased animals were destroyed by a protozoan. He also proved that the protozoan was transmitted by ticks (Smith and Kilbourne, 1893). This was the first time that an arthropod was found to be able to act as an intermediary host in the transmission of a protozoan disease. Further research showed that the ticks could be destroyed by certain dips and sprays, along with the vacating of pastures. Thus at an original cost of $65,000 a discovery was made that even thirty years ago was estimated to be worth about $40 million a year to the cattle industry (Harding, 1947).

[3]The hotel became known as Drovers' Cottage. Many years later the creamery in which Dwight D. Eisenhower worked as a youth was built on the site of this hotel.

the sides of which were large canvas signs advertising his cattle business. Along with cowboys and brightly dressed Mexican *vaqueros* riding and roping and throwing wild steers, the wild animals constituted a "Wild West Show." McCoy also organized a buffalo hunt for Illinois cattlemen. His advertising schemes were a great success and before winter all his cattle were sold. He estimated that 75,000 cattle were trailed into Abilene in 1868 and 50,000 of these were shipped east by rail. In 1869, 150,000 cattle reached Abilene, and Kansas City was developing a meat-packing industry rivaling those of St. Louis and Chicago.

In 1870 the number of cattle trailed up the Chisholm Trail again doubled. Mature steers purchased for $11 in Texas might sell for $31.50 in St. Louis or Chicago, or $55 in New York. Abilene had competition, as a shipping point, from other towns along the railroad, but retained its dominant position. Villages along the Chisholm Trail such as Fort Worth, Texas, were booming as outfitting and provisionary points for drovers. As had long been the case, some cattle were also trailed west or east to New Orleans or Natchez and some were shipped by coastal steamers. Nevertheless, the number of cattle reaching Kansas railheads was so great that the stockyards were swamped, loading pens couldn't handle them fast enough, and the surrounding prairies became "a sea of grazing cattle." Falling prices induced many Texas drovers to graze their herds until fall or winter, awaiting higher prices, but prices that year did not improve. Then came arctic blizzards that lasted for days and a thick layer of ice covered the grass. Many cattle and even some wolves and jackrabbits were frozen stiff. From 100,000 to 250,000 cattle and hundreds of horses were estimated to have perished in that terrible winter of 1870, and only their hides could be salvaged.

Abilene changed from town to city government and J. G. McCoy became the first mayor. However, a number of railway companies were extending their lines southwest to cow country—a portent of things to come.

Needless to say, Abilene and other cattle-shipping Kansas towns had to provide entertainment for the thousands of cowboys who were expected to compensate for their tedious months of deprivation on the long trail in a few nights of revelry in gambling saloons and bawdy houses. Mayor McCoy was moved to write that "Few more wild, reckless scenes of abandoned debauchery can be seen on the civilized earth than a dance house in full blast in one of the frontier towns." Cowboys always carried their revolvers with them and shootings were common. The sporting women carried a jeweled pistol or dagger for self-defense or to avoid fraud. When evicted from town by the board of trustees, these "soiled doves" built their own houses on the outskirts.

There were, however, lawmen who could subdue a town even as wild as Abilene. Among them was the famous James B. ("Wild Bill") Hickok, a handsome man with a flair for fancy dress and silver-mounted, pearl-handled revolvers. He established his reputation in Abilene by, among other things, shooting ten bullets into a fence post, "making only one small hole" (Gard, 1954). After eight months on the job, Wild Bill was dismissed for the reason that Abilene was "no longer in need of his services." His later adventures included a tour of the country with the theatrical troupe of Colonel William F. Cody (Buffalo Bill). On August 2, 1876, while playing poker in a saloon in Deadwood, Dakota Territory, Wild Bill Hickok was shot in the back and killed.

In February 1872 the Farmers' Protective Association and officers and citizens of Dickinson County, Kansas, signed a notice that was sent to Texas, asking prospective drovers who contemplated driving Texas cattle to Abilene to seek another point for shipment for they would "no longer submit to the evils of the trade." By that time the area south of Abilene had become too thickly settled for easy trailing. Although a few cattle were still shipped from Abilene, it became almost a ghost town. Abilene's newspaper noted that "hell is more than sixty miles away" (in Ellsworth, west of Abilene). More settlers came into Abilene and the town was able to make a new start as a local farm market (Gard, 1954).

By the summer of 1872, McCoy was traveling north and east to persuade cattle buyers to come to Wichita, which had been reached by railway in the spring. Wichita shipped 3,530 carloads of cattle (70,600 head) that year and Ellsworth, in second place, shipped 40,161 head. However, only 350,000 cattle were trailed north in 1872, compared with 600,000 in 1871. New developments were affecting the cattle industry. A large, decentralized packing industry was beginning to be developed in the Middle West and Southwest and the first shipment of meat in refrigerated cars had been made, from Salina, Kansas. The quality of the Texas Longhorn cattle was being improved through the use of Durham and Shorthorn bulls on the home range, particularly in northern and northwestern Texas. The steers were gradually becoming more squarely built and the horns of many of the animals seen on the Chisholm Trail were no longer so conspicuously long. Stockmen were fencing in the choicest pasture lands. This eventually induced many cattlemen to trail their herds to new pastures in northwestern Texas.

The Kansas legislature moved the cattle quarantine line farther west, putting Wichita in the forbidden area and bringing Dodge City into prominence as a cattle-shipping point. Cattle could still be driven up the Chisholm Trail and then pointed northwest somewhere in Indian Territory. However, a new route, called the "Western Trail," ran directly north to Dodge City, saving much distance and time when trailing cattle from western Texas. On this new trail was Fort Griffen, which engaged in lively competition with Fort Worth on the rival Chisholm. It is difficult to realize that only a century has passed since Fort Worth, now a modern city of over 400,000 population, was a tiny village with muddy, rock-strewn streets on which cattle occasionally went on a rampage as endless herds were trailed through the town on their long trek to the Kansas railheads and the Great Plains. In Fort Worth's Log Cabin Village can be seen six cabins of the Chisholm Trail era that were restored, moved to this one site, and furnished with appropriate antiques.

By 1877 Dodge City became mostly a stopping point for cattle moving farther north or west to stock new ranges. When the Santa Fe extended rail service to Caldwell, in June 1880, that town soon rivaled Dodge City as a cattle-shipping point. Although within the quarantined area of Kansas, it was so near the border of Indian Territory that the embargo did not affect it. It went through the usual cattle town development, along with the vice, crime, and shootings. Barbed wire fences and the ever-deepening penetration of the railroads into Texas virtually closed down the Chisholm Trail after 1884. Traffic on the Western Trail tapered off also after 1885.

All traces of the Chisholm Trail have vanished, but the approximate course of the original route is now followed by Highway 81. It passed near where Duncan, Oklahoma, is now situated. About thirty miles west of this point, in Lawton, is the Museum of the Great Plains, opened to the public in 1961 and dedicated to collect, preserve, interpret, and exhibit items of the cultural history of man in the plains. The museum maintains a special research library and archives; provides an educational program of tours, films, and demonstrations; conducts historical and archaeological research; and publishes the *Great Plains Journal,* available through membership.

The impact of the Chisholm Trail on the American cattle industry and the entire American economy was indelible. It is difficult to estimate how long the settlement of the northern ranges would have been delayed without it. Sometimes the Texans who went north with their Longhorns stayed and developed homesteads and herds of their own in Colorado, Wyoming, Montana, the Dakotas, and elsewhere. Throughout all the cattle ranges of the West, much of the cowboys' specialized skill was learned from transplanted Texans. In Texas, new ranches were developed, improved breeding stock was purchased, and many areas were fenced in, aiding the process for herd improvement—all made possible because Texans had made good profits on cattle trailed to Kansas and shipped to eastern markets. Perhaps as important as any other benefit was the fact that the Chisholm Trail, by bringing Texans and Northerners together in business transactions, helped to lessen the sectional animosity engendered by the Abolition movement and the Civil War (Gard, 1954).

Between 1866 and 1890, ten million cattle had been driven north out of Texas. As a result, post-Civil War reconstruction took place far more rapidly in Texas than in other southern states, laying the foundation for the fabulous growth and wealth of that giant state. It has been said that "civilization follows the plow," but throughout the great virgin prairie the plow followed the Longhorn. This hardy beast, along with the equally hardy Mexican vaquero and American cowboy, were the vanguard of civilization in the vast prairies and high plains of the West (Dobie, 1941).

## THE CATTLE KINGDOM

The cattleman possessed no land or grass; he had only cattle and what were recognized by his neighbors, but not by law, as "range rights," that is, the right to water sources. Control of water implied command of the surrounding range, for cattle could not exist without water. In this way the great "cattle kingdoms" were created. For example, the well-known cattleman John W. Iliff, by owning 105 parcels of rangeland totaling less than 16,000 acres, exerted considerable control on a range extending 100 miles east from the Colorado border and 60 miles north and south (Frink et al., 1956). This was an extreme form of the rugged individualism for which cattlemen were noted.

In *The Cattle Kings* (1961), Lewis Atherton pointed out that many of the famous cattle kings were self-made, starting "from scratch" to develop their cattle kingdoms. Such well-known cattlemen as Goodnight, Kendrick, Kenedy, King, Kohrs, Iliff, Littlefield, Pierce, and Swan started as cowboys or with other menial jobs. Atherton considered that the fortunes of the hero of Owen Wister's *The Virginian,* who rose from cowboy to ranch manager, financial success, and marriage to a blue-blooded eastern schoolmarm, had a solid basis in fact in the western cattle ranges, including even the usual admiration of successful cattlemen for educated women.

Perhaps the most colorful rise to success in cattle ranching could be claimed by Richard King, the owner of the enormous King Ranch of Texas. At the tender age of eleven, King had escaped from apprenticeship to a New York jeweler by stowing away on a sailing ship bound for Mobile, Alabama. He wangled a job as cabin boy and, upon arrival at Mobile, he entered into a new apprenticeship on southern riverboats, earning his pilot's license at age nineteen. During the Mexican War he became a pilot of a boat on the Rio Grande. When the war was over, King ran a grogshop and flophouse, and in 1849 he bought, for $750, a vessel for which the government had paid $14,000 three years previously. After only a brief career at hauling merchandise on the Rio Grande, King joined in a partnership with two bankers to operate riverboats. With one of his partners, Mifflin Kenedy, King bought cargoes of wool, hides, bones, tallow, or livestock to sell downriver. One of the original partners, Charles Stillman, withdrew from the river freighting business, and King and Kenedy continued the joint ownership until 1874, using the lucrative profits from the business to acquire ranches. Both became famed cattle kings. During the Civil War and the federal government's coastal blockade, King and his partners continued to operate under the flag of Mexico. They accepted pay only in gold and were thus able to survive the vicissitudes of Reconstruction. King had the financial means to make his Santa Gertrudis Ranch into a famed cattle empire (Atherton, 1961). In such ways the cattle kings (or cattle barons) of the West gained their immense lands, collectively referred to as the cattle kingdom.

The cattle kingdom had too many miles of running water—rivers, creeks, tiny brooks, and springs—to be guarded by lawful means; the cattle kings depended on the extralegal system of "range rights." The boundary between two ranches might be the crest of a ridge separating two river valleys. Up and down the stream the problem was not so simple, but it was considered to be "not good form" for ranches to crowd one another. The unwritten law of the cattlemen (Code of the West) preserved peace for a while. By 1880, in the enormous region between western Kansas and the Rockies and between the Texas Panhandle and Canada, range rights had been established. Ranches of a hundred thousand acres were "small outfits" and those of a million acres were not particularly large, for individual cattlemen and companies claimed range rights over four, five, and six million acres and extending over territories as large as the states of Massachusetts and Delaware combined (Terrell, 1972).

Many English, Scottish, and eastern investors became absentee owners of cattle ranges that did not legally exist and that contained "book counts" of cattle greatly at variance with the actual numbers. Fraud was rampant, but even the number of cattle actually on the ranges was sufficient to result in serious overgrazing. Also, beef prices that had steadily risen, until about 1885, went down even faster and the sources of foreign and eastern capital dried up.

### Barbed Wire and Windmills

Taking the law into their own hands, or with the connivance of unscrupulous lawmen, the big cattlemen (and their powerful associations) could more than hold their own against the small homesteaders or "nesters" until barbed wire fencing, patented in 1874, became so cheap that homesteaders could afford to fence in their small holdings.

Much smooth wire had already been used in the prairie region, but it contracted in cold weather, expanded in warm weather, and livestock learned that they could push through it without injury to themselves. The success of thorn hedges led to the idea of barbed wire. The first commercially practicable wire was made in 1873 by J. F. Glidden, a farmer of De Kalb, Illinois, and was patented the following year. Twisted fence wire was used to hold the barbs in place. In 1875, 600,000 pounds of barbed wire were manufactured; in 1880, 80,500,000 pounds, in 1901, 297,338,000 pounds. The price fell from $20 to $1.80 a hundred pounds, making it the least expensive form of fencing.

Cattlemen were against barbed wire from the very beginning. Argument arose as to whether cattlemen should fence their land to keep their cattle confined or whether farmers should fence their land to keep the cattle out. Violence and bloodshed followed in conflicts between "fence men" and "no-fence men." Fence-cutter wars broke out wherever men began to fence and make private what had been free land and grass. Webb (1931) pointed out that even the fertile prairie was but sparsely populated until the advent of barbed wire. Homesteads could not have been protected without this epoch-making innovation.

Despite shootings and lynchings, the nesters persisted and the cattlemen retaliated by cutting their fences and finally by fencing in what they considered to be their ranges and including the small but troublesome lands of the homesteaders. By this time considerable public indignation had developed and by presidential decree a famous

*Fig. 19. An American windmill pumping water for cattle.*

ex-Confederate cavalry raider was assigned the task of sweeping through the cattle kingdom, cutting fences on public lands and arresting cattlemen who had erected them (Terrell, 1972).

Barbed wire made the 160-acre homestead possible on the Great Plains, but it could not make it profitable, or even livable, without water. This is the point at which the windmill enters the picture. Windmills were of the "American-type" (fig. 19), familiar to the traveler on the plains even to this day. They were useful to large stockmen as well as to homesteaders. Used to pump water, they enabled livestock of the vast ranges to disperse themselves away from the streams and to more completely utilize the available forage. The homesteader could obtain enough water for his domestic needs and his livestock. By allowing the water to accumulate in a reservoir of sufficient size for several days and nights of continuous pumping, he also had enough water to irrigate a garden or even an orchard. He could then enclose his farm with a barbed wire fence, keeping out the livestock of the herdsman's range that had no legal basis. The homesteader often made his own windmill for as little as ten to fifty dollars, even when he bought new materials. Sometimes he used mostly scrap metal (Torrey, 1976).

Apart from the watering of livestock, the windmill allowed the homesteader family to subsist on produce grown in the irrigated garden. According to plains historian W. P. Webb (1931), "...it was the acre or two of ground irrigated by the windmill that enabled the homesteader to hold on when all others had to leave. It made the difference between starvation and livelihood. These primitive windmills, crudely made of broken machinery, scrap iron, and bits of wood, were to the drought-stricken people like floating spars to the survivors of a wrecked ship." Webb considered the evolution of the range to consist of four stages: (1) open range, (2) fenced range without windmills, (3) fenced range with windmills, and (4) fenced farms with windmills. Windmills also pumped water for the steam locomotives of the rapidly expanding railroads.

In 1889 there were seventy-seven windmill factories in the United States, in 1919 there were thirty-one, while in 1973 there were only two that were still well equipped to make complete old-fashioned farm windmills (Torrey, 1976).

Since about 1850, over six million windmills have been built in the United States (Metz, 1977). Windmills can still be seen in many areas of the Great Plains and elsewhere in the West. These are generally used for supplying water to livestock, commonly seen grouped around the tank or trough into which the water is pumped, as shown in figure 19. For example, in the Nebraska sand hills alone, a ranching country of some 18 million acres of grassland pasture, there are an estimated 30,000 windmills pumping water for cattle and other livestock. In some sections of the sand hills region of southeast North Dakota where the water table is high, windmills are common, but one may not see many unless one hikes cross-country, for few roads cross the rangelands. In Texas, along the 43 miles of Highway 6 from Stephenville to Meridian, 18 windmills can be seen. According to Hayes (1977), there are still 150,000 windmills being productively used in the United States.

In 1896, F. H. Newell, Chief Hydrographer from the U.S. Geological Survey, wrote: "Of the devices for operating pumps for irrigation upon the Great Plains, windmills are undoubtedly the most important, and they will always remain so from the fact that the winds blow almost incessantly over this vast country" (Newell, 1896). (In those days the other types of pumps were driven by steam or gasoline.) Could it be that modern wind-power technology will pluck Newell's prediction out of limbo and restore its credibility?

Windmills were not used only for pumping water. During the 1930s, prior to widespread rural electrification, tens of thousands of wind generators were used for generating electricity in rural areas. One machine, called the Wincharger, is currently being manufactured and obtainable at Dyna-Technology, Inc., Sioux City, Iowa. The Wincharger is one of the options suggested by James Bohlen (1975) in his *The New Pioneer's Handbook,* for disenchanted city folk who wish to try their hand at a solar-powered homestead with a considerably higher level of amenity than that which was available to pioneers of yore. Another successful machine of the 1930s, called the

Jacobs Wind Electric Plant, is being sold in reconditioned units by Northwind Power, a small company in Warren, Vermont. A number of other machines for producing electricity are now on the market (Metz, 1977).

## The Blizzards of 1885-1886 and 1886-1887

It is unlikely that the cattle kings would have succumbed to the reverses in their fortunes discussed to this point, but a more ominous threat moved down from the north in the winter of 1885-1886 in the form of the most severe storm that could be remembered by the oldest inhabitants of the High Plains. Temperatures of -50°F (-45.6°C) were common, and even -60°F in some places. Not only were hundreds of thousands of cattle frozen to death but families of settlers met the same fate in their cabins. Freighters died with their teams and for years human skeletons were found of presumed victims of the Great Blizzard. After the snow melted in the spring, the new grass started to grow but the rains needed to sustain its growth failed to come. The grass dried prematurely and there were many prairie fires. Large streams were reduced to trickles and small ones dried up completely.

Although many ranchers had lost half their cattle, the drought forced cattlemen to sell those they had left and thereby glut the markets with poor-grade animals. The next winter the snow came early. The tragedy of the previous winter was repeated with even greater severity. Some cattlemen lost 90 percent of their herds. There were reports that between 10 and 12 million cattle had been lost. Many cattle enterprises collapsed, particularly those financed by eastern and European capital.

A positive result of the two calamitous years of blizzard and drought was that ranges were no longer overcrowded. Overgrazing and the resulting erosion ceased and grass again became abundant where it was not destroyed by plows or sheep. The feud between the cattle kings and the homesteaders continued with renewed fury, but the flood of new settlers did not slow down. Economies became diversified, new political regimes came into power in the plains states, and by the turn of the century the cattle kingdom was history (Terrell, 1972).

Blizzards as severe as those described did not again occur until 1948-1949, when 200,000 head of cattle died of starvation and freezing. The greatest loss was to calves born during the blizzard. The Forest Service and the Bureau of Land Management gave prompt aid by providing road-clearing equipment so that feed could be transported to marooned and starving cattle. The melting snow helped both range and crops, so that during the summer following the blizzard the cattle that survived emerged in better condition than before the blizzard.

## Demise of the Longhorn

Rangemen were reluctant to exchange free-running Longhorns, able to rustle their own living, for the improved breeds that had to be fenced in and given constant care. But Longhorns had to be grazed until they were four to seven years old before they were fat enough for market. Shorthorn and Hereford breeds imported in 1873 from Britain were able to sire progeny that matured earlier and fattened to a desired market condition in two to four years when grazed on native grass (Cundiff, 1975). Buyers demanded the new breeds, which converted pasturage "into choice cuts of meat instead of into locomotor energy" (Dobie, 1941). Compulsory dipping of cattle against fever ticks, begun in southern Texas in 1922, finally sealed the fate of the last remnants of the Longhorns. Cattle had to be dipped every fourteen days or, alternatively, every one of them had to be kept out of pasture for from seven to eighteen months until the ticks that had fallen from the animals, finding no hosts to feed on, died without laying fertile eggs. The big ranches were forced to cut up and cross-fence[4] their pastures and Longhorns that could not be worked as draft animals were shot as nuisances.

The Longhorn, after playing such a prominent role in the development of the West, all but vanished from the American scene, giving way to more profitable cattle breeds. But in the Wichita Mountains National Wildlife Refuge in southwestern Oklahoma a thriving herd of 300 Longhorns may still be seen, along with buffalo, elk, white-tailed deer, wild turkeys, armadillos, coyotes, and other wild animals (Halpern, 1975). However, the Longhorns of that herd are not direct descendants of the wild, rangy Longhorns of the early years of the West. By the time their ancestors were introduced into the refuge in 1927 they had been "improved" by mixture with breeds that were commercially more desirable.

Some believe that the Longhorn might have developed into a superior breed by long, intensive breeding on well-controlled areas of land. But this did not happen; the magnificent Longhorn has become a part of history—a colorful and thrilling saga of the Old West (Dobie, 1941).

## IMPROVEMENT IN CATTLE BREEDS AND RANGE GRASSES

Long before the demise of the Longhorns, cattlemen began developing herds of standardized, uniform stock: first Durhams, then Polled Anguses and Devons, then Herefords and Charolais. The first Brahmans were imported in 1848 in South Carolina and began to enter Texas in the early 1880s. Cross-strains were developed, the most famous of which, until recently, was the Santa Gertrudis—a Brahman-Shorthorn cross. A more recent and more sensational cross is the "beefalo," a cross between the buffalo (American bison) and the modern beef cow. A "pure-bred" beefalo, as recognized by at least four beefalo associations, is three-eighths bison, three-eighths Charolais, and one-quarter Hereford (Ellis, 1978). The tendency of most crosses between different species to produce sterile offspring (like the mule) was overcome after years of intensive breeding. The beefalo has been said to have the best qualities of buffalo and beef cow, grazing as economically as the range-adapted buffalo and gaining weight faster than a cow. According to a California breeder, lower feed costs and higher fertility of the animals result in beefalos being 40 percent cheaper to raise than ordinary cattle. The meat has a higher protein and lower fat content. However, this is apparently because the objective of beefalo breeders is to rear bulls, whereas the objective of other cattlemen is to rear steers. With any breed of cattle, the bull grows larger and has leaner meat than a steer (Albaugh, 1976). Steers are reared because they have the "marbled" meat desired by the American consumer. Most cattlemen are taking a "wait-and-see" attitude toward the beefalo (Ellis, 1978).

Not only are cattle breeds being continually improved but so are the range grasses upon which they graze. On the High Plains the dominant vegetation still consists of native grasses. Introduced grasses are more common in the humid East than in the West, but some useful introductions have been made in the Great Plains. For example, crested wheatgrass, brought by plant explorers from Russia to the United States in 1898, resisted both cold and drought. It was admirably suited to the needs of the northern plains. For millennia, native grasses have protected land from wind and water erosion, binding the soil with fibrous roots. They use water efficiently, converting it to

[4]In Texas, the XIT Ranch of 3,000,000 acres, 150,000 cattle, and employing 150 cowboys, had 6,000 miles of single-strand fence (Sloan and Frantz, 1961).

economic plant growth. The successful stockman is the one who allows grazing at a rate that will permit the desirable plants to stay vigorous and productive and to reseed themselves. He must not only understand careful livestock husbandry, but also the equally careful management of grasses.

The deterioration of the prairie from overgrazing is one of the sad features of environmental degradation in this country. John Merrill, director of the Ranch Management Program at Texas Christian University at Fort Worth, and himself a lifelong rancher, has shown that three to four tons of high-quality dry matter per acre per year can be grown on Texas rangeland with no inputs except management. The Fort Worth prairie, with an annual rainfall of 31 inches, once supported native tall-grasses such as big bluestem, Indian grass, switch grass, and many other grasses and plants that cattle like to eat, and all well adapted to the soil and climate. In much of this region overgrazing thinned out these more productive plants and allowed them to be replaced by ones of lower nutritive quality, with more than 50 percent loss of productive capacity of the range.

Merrill operates a high-intensity, low-frequency (HILF) rotational system for grazing his cattle. This involves a half-dozen pastures, totaling 2,139 acres of his 4,000 acres of ranch and farmland five miles south of Fort Worth. Through a number of trials, Merrill found that during the growing season the optimum rotation was an average of 18 days of grazing followed by 90 days of rest for the grasses. This results in about 50 percent use of the bluestem grass. A small-grains pasture is grazed as winter protein supplement and grain is then harvested to cover farming costs. In the spring "transition period" Merrill practices "flash grazing" to take advantage of the lush growth of winter annuals and also to reduce their competition with desirable perennial grasses. Southern stockmen whose cattle graze pastures with introduced grasses such as fescue, Bermuda, and ryegrass, consider one cow to nine acres of pasture to be not too many. John Merrill, winner of the 1976 Rangeman of the Year Award, has achieved the same stocking rate on native grasses with no fertilizer, weed control, or hay feeding in winter, and with a great improvement in the grasses (Odle, 1977).

## Brush Control with Herbicides

One way of improving cattle range is to control brush. At the U.S. Southern Plains Field Station at Woodward, Oklahoma, experiments were made with a mist blower, mounted on a pickup truck, applying herbicide downwind at wind speeds of less than ten miles per hour. The drift of herbicide is usually effective for 100 feet and a single operator can treat 120 acres per hour.

Doses of herbicide are usually one ounce of active ingredient of 2,4,5-T or two ounces of active ingredient of 2,4-D in one gallon of water per acre. One treatment usually provides summer-long suppression of shinnery oak, sand sage, and most other weeds. Two annual sprayings thinned ragweed 70-90 percent, sand sage 34-50 percent, and shinnery oak 40-60 percent. Near-maximum grass growth can be obtained by a third annual spraying. Brush can be removed from brush-infested rangeland to such an extent that grass production can be increased 50-300 percent, at a cost of only 20 to 30 cents per acre per year.

Some herbicides of the terazine group not only control weeds on rangeland but also increase the protein content of range grasses and protect plants from drought. Atrazine, for example, increased nitrogen concentration in herbage by an average of 29 percent. When nitrogen fertilizer is added to treated rangelands, the protein content of the herbage increases in a way that indicates additive effects of fertilizer and herbicides.

*Fig. 20. A sprinkler-irrigated Oklahoma pasture that three years previously had been cleared of hardwoods and underbrush and planted to Coastal Bermuda grass (courtesy USDA Soil Conservation Service).*

Triazines control most annual weeds on shortgrass ranges, but are particularly effective against six-weeks fescue, which cattle will not eat (Anonymous, 1975*a*).

In some areas it pays to plant superior grasses in the cleared areas. Figure 20 shows a pasture near Ardmore, Oklahoma, that three years previously was heavily infested with hardwoods and underbrush. This is one of the many beautiful pasture and hay-producing fields that can be seen from Oklahoma's highways. The area was cleared and planted to Coastal Bermuda grass, thus becoming a highly productive piece of land. Sprinkler irrigation can be seen in action in figure 20. Sprinklers enabled ranchers to water land that otherwise would be unsuitable for irrigation and also made more economical use of the water.

## Ranch Farmers and Pasture Farmers

The development of the cattle industry involved transition from open-range herding to "ranch farming." Cattlemen grew fodder crops to supplement the range grasses. Some ranch farmers built silos and many also erected dams for water storage. In the severe blizzard of 1948-1949 it was the cattlemen depending mostly on the range who lost heavily. The ranch farmers suffered only a little inconvenience. Ranch farming sometimes evolved into "pasture farming," as illustrated in figure 20. Pasture farms were generally not as large as the ranch type, but an important feature was the new attitude of the cattlemen:

> Generally, they were not ranchers who reluctantly adopted a disliked makeshift. Neither were they just broken and disappointed dry farmers who hesitatingly accommodated to survive. Most pasture farmers embraced grassland farming, supplemental feeding, water development, advanced breeding, and pasture management as profitable goals in themselves. [Schlebecker, 1963]

## Increasing Dependence on Government Assistance

One of the most interesting phenomena in American agricultural history was the breakdown of the resistance of cattlemen—rugged individualists and most ardent champions of *laissez-faire*—to government interference in their affairs. Perhaps the first indication of their change of heart occurred in the early years of the twentieth century, when cattlemen's associations began in many ways to assist their old enemy, the Forest Service, particularly in range management, thus belatedly recognizing conservation as inevitable and for their own benefit. Then in the post-World War I years dependence on government assistance increased. Cattle-industry historian John T. Schlebecker (1963) writes:

> ...between 1919 and 1922 a host of troubles beset stockmen. Meat consumption declined; beef prices fell; drouth hit the Plains; grasshoppers, prairie dogs, and diseases destroyed crops and cattle;[5] and meanwhile bankers and packers hovered like Harpies to finish off any surviving cattlemen. In their extremity, cattlemen called for and got government assistance in combatting their adversaries. Farmers and cattlemen had changed their minds about government assistance. The change had been coming since the century began, but it reached a new stage after the war.

Again between 1929 and 1932, cattlemen asked for government help: price supports and tariffs, credit assistance, regulation of land use,[6] scientific research and education. During the darkest days of the Great Depression, federal emergency programs, such as the Federal Emergency Relief Administration, which ordered about eight pounds of beef each month for every family on relief, saved the cattlemen from complete disaster. Biochemical research contributed greatly to their later prosperity. Vaccination was shown to be effective against foot-and-mouth disease and brucellosis in the early 1950s. In 1955 the USDA announced success with the systemic insecticide phenothiazine against ox warbles. About that time the addition of antibiotics to cattle feed became widespread. As a feed additive, aureomycin proved effective against a complex of diseases ranging from foot rot to liver abscesses. Besides reducing dangerous bacteria, antibiotics also stimulated beneficial intestinal bacteria to produce more B-complex vitamins. Antibiotics also resulted in animals using protein more efficiently (Schlebecker, 1963).

[5]In 1921 and 1922 grasshoppers destroyed so much feed and grass that cattle in Montana and Wyoming had to be moved to other ranges. On the southern plains ten prairie dogs to an acre used almost all of one year's grass. Prairie dogs and predators of calves and sheep were combatted by United States Biological Survey personnel, but without decisive results. On the other hand, by 1922 the Plains were remarkably free of ticks because of federal eradication programs (Schlebecker, 1963). Currently (1978) grasshoppers are again causing serious damage on the Great Plains.

[6]Regulation of land use was the antithesis of laissez-faire, but was welcomed by cattlemen in the form of the Taylor Grazing Act of 1934, which gave the Department of the Interior authority to regulate grazing on the public domain in the West.

## WHEAT

Second only to cattle in the Great Plains is wheat. The Spaniards introduced wheat into the Western Hemisphere about 1520. The first introduction along the Atlantic seaboard by the English colonists met with little success, but the Mennonites were successful in growing wheat in Pennsylvania. Likewise, it was a later migration of Mennonites that put wheat growing in the West on a sound basis, as we shall see later.

The principal commercial species is "bread wheat," *Triticum aestivum* (= *sativum* = *vulgare*), of which there are many varieties. The other important species is durum wheat (*T. durum*).

In 1976, 58,565,636 metric tons of wheat were grown in the United States. The six leading states in wheat production were Kansas with 15.8 percent of the total, North Dakota 12.5 percent, Montana 7.6 percent, Oklahoma 7.0 percent, and Washington 6.8 percent and Texas 6.1 percent (USDA Agr. Stat., 1977). Five of these states are in the Great Plains. As in most of the world where wheat is grown, it is a region of cold winters; minimum winter temperatures are 0°F (-17.8°C) or below. Annual rainfall is less than 30 inches.

As one of the two leading bread cereals, the other being rice, wheat is the staff of life for peoples of the temperate zones of five continents. It is the world's most important food crop. World production is 360 million metric tons, compared with 320 million of rice, 300 million each of corn and potatoes, 170 million of barley, 130 million of sweet potatoes, and 100 million of cassava. These 7 crops total more than twice the tonnage of the remaining 23 that have over 10 million tons of annual production (Harlan, 1976).

### Origin of Wheat

What we now call "bread wheat" arose long after the three primitive "glume wheats" were initially domesticated from wild grasses. The glume wheats are so called because the spike, or seed-bearing head, breaks up when it is threshed and leaves each seed in a hard, shell-like glume, or husk. Further pounding in a mortar frees the seeds of the husks. The three kinds of glume wheat were einkorn, always a minor crop; emmer, from which Egyptians made their bread until it was replaced by bread wheat in the fourth century B.C.; and a third type of such trivial importance that it has no common name. Eventually a mutation occurred in emmer that caused the base of the glume to collapse at maturity, freeing the seed. Also the spike became tough, so that it did not fall off. This mutated, free-threshing emmer was the ancestor of present-day durum or macaroni wheats (Harlan, 1976).

Bread wheat is believed to be the result of emmer[7] receiving an extra set of chromosomes from a wild goat grass (*Triticum tauschii*), a hybridization that may have occurred near the southern end of the Caspian Sea. This may have made possible the raising of wheat on the dry steppes of the world. Thus a wild grass, in itself essentially worthless, contributed the genetic characteristics that "made a billion-dollar crop out of a million-dollar one" (Harlan, 1976).

Unlike its wild ancestors, bread wheat does not have kernels that can be spread by the wind; the spike is too tight to break up. Man and plant came to depend on each other, just as in the case of corn, as related earlier.

[7]It is not known whether the emmer parent of bread wheat was wild, tame, or even the mutated free-threshing durum wheat, but cytological behavior would be the same in all cases (J. R. Harlan, correspondence).

## Mennonites and Turkey Wheat

The ancestor of modern American wheat was the Turkey variety, introduced from Russia. American wheat is now a blend of modern varieties: Kaw, Lancer, Scout, Sturdy, Triumph, Warrior, and Wichita, but all have Turkey ancestry.

In the early days of the Great Plains, wheat farming was not a notable success. Winter wheat was continually threatened by drought, wind and duststorms, winterkilling, leaf and stem rust, chinch bugs, grasshoppers, "worms," and sometimes migratory ducks and geese. In the early 1870s the German Mennonites brought Turkey wheat with them when fleeing from Russia, the country to which they had been attracted by a promise of freedom of worship and freedom from military service, a promise broken in 1870-1871. The Mennonites called the wheat "Turkey" because they believed that it originated in a little valley in Turkey, but some authorities disagree, believing that it originated in the Crimea.

For the present story of Turkey wheat we are indebted to USDA agronomists Karl S. Quisenberry and L. P. Reitz (1974), who grew up on farms in south-central Kansas and began their interest in wheat breeding as student assistants in college. By 1875 many Mennonites had left Russia and migrated to the United States. Most of them settled in the Central Plains, mainly Kansas. Each family brought to this country anywhere from a few pounds to a bushel or two of seed, and this was the origin of our hard winter wheat. While most of the Mennonites settled in Kansas, particularly Marion and Harvey counties, some settled in Nebraska, Dakota, Minnesota, and Manitoba, and it is likely that Turkey wheat was carried to those areas also. No doubt, by far the most of this famous variety reached Kansas.

Farming on the Central Plains was a formidable task that had driven many a farmer to desperation. The Mennonites were much better at it than most other American settlers. As is usually the habit of immigrants, they deliberately picked out the prairies of Kansas and Nebraska as having land and climate similar to that which they left behind in their homeland in Crimea and Ekaterinoslav. They had experience in farming on prairie land and had seed that was well adapted to the harsh Great Plains environment. It was the Mennonites who got the gigantic wheat empire of the Great Plains underway.

Unfortunately, it was not until the Hatch Act of 1887 and a subsequent period required for organization of test plots, as well as experiences during some severe winters, that Turkey wheat was proved to be the best for yield, winterhardiness, and quality. It was probably not until 1903, when the Kansas Experiment Station at Hays was established in the rigorous climatic conditions of the High Plains, that winterhardiness was shown to be the most important factor in a wheat variety. By 1919 Turkey comprised 99 percent of the hard red winter wheat acreage of the United States. It was not until 1944 that the lead in acreage was taken by Tenmarq, one parent of which was a Turkey type. Even for the famous Japanese variety Norin 10, giving the semi-dwarf characteristic of Green Revolution wheats, one of the parents was Turkey, obtained by the Japanese sometime before 1892.

## Other Ethnic Groups

The Mennonites should not be confused with the Volga Germans, who were refugees from the Volga River region of Russia. Like the Mennonites, they were lured to Russia during the reign of Catherine the Great, who wished to increase the ratio of free citizens to serfs and set an example of industriousness and advanced agricultural methods, for which German farmers were well known (Tooke, 1800).

Having left their wealth in Russia, the Volga-German refugees could not afford to purchase expensive land and therefore they tended to settle on the poorer lands of the western fringes of the Great Plains. Nevertheless they had the necessary human resources to succeed where others failed, and they were therefore an important factor in the agricultural development of the High Plains. In part, their success was the result of their substitution of communal village life for the isolated homesteading existence of most frontier farmers, affording greater opportunities for supplementing their meager farm income with nonfarm work (Forsythe, 1977).

The Russian-Germans worked in family units and, because the families were large, the farmers could earn and save enough money to become independent. The sugar beet companies wanted resident labor and often provided generous terms to those wishing to buy or rent company land. Families that had climbed the "agricultural ladder" were replaced by newly arrived immigrants and the sugar beet acreage increased until World War I. Later, Japanese arrived from California and they too were anxious to become growers. In northern Colorado, Spanish-Americans from New Mexico and Mexican nationals were employed, but most of them had no interest in acquiring farms and seemed content to remain as farm laborers (Schwartz, 1945).

Another group of agriculture-oriented German-speaking people that sought sanctuary in the Russia of Catherine the Great, and then emigrated for the same reason as the Mennonites, were the Hutterites. They are followers of Jacob Hutter, a Christian martyr who founded the sect in Austria in 1528 and was burned at the stake in Germany in 1536 for his pacifist beliefs. The Hutterite faith centered on adult baptism, nonviolence, and common ownership of property. Hutterites first emigrated to Dakota Territory in 1874 and founded three colonies (*Bruderhofs*) totaling less than 500 adults and children. Now there are more than 200 colonies in the United States and Canadian West, with a total population of about 25,000. They have the highest recorded birthrate in the world, an average of thirteen children to a family.

In the Dakotas, Minnesota, and Manitoba a colony of ten to fifteen Hutterite families occupies about 4,000 acres, whereas in Montana, eastern Washington, Alberta, and Saskatchewan, such a colony will own more than 10,000 acres. The colonies raise grain, cattle, and a wide variety of other crops and livestock. They have machine and other shops and technicians to make their colonies highly self-sufficient. They are said to be very progressive and innovative in agricultural technology and experimentation (Hostetler, 1974; Easton, 1976).

## Campbell Revolutionizes Wheat Farming

It is not surprising that someone took advantage of World War I and the resulting food crisis to stage the kind of show that Americans enjoy most, a mechanization extravaganza—this one in wheat farming. The principal role was played by Thomas D. Campbell, who grew up among the giant bonanza farms of the Red River Valley of North Dakota and had dreamed of some day having the biggest of them all. Characterized as "a natural born promoter, a wonderful personality, smart as hell, and not afraid of anything," Tom Campbell utilized these useful traits to obtain a $2 million loan from J. P. Morgan and other financiers and joined them in forming the Montana Farming Company (MFC), with himself as manager, thereby becoming responsible for many practices relating to the mechanized, industrialized, large-scale farming that eventually changed American and world agriculture (Drache, 1977).

Campbell obtained permission from the Indian Bureau to farm Indian reservation lands in Montana, with the Indians receiving one-tenth of the crop, delivered to the railroad, for five years, then one-fifth of the crop for the second five years. Convinced that the big tractor gave the farmers fullest economy, he ordered thirty-four 30-60

Aultman Taylor 35 horsepower tractors. He also ordered several large Holt caterpillars and several medium-sized tractors, 40 ten-foot binders, 10 threshing machines, 4 combines, 100 grain wagons, 60 grain drills, 50 discs, 50 ten-bottom plows, and 10 trucks. Nearly 10,000 spools of barbed wire were used on 625 miles of four-strand fence. There was need for bunkhouses, cooking and dining facilities, grain storage, fuel depots, a machine shop, roads, wells, and a main office at Hardin. The expenditure for these items added up to over a million dollars. The cost of labor, fuel, repairs, and seed exhausted the remainder of the $2 million.

Seven thousand acres were plowed, seeded, and fenced in 1918, the first full year of operations, 45,000 acres by 1919, 110,000 by 1923, and eventually 150,000 acres were fenced.

The extended drought in Montana resulted in no profits during the four years of MFC operations. In 1920 wheat prices fell from $2.75 to $1.05 per bushel during a period of a few months. The corporation, which had been formed as a part of the overall war effort, discontinued its operations, charging some losses to the war cause and recapturing others via tax refunds. Campbell paid $150,000 for the machinery and existing leases. Having settled the Morgan account for seventeen cents on the dollar, Campbell awarded his Wall Street benefactors the dubious distinction of being the most reasonable group of men he had ever known.

At the age of forty, Campbell had become a veteran of industrialized, mechanized farming on what was reported to be the world's largest wheat farm. He organized the Campbell Farming Corporation (CFC), retained 75 percent of the corporation stock, and distributed remaining shares to managers, foremen, and other key personnel. Some of the poorer lands were abandoned and others were developed. Economies were effected and fallowing was changed from every second to every third year. Thus two-thirds of the farm was always in production. This decision was based on Campbell's belief that moisture conditions would improve, which happened to be the case. The CFC changed from deep to shallow plowing, and later to stubble mulching; both changes helped to retain winter moisture and reduce wind erosion.

The CFC established several records. In 1924, fifteen 30-60 Aultman Taylor tractors, each pulling a plow, a disc, a drill, and a packer, plowed and seeded a 640-acre field in 16 hours. The tractors were refueled while moving. To do the same job in the same number of hours would have required 500 men with 1,800 horses. Campbell started using 24-foot combines, each of which would combine 60 acres of grain a day.

CFC doubled its number of grain wagons, for the sacking of grain in the field had been discontinued. Twelve wagons could be pulled by a single Holt Caterpillar controlled by a two-man crew, effecting great economies in hauling a half-million-bushel grain crop as far as forty-two miles.

At first enamored with bigness of all kinds, including big tractors, Campbell did not hesitate to reverse his "bigger is better" philosophy when economics and efficiency so dictated. After 1923, his huge Aultman Taylor tractors were gradually replaced by less costly, much lighter, more maneuverable types using only one-fourth to one-half as much gas, particularly the McCormick-Deering 15-30 and 22-36 models.

Although "corporate giants" have come in for much criticism on sociological grounds, it must be said on Campbell's behalf that he always proudly bore the title of "farmer," remained on the farm, and dedicated his considerable energy and talent toward innovation in methods and equipment that were to transform Great Plains agriculture. In doing so, he proved that through proper conservation practices the plains could profitably produce grain; that mechanized farming was the most economical and productive procedure; and that farming with hired labor and management could compete with farming based on family labor.

The huge corporate farming so fervently defended by Campbell was not without its adverse externalities. For him the new technology and improvements were a boon, but they were beyond the means of thousands of homesteaders; their incomes were reduced by expanding supply and, in postwar years, of greatly diminished demand. Although Campbell was an avid believer in supercolossal farm operation, its presumed merit was not fairly tested, for one is left to surmise how well his corporation would have fared if it had to cover the entire $2 million original investment instead of liquidating at 17 cents on the dollar (Anderson, 1977).

## Winter and Spring Wheat

In the southern Great Plains wheat fields are green even before trees begin to bud, because wheat planted in autumn has already developed a strong root system before growth stops in winter. Remaining dormant during winter, the plants resume their growth in the early spring. This is called "winter wheat." It survives cold winters remarkably well. Except under most severe conditions when a snow blanket is lacking, wheat planted in the autumn "tillers" well in the spring, that is, it produces many stalks and therefore many seed heads. The grain is ripe and ready to be harvested in a few months.

In the northern Great Plains winters are too severe for winter wheat. Wheat is planted in the spring as soon as the ground is dry enough to be worked with power equipment and the threat of frost is over. This is called "spring wheat." It is usually sown in March for harvest in autumn. On the North American continent spring wheat is grown in the northern wheat belt of the United States (fig. 3, Region F) and in Canada north to the Arctic Circle. Durum is one of the spring wheats. It has a high gluten content and is used as a semolina for preparing macaroni and spaghetti. In 1976, 67 percent of the 3,672,800 tons of durum wheat grown in the United States was raised in North Dakota. Surprisingly, Arizona raised 54 percent of the remainder (USDA Agr. Stat., 1977).

Thus harvesting of wheat begins as early as May in the southernmost wheat-growing areas, as in Texas, and sometimes the same group of giant self-pulled combines that begin to harvest wheat in May will move northward at the rate of about ten miles a day and work continuously until they finish the harvesting of spring wheat in the fall in the Dakotas or in Canada.

The self-pulled or self-propelled (SP) combine (fig. 21) has gradually replaced the "pull-type" machine, having the obvious advantage that no tractor is needed to pull it. It is also more maneuverable and can be easily transported by truck and, with interchangeable heads, can be adapted to multicrop harvesting more easily than the pull-type. Despite costing twice as much as pull-type combines, by 1967 over 96 percent of the 42,000 combines sold were SPs. The SPs are also becoming larger, with "cutting heads" 20 feet or more in width. Besides cutting and threshing small grains, the newer units can handle most seed-yielding crops, including beans, soybeans, sorghums, sunflowers, and corn. A single machine can be used to harvest three or more crops with only minor changes (Hilliard, 1972). The bold entrepreneur and clever inventor are second to none in importance in the development of the American agricultural miracle.

## How Wheat Is Milled and Used

When wheat is milled, the outer portions of the grain (pericarp and aleurone) become bran. This amounts to from 8 to 9 percent of the grain; the starchy endosperm, 82 to 86 percent; and the oily embryo or "germ" at the lower end, 6 percent. Wheat is a little

*Fig. 21. A self-propelled (SP) combine harvesting wheat (courtesy Deere and Company).*

higher in protein than corn and a little lower in fat and carbohydrate. The wheat flours with the higher percentage of protein result in the more sticky and tenacious doughs for baking. There are various types of wheat grown in the United States, each chosen either for certain baking purposes, that is, for bread, pastries, and breakfast foods, or for making macaroni and spaghetti. Much wheat is also used in making alcohol. Some wheats are used for feeding livestock, but comparatively little when compared with corn. Great quantities of bran are left over from the milling operations and this is used as a high-grade animal feed. The straw is used as a mulch in agriculture or as a stuffing or weaving material (Schery, 1972).

One is impressed by the complete utilization of agricultural products except in one important respect—the excessive refinement and devitalization of foods during their preparation for human consumption. The germ in wheat, for example, is screened off in the milling operation in the preparation of white flour. Also the outer grain coatings are rejected, depriving consumers of fats, proteins, vitamins, and minerals.[8] Flour or

[8]The following percentages of the nutrients are lost: 22 percent protein; 51 percent fat (USDA, 1975); an average of 72 percent B complex vitamins—thiamine, riboflavin, niacin, vitamin $B_6$, and pantothenic acid; 86.3 percent of vitamin E; an average of 73 percent of calcium, magnesium, phosphorus, and potassium; 76 percent of the iron; and an average of 63 percent of seven essential minor elements: chromium, cobalt, cop-

bread does not keep as well with the wheat germ present, but with practically universal refrigeration in the American home, this should not be a problem. The fact that genuine whole wheat bread is being sold in some stores shows that it can be done. However, generally the most nutritious parts of wheat are being fed to cattle (Schery, 1972). During the question period after he had lectured at UCLA, the well-known nutritionist Jean Mayer was asked what could be added to further improve "enriched" bread. After listing the many nutritional factors, he added, "but then you would have whole-wheat bread, which is already available."

The ultimate in processing is reached with breakfast foods, particularly those based on corn, in which the sugar added may amount to as much as 50 percent of the finished product (Yudkin, 1972). The cost to the consumer per unit of basic food can rise to absurd levels (Lappé and Collins, 1977).

### Variety Improvement

Most of the factors that have led to increased production of food crops—for example, fertilizers, irrigation, herbicides, pesticides, and mechanization—have been energy-consuming. As in all aspects of modern industry, the ever-increasing demand for energy is a matter for concern. Only the plant breeders have made significant advances in agriculture without increasing energy demands. This they have accomplished by sophisticated procedures for incorporating new and useful genes into the genotypes of cultivated varieties or by selecting individual plants in which the desired trait has arisen by spontaneous or induced mutation (Nabors, 1976). (As discussed in chapter 3, in the case of tree fruits only a portion of the plant need show a desirable trait and it can be thenceforth propagated by grafting or budding.)

With about a third of the world's people depending upon wheat as a major source of calories and protein, a wheat variety combining increased yield with increased protein content would be a development of utmost global significance. About twenty years ago the U.S. Department of Agriculture and the Nebraska Agricultural Experiment Station in Lincoln began a cooperative breeding program for improvement of the nutritional value of wheat. The result was the new hard red winter wheat variety Lancota. This is the first hard winter wheat variety adapted to the Great Plains that has the potential for both high yield and increased protein content, the latter deriving from genes carried by the soft wheat variety Atlas 66. Lancota excels its male parent, Lancer, in yield and in resistance to leaf rust and Septoria leaf blotch, and it has excellent milling and baking qualities. The added protein is in that portion of the wheat grain that is made into white flour, so it will not be lost in the milling process (USDA, 1975*a*). Limited quantities of certified seed of the new variety became available in 1976.

At one time it appeared that hard red spring wheat was doomed by black stem rust. A cooperative research program of the U.S. Department of Agriculture and the Minnesota Agricultural Experiment Station, beginning in 1907, led to the development of rust-resistant Thatcher wheat, of which eighteen million acres were grown by 1941. Likewise the crossing of the Victoria and Richland varieties of oats resulted in high-yielding, strong-strawed varieties, resistant to black stem, crown rust, and smuts. Cooperative work of the agricultural experiment stations of Iowa, Wisconsin, South

per, manganese, molybdenum, selenium, sodium, and zinc (Schroeder, 1971). "Enriched" white flour, in which thiamine, niacin, riboflavin, and iron are added to approximately the amount present in whole wheat flour, still deprives the consumer of most of the above nutrients as well as fiber, likewise important for health (Abrams, 1978).

Dakota, and Nebraska led to varieties now grown on most of the oat acreage in these states. Barley production was increased by the development of high-yielding smooth-awn varieties.

Plant breeders have also done much to avoid losses from lodging. Lodging refers to the permanent displacement of the stems from the upright position. It may be caused by wind, rain, or hail. Conditions that are most favorable for growth and grain yield will aggravate lodging tendency and increase its severity. It has been regarded as an "abundance disease" that restricts the exploitation of factors that would otherwise increase yield. Lodging increases the difficulty of harvesting and results in considerable loss of grain and a reduction in its quality. Yield-promoting factors such as increased nitrogen fertilization or irrigation are not useful unless lodging can be prevented. Considerable progress has been made by plant breeders, generally in the direction of developing short-strawed varieties, less susceptible to lodging but with increased yields. Varieties have been developed that have strong stems and many adventitious roots, for they are less likely to lodge when heavily fertilized. As a further insurance against lodging, foliar sprays of 2-chloroethyl trimethylammonium chloride (CTC) at rates of 1 to 5 kilograms per hectare are commonly applied, especially on very fertile land where high grain yields are attainable (Pinthus, 1973).

## BARLEY

Barley is one of the oldest of cultivated cereals and is presumed to have been first domesticated in the Near East, with Ethiopia and Tibet also possible loci of domestication. Cultivated species are presumed to have descended from one of the wild barleys (*Hordeum spontaneum*) with which they readily cross. There are two "six-row" and two "two-row" cultivated species, referring to the number of rows of grains in a head. The two-row species are the more primitive. In the United States the six-row species are the ones most generally grown and all four species are usually lumped together as *H. vulgare*. Spring barley, grown mostly in the North Central States and in California, accounts for a major portion of the tonnage. Most of the winter barley is grown in the Southeast as a cover crop or for autumn and spring pasture. Average yield of barley in the United States is about a ton per acre, but the hybrid variety Hembar has yielded more than four tons per acre in test plantings in Arizona.

The brewing industry consumes over 30 percent of the barley crop, germinating it for malt. About 3 percent of the crop is made into pearl barley, by dehulling the grains, and is used for soups. Much use is made of barley as a substitute for corn in animal feeding. Some of it is rolled or partly crushed and fed to livestock. Barley was used as a staple food cereal in southern Europe until the discovery of the utilization of yeast for making bread, for which wheat flour is more satisfactory. A very small quantity of barley is still made into flour, for use by invalids and infants (Schery, 1972).

## OATS

Oats (*Avena* spp.) are widely cultivated, particularly in North America and Europe, and are adaptable to a wider range of climate, soil types, and cultural techniques than are most other cereals. They are believed to have been domesticated in northern Africa, the Near East, and temperate Russia, perhaps as late as 2500 B.C. They were introduced into North America in 1602. World production of oats is about 50 million tons a year, with the United States as the principal producer. The common cultivated species (*Avena sativa*) is often seen as an "escape" from cultivated fields. Other species of oats are common in the wild over large areas in fields and waste places, as in the case of

*A. fatua* and *A. barbata* in California (Jepson, 1925). *A. fatua,* the common wild oat, believed by many to be the ancestor of *A. sativa* and *A. barbata,* is a tall, weak range grass in the western United States (Schery, 1972).

Oats are broadly classified into spring and winter types, depending on the season of planting. Most of the domestic production comes from the spring type of the northern states. The grain is fairly rich in protein and fat (13.8 percent protein, 4.3 percent fat, and 66.4 percent carbohydrate). Only 3 or 4 percent of the crop is used for human consumption, mostly as rolled oats; thus it is generally used as animal feed.

## RYE

Although known to the ancient Greeks and Romans, rye (*Secale cereale*) is believed to be of more recent origin than most cereals. It is not found among Egyptian ruins or Swiss lake dwellings (Schery, 1972). As related in the introductory chapter, rye and oats are believed to have entered northern Europe as weeds that accompanied wheat and barley. In the acid soils of the region, and the colder and wetter summers, the two "weeds" were better adapted than wheat and barley and eventually became cultivated crops. For centuries rye flour served as the main ingredient in *Schwartzbrot* for much of the European population. This black, soggy, and rather bitter bread is still common in rural Germany, Poland, and Russia.

To this day northern Europe is the only region in which rye is an important crop. It is understandable that in the United States also, rye is grown principally in northern regions; South Dakota, North Dakota, Minnesota and Nebraska usually account for approximately 65 percent of the annual crop of about three million metric tons. In this country rye bread is made of a combination of wheat and rye flours. Most of the rye crop is fed to stock.

## TRITICALE

There is good reason to believe that triticale, the world's first man-made crop, will have certain unique advantages not possessed by wheat and other grains. The word *triticale* derives from the wheat genus *Triticum* and the rye genus *Secale,* the two grains from which the new grain was derived as a hybrid. Hybrids are generally sterile and, as in the case of the beefalo related earlier in this chapter, an enormous amount of patient crossing, selection, and careful observation was required before a strain was found that could reproduce itself. Triticale combines the high yield and high protein content of wheat with the ruggedness of rye.

Scientists in nearly every state are doing some research with the new grain. Triticale flour and pancake mix are now being marketed. Hardy spring and winter varieties are being developed for even higher yields, greater disease resistance, and wider adaptability to different regions (Friggens, 1975).

## SORGHUM

The sorghums are a group of closely related grasses presumed to be native to Africa, but are cultivated extensively in tropical and temperate regions. Sorghum is an important part of the diet in Africa, India, and China. The plant is utilized in much the same way as maize (corn) and has the advantage of being suitable for cultivation in areas too arid for the latter crop. Sorghums are also grown as animal-feed grain over a much wider area, as well as for forage, silage, syrup, brooms, and other purposes. Sorghums, particularly those with sweet stems, are highly nutritious feeds, being rich in proteins and vitamins. In North America sorghum is grown only in the United States, principally in the southern Great Plains. It withstands drought well and grasshoppers feed on it only as a last resort (Schlebecker, 1963).

This country produced 19.7 million bushels of sorghum in 1976. Texas led with 40 percent of the total, followed by Kansas with 24 percent and Nebraska with 17 percent. There were also 7.3 million bushels grown for forage, of which Kansas produced 34 percent (USDA Agr. Stat., 1977).

The sorghums are most closely related to sugarcane among the grasses of economic importance. However, the plant looks much like corn, also in the grass family. In the United States much breeding and selection have been done to develop strains having early maturity, dwarf stature, and disease resistance. Dwarf varieties with easily threshed seeds are especially suitable for mechanized cultivation. In fact, grain sorghum provided the first example of the ability of plant breeders to modify the characteristics of plants so as to make harvesting with mechanical equipment (combines) possible. This occurred when plant breeders at the Oklahoma and Kansas experiment stations developed the double-dwarf milo.

Breeders at Funk Seeds discovered a new germ plasm known as 3-Dwarf-3, which has excellent resistance to maize dwarf mosaic virus (MDMV) and sorghum head smut, as well as excellent drought tolerance. Many of the forage sorghums carry this germ plasm, and could be useful in grain-sorghum breeding. The new germ plasm was used in developing the G-404, first available for limited planting in 1976. It consistently outyielded hybrids of the same maturity in those areas where it was adapted and has done exceptionally well in dryland areas. Germ plasm 3-Dwarf-3 should help breeders to more rapidly utilize genetic resistance to conventional disease and insect attacks (Anonymous, 1977*d*).

The United States is also interested in new strains of sweet sorghums for the production of sorghum syrup, a sizable industry. Approximately 200 million brooms are made annually in this country from broomcorn, a sorghum with long, loose, stiff panicles.

Sorghums are planted and cared for in much the same way as corn. About four times as many plants are grown per acre when grown for forage or silage than for grain. If utilized for forage, sorghum is cut and allowed to wilt for several hours before cattle and other animals are allowed to feed on it, for the fresh plant parts, particularly young leaves, contain a glucoside that releases hydrocyanic acid when eaten.

As with sugarcane, sorghums produced for syrup should be harvested when vegetative development ceases and before blooming. The leaves and upper, immature portions of the plant are removed and left in the field to rot. Juice may be extracted from the remainder of the cane, either in primitive mills or in factories with crushing equipment similar to that used for sugarcane. The crude juice is boiled in large kettles until syrupy. To prevent crystallization of sugar, a small amount of malt diastase is added directly after pressing. The starch, sucrose, and maltose are hydrolyzed by the enzyme to glucose, which does not precipitate out. Sorghum syrup is usually not filtered or otherwise treated to remove the nonsugars and that is why it has its characteristic taste and aroma (Ochse et al., 1961).

## MONOCULTURE AND GENE BANKS

As related in chapter 2, one of the reasons for the high efficiency of agriculture in developed nations is the practice of monoculture. Vast areas are planted to just a few varieties of one or two crops that happen to yield heavily under the conditions of soil and climate in these areas. Even within a crop species, uniformity is encouraged by the requirements of modern agricultural operations and the customers' demand for agricultural products of uniform characteristics. However, monoculture is in some ways hazardous.

### Plant Resistance to Insects

High-yielding varieties may not be the ones most resistant to insect pests. Yet monoculture puts the natural enemies (parasites and predators) of these pests at a disadvantage, for it does not provide the great diversity of ecological niches required to develop high populations of those natural enemies. This results in a greater need for pesticides than would be the case in a more diversified agriculture.

The ecologically ideal way to control insect pests is to breed crop plants that are resistant to them. Breeding plants for resistance to disease has been successful more often than breeding for resistance to insect pests, nevertheless some outstanding successes have been recorded. Possibly progress is as rapid as can be expected in view of the fact that at least six to 10 years are required to produce a new variety, involving hybridization followed by selection. Plant resistance can take three forms: (1) the crop may produce well despite insect infestations seriously injurious to susceptible plants; (2) it may contain an insecticidal ingredient or lack some nutrient or vitamin necessary for insect development; or (3) it may not attract the insect or may contain a substance that repels it. An insect pest may refuse to lay eggs on such a plant and the parent or her progeny may starve to death rather than feed on it (Painter, 1958).

### Examples of Economic Value of Resistant Varieties

Some examples of what breeding of resistant plant varieties can mean, in terms of the current food crisis, are the following: The Kansas Agricultural Experiment Station released or approved ten different wheat varieties resistant to the Hessian fly (*Mayetiola destructor*) and also improved in other respects. One of these varieties, the Pawnee, released in 1943, had by 1949 become the leading wheat variety in acreage in the United States. Increased annual yield of wheat in Kansas alone, as the result of Hessian fly resistant wheats, has averaged about 5 million bushels or about 150 thousand tons. In 1950 it was estimated that the increased annual yield of wheat resulting from the Pawnee variety alone would pay for all the expenses of operating the Kansas Agricultural Experiment Station for over twenty years (Painter, 1968). The development of varieties resistant to the Hessian fly, sawfly, alfalfa aphid, and European corn borer by federal, state, and private agencies was estimated to cost about $9.3 million. These resistant varieties have resulted in savings to farmers of about $308 million per year, an amount that would otherwise be reflected in higher food prices to consumers. Assuming that each of the resistant varieties can be successfully grown for about ten years before it is replaced by another variety, the annual return for each dollar invested in research and development of the resistant crop plants was $300. An additional bonus was the elimination of pesticide residues (Luginbill, 1969).

An important thing to bear in mind when evaluating the contributions of plant breeders is that resistance to insects bred into a plant crop species may not in itself be sufficient to solve a pest or disease problem but might be one of the factors in an integrated pest-management program utilizing cultural, biological, and chemical measures. The status of resistant varieties among the different families of plant crops has been discussed by Maxwell et al. (1972).

## Plant Resistance to Disease Pathogens

Monoculture may also favor the spread of disease pathogens. For example, when crops are grown year after year on the same land, root rots and nematodes may become severe problems. Reservoirs of disease viruses may build up in the native vegetation or an insect vector of viruses may increase in numbers, although it may not be a serious pest itself in the particular crop affected. A virulent new rust destroyed 75 percent of the Durham wheat crop in the United States between 1953 and 1954.

There are many races of wheat stem rust and new ones are constantly developing. Wheat strains may be developed that are resistant to one race of stem rust, but may be susceptible to others. Nonvirulent rusts may hybridize and thereby develop a new virulent strain. This is just one example of the fact that nature is continuously in a state of flux, and research on any given agricultural problem must be continued and implemented indefinitely.

Corn became susceptible to southern leaf blight in 1970. By the end of the season about 15 percent of the corn crop had been lost nationwide and more than 50 percent in some sections of the South. A new race of the fungus (race T) provided a phytotoxin specific to sms-T cytoplasm for male-sterility.[9] Many hybrids were produced by the male-sterile method in the United States by 1970 and this, plus favorable weather conditions in many areas, resulted in infection to epiphytotic proportions. There are alternative sources of male-sterile cytoplasm, but they will now all be suspect because of potential susceptibility to other races of the blight. In 1971 hybrid seed was planted that had been developed by methods other than T-cytoplasm (Hooker, 1972; Wilkes, 1977).

Under natural conditions plants and the diseases that attack them are continually adapting to each other through the evolutionary process, the disease mutating new forms of attack and the plants new forms of resistance. In modern agriculture crop plants are grown from new seeds each year for continual high yields and are no longer subject to mutation, whereas the mutation of diseases is continuing.

Plant breeders have usually been able to develop resistant varieties, when needed, by hybridizing high-yield varieties with others that can contribute characteristics toward resistance to the disease. For wheat and corn, that usually means a four- to five-year wait for a resistant hybrid. If weather conditions remain favorable to the disease, years of heavy losses may follow, a situation increasingly more precarious as the world's population steadily outstrips its gains in food production. Chemical pesticides are usually ineffective or lack economic feasibility against plant diseases except as preventives.

## Importance of Genetic Complexity

Until about 1950 only the United States and portions of western Europe had restricted their important food crops to a few high-yield varieties. In the rest of the world, agriculture remained a family affair and crop variation differed from village to village or even from household to household. Such primitive varieties are now being lost as farmers adopt a few new high-yield varieties of "miracle" grains developed specifically for tropical climates and soils. Yet the primitive varieties remain very important as sources of germ plasm for hybridization. The United States lacks any substantial indigenous genetic base of its own; every major food crop grown in this country has its base in other regions of the world.

[9] A "male sterile" strain of corn had been developed that was controlled by a cytoplasmic factor, eliminating the necessity for removing tassels from all the ear-producing parent plants in the hybridization process. Male-sterile lines have also been useful in improving other crops, for example, by eliminating the tedious emasculation process in sorghum and sugar beet. It is widely used in the production of hybrid barley.

The possibility of success in the search for resistant genes is generally enhanced by the great number and diversity of plants to study. The extensive monocultures that characterize modern agriculture—thousands of acres of a single crop, such as wheat, corn, sorghum, alfalfa, or cotton—have been criticized by ecologists because of the danger of losing gene diversity. If a crop develops susceptibility to a disease or pest, there might be a critical need for a wide variety of genes for certain physical or physiological characters that might be bred into the commercial variety by hybridization. The same may be said of animals. This is only one reason, of course, for avoiding the extermination of plant and animal species, but it is an important practical reason. That is why the collection of plant species and varieties from all over the world, as in the Pacific Tropical Botanical Garden on the island of Kauai in Hawaii, or the protection and care of wild animal species and varieties in the game refuges of the different continents, is so important.

Preservation of domestic animal or crop varieties is also important, for a strain abandoned at one time may have certain characters that prove useful for hybridization when ostensibly superior types that replaced it begin to manifest certain serious weaknesses. Another possible procedure that has already shown some promise is the development of varieties that possess both a reasonable measure of genetic diversity and high productivity.

Even after exotic plants are imported, much work must be done to convert them to suitable types before their germ plasm can be used, for most are poorly adapted as crop plants. Only limited support has been given for this kind of research, and seldom on a sustained and systematic basis.

Fortunately the USDA operates the world's largest "gene bank" or "germ plasm bank" in Ft. Collins, Colorado, with 100,000 plant varieties, mostly grain. The USDA has also assembled a team of geneticists in Beltsville, Maryland, for development of additional collections. More than 2,200 wild wheats and their near relatives have been assembled at the University of California, Riverside (Johnson and Waines, 1977).

The Consultative Group on International Agricultural Research (CGIAR), supported by the Rockefeller Foundation, the World Bank, and other international agencies, has recommended a worldwide series of gene banks. This organization has developed a plan for systematically cataloging and storing the remaining varieties of most major crops, analyzing their favorable characteristics and maintaining a breeding program to keep the seed stock viable. Unfortunately the USSR and China, which contain enormous seed resources, are not members of CGIAR. The cost of such lack of cooperation can be illustrated with soybeans. The yields of most soybean crops in the southern United States are annually reduced about 20 percent by nematode infestation, a loss that could probably be avoided by hybridization with plant varieties from China, where the genetic base for soybeans is located.

## THE DUST BOWL

On the Great Plains during the 1930s the tragedy of the Great Depression was further aggravated by man's careless assault on the environment. Millions of acres of soil that had been anchored down by native grasses were plowed up and planted to wheat in response to the high grain prices following World War I. This in itself would have made the soil more subject to wind and water erosion, but in addition none of the soil-conservation measures that might have been taken to minimize erosion were practiced. The resulting devastation was nowhere as complete as in the Dust Bowl, a 150,000-square-mile area that included southwestern Kansas, southeastern Colorado, northeastern New Mexico, and the Oklahoma and Texas panhandles. The long drought from 1933 to

1939 rendered the land defenseless against high winds that piled sand in dunes as high as thirty feet. The wind picked up the lighter silt to form black dust clouds five miles high and carried them to the Atlantic Coast and beyond (Sears, 1971).

The dust clouds hovering over Washington, D.C., brought to Congress physical evidence of the tragedy that befell the Great Plains. The Soil Conservation Service was formed in 1935, headed by Hugh H. Bennett of the USDA, one of the world's leading soil conservationists. The federal government began to encourage, teach, and financially support sound soil conservation practices (see chapter 4) not only in the Dust Bowl area, but throughout the nation (GPC, 1936; Eckholm, 1976).

Now again surging world grain prices have resulted in the conversion of about ten million acres of former pastures, woodlands, and idle fields into cropland, about half of it without adequate conservation treatment and losing an average of over ten tons of topsoil per acre per year to water and wind. Surveys in 1974 showed that in the southern Great Plains soil loss on about 50,000 acres of newly planted land ranged from 14 to 127 tons per acre (Grant, 1975). But the worst was yet to come.

In late February 1977, following a prolonged drought, gusts of wind up to 90 miles per hour scooped up dry topsoil at the rate of 5 tons per acre during a 24-hour storm in eastern Colorado. From Nebraska to the panhandles of Oklahoma and Texas dust clouds were billowed up to 12,000 feet and brought visibility to near zero. It was believed by officials of the National Oceanographic and Atmospheric Administration that a calamity as severe as that of the Dust Bowl era could occur if the drought lasted another thirty days.

In the early 1970s Great Plains farmers were again sacrificing sound soil conservation practices to increase grain production. They were changing from wheat-fallow or wheat-sorghum-fallow rotations to continuous planting of wheat, to take advantage of high wheat prices. In addition, they were plowing with moldboard plows to bury the seed of downy brome and were keeping the surface free of the harvest residues that would help to prevent erosion (Carter, 1977*a*).

Droughts may be related to sunspot activity, which has an average 11-year peak-to-peak cycle of intensity; thus they have been occurring at roughly 20- to 22-year intervals. For long-term planning, it would be wise to take into consideration that the weather of the past forty years has actually been unusually stable for the Great Plains and will probably become more variable in the future (Schneider, 1976).

## Sound Tillage Practices

Tillage practices are important in determining soil loss through erosion (Schwab et al., 1966). Conventionally, soils are plowed with either a disk or moldboard plow, followed by one or more disk, spring, or spike-tooth harrowings. "Strip" or "zone" tillage conditions the soil for planting in a narrow strip, leaving the remainder untilled or tilled in a different manner. In the practice known as "listing," seed is planted in the bottoms of furrows at the same time the furrows are made. This tends to avoid excessive tillage, an important factor in soil compaction and in soil erosion. "Mulch tillage" or "stubble mulching," described in chapter 4, is a soil management practice of special importance on the arid, windy plains.

Many farmers are now practicing "no-tillage" (chapter 4) wherever possible, thereby improving water penetration and soil texture. The possibilities have been enhanced through water-soluble fertilizers.

Tillage and other soil-conservation practices are important in protecting soil not only from water erosion but also from wind erosion, of particular importance on the Great Plains. Only a half-ton of straw per acre, anchored to the soil with a straight-disk, is required to reduce soil loss by 92 percent. Tilled-in manure is also effective in reducing wind erosion, aside from adding organic matter and nutrients to the soil and removing the manure as a potential source of pollution. After the fall harvest of wheat, a farmer might plant a winter cover crop of rye, kill the rye with herbicides in the spring, then seed the main crop into the resultant mulch with a special planter. No furrows are made, and no further work need be done until harvest time, when the cycle is begun anew. This type of culture encourages insects that were formerly kept in check by fall deep-plowing; they must now be controlled with insecticide. However, the unturned soil retains 50 percent more moisture than a plowed field and erosion and runoff are reduced by up to 90 percent. More than 6 million acres in the United States are now devoted to plowless farming and the USDA predicts that by the year 2010 only 5 percent of the American farmers will practice conventional tillage (Kaniuoka, 1976). Stubble mulch tillage that leaves a rough, cloddy surface is also effective. Such a surface can be obtained by performing the primary tillage as soon as practicable after a rain and delaying the secondary one for seedbed preparation as long as possible. Appropriate stripcropping systems and the planting of windbreaks are likewise effective measures in reducing erosion caused by windstorms.

### Government Aid

A program designed by Congress in 1956 enabled farmers and ranchers of the Great Plains, working through their local conservation districts, to develop and apply conservation plans on their operating units. Called The Great Plains Conservation Program,[10] it was facilitated through federal cost sharing to the extent of 50-80 percent as the farmer completed each conservation step, and by getting high priority in technical help from the federal Soil Conservation Service.

Since 1957, the program has protected and improved cropland, reduced wind and water erosion, improved rangeland, developed dependable water for livestock, stabilized income, saved scarce irrigation water, strengthened community economy, and made the Great Plains countryside more attractive. Many of the nation's approximately two million water-conserving farm ponds, often containing fish, were developed in cooperation with the Soil Conservation Service.

Congress extended the program to December 31, 1981, making additional provisions for dealing with special problems and opportunities, such as disposal of animal wastes that have become of great concern in hundreds of rural communities, developing recreation and fish and wildlife resources, and promoting conservation work on nonfarm land adversely affecting a farming area.

## WINDBREAKS

A windbreak is any protective shelter against the wind, but is generally thought of in agriculture as a row or more of trees or shrubs that serve that purpose. Windbreaks are needed most in areas where wind erosion is a problem, as in parts of the Great Plains; in the irrigated valleys of the Pacific Southwest and the Rio Grande, Colorado, and Columbia river basins; on the sandy and muck soils in the Great Lakes region; and on sandy soils along the Atlantic and Gulf seaboards. Windbreaks are useful when planted around country homes and farmsteads (fig. 22), protecting people and livestock from high winds, airborne dust, and drifting snow. They reduce pollution from dusty air and

[10]Farmers and ranchers can obtain information concerning the Great Plains Conservation Program from the local Soil Conservation Service representative, county agent, or a member of the local conservation district (USDA, 1970*b*).

*Fig. 22. A North Dakota farmstead protected by windbreaks (courtesy USDA Soil Conservation Service).*

noise from highways or other sources. They increase comfort by reducing the "wind-chill index"—a number expressing the combined effect of wind and temperature. For example, a 30-mile-per-hour (mph) wind at a temperature of 10°F can cause the same heat loss from exposed skin as a temperature of -33°F with no wind. A wind of that velocity would be reduced to 10-15 mph by a dense windbreak (Ferber, 1969).

Fine soil particles, both organic and mineral, are easily moved by wind and often contain ten to twenty times as much humus and phosphate as the heavier particles that stay behind in the windswept fields. The great clouds of the Dust Bowl in the 1930s were black because they contained the dark humus of the topsoil. The "threshold velocity" for movement of soil particles is about thirteen miles per hour one foot above ground. Beyond that, the carrying capacity of wind is proportional to the cube of its velocity. Therefore a small reduction in wind velocity has a proportionately great effect in reducing airborne dust particles.

Field windbreaks can take some of the risks out of farming. Strong winds can lower yields or blow crops to the ground. Windblown soil particles can damage leaves of tender plants or break their stems. Windbreaks also protect orchards, especially during pollination and when the fruit is ripening. Hot winds can desiccate foliage, scar fruit or cause it to fall. Reducing wind velocity also reduces evaporation of moisture from soil and vegetation. Where rainfall is adequate, one or two rows of trees may suffice for a windbreak, whereas in areas where tree-growing is more difficult, a three- to five-row planting is better (Ferber, 1969). In the northern Great Plains, an eight-row windbreak may be desirable for protecting a farmstead or feedlot, with a combination of deciduous and evergreen coniferous species. In some areas a fence may be required to prevent livestock from browsing on lower branches and bark on trunks and trampling on and packing down the soil, keeping it from readily absorbing moisture (George, 1966). A varying combination of species is appropriate for different areas of the country. Information can be obtained from the local county agricultural agent or local Soil Conservation Service representative.

### The Federal Shelterbelt Project

During the Great Depression of the 1930s a series of windbreaks for the Great Plains was proposed by President Franklin D. Roosevelt and became known as the Shelterbelt Project. The project as originally instituted was to include a zone extending from the Canadian border in eastern North Dakota to a point just north of Abilene, Texas. It was to be 100 miles wide and 1,150 miles long and limited to the area of transition between the tallgrass and shortgrass prairies. The windbreaks were to be wide enough to permit the use of slow-growing, shrubby trees on the outside and those with a more rapid rate of growth in the center. The Shelterbelt was intended not only to control wind and water erosion, protect crops, provide food and cover for wildlife, but also to supply wood for lumber, fence posts, or fuel, and to provide a large measure of employment (Potter, 1976).

The project was financed by an allocation of $5 million to the Federal Relief Administration and $10 million to establish Civilian Conservation Corps (CCC) camps to supply manpower. After 1937 the Forest Service continued the Shelterbelt as the "Prairie States Forestry Project," using emergency funds from the Works Project Administration (WPA). In 1942, when the project, again named "Shelterbelt," was transferred to the Soil Conservation Service, 18,600 miles of windbreaks had been planted. The Forest Service had planted over 217 million trees on 232,212 acres surrounding over 30,000 Great Plains farms (Wessel, 1969).

As the memory of the "dirty thirties" faded among the older farmers, and young people were not told why the trees were planted, the trees were gradually removed to provide more land to cultivate and to make room for center pivot sprinkler irrigation systems increasingly used by farmers in the plains. Most of the trees will be removed within ten years unless there is an educational campaign to emphasize the need for them. Potter (1976) suggests that the Shelterbelt Project should be reinstituted and expanded through the use of a revived Civilian Conservation Corps. As in the Great Depression, afforestation could be accomplished and meaningful employment provided for the nation's youth. Confronted in 1977 with the worst drought and biggest dust storms in twenty years, governmental agencies were taking another look at this sound policy.

Prompted by a 1974 article in the *New York Times* the following year the General Accounting Office (GAO), declaring that an important resource was disappearing in the Great Plains, called on the Secretary of Agriculture. Besides economic benefits to farmers in reducing loss of soil, the GAO called attention to the benefits of beautification in an otherwise largely treeless region and improvement of living conditions for humans and livestock. The windbreaks are also beneficial to wildlife. According to an Oklahoma wildlife official, the Shelterbelt trees served to bring deer and turkeys back to many sections of western Oklahoma and even into areas where they never occurred before (Nelson, 1975*a*).

### Weather Information

Increasing technology has resulted in management systems increasingly sensitive to weather events and demanding more meteorological information than formerly. Accurate and timely weather information is needed for such activities as planting (soil temperatures); harvesting (grain moisture content or threatening storm); accurate interpretation of the micrometeorological conditions favoring weed, spore, or insect development or the dispersion of pesticides into a turbulent atmosphere; drying fields or drying hay or grain; irrigation; freeze protection; and sheltering of livestock. There is need for an expansion of the agricultural forecasting program of the National Weather Service to a nationwide coverage. In the arid lands of the West, climate and weather variation are particularly great, and improved long-range and short-range weather forecasting are especially urgent to help ranchers and farmers to devise strategies that take into account risks and benefits (USDC, 1971; BARR, 1975).

## IRRIGATED FARMS

Irrigated farms are by no means without their own kinds of difficulties and problems, but everywhere in the West irrigation has contributed the element of stability to agriculture, even in areas of considerable rainfall, but with dry summer seasons. The antagonism cattlemen showed toward dry-farm "nesters" was not extended toward irrigation farmers. The latter required too little land to pose a problem. What land they used was highly productive and helped finance municipal services needed by ranchers. Eventually irrigated farms were found to be a great boon to cattlemen because irrigated forage crops greatly supplemented the scanty natural forage of the High Plains, particularly in winter. Eventually a complete symbiosis between irrigation farmer and rancher developed. Throughout the West, agriculture became largely a complex of interdependent crop-farming and grazing of rangeland.

### A Colorado Cooperative Venture in Irrigated Agriculture

Second only to the Mormon experience in Utah in its influence on the development of irrigation in the arid regions of the West was the brilliant success of Union Colony in Greeley, Colorado. Chief credit for the success belongs to Horace Greeley, the New Hampshire farm boy who founded an influential newspaper (the *Tribune*) in New York City, and wrote as often about farming as about politics. Greeley was lured to Colorado in 1859 by the abortive Pikes Peak gold rush but became more interested in what he considered to be the agricultural possibilities of the High Plains country he passed through on the way to the mountains. Through Greeley's efforts Union Colony was founded in 1869. Greeley sent his agricultural editor, N. C. Meeker, to find a suitable area in Colorado where the colony could establish a cooperative irrigated farming venture, taking advantage of what they could learn from the Mormons. Meeker selected a site on the South Platte River about fifty miles north of Denver and on the Denver Pacific Railroad. He obtained 12,000 acres from the railroad's land department and 100,000 acres from the government (Holbrook, 1950).

Membership in Union Colony cost $155. Each member was entitled to a parcel of land outside the colony and the right to purchase a town lot for $25 to $50. Every deed had an antiliquor clause, the violation of which meant loss of the land. And there was to be no land speculation. Great care was exercised in selecting only men of demonstrated intelligence, industry, and good character.

Oberlin-educated Edwin Nettleton was selected to survey the site and lay out the irrigation ditches. By early spring of 1871 the Greeley irrigation ditch was 27 miles long and carried its first water, becoming Colorado's first large-scale organized irrigation project.

Nettleton went on from Greeley to become one of the greatest figures in western irrigation. He laid out the vast system of an English syndicate that began farming on a huge scale near Fort Collins. After studying irrigation in Spain, he returned to lay out immense irrigation systems in Idaho and Wyoming.

Greeley's citizens fenced in some 35,000 acres of land, much to the dismay of the region's cattlemen. By May 1, 1871, the town of Greeley had a population of more than 1,500, including 700 shareholders in Union Colony. Thousands of maple trees had been planted around the settlers houses. The first season's crops demonstrated the fertility of irrigated land on the High Plains. The town flourished, first on potatoes and then on sugar beets (Holbrook, 1950).

Greeley and the surrounding farm area still reflect the planning of the Union Colony. The canals carrying irrigation water to the area are essentially the same as originally laid out by the Union Colonists. The present population of Greeley is 38,900. There are about 337,000 acres of irrigated land in Weld Country that surround the city of Greeley. The crops of the area, in order of importance, are corn for silage, sugar beets, wheat, corn for grain, all hay, dry beans, "other crops" (includes sorghum, rye, fruits, and vegetables), and potatoes. In 1973 these crops were valued at $116 million (V. W. Youngeman, correspondence).

### Center Pivot Sprinklers

Even with the most ambitious land-leveling efforts with bulldozers, there is much that can never be irrigated by gravity systems. Under the best of conditions water running down furrows saturates the end of the field where it is introduced and tends to supply too little to the far end of the field. Sandy soils, which absorb water quickly, are particularly subject to uneven water distribution.

The drawbacks to gravity irrigation are overcome by the installation of sprinklers, of which the center pivot type has attracted greatest interest. A prototype of this device was built in 1949 by Frank Zybach, a Colorado tenant wheat farmer, who called his device a "self-propelled irrigation apparatus" (CRA, 1976). As related in chapter 2, the device as now constructed consists of a quarter-mile-long pipeline with sprinkler nozzles located at intervals along its length. It is anchored in the center of the field and rotated around the central anchor (fig. 23). It is propelled by oil or water hydraulics or electric or air drive and rides on steel wheels or, more frequently, on rubber tires. One sprinkler can irrigate 133 acres and its speed can be adjusted to complete its rotation in periods ranging from less than one day to up to as long as seven (Splinter, 1976; CRA, 1976). The typical road alignment of the Great Plains suits the use of center pivot sprinklers. The land in the corners not reached by the sprinklers is commonly used for pasture, feedlots, grain storage, trees, or dry land crops.

Center pivots have found their greatest acceptance in Nebraska. Most soils there, particularly in the central and western parts of the state, consist of sand, or sand and silt, and receive an annual precipitation of less than 25 inches. Much of the area is underlain with a great thickness of permeable water-bearing sand and gravel. This is the Ogallala Formation, one of the nation's richest aquifers and generally close enough to the surface to permit massive pumping from relatively shallow wells. The region includes the vast Nebraska sand hills, with marginal soils not suited for gravity irrigation and risky for dryland farming. The land is relatively cheap and available in suit-

*Fig. 23. Center pivot sprinkler in a Nebraska cornfield; this system pumps 1,400 gallons of water per minute (courtesy USDA Soil Conservation Service).*

ably large parcels. Pivot systems were commercially pioneered in this region and most of the state's 10,000 center pivots are located there. They have been a great economic boon to the state.

Nebraska now irrigates about 5.5 million acres, exceeded only by California and Texas. With current growth rates, total irrigated acreage in the state could exceed 7 million by 1980. Nebraska has approximately as much dry-farmed as irrigated corn acreage. In the drought year of 1974, the dryland corn crop was worth only $229,942,000 compared with $1,313,818,000 for the irrigated corn crop (CRA, 1976).

Coarse-textured or sandy soils hold less than an inch of water per foot of soil depth compared with two or more inches per foot for fine-textured or loamy soils. The coarse or sandy soils can usually serve only for rangeland or marginal farming. The center pivot system, by applying water lightly and frequently, replenishes the moisture in the root zone sufficiently to allow intensive cropping on such soils. University of Nebraska researchers found that pasture irrigated by a center pivot system produced 700 and 900 pounds of live beef per acre per year in cool-season grasses, compared to a normal production of 27 pounds per acre per year for open range in the Nebraska sand hills (Splinter, 1976).

Besides the advantage of being adaptable to a wide range of soils and terrains, the center pivot eliminates much of the labor associated with irrigation. Once programmed to deliver proper amounts of water, labor is virtually reduced to maintenance. An experienced center pivot operator can operate ten or more pivots (Hagood, 1972).

In Nebraska corn is the principal crop grown in the pivot sprinkler circles. It is used for grain, but also for forage (silage), for cattle-raising is important in the area. Fertilizers and herbicides are often applied through the sprinkler systems, but only limited amounts of insecticide. Moisture-sensitive crops like corn grow well in sandy soils when sufficiently fertilized. When fertilizer is applied only as the crop utilizes it, as with the sprinklers, the hazard of contamination of underground water by nitrates and other mobile nutrients is greatly reduced. Yields range between 120 and 175 bushels per acre. Grain elevators and feedlots are suddenly appearing in regions once limited to breeding livestock.

## Some Sociological and Environmental Costs of Center Pivot Systems

It was not very long ago that a farmer could install a center pivot system, necessary equipment, and a well for about $20,000. Today the well, casing, pump, power line, pipe to the sprinkler, and the sprinkler cost about $60,000 and this does not include land, land preparation, interest, taxes, and insurance. Many farmers cannot afford center pivot irrigation and the financial risks involved. Thus pivot irrigation has become an investment novelty in Nebraska. (In Wyoming, the state supplies low-interest loans to farmers wishing to install sprinklers.)

In a report published in January 1976, the Center for Rural Affairs[11] concluded that from one-fourth to one-third of the land irrigated by pivots, in a six-county region of Nebraska that they extensively studied, is not owned by farm operators. Farm managers are using a multitude of investors—many of them not Nebraskans and who will probably never see their farms—as a financial base for their own industrialized farming operation. To the extent that investments are made by absentee landlords and corporate investors, the primary benefits are siphoned out of the local community or even the state. One custom farm manager boasted of having made a New Jersey investor a millionaire with center pivot systems in Nebraska (D. Morgan, 1975).

The increase in land values spurred in parts of Nebraska by the rapid growth in irrigation development cannot fail to contribute to the decline of the operator-owner. Professor Harold Breimyer, agricultural economist at the University of Missouri, has this to say on the subject:

> When land values rise fast and conspicuously they become an attraction of their own. No longer is ownership of a farm seen as just a place of employment for a hard-sweating operator. Ownership is viewed as a ticket for further speculative gains. Although the operating farmer may still want to increase his ownership, he will be swamped by the large numbers of non-farm investors seeking the opportunity to invest for speculative profit. The overall effect of steadily large capital appreciation in farming will be to move land out of the hands of operating farmers. [Breimyer, 1975]

## Center Pivot Sprinklers on the Texas High Plains

In a 50-county area of Texas—the Texas High Plains—where rainfall averages from 8 inches in the southern counties to 21 inches in some of the northern counties, irrigated agriculture has made the area one of great economic importance. In the late 1960s there were 45,000 wells in the region, delivering water for 5.5 million acres of irrigated land. The area produces a fifth of the cotton and over a third of the sorghum grown in the United States. It also produces about two-thirds of the wheat raised in Texas. The

[11]P.O. Box 405, Walthill, Nebraska 68067.

rapidly expanding cattle and swine feedlot industry depends on feed grains produced in the area. The irrigated area supports a million people and a billion-dollar economy. The use of groundwater for irrigated agriculture, including center-pivot sprinklers, has become an important factor in the regional economy. The major source of water for the Texas High Plains is from a part of the Ogallala aquifer of water-bearing sand. Estimates of rate of recharge of the aquifer range from only 0.05-0.5 inch per year. Water is being pumped out of the wide but relatively thin aquifer at a rate that far exceeds recharge and recirculation. According to estimates, the economic activity attributable to crops, feedlots, and meat processing will continuously decrease after 1980 and is expected to be terminated in from 30 to 40 years (Osborn, 1975; Bowden, 1977).

## A WYOMING SHANGRI-LA

As related in chapter 2, millions of family-farm enterprises have been terminated during the past century because potential heirs chose to abandon them and move to cities. But our nation is experiencing a rebirth of love for the land. A striking example was seen in a picturesque valley reached by a ten-mile gravel road off Highway 287 near the tiny town of Crowheart, Wyoming. In this valley is located a ranch inherited by Nick and Neal Anderson. The ranch follows the course of the Wind River for about five miles and averages about a half-mile in width. The lofty Wind River Range, a massive spur of the Rockies, rises majestically in the west.

The Anderson brothers are busily engaged in expanding and intensifying the farm enterprises. Obtaining water by gravity flow from the river, they are farming 350 irrigated acres of malt barley, oats and alfalfa on the 1,400-acre ranch. A herd of 135 cattle and calves graze in grassy open areas among the cottonwoods along the river. Beavers periodically float logs and brush down one of the irrigation ditches to form a dam that provides them an earthen slide for fun and frolic. This obstruction is patiently removed by the Andersons when necessary. Deer easily jump over the barbed-wire fences and are welcome to their small share of the alfalfa. (The Andersons occasionally hunt for deer—back in the mountains.) Migrating elk cross the ranch. Raccoons, mink, and porcupines live on the ranch and foxes, bobcats, and coyotes are visitors—the coyotes being the only ones that are not welcome. Among the feathered visitors are wild ducks, geese, pheasants, and grouse. An eagle has a nest near the ranch.

At the time of our visit (August, 1978) Nick's attractive young wife, Nancy, was carrying her four-month old "natural birth" baby in a rucksack, designed for that purpose, while sharing in many of the farm operations, with which she was thoroughly familiar. She invited my wife and me to a lunch of lean and remarkably palatable grass-fed beef and home-grown vegetables.

There are other such potential Shangri-las in the West awaiting transformation to "humanized nature" by other young couples who hopefully would share the enthusiasm and environmental concern of the Andersons, along with the same degree of experience and practical knowledge. How could one find a life of greater challenge and adventure?

## WIND, AN INEXHAUSTIBLE HIGH PLAINS POWER RESOURCE

Wind in the Great Plains, generally thought of as a curse, has been used as a source of power. We have seen how windmills, together with barbed wire, were a key factor promoting the changes that took place in the cattle industry as well as the general agricultural development of the plains. Yet in view of the potential provided by modern materials and technology, the windmill's usefulness to mankind has been sadly neglected.

### A Modern Wind-Energy Conversion System

There is great incentive to strip-mine the coal and develop the oil-shale resources of the West, exploiting a vast but exhaustible and pollutive natural energy resource. Yet there has been little economic incentive to exploit the wind, an even greater energy resource—inexhaustible and nonpollutive—which Abraham Lincoln once referred to as "The greatest untamed force of nature." Government promotion and subsidy of wind power are miniscule compared with other sources, particularly nuclear energy. But each source of energy should be given a fair chance to prove its worth.

In early 1975 all the solar research programs were shifted from the National Science Foundation (NSF) to the newly formed Energy Research Development Administration (ERDA), which is now the Department of Energy (DOE). The program grew rapidly because Congress, responding to great public enthusiasm, authorized funds as much as 80 percent above what ERDA requested. So government solar research was funded in 1977 in the respectable amount of $290 million, with $320 million recommended by President Carter for 1978. Hammond and Metz (1977) comment:

> Wind-power research, although it is the solar electric technology closest to being economically competitive, receives only about 8 percent of the solar budget.... Still less money presently goes to the solar resource that would be most versatile of all—plant matter or biomass, which can be converted into either heat, fuels, or electricity. ERDA officials are generally agreed that biomass is one area in which they have yet to get a strong and coherent program underway.

The states where the greatest coal- and oil-shale deposits occur are among those with the greatest area of high wind-power potential. The greatest efficiency from solar energy is said to be obtainable from mechanical energy indirectly derived, as in the case of the windmill. Furthermore, other solar-related processes require much more research and development before their initial large-scale development as sources of power, but wind power does not. Large modern wind-electric machines have been sufficiently demonstrated in the United States and Europe, and full-scale units could be designed and built now to provide the operating experience and data for dependable cost analyses needed before large numbers of these machines would be deployed. As petroleum prices steadily increase, the practicability of wind power increases correspondingly (Inglis, 1975).

At 20 mph the power through a surface perpendicular to the wind is 45 watts per square foot or nearly three times the average solar insolation on the ground. A windmill can convert to electricity about a third of the energy incident on the area swept out by the propeller. At 20 mph a wind would yield nine times more electric power per unit area than a 10 percent efficient solar thermal unit. Since power developed increases as the cube of the wind velocity, the power available at 25 mph is about double that available at 20 mph (von Hippel and Williams, 1975).

Wind-power electric generators are already in operation in some regions of the world, particularly Australia and Switzerland. The NSF considered wind power to be among the most amenable to commercial exploitation of all its research projects. An NSF wind-power generator will have two blades 125 feet in diameter and is expected to produce 100 kilowatts in an 18-mph wind. This was only one of six research programs that were in progress in 1974. About 50 percent of the electricity needed annually in the

United States by 1985, equivalent to about 140 billion gallons of gasoline, could be produced in a good location by wind, such as on the Great Plains, according to a NASA report to Congress (Spaulding, 1974).

Added interest in wind-power machines is being generated by the vertical-axis wind turbine. Its cost is said to be about one-sixth of the conventional windmill and it should be able to produce power for 1.3-2.0 cents per horsepower-hour at a wind velocity of 10 mph. If produced in quantity it could supplement the conventional sources of power at comparable cost where the annual mean winds are 10 mph or higher (South and Rangi, 1974).

A report produced under a contract with the NSF states that the power that could be extracted from winds within the 18-mph contour surrounding the high Great Plains region, with conservative assumptions regarding the operating efficiencies and proper spacing of wind machines, "is several times the present United States electrical power demand" (Mitre Corp., 1975). Many large wind generators widely spaced in sparsely populated regions need not interfere appreciably with the principal present land use in the area—grazing or agriculture. Offshore floating units, in certain areas of high wind velocity, likewise are an attractive possibility (Inglis, 1975).

Vannevar Bush, who pioneered the electronic computer and was head of the nation's Weapons Development Program during World War II, wrote in the "Introduction" to P. C. Putnam's *Power from the Wind,* that the cost of electricity generated by the wind turbine described in the book was close to that of electricity generated by the supposed more economical conventional means. This was in an era when the cost of oil and coal was only a fraction of what it is today (Tarver, 1976).

D. R. Inglis, professor of physics at the University of Massachusetts and recipient of the 1974 Leo Szilard Award for Physics in the Public Interest, reported (in 1975):

> The prospects of wind power should be compared with the program now being promoted in Washington to build, probably largely at government expense, 200 nuclear reactors in nuclear parks by 1990. There is no apparent limit to the harnessing of wind power that would preclude its generating as much power as these nuclear reactors would by that time or supplementing other sources of power as reliably as can be expected of nuclear power. It will involve a huge construction effort, but not one beyond the capacity of industry.

Regarding the "huge construction effort," Inglis remarks:

> It would seem socially irresponsible to waste the opportunity [to provide a new source of power] by meeting the need with compact, high-technology nuclear plants when low-technology and labor-intensive wind-electric systems could meet the need, perhaps more economically, and provide considerably more employment in large-scale construction.

Professor Inglis believes that government needs to back wind power as it did nuclear power, although the financial backing needed is much less. He believes wind power would then be able to supplant nuclear power completely.

Steven Tarver (1976) pointed out that a wind-powered turbine that operated successfully as a component generating unit in the grid of the Central Vermont Public Service Corporation, had a rated capacity of 1,250 kilowatts (equivalent to nearly 1,900 hp) and that 844,000 such turbines would produce electricity equal to our entire present national output. Homer J. Stewart, who spent his youth on a Kansas farm surrounded by neighboring windmills, was a consultant on the famed Vermont windmill in 1940. He points out that the windmill missed by only 20 percent of being economically competitive even in those days of cheap gas, oil, and coal. Stewart has spent most of his life in wind-related work. Now a professor of aeronautics at Caltech[12] in Pasadena, California, he has written a research paper on designing blades to make windmills 20 percent more efficient, which he hopes will be useful in designing a large model at General Electric's Space Division at Valley Forge, Pennsylvania, under a $10 million contract with DOE and NASA (Austin, 1965).

There is great interest in wind turbines with technology of design and construction similar in many respects to that of aircraft. DOE officials anticipate that the size of the wind-turbine industry may equal that of the aircraft industry in the 1980s. If solar energy were derived from both solar radiation and wind, the two sources would be complementary, for the former is most abundant in summer and the latter in winter (McCaull, 1976).

Wind power has the disadvantage of not being uniform, but Tarver (1976) proposed that the electricity generated be fed into a national grid of transmission lines operated by a government corporation. There are already good reasons for developing such a nationwide grid. Legislation for establishing a national power grid had already been introduced in Congress before the power crisis during the record cold wave of January and February 1977 added further emphasis. Tarver suggests a continually growing number of pumped storage hydroelectric plants,[13] some built and owned by a national power grid corporation and others built and owned by utilities. Coal-fired generators could be retired from continuous operation but kept in operating condition for backup purposes (Tarver, 1976).

There are other ways of using the electricity generated by wind. Water can be electrolyzed into hydrogen and oxygen and the hydrogen used for fuel to run automobiles much the same as gasoline does now. There would be no exhaust except water vapor. Usually existing automobile engines can be converted to burn hydrogen, methane, or alcohol. Utilizing air and water, the electricity can also be used to produce fertilizer—anhydrous ammonia and ammonium nitrate—now usually synthesized by using natural gas.

"Wind power permits humanity to participate in cosmic economies and evolutionary accommodation," says R. Buckminster Fuller, "without in any way depleting or offending the great ecological regeneration of life on earth" (Torrey, 1976).

## CONVERSION OF BIOMASS TO FUEL

The earth's mass of organic material (biomass) received its energy from the sun and in turn it offers prodigious potential as a source of energy for man. Much of the potential exists in what is now waste and pollutive material. So much progress has been made in the utilization of pyrolysis for the recovery of fuel gas and chemicals from agricultural crop wastes and manure that operations featuring plants of scale size could be started now, and without environmental impact. Pyrolysis has been defined as "a thermal process carried on in a non-oxidizing atmosphere where the applied thermal energy

[12]A course in windmills taught by distinguished aeronautical engineers, was added to the curriculum at Caltech in the spring of 1974 (Torrey, 1976).

[13]Two water-storage reservoirs on different elevations with a connecting pipe and a turbine pump/generator. Electricity supplied to the system pumps water from the lower to the upper reservoir. When there is a shortage of electricity the water is allowed to flow back down the pipe through the turbine pump, causing it to run in the opposite direction and making an electric generator out of what was an electric motor when surplus electricity was fed into it. Thirty such units have already been built in various parts of the United States and more are being planned and built (Tarver, 1976).

is sufficient to disrupt the original chemical bonds of the feed stock." It is not an energy-intensive process; it produces export energy as well as chemical products (Crane, 1976).

Methane gas could be obtained in enormous quantities, particularly from the approximately 1.5 billion tons per year of animal and human wastes in this country. If two-thirds of the waste were processed into methane and fertilizer, the former would approximate the nation's current consumption of natural gas. Manure from feedlots, particularly abundant in the Great Plains, could be an important source (Tarver, 1976). In his "Sun Day" speech at the Solar Energy Research Institute in Golden, Colorado, on May 3, 1978, President Carter announced the reallocation of $100 million from funds already in the administration's fiscal 1979 energy budget so that the money could be spent entirely on solar and other renewable energy projects. He also said that he had directed the USDA to lend Lamar, Colorado, $14 million for a project to turn livestock manure into methane gas.

Through anaerobic digestion with bacteria, manure can be broken down into organic matter, ammonia, and carbon dioxide. Another kind of bacteria breaks the organic acids into 70 percent methane and 29 percent carbon dioxide, a combination that burns readily. During World War II rural methane plants were developed in Germany and the gas was used to run tractors. The ammonia can be used as a fertilizer. Some cities in the United States are employing a similar system in human-sewage treatment plants, using the gas to run the engines of the sewage-disposal equipment. The anaerobic digestion of the daily manure from 10,000 cattle, amounting to 135 tons, could produce 220,000 cubic feet of methane gas or the equivalent of 1,000 gallons of gasoline. Biomass farming—growing trees and other plants for conversion into methanol—is in its early planning stages in some states.

Ethyl alcohol (ethanol) is superior to gasoline as an engine fuel. It delivers as much power with much less pollution in a properly tuned motor. Only about half as much carbon monoxide and oxides of nitrogen are emitted. Alcohol is not a hydrocarbon and does not require tetraethyl lead additives; so these substances are eliminated altogether. Brazil is already well on its way toward an energy strategy based on conversion of biomass to alcohol. Its government has already committed more than $400 million toward an alcohol production program. Fifteen new distilleries for producing alcohol from sugarcane have already been funded, but a far larger potential could ultimately be provided by manioc (cassava). Brazil is already the world's largest producer of this root crop. Its Ministry of Agriculture has established a research institute devoted to manioc and the "new agricultural frontier" (Hammond, 1977).

A hydrogen-methane-alcohol economy based on solar energy would appear to be more appealing than a plutonium economy, which seems to be the direction in which we are now heading. Solar energy reaches our planet at a rate more than ten thousand times larger than our current annual consumption of all other forms of energy (Wilcox, 1975). Energy derived from the sun directly, or indirectly as in wind power or conversion of biomass, would result in minimal air pollution, and another advantage is that it would not add to the heat in the atmosphere unless the energy were transmitted via shortwaves to the earth from space satellites.

Major importance should also be attached to the social implications of solar versus nuclear power. Large-scale development of the various types of solar energy is the more democratic alternative. It can be undertaken by many relatively small businesses and can employ by far the larger number of workers. The nuclear route takes us in the opposite direction, to a concentration of economic and political power on a scale of almost unimaginable magnitude.

# 7 The Pacific Northwest

*The American farmer, living on his own land, remains our ideal of self-reliance and of spiritual balance—the source from which the reservoirs of the Nation's strength are constantly renewed. It is from the men and women of our farms living close to the soil that this Nation, like the Greek giant Antaeus, touches Mother Earth and rises with strength renewed a hundredfold.*

—Franklin D. Roosevelt

The Pacific Northwest as here defined consists of the states of Washington, Oregon, and Idaho. It is an area of a great variety of topography, climate, and vegetation. From the Cascade Mountains of Washington and Oregon to the sea, and to a lesser extent on the west slopes of the Rocky Mountains of Idaho, are areas of heavy rainfall and dense forests. Between the two mountain ranges is a large semiarid region, but it is traversed by the great Columbia and Snake river systems and provides extensive opportunities for irrigated agriculture.

The humid, cool, northwest coast of North America, extending from Alaska to northern California, contains the densest, tallest, most magnificent stands of timber on earth. Despite being a small proportion of the area of the United States, the Pacific Coast has about half the nation's timber. Half of this timber is Douglas-fir (*Pseudotsuga menziesii*). It is the leading commercial forest tree not only in the Northwest but from a world standpoint. In Washington and Oregon, from the crest of the Cascades to the sea, Douglas-fir is the dominant species in a mixed forest of hemlock, spruce, fir, cedar, a few pines, a scattering of hardwoods, and with an undergrowth of yew, alder, maple, cottonwood, willow, dogwood, cascara, vine maple, manzanita, salal, hackberry, ceanothus, and with the ground often carpeted with ferns (Eliot, 1948).

The Douglas-fir is not a true fir; in botanical relationship it stands in intermediate position among the spruces, hemlocks, and firs. The coastal form is a fast-growing tree, the largest of the Northwest conifers and not greatly exceeded in size even by the redwood of California. Mature trees can be up to 300 feet tall and 6 feet or more in trunk diameter. Trees with trunks up to 17 feet in diameter have been recorded. This huge tree often grows in extensive pure stands in Pacific Slope forests.

The tourist is at times shocked at the sight of vast "clear-cut" areas in the Douglas-fir regions (fig. 24). However, the controversial practice of clear-cutting is regarded by

*Fig. 24. Typical coniferous forest of the west slopes of the Cascade Mountains; bare areas have been harvested by "clear-cutting" (courtesy Larry L. Streeby, Oregon State University).*

some authorities as best suited for the tree because of the need of this species for sunlight for regeneration under artificial reforestation (Anderson, 1977). In virgin forests hemlock seedlings, which thrive in deep shade, tend to crowd out the more desirable species, which requires more sun (Eliot, 1948; Morgan, 1955).

On the eastern slopes of the Cascades are the great forests of ponderosa pine in almost pure stands. Then, except for the Blue Mountains, which repeat the forestation of the Cascades, no important stands of timber are found until one reaches the Rockies and their secondary ranges in Idaho. Here again the Douglas-fir is the most important commercial forest tree, but it is the smaller mountain or inland form of the species.

The welfare of the inhabitants of the Pacific Northwest has always been closely dependent on its forests. Like dense forests anywhere, those of the Northwest provided little in the way of game; the aboriginal inhabitants of the area turned to the sea and the rivers, teeming with salmon, for their principal food supply. However, the forests were important in many other ways. Probably the tree of greatest value to the Coast Indians was the western red cedar, providing them with material for homes, utensils, canoes, weapons, nets, mats, clothing, rope (from bark), and even food. However, where this species did not occur, Alaska cedar, Port Orford cedar, or whatever species that came nearest to being satisfactory, were utilized (Eliot, 1948).

Many American Indian tribes planted crops to supplement their diet, but the Northwest Indians appear to have gained their food only from hunting, fishing, and gathering. In summer they moved around the country in quest of berries and nuts. They needed summer tents that could be moved around easily. These were made of a frame of light poles, preferably lodgepole pine, that was covered with sheets of bark or mats of cedar, bark fiber, or hides. Lodgepole pines were an item of Indian trade, floated down the Columbia River from The Dalles to tribes that lived where this species was not available.

Between the outer bark and the wood of the cedar was a thin sheath of material, the cambium, used for food. It was scraped away and the scrapings were pounded and might be mixed with roots and berries, then molded into cakes about a foot square and an inch thick. The dried cakes were put away and eaten, with oil, in winter. The cambium of spruce, hemlock, and pine was sometimes used in the same way. Acorns, chinquapin nuts, hazelnuts, and seeds of fir and pine, particularly the sugar pine, were used as food. Crab apples, the Pacific plum, blackberries, serviceberries, madrona berries, and elderberries were eaten or stored away in fiber baskets or skin bags for winter use.

East of the Cascades, the high desert of the Columbia Plateau was the home of the Plateau Indians, including the Nez Percé, Yakima, Wenatchee, and Cayuse. They caught great quantities of salmon when these fish were running heavily in the Columbia and Snake rivers and their tributaries. Many of the salmon were dried in the sun or over a fire and stored away for winter food. The diet of salmon was supplemented with game and wild plant foods.

## EARLY EXPLORATIONS

The principal features of the coastline of the Pacific Northwest were revealed by incidental voyages made by the Spaniards in the late seventeenth century, but even such a prominent feature as the Columbia River was not discovered until an American, Captain Robert Gray, entered it in 1792, in the vessel *Columbia,* in search of sea otter furs. The great British explorer, Alexander Mackenzie, made a successful overland journey through Canada to the Pacific in 1793, reaching the ocean north of Vancouver Island.

The first American overland exploration and the first clear incentive for Americans to push the frontier westward to the Pacific was provided by a geographical and scientific expedition from 1804 to 1806 headed by Captains Meriwether Lewis and William Clark, who led a band of soldiers and civilians up the Missouri River, across the Rocky Mountains, and down the Columbia River to the ocean, returning by a similar route. Lewis and Clark were commissioned by President Thomas Jefferson to explore the Missouri River and such tributaries of it as might offer the most direct and practicable water route across the continent. (The prevalent belief at the time was that there was probably only one portage of less than a day between the Missouri and Columbia rivers.) An implied purpose of the trip was to buttress the American claim to the "Oregon Country" (what is now Oregon, Washington, and parts of Idaho and Montana), which rested on Gray's discovery, by establishing a practicable overland route to the Pacific before the British could do so.

The diaries of Lewis and Clark and others in their party give an account of two years of struggle, sacrifice, and backbreaking toil against what frequently seemed intolerable obstacles encountered under conditions ranging from oppressive heat and serious illnesses to bitter cold and starvation (De Voto, 1953). Woven into the voluminous diaries and notes of the expedition is an occasional laconic statement of appreciation and admiration for the remarkable young Indian guide and interpreter Sacajawea who, with a tiny infant on her back, accompanied her French-Indian husband on the expedition and uncomplainingly shared all its hardships and perils and contributed in important ways to its success. Sacajawea was destined to have more monuments, markers, and shafts erected in her honor, and to have more parks, lakes, and mountain peaks named for her, than any other American woman.

Fortunately there is a 155-mile stretch of the river route taken by Lewis and Clark, the upper Missouri Wilderness Waterway, which extends from Fort Benton to Kipp Park, Montana, that has remained much the same as when the two explorers and their expedition ascended this part of the river in their canoes and pirogues, with locomotion variously provided with towline, sail, and oars, during the spring flood, approximately from May 23 to June 10, 1805. Today the canyon walls, landmarks, and vegetation that Lewis and Clark and other early explorers described are still practically unchanged.

## THE FUR TRADE

The treatment of the Indians by Lewis and Clark during the entire two-year period of their trip, combining fearlessness and firmness with understanding, courtesy, and respect, accounted in large measure for their success. It also established a feeling of goodwill among the Indians of the region traversed that accounted for an era of far less violence in the early years of the Northwest fur trade than otherwise would have been possible. That trade always preceded agriculture, establishing the best land routes for settlers. The Lewis and Clark expedition revealed the enormous potential for the fur trade in the vast area they traversed. Trader Manuel Lisa organized a fur brigade the year following the expedition's return, hiring three of Lewis and Clark's veteran frontiersmen and persuading another, the redoubtable John Colter, to return to fur country (De Voto, 1953). Lisa's trip up the Missouri marked the beginning in earnest of the western fur trade. He left for the country Lewis and Clark most highly recommended, the Three Forks and the valley of the Yellowstone. Then in 1811 the founding of Fort Astoria by the influential fur merchant Jacob Astor resulted in the Americans winning the British-American race to the Pacific.

## THE OREGON TRAIL

One of the principal routes to the Pacific was the Oregon Trail, first used by fur trappers in the 1820s and 1830s. The first wagons were said to have been taken through the South Pass of the Rockies by Benjamin Bonneville in the 1830s. Settlers began using the trail about 1841. The big trek was started in earnest in 1842, by thousands of settlers seized by "Oregon Fever." The covered wagons were generally pulled by oxen because they did not stampede or stray as readily as horses and their feet stood the long journey better. Indians did not try to steal them, preferring horses; also, oxen were better for eating in case of extreme hunger.

Beginning at Independence, Missouri, the Oregon Trail followed the Platte River past Fort Laramie, Wyoming, traversed barren country to the Sweetwater River, then up that river and across a gap in the Rocky Mountains known as South Pass (elevation 7,450 feet); one branch of the trail extended southward to Fort Bridger while a shorter route (Sublette's Cutoff) led directly to the lush Bear River Valley in southeast Idaho, thence to Fort Hall, a Hudson's Bay Company post on the Snake River. The trail then led through broken, arid land, following the Snake River westward and northward, descended into the delightful Grande Ronde Valley in what is now northeast Oregon, scaled the Blue Mountains, and finally emerged at The Dalles on the Columbia River. From that point, in the early years of the trail, some settlers floated down the Columbia River on rafts built from their wagons, while others drove the livestock overland. The trip from Independence to Fort Vancouver usually required four or five months. The weary travelers then turned southward to settle in the rich Willamette Valley.

By the end of 1842, "Oregon Fever" raged through the Mississippi Valley and, in the spring of 1843, a thousand farmers in two parties left Independence for Oregon, having prepared for the trip all winter. They reached their destination late in October, having pioneered the use of wagons over the entire trail. In 1844 another 1,000 made the trip; in 1845, 3,000, but in smaller parties; in 1846 a total of 3,500; and numbers continued to increase.

The most intimate acquaintance with what life was like on the Oregon Trail can be obtained from certain published accounts of pioneers who made the trip. One was published by Ezra Meeker in 1912, sixty years after the trip was made (Meeker, 1912). In May 1850, at the age of twenty-two, Meeker and his wife and seven-week-old baby boy set out on the Oregon Trail, lured by the 640 acres of virgin land that was offered to married men during that period to stimulate the migration of settlers to Oregon. The family left Eddyville, Iowa, with a senior partner by the name of William Buck, a covered wagon, two yoke of steers, and a yoke of cows. At the time of Meeker's emigration, "for long distances the throng was so great that the road was literally filled with wagons as far as the eye could reach." Many families suffered from an inadequate diet. Cholera struck, and fresh graves along the route became a common sight.

On June 15, 1846, the Oregon Treaty was signed in Washington. Largely on the basis of the Lewis and Clark expedition, the treaty extended the international boundary along the 49th parallel to Georgia Strait, then to Juan de Fuca Strait, and through the middle of the channel to the Pacific. The treaty was international recognition that the American agricultural frontier had been pushed all the way to Oregon (Smith, 1950). By 1848 the population of the Pacific Northwest was sufficiently large to allow Congress to create the Oregon Territory.

## CATTLE AND SHEEP
(Oliphant, 1968)

As elsewhere on the continent, cattlemen followed the trappers and fur traders and preceded the sheepmen, who in turn preceded the grain farmers and growers of specialty crops. As early as 1792, English explorers saw cattle, hogs, goats, sheep, and poultry belonging to the Spaniards in the Pacific Northwest. After the Spaniards left, the citizens of the United States and Britain shared the disputed territory and laid the foundations of agriculture and animal husbandry. By 1837 the Hudson's Bay Company was reported to have 1,200 cattle among its various posts in Oregon Country, as well as many horses and other livestock, milk and butter, corn, potatoes, and other garden vegetables.

The Hudson's Bay Company developed agriculture to support the fur trade. Many cattle as well as horses and sheep were driven up from California in the 1830s and 1840s. During the same period American missionaries developed agriculture with the object of teaching the art to the Indians, believing this to be the prerequisite to teaching them the white man's way of life and religion. By 1843 colonists began to stream in from the American Midwest, bringing cattle with them which were more gentle and gave more milk than those driven up from California. The Pacific Northwest rapidly gained a reputation as choice country for livestock, abounding in nutritious grasses, and with a climate sufficiently mild so that cattle and other livestock could subsist throughout the winter without having to be fed or sheltered.

The part of the Northwest that lies between the Cascades and the Rockies went through certain stages of development similar to those of the Great Plains. Initially referred to as the Great Columbia Desert, it was generally considered an inhospitable wasteland that had to be passed through to reach the rich, verdant lands west of the Cascades, particularly fabulous Willamette Valley. The vast area was later to be called Transcascadia or the Inland Empire. The routes of the pioneers, including that of Lewis and Clark, usually did not bring them into contact with all its variations of sun and rain, slope and elevation, and corresponding variety of bunchgrass, ranging from scattered tufts on the desert margins to the luxuriant pastures of the eastern hills. But even Meriwether Lewis was prompted to note in his diary the splendid condition of the Indians' horses that had wintered on the short brown grass of the plains along the route traveled by the expedition (Meinig, 1968). One is reminded of the observations of certain astute individuals, soon to become known as cattle kings, regarding the obviously nutritious quality of the bunchgrass that supported the great herds of buffalo of the Great Plains.

Despite the early reputation of the Northwest as a region of reasonably mild winters, cattlemen in Transcascadia suffered severe losses during "unusual" winters when the chinook wind from the Pacific failed to appear to temporize the chilly gales from the north and east. Losses of cattle, sheep, and horses of up to 90 percent were then experienced, the animals dying of cold, hunger, and thirst. A few provident stockmen might collect one or two hundred tons of natural hay that could remain unused for two or three years, but then would serve to prevent severe losses (Nash, 1882). But such stockmen were in a small minority, an example of man's persistent failure, as Schneider (1976) has pointed out, to anticipate and prepare for the vagaries of climate. One feels less remorse for the periodic financial reverses of stockmen who chose to gamble with climate (and figure on averages) than for the dreadful suffering of thousands of helpless animals—pawns in a cruel game of chance.

"Unusual" weather was not the only hazard to livestock. There were diseases and predatory animals—wolves, mountain lions, and bears. In May 1880 four grizzly bears killed more than 100 calves belonging to one rancher in south-central Oregon, and in 1882 grizzlies killed from 75 to 100 head of cattle of another rancher. Trains killed cattle in considerable numbers and laws were enacted to enforce fencing of the tracks. Cattlemen formed associations to stop cattle and horse stealing and to "see that no guilty man escapes." Stockmen's associations and laws were unable, however, to prevent range fires set by careless white men or malicious Indians. Finally, many cattle died from eating poisonous plants, particularly in the early spring.

In 1850 the federal census reported 41,729 head of cattle in what was then called Oregon Territory: 24,188 beef cattle, 9,427 milk cows, and 8,114 working oxen. During the 1850s thousands of cattle and sheep were driven into California, but by the end of the decade that market had declined. New markets were then developed as the result of gold strikes in many areas throughout the Northwest and into British Columbia. By the late 1860s placer mining in the Northwest had subsided, but by 1869 cattle and sheep could be driven to Winnemucca, Nevada, and shipped to eastern markets, for the transcontinental railroad had been completed. Simultaneously, livestock was giving way to plant crops in the valleys west of the Cascades, but the vast area east of the range (Transcascadia) was to enter a couple of decades of rapid expansion in cattle- and sheep-raising.

Much of Transcascadia was sagebrush country, which lay mostly east of the Blue Mountains, over nearly all the main tributary valleys of the Snake River, and in southeastern Oregon. In this area bunchgrass grew at high altitudes and in the lower bottoms along the streams rye was grown, or a native redtop and clover. Scattered about through the black or gray sage was a "white sage" or "winterfat" (*Eurotia lanata*). In some places it occurred in large tracts, as on the north side of the Yakima River. White sage was found to be a remarkably nutritious winter forage for cattle. According to a General Land Office report in 1872, "after an occurrence of frost, horses, cattle, and sheep feed upon it with great avidity. It possesses great value for winter feeding." The Surveyor-General of Idaho wrote in 1870 "that the white sage, after the maturity of its seeds in the fall, is sought for by cattle in preference to grass." He considered more than half of the wild sage in the center and southern portions of Idaho to be of this species (Oliphant, 1968). White sage still occurs quite extensively in the drier areas of deep silty, slightly alkaline and calcareous soils and where precipitation is less than nine inches per year in southern Idaho, Utah, Wyoming, Nevada, and other areas. In the same plant community with white sage are the bunchgrasses (*Poa secunda, Sitanion hystrix,* and *Oryzopsis hymenoides* (M. Goodberg, correspondence).

By the early 1880s the grazing of cattle on the open ranges was already being hampered by farming in the Yakima, Klickitat, and Palouse areas and in the Walla Walla Valley. As early as 1879, 225,000 bushels of grain were harvested in Yakima County alone. Whereas in the 1870s the luxuriant growth of bunchgrass in the Palouse country had made it popular among cattlemen, the fertile soil and unfailing yield of grain caused a rapid displacement of cattle, which were driven farther eastward and north. In 1881, Idaho was reported to have 450,000 cattle, 60,000 sheep, and 60,000 horses. Bunchgrass and white sage in Idaho had taken the place of placer mines as valuable natural resources.

With the completion of the Northern Pacific Railroad in 1883, the ranges of the Oregon Country had access by rail to both the Atlantic and Pacific seaboards. By 1887 eastern Washington was directly connected by rail to Puget Sound. By the end of the decade all the Northwest except south-central and southeastern Oregon had relatively easy access to railway transportation. Products not able to "walk to market" could then be raised in great abundance. Agricultural expansion was particularly notable in Transcascadia (Oliphant, 1946).

Livestock raising went through the same changes in the Northwest as on previous American frontiers since earliest times. At first custom or local law gave an implied license to stockmen to graze their cattle or other livestock on all unenclosed land, whether publicly or privately owned. (Likewise cattlemen continually encroached on Indian reservations.) Inevitably the time came when it cost less to fence the pastures than to fence the grainfields. The new settlers, particularly those far removed from timber for fence rails, campaigned for "no-fence laws" or "herd laws" that placed the responsibility for keeping herds out of their fields of wheat and other plant crops up to the stockmen, who were expected to herd their stock.

When the issue came to a vote in eastern Washington in November 1880, the "no-fence" forces were badly defeated. This was considered to be a victory for "free enterprise" and an indication that stockmen would not yet have to move into the free territory of Idaho. A few years later sentiment had changed in favor of "no-fence," but by 1875 barbed wire fencing had made its appearance and the controversy died down. Barbed wire was relatively inexpensive and the grain farmers found some advantage in fencing in their grainfields in which, after harvesting, they could graze some livestock themselves.

Another lively issue arose between cattlemen and sheepmen. Cattlemen claimed that cattle would not graze on pasture where sheep had grazed, for the latter "poisoned and scented the grass." Sheep raising happened to be a good business, however, and by 1885 many cattlemen were fleeing eastward from Oregon and Washington. Others swallowed their pride and moved "downward" to the status of sheepmen.

In the shrinking cattle ranges the tendency to settle controversies between cattlemen and sheepmen and between stockmen and farmers by intimidation and violence continued, reaching a peak in 1904 and 1905, then virtually ending in January 1906, when the government began to lease grazing lands within its forest reserves and to regulate grazing on its summer reserves. It was not until the Taylor Grazing Act was approved by President Franklin D. Roosevelt on June 28, 1934, however, that the orderly use, improvement, and development of public grazing lands began. An effort was then made to prevent injury to those lands from overgrazing and soil deterioration, and a legal basis was provided for the stabilization of the livestock industry that depended upon the public range (Peffer, 1951; Dana, 1956).

Despite the near elimination of the cattle ranges of the Northwest, the production of beef increased, but from the fenced-in farms and not from the range. Cattle and sheep to this day are an important part of Northwest agriculture, Washington leading in dairy cattle, Oregon in beef cattle, and Idaho in sheep (USDA Agr. Stat., 1977).

## THE ROLE OF WHEAT IN THE HISTORY OF THE NORTHWEST

The principal goal of those infected with "Oregon Fever" was wheat as it was said to grow in the rich soil of Oregon. Three of the famed mountain men of the early days of the West, Jedediah Smith, David Jackson, and William Sublette, visited Vancouver (Washington) in 1825 and reported that the British of Hudson's Bay Company had raised 700 bushels of an exceptionally fine grade of wheat, making good flour. They had noticed a hand-worked gristmill at the fur company's post. By 1830 a gristmill to be run by oxen was built. The first mill in the Willamette Valley was built in 1834 for new settlers on French Prairie. In 1837 these settlers grew 5,000 bushels of wheat for their new mill. Wheat was sold to the Russians at Kamchatka until the Oregon country was ceded to the United States. After the big wave of immigration started in 1842, the location of nearly every town in western Oregon was decided by the presence of water and a mill site. Nevertheless, not all farmers raised wheat. Freight charges down rivers such as the Willamette were high, and the earliest settlers mainly raised cattle and sheep; they would walk them to market.

Prior to the construction of the mills, flour was brought to the fur traders from Chile on ships that made the return journey with furs from Oregon and hides from California. After the discovery of gold in California, the miners could take all the flour the Oregon flour mills could produce. Flour was sent to California by ship and mule pack train.

Mining was a stimulus to the raising of grain in Transcascadia, just as it was for the livestock industry. Oats were the first grain crop of importance, needed for hay for the hundreds of pack mules, freighting teams, and cavalry horses. Grain was first cut with cradles and the few simple McCormick reapers in the area. The first large threshing machine arrived in 1861. There was a sudden great increase in demand for wheat for the mines in 1862, but very little was then raised in Transcascadia. By the end of the year three grist mills were in operation. By 1863 farmers had found that not only the river valleys, but also the "benchlands" were suitable for growing wheat. However, the waning mining enterprise, plus the increased production of wheat (and oats) in mountain valleys to supply the miners, resulted in a glut of wheat on the market in the mid-1860s.

There was no sign of relief until wheat began to be exported in ocean vessels via Portland (Meinig, 1968). In 1867 Oregon flour was reported to be the best flour on the New York market. The first British vessel to carry a full load of wheat and flour from Portland to Liverpool did so in 1868. In 1870 twelve vessels sailed from Portland with a total of 242,579 bushels of wheat. Settlers arrived in Oregon with practically nothing in the way of possessions, depending on income from wheat to pay for such essentials as stoves, harness, farm tools, household utensils, furniture, clothes, coffee, and spices (Brumfield, 1968; Meinig, 1968). The grain trade ensured a steady supply of such essentials. Grain accounted for more than two-thirds of the value of exports from Pacific ports between 1871 and 1895; it took up about 90 percent of the shipping space. By attracting such a large volume of shipping, the grain trade assured a constant flow of scarce materials into the Northwest at relatively low freight costs, thus greatly aiding its economic development (Rothstein, 1975).

Even during the 1870s there was already so much grain and other agricultural products being grown in Transcascadia that the Oregon Steam Navigation Company had difficulty in providing enough boats to carry the products down the Snake and Columbia rivers (Oliphant, 1968). Then after the Union Pacific Railroad reached Portland in 1883, ever greater numbers of wheat growers were lured to the larger and cheaper lands of eastern Oregon. Wheat was still grown in western Oregon in some rotation systems, but only about 10 percent of the state's total. The same thing happened in Washington; wheat was first grown west of the Cascades, then around Walla Walla, then through the vast areas of the Snake and Columbia river basins wherever this highly adaptable crop could thrive (Brumfield, 1968).

In the 1880s when the Northern Pacific was building its railroads into the Walla Walla and Palouse regions, its pamphleteers could say, based on the experiences of wheat farmers, that wheat would grow and mature wherever bunchgrass grew. The bunchgrass country, they said, would be turned into the "finest wheat country on earth." The railroads resulted in rapid growth in wheat towns such as Colfax, Palouse City, Pullman, and Moscow, the last named in the Idaho corner of the Palouse. South of the Snake River, railroads and wheat spurred the growth of towns such as Pomeroy,

Dayton, Walla Walla, and Pendleton, the last named in Oregon and to this day the home of the well-known annual "Pendleton Roundup" festival and rodeo.

Adams County in the Big Bend area got a late start, with disappointing initial results when compared with the thriving Walla Walla and Palouse districts. Then in 1883 an event occurred that had become familiar in the early development of the bleak and unpromising areas of the Great Plains. A group of Volga Germans settled a short distance west of Ritzville, each purchasing a half section of land. Living in dugouts and gathering cow dung and sage brush for fuel, through hard work, thrift, and good management they achieved unusual success in farming. Soon they had good homes, orchards, windbreaks, and windmills, marking a distinctive oasis in the semiarid pioneer landscape. These farms were a sharp contrast to those elsewhere in the area. With this example, the Big Bend got its start as one of the great farming areas of the Northwest's Inland Empire (Meinig, 1968).

## CROPS OF THE SNAKE RIVER PLAINS

The Snake River Plains are in the southern part of the high desert of the Columbia Plateau as defined by Nuttonson (1966). The Plateau is bordered on the east by the Rocky Mountains and on the west by the Cascade Range.[1] It includes all of Land Resource Region B in figure 3. In this important agricultural area, precipitation ranges from less than ten inches in the south to more than fifty on the slopes of the higher mountains. Like much of the United States west of the Rocky Mountains, there is little or no summer rain. The factor governing the amount of land that can be devoted to agriculture in such regions is the amount of mountain watershed in relation to the amount of arable land in the valleys. In this respect the entire Northwest is fortunate. The Cascade Ranges take most of the moisture out of the air mass coming in from the Pacific, providing an abundance of water for the coastal valleys of Washington and Oregon and causing the region east of the mountains to be arid. Yet enough moisture remains in the air mass reaching the high ranges and forested areas of the Rocky Mountains to provide the heavy precipitation that eventually finds its way to the giant reservoirs and irrigation canals of the lower plains. Two great rivers flow through the Columbia Plateau, the Columbia, which enters from the northeast, and its greatest tributary, the Snake, which enters from the southeast. They meet in southeastern Washington, about fifteen miles from the Oregon state line, then form the major part of the border between Washington and Oregon.

The Snake River flows through southern Idaho from east to west in a great semicircle, partly through broad valleys and partly through box canyons with precipitous sides rising a thousand feet above the river bed. Extending for miles on either side of the river are the Snake River Plains, interrupted here and there by detached buttes of ancient volcanic origin that rise abruptly, high above the level of the plains. The plains vary in elevation from around 5,000 feet at their eastern extremity, not far from Yellowstone National Park, to 2,114 feet at Weiser, the western end (Nuttonson, 1966).

Irrigated land comprises about a fourth of the 16,200 square miles of the Snake River Plains region, where average annual precipitation is only 7-13 inches, with little or none in summer. In addition in some of the deeper alluvial deposits throughout the region and the lavas north of Snake River in eastern and central Idaho, groundwater is plentiful and has been used extensively for irrigation. These two regions make up the bulk of

[1]The term "Columbia Plateau" has also been used by the U.S. Soil Conservation Service to designate a much more limited area in eastern Washington and north-central Oregon (Austin, 1965).

that portion of Region B (fig. 3) that extends into southern Idaho. The eastern part of the region is referred to as the Eastern Idaho Plateau.

The irrigated plateau plain of southern Idaho is at a high enough altitude to have cool summers, favorable to the growth of potatoes of high quality. The average freeze-free period in this area is only 90-170 days; yet, besides potatoes, large crops of grain, hay, sugar beets, beans, fruits, and vegetables are raised.

At some of the higher elevations in the eastern part of southern Idaho the land is used mainly for livestock and wheat. At somewhat lower elevations, one of the principal crops is the Idaho russet potato, the crop for which the state is best known. About 100,000 acres, in the area between Gooding and Minidoka counties in south-central Idaho, are devoted to raising dry edible kidney beans. Another important crop is the sugar beet, and Idaho is exceeded only by California in annual tonnage. In the upper portion of the Snake River Basin of Idaho more than a third of the crop area is devoted to hay, mainly alfalfa, and dairy products are the chief source of income for many farms (Nuttonson, 1966).

In southwestern Idaho the 750,000 acres of irrigated farms constitute a comparatively small percentage of the area's total acreage of farms and stock ranches, but a large variety of crops are grown. Dairy products, sugar beets, onions, lettuce, hops, spring wheat, alfalfa, various seed crops, cherries, prunes, peaches, and apples all contribute to agricultural income. In Canyon, Ada, and Payette counties of Boise Valley there is considerable double cropping, not possible at higher altitudes where the growing season is too short. Early crops include peas, intermediate potatoes, turnip seed, and spring lettuce, followed by fall crops such as fall lettuce, late seed potatoes, late grain, or a green manure crop such as Hubam clover (Nuttonson, 1966).

### Irish Potatoes

The quantity of root crops, worldwide, probably rivals that of the cereal grains, but the former have a relatively high moisture content so the dry nutrient weight would be much lower. Nevertheless, with an annual world production of about 300 million tons of white potatoes and over 100 million tons of sweet potatoes and yams, the root crops loom large as foods for mankind (Schery, 1972).

The Irish or white potato (*Solanum tuberosum*) came to be intentionally cultivated probably more than 4,000 years ago in the Andes, where the wild plants were common. Potatoes do well at elevations as high as 15,000 feet. Where temperatures were too low for potatoes to keep well, the Indians learned how to "dry freeze" them. They allowed the potatoes to freeze at night and the next day they would stamp on the thawing tubers. Repeated for several days, this process would eventually remove the water, resulting in a desiccated potato called *chuño*. The product could be kept almost indefinitely and used as required (Heiser, 1973).

The Spaniards probably discovered the potato in 1537 when they occupied what is now Colombia (Morrison, 1945). Although distributed throughout continental Europe by 1600 and into the British Isles by 1663, the potato does not appear to have been widely planted until after 1700. It then soon became an important food staple, especially in Europe, which now grows by far the largest quantity. Yet in America of the eighteenth century many people feared that eating potatoes would shorten a person's life. Some apprentices refused to be indentured unless they were assured by their masters that they would not be served potatoes.

How important potatoes eventually became in some areas is indicated by the fact that in Ireland two potato crop failures in succession, in 1845 and 1846, caused by the potato blight (*Phytophthora infestans*), resulted in such a severe famine that about one million Irishmen died of starvation and one million migrated to the United States, while many of the survivors remaining in Ireland were left in abject poverty. Potato blight is found wherever the tubers are grown. The first symptoms consist of dark brown spots on the foliage that rapidly increase in size. The disease not only destroys the foliage but also attacks the tubers. The latter become reddish brown and in later stages they rot. Resistant varieties and fungicides have reduced the potato blight problem to manageable proportions. The federal government is pursuing an intensive breeding program, utilizing breeding stock principally from South America in an attempt to develop varieties with increased resistance to potato blight.

Irish potatoes are grown throughout the United States. The earliest potatoes to reach the market are grown principally in California and Florida, late spring and summer potatoes farther north, and the fall potatoes in the northern half of the country. Fall potatoes comprise 86 percent of the annual production of about 16 million tons in the country. In 1976 Idaho produced 25.5 percent of the fall crop. Washington, Maine, and Oregon are also big producers (USDA Agr. Stat., 1977). In favorable areas for growing potatoes, such as the mesas above the Snake River in Idaho, yields average 15 tons per acre, but may be as high as 35 tons in some fields. As with other major food crops, greatly increased yields are expected in the future by agronomists, possibly 100 tons per acre (Schery, 1972).

About 7 percent of a potato crop is used for planting new fields. These "seed" potatoes are cut into four or more sections, each containing an "eye"—a dormant bud. Largest crops are produced where soils are well drained and where days are warm and nights cool. About two weeks before digging, the plants may be sprayed with a growth retardant to inhibit premature sprouting. Just before harvest, plant tops are usually cut off or killed chemically to facilitate digging. In the United States, most of the crop is mechanically harvested, then washed to remove soil residues. The potatoes are then stored in a cool place to inhibit sprouting.

Americans spend only 2 percent of their food budget on this most economical, nutritionally balanced staple food. Mistakenly shunned by many weight watchers, it actually has no more calories per unit weight than the apple. According to the USDA, if one's diet consisted entirely of potatoes they would supply 150 percent of the iron, all the riboflavin ($B_2$), three or four times the thiamin ($B_1$) and niacin ($B_3$), and more than ten times the vitamin C that the body needs. Potatoes contain good quality, easily digested protein, and combined with whole milk they would supply almost all of the required food elements. The Incas ate potatoes with other foods to prevent indigestion. They are now known to be high in potassium, a valuable aid in treating digestive disorders (Scott, 1976).

With about 20 percent starch, the potato is a fairly economical source of this substance. It has only about 2 percent protein, but because of high yield, it provides an amount of protein per acre exceeded only by soybeans and brussel sprouts among food crops. It is exceeded (slightly) only by rice in yield of food energy per acre while greatly surpassing the latter in food-energy, in relation to fossil-energy, input. However, potatoes require a high input of labor (see table 1).

Of the Irish potatoes sold in the United States in 1975, 52 percent were processed in some way, 7 percent were used for seed, 1.5 percent for livestock feed, 0.7 percent for starch, leaving only 38.8 percent for "table stock" (USDA Agr. Stat., 1977). Processing for "ready-mix" mashed potatoes is becoming increasingly common. Consequently, plant breeders are inclined to emphasize high yield and high-starch content rather than appearance. Also the potatoes of the future may be larger, heavier, and more solid (Schery, 1972).

The potato is a basic, inexpensive, nutritious food. However, the more it is processed, the more the price goes up (as much as 23-fold) and the nutritional value declines —that is, the fat, salt, and other chemicals increase (Lappé and Collins, 1977).

### Seed Corn

One of the things that enabled plant breeders to make new sweet-corn hybrids so quickly available to the public was the discovery that southern Idaho provided a nearly ideal area for seed production. A little-known fact about this area is that about 90-95 percent of all sweet-corn seed planted in the United States is grown in the region of Treasure Valley about 30 miles southwest of Boise, amounting to about 18,000 acres. Credit belongs to George Crookham, Jr., and Don Baldridge, two young men who got out of college about the time the Golden Cross Bantam variety of sweet corn was developed. These two men foresaw the tremendous demand for Golden Cross Bantam seed, and as a result went into the business of producing this and other hybrid sweet-corn seed. They discovered that southern Idaho land, properly irrigated, brought together a number of conditions that made it remarkably well suited to the production of many types of seed. Since then practically all the large growers of hybrid seed have transferred their production to southern Idaho.

Idaho's prominence in sweet-corn-seed production is believed to be the result of the dry climate and use of irrigation. There are very few pathogens; the potentially most serious problem, *Fusarium* stalk rot, is controlled through planting of tolerant varieties. Excellent seed quality is ensured by the long, warm days and cool nights of the growing season.

## THE MOUNTAINOUS REGION OF IDAHO

North of the Snake River Plains is a labyrinthine mass of mountains and canyons. Most of it is used as wildlife habitat, or for recreation, mining, and production of timber. (Whatever can be said of this area is equally true of the contiguous mountainous area of western Montana.) On the upper mountain slopes and crests above the timberline there are meadows partly used for grazing. There are also occasional open valleys where there is farming, including the Big Wood River Valley, the Big Lost River Valley, the Salmon River Valley, the Pahsimeroi Valley—all named for the rivers that traverse them, and Long Valley, traversed by the north fork of the Payette River.

In the valleys, nearly all the area is given to farms and ranches. In some valleys as much as a third of the land is irrigated. Potatoes, sugar beets, and peas are grown, but a much larger acreage is used for hay and for pasture for livestock. Where there is enough rainfall, some areas are dryfarmed to wheat and peas. There are large areas planted to native grasses and shrubs, with beef cattle and sheep as the principal livestock. Near the larger towns, dairying is important.

Precipitation is fairly copious over most of northern Idaho, and generally sufficient for local crop needs. In some areas, such as the valleys of the Clearwater and Spokane rivers, irrigation is necessary. In the western portion of northern Idaho, the lower valleys and the gently rolling prairies are well adapted to rain-fed agriculture. Native grasses grow luxuriantly, and cereals, alfalfa, timothy, beans, peas, corn, and all common garden vegetables are grown successfully. Hardy deciduous fruits, mainly apples, are grown generally throughout northern Idaho but commercial orchards are almost exclusively in the more favorably situated valleys (Nuttonson, 1966).

## THE PALOUSE REGION

Around the intersections of Washington, Oregon, and Idaho is the Palouse region. About 40 percent of the land is dryfarmed to wheat and peas. Only about 1 percent is irrigated and used to grow vegetables, seeds, and other specialty crops. Much of the publicly and privately owned land is in range and there are small wooded areas on the steep slopes (Austin, 1965).

The Palouse proper is mostly in southeast Washington, dipping only slightly into Oregon east of Pendleton and far enough into Idaho to include the region around Moscow. Thus it includes the state agricultural colleges and agricultural experiment stations of Washington at Pullman and Idaho at Moscow. The Palouse proper is often considered together with the Camas and Nez Perce prairies in discussion of dryfarming wheat production, thus extending the area farther into Idaho and northeastern Oregon.

The rainfall in the Palouse varies from about 18 inches at its western margin to about 22 inches at the Washington-Idaho border and 30 inches in portions of the Idaho wheat-pea area in its eastern margin at the foot of the northern Rockies, where the small amount of moisture remaining in the air mass that passes over the Cascades is finally precipitated as the air rises and cools with the increasing elevation of the Columbia Plateau and finally the Rocky Mountains. Here the pioneer farmers, who began to settle the Palouse region about 1864, found another grass land with soil rich in organic matter, as in the vast grasslands of the Midwest.

"Palouse" means "grassland" and is a name believed to have originated with the early French explorers. However, the grasses were mainly bunchgrass types, as is generally true in the West where rainfall is low and occurs principally during winter months. Bunchgrasses have roots that penetrate deeply and can sustain the plants during the dry summer months. Wheat is an annual bunchgrass and its roots can penetrate down as far as six feet in search of moisture. Therefore wheat was the logical crop for the Palouse.

Thus the Palouse is the heart of the northwest wheat country. Much of the wheat is growing on the slopes of silt dunes with grades up to 55 percent. While the region has by no means been free from soil erosion, this has been minimized because the region has had the good fortune of having rains of remarkably low intensity, 92 percent of the rains amounting to less than half an inch in 24 hours. Although moisture is rather deficient, there is considerable storage of water in the soil during the winter rest period. Like the short grass of the Great Plains (fig. 3, Region G), the grass of the Palouse passes through a drought rest period much earlier in the summer than the grasses of the tallgrass prairie of the Midwest (Region M). In the Palouse the "wheatgrass sod" (*Agropyron spicatum*), little bunchgrass (*Festuca idahoensis*), little bluegrass (*Poa sandbergii*), and balsam root (*Balsamorrhiza sagittata*) form a luxurious and varied sod which supports plants with large showy flowers and leaves. Also the silt loam soils, high in organic matter, are absorptive and store the winter rains efficiently for summer plant growth. If it had not been for these fortunate circumstances, the steep-sloped dunes would by now have been severely eroded into an unplowable condition, for the wheat-pea and wheat-fallow rotations have been continuously practiced for over a century.

Agricultural practice is continually changing. Recently farmers in the Palouse have found that, with improved tillage equipment and chemicals for weed control, continuous cropping is not only more profitable, but reduces erosion, when compared with summer fallowing. Best protection of the soil is obtained with a recrop system that includes only grains, but some systems might include peas or lentils somewhere in the rotation. *Cercosporella* foot rot[2] might become a problem in land recropped continually to grain.

The secret of success seems to be the incorporation of wheat stubble (or pea and lentil plant residue) into from 4 to 6 inches of soil, using a chisel plow or other cultivator-type implements (rather than a moldboard plow and disk). The land is chiseled, fertilized, and drilled on the contour. The plant residue keeps soil open to moisture intake and binds the soil against running water. Soil becomes more mellow and moisture retention for germination is good (Anonymous, 1974).

In view of the hilly nature of the terrain in the Palouse country, the leveling device developed by Holt Brothers in 1890, which adapted the combine for use in harvesting and threshing wheat on hillsides, was an important development.

The Gaines wheat variety, developed from dwarf Japanese "Norin-10" germ plasm, resulted in great increases in wheat yields in the Northwest during the 1960s. When moisture is not a limiting factor, this strain will produce 100 or more bushels per acre and it does not lodge (fall over). Norin-10 germ plasm also was the basis for the miracle wheats, bred in Mexico, for which N. E. Borlaug received the Nobel Peace Prize in 1970 and which played an important role in the Green Revolution.

In 1976 Washington produced 149,050,000 bushels (4,211,000 metric tons) of wheat. Although this quantity was exceeded by Kansas, North Dakota, Montana, and Oklahoma, in yield per acre Washington greatly excelled, with an average of 45 bushels compared with yields ranging from 24 to 31 bushels for the other states. In the Palouse area 60 to 90 bushels per acre are not uncommon. This high yield is not the result of irrigation. In 1976 Washington wheat farmers harvested 3,285,000 acres of wheat, but only from 150,000 to 200,000 acres were irrigated (K. J. Morrison, correspondence).

### Peas

The dominant rotation in the Pacific Northwest is between wheat and peas. The area is by far the nation's most important region for the production of peas. In 1971 Washington, with 98,303 and Idaho, with 66,071 metric tons, produced 93.7 percent of the dry edible peas grown in the United States. Green peas are also grown extensively in the Northwest for processing, mainly freezing, and are the principal cash crop in some areas. They are an early-maturing crop and some farmers grow a second one on the same land. Peas are also grown as a companion or nurse crop for spring-seeded alfalfa, pasture, or clover. In 1976, Washington, with 106,400 metric tons, was exceeded only by Wisconsin in green-pea production. Oregon and Idaho also produce heavily (USDA Agr. Stat., 1977).

### Lentils

The lentil (*Lens esculenta*), believed to be indigenous to southwestern Asia, is one of the oldest of legumes, as well as one of the most nutritious. Lentils are an important food crop in India, Pakistan, southwestern Asia, northern Africa, and southern Europe. In North America, the Palouse area of Washington and Idaho, where 3,000 growers produce about 40,000 tons a year, is the principal area for the production of

[2]Foot rot is a soilborne disease that can reduce grain yields as much as 75 percent. If winter wheat is seeded early it provides leaves and roots to protect and hold the soil. However, the earlier it is seeded the more susceptible it is to foot rot. Plant breeders are now attempting to develop a high-yielding wheat variety resistant to foot rot and other diseases.

lentils and yields are about a half-ton per acre. Lentils are generally rotated with wheat. In the Northwest the vines are permitted to dry before being harvested with combines, and are then cleaned at local processing plants. Lentils are a good meat substitute, consisting of about 25 percent protein, 1 percent oil, and 5 percent carbohydrate (Schery, 1972).

## THE COLUMBIA BASIN

The Columbia Basin as defined by Yates et al. (1966), comprises the northern part of the Columbia Plateau. It includes most of eastern Washington, forming a part of the drainage basin of the Columbia River. The Columbia is one of North America's greatest rivers. Rising in the Canadian Rockies of British Columbia, it is 1,214 miles long and on this continent it is exceeded in volume of water only by the Mississippi and McKenzie rivers. Its flow exceeds that of the Colorado tenfold. It winds through the Columbia Basin amid narrow valleys and steep-walled canyons. The basin extends from the Cascades to a series of relatively low north-south ranges (The Okanogan Highlands) in the northeast and even as far as the beginning of the Rocky Mountains, while in the extreme southeast it extends to the Blue Mountains. The region is treeless except for the mountains, which have considerable timber. The climate is continental in character except to the extent that it is tempered by the mild westerly air currents from the Pacific Ocean. In eastern Washington the January mean temperatures range from 17 to 35°F (-8.3 to 1.7°C) and July mean temperatures generally range from 59 to 79°F (15-26°C). If a hot spell continues for more than two days before grain is fully matured, the hot, dry, north to east winds cause great damage to grain and become a serious fire hazard to grain and mountain timber. The frost-free period varies greatly with altitude and latitude, ranging from 68 to 218 days, the latter in the region around Walla Walla, near the Oregon border (Nuttonson, 1966).

### The Grand Coulee Dam

The relatively steep gradient of the Columbia River, the low percentage of silt in the water, and the strength of bedrock have made possible the construction of many dams. The greatest of these is the Grand Coulee, about 80 miles west of Spokane. Completed in 1942, it is one of the largest concrete structures in the world. It creates a reservoir, called Franklin D. Roosevelt Lake, that extends 151 miles and reaches the Canadian border. Some of the water stored behind the Grand Coulee Dam is pumped onto the plateau with energy created by the dam. The water then flows southward by gravity to irrigate 1,029,000 acres of land previously used only for dryfarming and grazing. When full capacity is reached, Grand Coulee will be the largest hydroelectric installation in the world.

The Northwest has been fortunate in its abundance of clean, renewable, low-cost hydroelectric power by means of the scores of dams that now control the Columbia River. However, nuclear plants already operate at Hanford and on the lower Columbia River near Portland. Predictably, "growth" demands further reliance on expensive thermal generating plants. But there is strong public support for a quite different vision of the future that would involve a transition to innovative, smaller-scale, appropriate energy sources. Time would be gained for this transition through serious energy conservation on a region-wide basis (Scott and Blomquist, 1978). There has been an unconscionable neglect, for example, of thermodynamically appropriate use of the various energy sources (Commoner, 1976). The thermodynamic efficiency of industry in the

United States is only about 25 percent (Georgescu-Roegen, 1971; USFEA, 1976). Saved energy has become "the world's most promising energy resource" (Hayes, 1977).

### Effect of Irrigation on Climate in Arid Regions

Climate in arid regions can be remarkably influenced by irrigation in and beyond the immediate areas. In the early 1950s, irrigation water became available from Grand Coulee Dam and was applied to about 400 square miles of formerly near-desert land in the center of the Columbia Basin in Washington. In the period 1955 through 1973, July and August rainfall in that area was 50 percent higher than in the period 1931 through 1950. This compared with an increase of only 23 percent for the region outside the basin but within 150 miles of its center. The higher percentage increases were in the foothills inside the basin. Stidd (1975) believes that irrigation water in the Columbia Basin is recycled at least once as rainfall, with benefit to the rich wheatlands in the foothill areas.

In the semiarid agricultural lands east of the Cascades, just as in the Great Plains region, huge automatically controlled center pivot sprinklers are being increasingly brought into service for irrigation of grain, forage, and other field crops. Pivot-irrigation circles (fig. 25) have become a familiar sight to air travelers. This development, by extending the irrigated areas, might be expected to accentuate the climatic influences of the other types of irrigation.

The center pivot systems are most commonly propelled with electricity, though some are water-driven. Ten or more of these units may be operated by one irrigator and a high degree of uniformity of water application is obtained. A disadvantage is that if several types of crops are planted in one circle, scheduling, cultural, and harvesting problems are created. It is best to have one crop per circle. The entire circle is dependent on one lateral, so lengthy breakdowns may cause more problems than with systems involving several irrigation units. The center pivot systems are not being used in orchards and vineyards, where other sprinkling systems are more appropriate (Hagood, 1972).

### Wheat

In Washington, the principal wheat-producing section is on the rolling-to-hilly plateau of the east-central and extreme southeastern part of the state. The average annual precipitation averages only from 10 to 20 inches where dryland wheat is grown, but the soils have a relatively high moisture-holding capacity. The fall, winter, and spring precipitation favors germination of both fall and spring grain, and ample June rains with moderate temperature are favorable to heavy wheat yield. Furthermore, the generally long, comparatively dry spell in July, August, and September is nearly ideal for grain-harvesting. Green peas are often grown in rotation with wheat.

### Tree Fruits

Precipitation in eastern Washington ranges from over 60 inches in the higher eastern slopes of the Cascades down to about 6 inches in the lowest central valleys. Agricultural development in the valleys depends on irrigation. Apples and other fruits of good quality and color are grown on an extensive scale, under irrigation, in the Yakima and Wenatchee valleys on the lower east slopes of the Cascades as well as in the Columbia, Okanogan, and Walla Walla valleys.

Yakima Valley may be given as an example of the types of fruit grown in the irrigated

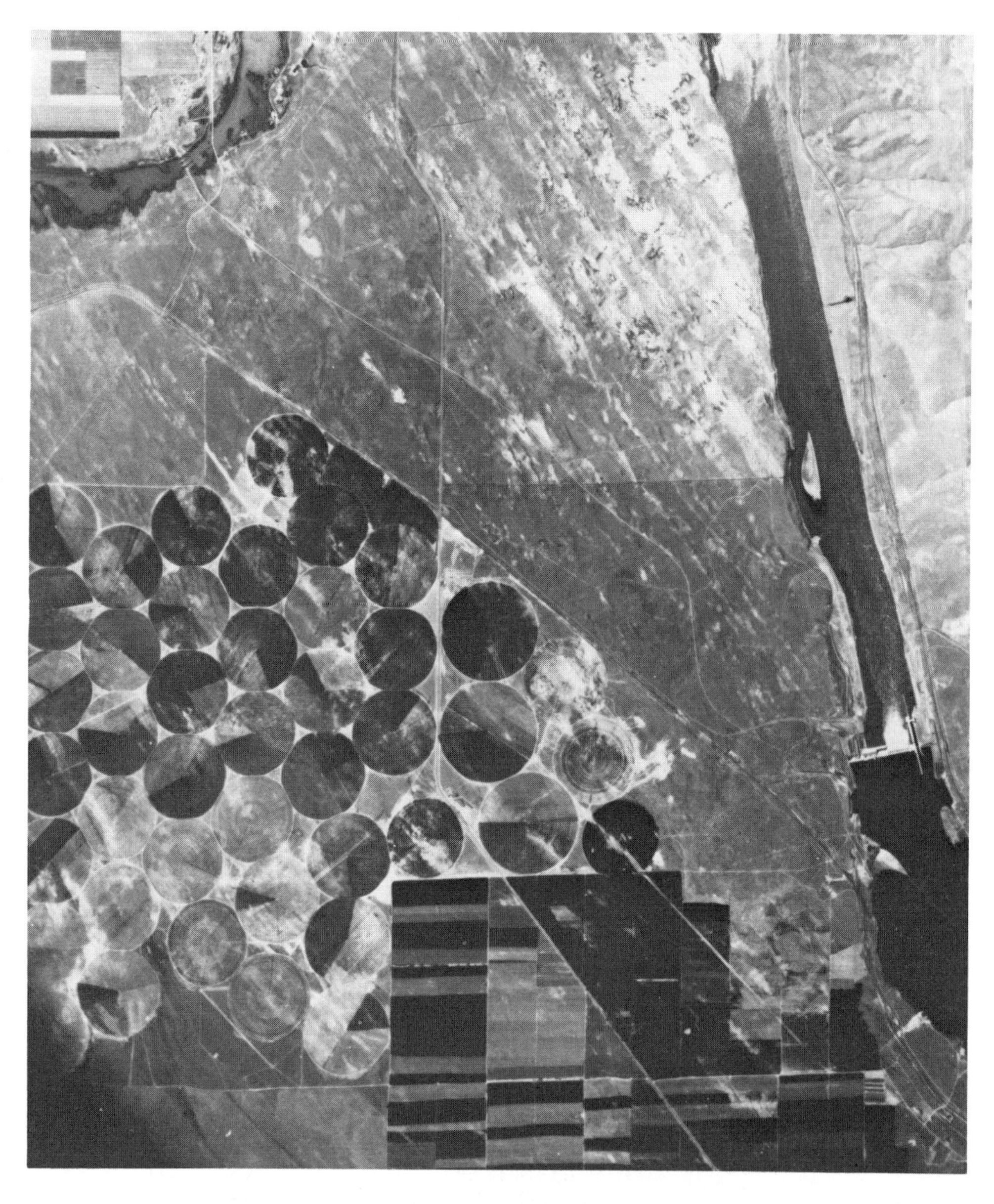

*Fig. 25. Pivot-irrigation circles made by center pivot sprinklers (courtesy USDA Soil Conservation Service).*

valleys. The 41,226 acres of apples in the valley had a value of $66 million in 1975. The next five fruit crops, in descending order of importance, were cherries, pears, grapes, peaches, and prunes, totaling $45 million.[3]

In 1826, Marcus Whitman planted apple seeds at his mission near Walla Walla,

[3]Although Yakima Valley is best known for its fruits, the value of hops from the valley's 20,922 acres of the crop in 1975 was $30 million. The Yakima and Moxee valleys provide more than 65 percent of the nation's hops. The value of Yakima Valley's beef cattle was $75 million and dairy products $12 million. (Data from "Facts About Yakima, Washington, and Yakima Valley," Greater Yakima Chamber of Commerce.)

Washington, having carried them across the continent on horseback. An apple seed brought from England was planted at Fort Vancouver, Washington, in 1826 and is still bearing fruit. The apple is a long-lived tree (Larsen, 1975).

About half of the apples produced in the United States are processed into apple sauce, dried, or converted into cider and vinegar. Pectin gum is an important by-product obtained from the fruit residues at the cider mills. Washington, which produced 1,021,000 metric tons in 1976, over 35 percent of the nation's apple crop, usually sells about 80 percent of its fruit fresh. This is because Washington apple growers produce mostly Red and Golden Delicious, which are of very high quality and are grown specifically for the fresh market. Most of the apples processed in Washington are those sorted out from the packing lines.

Most fruit trees do not "come true" from seed, so a bud or piece of shoot of the desired variety is grafted onto a small tree known as the rootstock. The rootstock has no effect on the size, shape, or quality of the apples, but has an important effect on ultimate tree size. Rootstocks resulting in semidwarf trees have become popular. As many as 200-300 can be planted to an acre, compared with only 50-100 full-size trees. The smaller trees are easier to manage, yet produce more fruit per acre and the fruit is of better quality. The semidwarf trees become full-bearing in 6-8 years after planting compared to 12-15 years or more for standard trees (Larsen, 1975). As recounted in chapter 3, dwarf or semidwarf characteristics can also be obtained by grafting mutations or sports onto standard rootstocks. It is important for the commercial apple grower to be fully aware of his soil fertility when he decides what variety and what rootstock he is going to use.

The apple tree can withstand subfreezing temperatures well and requires a winter conditioning in order to bear fruit. Most orchards are located in rolling country above low-lying pockets so as to avoid loss of blossoms from frost. The tree can bear fruit for as long as a century, but most orchards are renewed much sooner. Sprays are used to protect the fruit from many insect and fungus pests and to control the set of fruit, its retention on the tree, or its fall where thinning is needed.

The pear is closely related to the apple, and like the latter it is indigenous to western Asia. The pear tree is similar to the apple tree in appearance, but is less tolerant of either extreme heat or cold. The winter chilling requirement of these two pome fruits is similar. Care and culture of apples and pears are similar, but some varieties of pears, such as the Bartlett, are susceptible to fire blight, a serious threat to pear-growing. It pays to plant more than one variety of either apples or pears to ensure cross-pollination, for some varieties are self-sterile.

In the Pacific Northwest about 52 percent of the pear crop is of the Bartlett variety, the remainder being various other varieties, mainly the Anjou, but also a few Comice, Hardy, and Winter Nelis. This differs sharply from the situation in California, where, in 1976, 97.6 percent of the pear crop was of the Bartlett variety (USDA Agr. Stat., 1977). In the Northwest pears are grown principally in the Rogue River Valley of southwestern Oregon, the Hood River Valley of north-central Oregon, and the Yakima and Wenatchee valleys of central Washington.

The Bartlett is a prolific bearer and is unsurpassed in flavor and texture among the large commercially grown pears. Nearly two-thirds of the crop is canned and about one-third is sold as fresh fruit. It is the first of the important commercial varieties to ripen. The Anjou is a large green pear, the most important winter variety, and keeps well in storage until March or April. The entire commercial crop is marketed as fresh fruit. Of all the large high-quality varieties, the Anjou is the most resistant to blight (Magness, 1957).

The pear crops of Washington and Oregon are exceeded in volume by that of California, and the three Pacific Coast states produced 96 percent of the 826,700-ton pear crop of the United States in 1976 (USDA Agr. Stat., 1977).

## WEST OF THE CASCADES

West of the Cascades of Washington and Oregon (fig. 3, Region A), the average rainfall is from 50 to 90 inches in most areas, but can be as high as 200 inches, increasing with elevation and from south to north, falling mostly in autumn, winter, and spring. Most of the land is in forest, but even in mountainous areas there are narrow valleys and coastal plains where some land is cleared and hay for dairy cattle is the chief crop, with some grain and improved pasture also grown. In the Puget Sound Valley there is a highly diversified agriculture, with deciduous fruits, berries, vegetables, seed crops, and grain, all grown under intensive management. Much of the land is planted to hay and grain for dairy and poultry feed. Most of the crop depends on rainfall, generally 30-45 inches per year, but some high-value crops are irrigated for there is little rainfall in summer.

In Washington there are still 1.5 million acres of high-quality agricultural soils west of the Cascades, located on the valley floors of major streams. These are highly fertile soils of good texture and lie on gentle slopes, not readily erodable. The soils are well suited for the production of row crops; some types producing greater yields than soils anywhere else in the world (Gray, 1975). The Skagit River Valley of northwestern Washington is a large producer of dairy products as well as peas and beans. In Washington the dairy industry is second only to wheat in importance.

The largest and richest of the mountain valleys west of the Cascades is the Willamette Valley of Oregon, following the river by that name that flows from south to north and empties into the Columbia near Portland. At the time the gold rush was attracting most people to California, the 1850 Oregon donation law granted a square mile (640 acres) of land to any married man willing to homestead in the Willamette Valley. Although this liberal land policy lasted only five years, it was long enough to fill the valley with large farms that the settlers used as range and for growing wheat.

The Willamette Valley contains about 1.7 million acres of tillable riverbottom and bench-terrace soils and over a million acres of usable hill lands in the central portion of the valley and along its piedmont borders. Here cane and tree fruits can be grown or improved meadows can be developed. Despite the annual average of about forty inches of rainfall, the dry summers make irrigation valuable for intensive land use and much of the valley is irrigated. Gravity-flow irrigation ditches sometimes serve for drainage during the rainy season. Intensity of farming may be illustrated by green (snap) bean production for processing. In 1976, Oregon produced 136,250 tons on 29,300 acres—4.6 tons per acre. Wisconsin produced 170,250 tons on 69,500 acres—2.5 tons per acre (USDA Agr. Stat., 1977).

In the Willamette Valley, as elsewhere west of the Cascades, dairying is the principal occupation of farmers, but the valley is also well known for its diversified farming: fresh vegetable and canning crops, cane and orchard fruits, filberts, grass seeds, and oil of mint—almost everything that can be grown in the humid eastern United States.

The production of pears has increased in Oregon while apple production has declined. The pear varieties of the state, in order of importance, are Bartlett, Anjou, Bosc, Comice, Seckel. In western Oregon the Rogue River Valley is the principal pear-growing region, although there are some plantings in the Umpqua and Willamette river valleys. Much of the soil in the Rogue River Valley is a rather heavy adobe that is better suited to pears than other fruits. In all but a small percentage of the orchards, the rainfall is supplemented by irrigation. Pear orchards are protected from spring frosts by artificial heating.

## FARMLAND IN AN URBAN SYSTEM

An increasingly pressing modern problem is the function of farmland as it is encroached upon by expanding cities. Its function as the principal source of food must compete with its use as the most easily developed space for further urban expansion. The latter nearly always brings the greatest economic returns. This is a nationwide problem that is intensified in areas where the economic base has shifted from resource-dependent production to more technological industries. Some of the economic and social issues involved, with special relation to the state of Washington, have been discussed by Gray (1975).

Where land may have a market value of approximately $3,500 an acre, high property taxes, and a capital gain and federal inheritance tax structure that limits the transfer of land from one generation to another, the family farm is fast passing out of the picture. Yet much of the charm of these areas for residents and visitors derives from the aesthetically pleasing nature of the agriculture—berry farms, vegetable acreages, fields of flowers, wheat fields, and other crops that all contribute to an ever-changing panorama.

At present a vital consideration, when considering the conversion of fertile agricultural land to alternate uses, should be the high output of such land per unit of energy input. The extra nitrogen required to produce 100 acres of corn in poor Class IV soil, when compared with prime Class II without fertilizer, is more than 2 tons per year, to say nothing of the phosphorus, potassium, and lime also required. To produce 2 tons of nitrogen requires the burning of 76,500 cubic feet of natural gas. (Ammonia nitrogen is synthetically derived by combining atmospheric nitrogen with hydrogen from fossil fuel.) When prime agricultural land is converted to other uses, the yield lost must be replaced through additional application of energy to less productive land (where the housing should be). Generally irrigation is required for the latter, incurring additional energy expenditure, particularly when electrical power is needed to lift water to the new land to be cultivated. Also, in such areas salinity problems are often encountered.

In urban areas, possibly too little attention has been directed toward the ability of agricultural land to degrade certain organic wastes, including both animal wastes and urban sewage sludge. A thousand acres of agricultural land has been estimated to be able to absorb the current volume of sewage sludge from the Seattle metropolitan area. Farmlands could become a valuable resource for disposal of biodegradable waste materials of urban areas (Gray, 1975).

## PUBLIC UTILITY DISTRICTS

People in the state of Washington have accomplished much to prevent environmental deterioration and impoverishment of land, and to restore those formerly impoverished, through the creation of "public utility districts" (PUDs). A PUD can embrace areas ranging from a part of a county to several counties. It has the right to own land, sell revenue bonds, generate and sell electricity, and devote the profits in part to bond retirement and the remainder to finance schools, hospitals, libraries, land

*Fig. 26. Deciduous-fruit orchards in Hood River Valley, Oregon, with Mount Hood in background (courtesy Joseph E. Kollas).*

reclamation, reforestation, and other essential facilities and services. Formerly depressed areas (e.g., Chelan and Grant counties) have become areas of environmental restoration, opportunity, and growth. The suggestion has been made that Appalachian states, plagued by despoliation by absentee corporations, could benefit by establishing PUDs in their own areas (Caudill, 1975).

## EAST OF THE OREGON CASCADES

On the east slopes of the Cascades, in Washington and Oregon, more than three-fourths of the land is in forests which serve for wildlife habitat and recreation and there is also much lumbering because of the large stands of Douglas-fir and ponderosa pine. The parklike woodlands of the lower elevations have an understory of grass, sagebrush, and bitterbrush where there is some grazing. In the valleys, perhaps 5 percent of the area, there is irrigated cropland with mostly grain and forage crops. The Hood River Valley of north-central Oregon (fig. 26) is famous for its pears and apples. This beautiful valley is in the Cascades, about 120 miles from the coast and at the extreme northern boundary of Oregon. During the summer, rainfall is light and practically all orchards are irrigated. The valley cannot be considered strictly either west or east of the mountains. However, the mountains on the east side are somewhat lower than those on the west and the valley has more warmth and sunshine in summer than the areas west of

*Fig. 27. A center pivot sprinkler in a wheat field near Klamath Falls, Oregon (courtesy USDA Soil Conservation Service).*

the Cascades. (As already related, equally well known for fruit-growing are the Yakima and Wenatchee valleys on the east slopes of the Cascades in Washington.)

About two-thirds of Oregon lies east of the Cascades. Except for the Blue Mountains in the northeast, this area is relatively arid, with annual rainfall mostly from 5 to 15 inches. Oregon's portion of Region B (fig. 3) is largely devoted to range or cattle production. In some of the higher rainfall areas, dryland wheat is grown in a rotation of wheat one year and fallow the next. At higher elevations in the eastern portion of the region, wheat is grown on an annual basis. Some irrigated lands along the Columbia River are now producing alfalfa and potatoes. Oregon's portion of Region D (fig. 3) is a higher plateau, largely devoted to range supporting cattle and some sheep. Meadow hay is harvested on some lowlands that are flooded by runoff from melting snow. Included within Region D are some productive irrigated regions in the Klamath basin that produce a wide variety of crops, including spring barley, winter wheat, potatoes, and alfalfa (Harold Youngberg, correspondence).

Figure 27 shows a "circle sprinkler system" irrigating wheat ten miles southeast of Klamath Falls, Oregon. There are thirteen "stations," each with its own drive system and wheels. The sprinkler system can be set to irrigate 138 acres either in 11 hours or 22 hours. The water flow is a heavy mist.

# 8 The Great Basin and the Southwest

*Our story has been peculiarly the story of man and the land, man and the forests, man and the plains, man and water, man and resources... the ever beckoning and ever-receding frontier left an indelible imprint on American Society and the American character.*

—John F. Kennedy

The Great Basin is that large, roughly triangular area of interior drainage that includes western Utah, Nevada, eastern California, and small parts of Wyoming, Idaho, and Oregon. It is bounded on the east by the crests of the Wasatch Range in Utah and on the west by the crests of the Sierra Nevada in California. It is about 800 miles wide and covers more than 200,000 square miles. It was so named, following John C. Fremont's survey of 1843, because it has no outlet to the sea. Many streams that flow down from the mountains may reach shallow, permanent salt lakes, the largest of which is the Great Salt Lake of Utah into which the Bear, Weber, and Jordan rivers flow. The smaller streams terminate in *playa* lakes that contain water in the spring but dry up in summer. The salts deposited by the evaporating water form a thin white crust of alkali on the surface of the ground. Many streams simply disappear into the desert without reaching even a temporary lake.

The rest of the area to be considered in this chapter includes Arizona and the part of New Mexico west of the Rocky Mountains. In Arizona the "Mexican Highland" stretches diagonally from northwest to southeast and varies from 60 to 150 miles in width. South and west of this mountainous region is the Sonoran Desert, which has many mountain ranges but of lower altitude than those of the Mexican Highland. An important physical feature of the Sonoran Desert is the Gila River, which arises in the highland and flows into the Colorado. Ninety-five percent of Arizona comprises a drainage basin of the Colorado.

The entire northeast section of Arizona—about 40 percent of its area—comprises the Colorado Plateau. The plateaus of this region range from 5,000 to 9,000 feet in elevation. The Colorado Plateau includes the drainage basin of the Little Colorado River and the Grand Canyon area. Much of its southwest edge is marked by the Mogollon Rim and the Mogollon Mesa. The colorful plateaus, mesas, and canyons of the Colorado Plateau have a unique beauty and are of great geological interest.

The part of New Mexico west of the Rockies resembles in topography and climate the adjacent areas of Arizona and need not be treated separately.

## CLIMATE

Moisture-laden air in the region discussed in this chapter comes from two sources: the Pacific Ocean and the Gulf of Mexico. In the winter, moist air passes inland from the ocean, depositing considerable moisture in the form of rain and snow on the high mountains as it rises against the many escarpments of the rugged terrain, but very little to the leeward of each escarpment, which therefore is a true desert.

In the Great Basin the infrequent rains come mostly in winter, although there are also some summer thundershowers, especially in the mountains. Annual rainfall is usually less than 10 inches and seldom more than 15 inches. In most of the region the valleys are at 3,000-6,000 feet elevation and relatively cool, when compared with the hot Sonoran desert lands to the south.

Moist air from the Gulf of Mexico flows westward across southern Texas and occasionally penetrates the deserts of New Mexico, Arizona, and even southeastern California. Short-lived, torrential rains result from the convectional currents that develop as the result of expansion of the air mass as it passes over the hot desert surfaces. The driest area is southeastern California and southwestern Arizona, a large region that lies between the two rainfall zones and receives very little precipitation from either (Logan, 1961). This is the region that includes the agricultural lands of the Salt River and Gila River valleys of Arizona, the lower Colorado River valley, and the Palo Verde, Imperial, and Coachella valleys of California. It is the only agricultural area of the western desert that comes close to being tropical, with January minimum temperatures normally above freezing. The plateaus and higher basins and valleys of the remainder of the desert country have cool to cold winters, with dawn temperatures often approaching 0°F (-18°C).

## VEGETATION

Most of the Great Basin ranges in elevation from 2,000 to 6,000 feet above sea level, the mountains, of course, reaching much higher elevations. Latitude and elevation result in the region being a "cold desert." There are not many cacti. The common sagebrush comprises the bulk of the vegetation. This is an erect, much-branched evergreen shrub three to six feet high, with gray, aromatic foliage. Other common plants throughout the region are greasewood and several species of shad scale (*Atriplex*). The grayish aspect of the desert results from the predominant vegetation, sagebrush and shadscale being gray-green rather than bright green. In the southern reaches of the Great Basin, sagebrush has a grayish lavender tint, accounting for the title of Zane Grey's now ancient but well-remembered *Riders of the Purple Sage*. The height of sagebrush is a good indicator of the quality of the soil. Where it is less than knee-high, the land is said to be not worth farming even if water is available (Gleason and Cronquist, 1965).

Wherever there is sagebrush there are likely to be some perennial bunchgrasses, the blue-bunch wheatgrass (*Agropyron spicatum*) and sheep fescue (*Festuca ovina*). Where an area is overgrazed the grasses tend to diminish and the sagebrush increases. In some areas, particularly in the northern part of the basin, there are places where there is no grass left. There are many perennial flowers in the spring and early summer in the sagebrush regions, but not the brilliant display that one may see farther south in the

Sonoran Desert. The most common species are arrowleaved balsamroot (*Balsamorrhiza sagittata*), with sunflowerlike flowerheads; sego lily, the bulbs of which were used by the early Mormons as an emergency food; many species of Indian paintbrush (*Castilleia*) and beardtongue (*Penstemon*); and lupines and phlox.

Precipitation increases and evaporation decreases with increasing elevation. On the foothills of the mountains of the Great Basin are, first, "desert forests" of junipers; these are shrubs or small trees in the cypress family. There are many species, extensively distributed over the West at intermediate altitudes. Browsing animals, especially goats and deer, frequently feed on the "berries" of the widely distributed "white cedar" (*Juniperus utahensis*), and also feed on its foliage to some extent in extremely cold weather, probably because of its high oil content. Higher up are the nut pines (*Pinus monophylla* and *P. edulis*) with large seeds that the Indians used to supplement their diet and which can now be purchased locally as piñon nuts. In Utah at these altitudes the land is often covered with dense thickets of scrub oak. At elevations above the desert-forest zone the mountains support coniferous forests and aspen groves similar to those of the Rocky Mountains.

The Sonoran desert region extends from the Great Basin region southward into Mexico and also eastward to include a considerable portion of western Texas. In the lower elevations there are forms of plant life strange and bizarre to persons who may be seeing desert flora for the first time, but the plants are all well adapted to the hot, dry climate. Characteristic vegetation throughout this vast, sparsely inhabited region includes creosote bush, saguaro, cholla, mesquite, paloverde, ocotillo, and other desert plants with melodious Spanish names. The saguaro can be as much as 50 feet in height —the largest of all cacti—and is most abundant in southern Arizona and northern Sonora, Mexico. The ocotillo is a tall shrub, 8-25 feet, with a cluster of slender branches arising from a common root crown and terminating in conspicuous flame-red flower clusters in March and April. Some years, when there is a heavy rainstorm at the right season, seeds will sprout that may have remained dormant in the ground for months, or even years, and the desert floor is then covered with a spectacular carpet of blossoms. If more than one inch of rainfall has accumulated in November and December, one may also be sure of splendid displays of blossoms of perennial plants. The North American deserts support more than 5,000 kinds of wild flowers.

A previously little-known desert shrub, the jojoba (*Simmondsia chinensis*), is now prominently in the news. The jojoba, also known as goat nut, is a rigid much-branched shrub ranging from two to seven feet in height and is common in the arid hills of the deserts of southern California, Arizona, and south into Mexico. From the acornlike seed of the female plant may be obtained an oil which, when refined, is remarkably similar to a heretofore unique lubricant for which the sperm whale has been hunted almost to extinction. Sperm-whale oil has been indispensable for making steel, leather, and textiles and for lubricating high-speed machinery and precision instruments. Treated with sulfur, it can survive the tremendous pressures and the heat and wear caused by friction between metal parts and grease. It is used for making automobile transmission fluid. Jojoba might become a useful crop in areas of the world where agriculture is at best marginal (Wolff, 1975; Yermanos, 1974).

To many Easterners the term *arid Southwest* brings visions of an immense, inhospitable desert with here and there an oasis artificially created with the aid of irrigation. Yet the Southwest has millions of acres of forests. Compared with certain other regions, their commercial value for wood products is relatively small, but from the standpoint of irrigated agriculture their value as watershed is enormous. Enough water to irrigate an acre of desert lowland is supplied by about seven acres of high-mountain watershed. Irrigated land in the arid West produces about 150 times more forage per acre than open range. Therefore it is particularly important here to avoid abuse of high-mountain vegetation. The trees, shrubs, and grass hold back much of the water, allowing it to percolate to subterranean streams and storage areas from which it can be pumped or obtained from artesian wells. Deforestation and overgrazing lead not only to decreased value of the watershed but also to erosion of lower rangelands and increased silting of reservoirs. Moreover, mountain soils are thin and the infrequent brief but pelting rains of the Great Basin and Southwest remove topsoil not protected by vegetative cover (Higbee, 1957).

In central Arizona a lofty 200-mile-long barrier known as the Mogollon Rim extends from around Flagstaff in a southeasterly direction. It rises up from the desert to form an immense plateau 7,000 feet or more in elevation. In New Mexico it merges with the southern Rockies. Throughout this plateau and mountain area there is enough rainfall to sustain fine stands of timber. The predominant forest tree is the ponderosa pine. Although this is a large tree, relatively limited rainfall results in it being widely spaced in pleasant, sunny, parklike woods. At higher elevations, and with greater rainfall, there are also stands of Douglas-fir, Engelmann spruce, white fir, and aspen.

## INDIAN AGRICULTURE

The Indians of the Southwest date back at least as far as the Folsom Man and Sandia Man of New Mexico—about 23,000 years. The pre-Columbian inhabitants of the region used at least 225 different species of wild plants for food and in other ways, about 15 percent of them being major food sources. Mesquite beans were highly nutritious. Those of *Prosopis juliflora* had a protein content of from 34 to 39 percent, equivalent to that of soybeans. The Indians also used screw beans, agave, yucca fruit, prickly pear fruit, piñon nuts, and the seed of chia sage, the latter ground and cooked into a porridge called *pinole* (Meyer, 1975). The giant saguaros served as water towers. Rats, rabbits, deer, lizards, and insects were consumed (Bowden, 1975).

The Indians used to twist into thread the fibers from the pods of a plant we now call cotton. They wove the thread into cloth for clothing and blankets. Fibers of other plants, such as yucca, were used in the same way. Reeds were woven into baskets used to transport seeds, nuts, and fruits. Some were woven so tightly and perfectly that they could hold water. Clay was utilized for building houses and for making pots and dishes used for cooking, and for storage of food and water (Mann and Harvey, 1955).

Indians were the first agriculturists in the area. When corn arrived from the south, it was the first crop that could be planted and cultivated and the first that was sufficiently productive and nutritious to provide much of the food for sustaining life. Corn, beans, and squash are believed to have been in New Mexico as early as 2,000 B.C. and were being raised as food crops in that region (Higbee, 1957; Fergusson, 1964).

Something had to be devised for grinding corn. An early device was a stone with a depression on the surface and a smaller stone that could be held in the hand to do the pounding and grinding. Later these evolved into the flat metate and mano in use today. A planting stick was shaped for making holes in the ground in which to put seed; the Indians had become farmers. No longer needing to follow game herds, they could develop permanent villages. In fact, when man started planting seeds, he had to stay in the area until time for harvest; he had to stop wandering and start building homes. He continued to hunt, but had a home to which he could return. The first homes built by Southwest Indians were circular pits dug into the ground and covered with logs, brush, and earth. Remains of these pit houses can be seen in many places in Utah, southern

Colorado, New Mexico, Arizona, and as far south as Chihuahua in Mexico. Beams that were cut to support these pit houses date back to as early as A.D. 217 (Peck, 1961; Fergusson, 1964). As early as A.D. 1000 groups of Indians progressed from pit houses to large permanent apartment dwellings in genuine cities (Mann and Harvey, 1955).

The importance that corn assumed in the lives of the Indians is indicated by the role it played in their religions. Their myths relate that in the beginning First Man was created along with an ear of white corn and First Woman was created along with an ear of yellow corn (Terrell, 1972). To this day corn is the most sacred plant of the Pueblo Indians. Once-ground coarse cornmeal is strewn on and around the participants of ceremonial dances, over prayer sticks and feathers, tossed toward the sun at dawn, rubbed on the newborn child, and dropped on the faces of the dead. When the religion of the Indian became blended with that of the Spaniard, the Pueblo child might be sprinkled with holy meal according to tribal custom and sprinkled with holy water by a Catholic priest (Mann and Harvey, 1955). Estimates of the amount of corn now used for ceremonial purposes range from 50 to 80 percent of the crop!

The religious significance attached to corn by the Hopis was characteristic throughout the New World, and was prominent in the rites of the early Mayas. The notoriously ruthless Franciscan Bishop de Landa failed to stop the Mayas from worshiping the "Earth Mother" and her son, the young "Green God," a deification of the maize plant, despite a campaign of cruel suppression. Never was worship of corn extirpated among the Indians, whose word for corn everywhere signified "Our Life," or "Our Mother," or "She Who Sustains Us." Linnaeus was aware of this when he chose the word "zea," meaning "the cause of life," for the genus of the fabulous New World plant. *Zea mays* might be translated "That which caused the Maya to live" (Giles, 1940).

## Hohokam Indians and Irrigation

The Hohokam Indians were believed to have migrated from Mexico into what is now Arizona as early as about 300 B.C. They brought with them water-control technology, including well-digging, and an associated tillage economy involving corn, beans, squash, and cotton. They built large sunken-floored houses for extended-family living in established villages (Haury, 1976).

In the dry but fertile valleys of the Salt and Gila rivers of Arizona, the Hohokams, with neither draft animals nor tools other than those they could fashion from wood and stone, eventually excavated great irrigation ditches 30 feet wide and 10 feet deep. In the Salt River Valley at least 150 miles of irrigation canals were constructed. Some of the modern irrigation canals follow courses set by these ancient ditches, for even with his best surveying instruments, modern man cannot find a better course to follow. As in other areas of the world, so in the American Southwest, it was irrigated agriculture that enabled prehistoric peoples to develop the populous towns, and to obtain the security and leisure time necessary for the development of ceramic art, textile design, trade, and government (Turney, 1929; Higbee, 1957; Terrell, 1971).

Shortly after A.D. 1400 the canals were abandoned and the Hohokam people disappeared, no one knows why. They are known to the present Salt River Indians only as legendary people, and in fact Hohokam means "Those Who Have Gone" (Terrell, 1971). It is likely that problems of salinity and waterlogging resulted in the abandonment of the Hohokam region. The same problems occur in the same area today, but modern farmers have pumps and other means of getting rid of surplus water (Peck, 1962).

### The Hopi Indians

Among the most progressive agricultural tribes were the Hopis. In the deep and intricate canyons far to the north of the present Hopi Indian reservations, large, shallow caves occurred in the canyon walls, with level floors and arched roofs. On the floor of the canyon was tillable land and there were springs nearby. These caves provided high and dry locations for groups of houses or even sizable villages—the cliff dwellings. A prolonged drought that prevailed in northern Arizona for 25 years, ending in A.D. 1299, forced the cliff dwellers to their present location farther south.[1]

Nearly three-fourths of all cultivated land on a Hopi Indian reservation visited by retired entomologist Walter C. O'Kane in 1950 was planted to corn (O'Kane, 1950). The main corn crop was likely to be the white-eared variety, but yellow-eared, red-eared, and blue-eared corn was also raised, the latter for making piki. When preparing piki, ashes made from sage are mixed with the cornmeal. Sometimes tumbleweed ashes are allowed to steep for a while in water and the latter are used in preparing cornmeal batter. This is poured on a flat stone heated to just the right temperature and greased with burro fat. (Mutton fat would cause a bad flavor!) The batter is spread over the stone by hand into a very thin layer to bake. When an edge of the baked bread begins to curl up, the sheet is transferred to a cloth and rolled up into a cylinder. The different colored corn kernels have different flavors and each is adapted to its own cookery. A young Hopi woman had to become proficient at making piki before any man would marry her.

The Hopis developed a special variety of corn that could be planted at depths ranging from 8 to 18 inches and still reach the surface. The seed must be planted so deep that it will be in contact with enough moisture to germinate at the time of planting (May or June), when there is little or no rain. Like their remote ancestors, the Hopis visited by O'Kane used a planting stick, thrusting it into the ground to the desired depth, then dropping twelve kernels (a multiple of the magic four of the Maya) into the hole. The Hopi corn shoot could reach the surface and proceed with normal growth. Moisture is best preserved when the surface of the soil is covered with sand. Flood irrigation was an ancient art for the Hopi Indians, but only in a few lateral canyons is it now possible. The main watercourses have become too severely eroded because of overgrazing of the watersheds since the early 1900s.

O'Kane found that Hopi farms might not be impressive to some observers, but actually were productive. Corn plants were widely spaced and not more than three feet tall, but bore many ears within a few inches of the ground. The low, spreading plants shaded the ground and preserved moisture. O'Kane was given an ear twelve inches long and seven inches in circumference. It had sixteen rows of deep, heavy kernels and more than fifty kernels to the row. Squaw corn on the Navajo Indian reservation in Arizona, and severe soil erosion, are shown in figure 28.

O'Kane also observed good harvests of beans, squash, melons, and peppers, and peach and apricot trees with excellent fruit, in the land of the Hopis. The prehistoric Hopis developed a distinct type of cotton that matured in a short growing season. They ceased to grow cotton when goods became available at the trader's. Land formerly planted to cotton could then be better devoted to raising food.

In early May of 1977 a Hopi Indian demonstrated to me his method of planting corn, using a pipe, flattened on one end, as a "planting stick." He made a hole with an area

[1]The archaeology of the Indians of the Southwest, from the time of the archaic Anasazi hunters and gatherers, through the various Basketmaker and Pueblo cultures, and up to the time of the arrival of the Spaniards in 1540, is excellently depicted in the dioramas in the Museum of Northern Arizona in Flagstaff.

*Fig. 28. Squaw corn on a Navajo Reservation; note severe erosion (courtesy Branch of Land Operations, Bureau of Indian Affairs).*

of about 4 square inches at the bottom and dropped in 12-15 seeds. (The Museum of Northern Arizona says 6-12 seeds are commonly used.) This is a considerable departure from the traditional 4 seeds used by Indians in most places, a practice dating back to the ancient Mayas. However, traditionalists use some multiple of the magic number 4, generally 8 or 12, as was also the practice among the Pimas and Papagos (Castetter and Bell, 1942). My host explained that many plants are destroyed by cutworms but that the larvae are then dug out by hand and destroyed.

The wind blows much of the time on the northeast Arizona plateau. During my visit, wherever a little patch of land was cultivated one could see a plume of dust rising. In areas of considerable cultivated land, as at Many Farms along Highway 63 in the Navajo Reservation, the dust cloud was so dense that headlights were required in mid-day. Probably the traditional way of dropping seeds in holes made with a planting stick, then weeding with a wooden hoe, caused far less environmental degradation and accounted for the centuries of sustained corn culture. Livestock are now the only agricultural money crop. No dust rose from the ranges, even though they were said to be severely overgrazed, principally by sheep.

## Apaches and Navajos

The Apaches and Navajos came down from northwestern Canada and Alaska about 2,000 years ago, but arrived in the Southwest only a short time before the first Spanish explorers. They learned to grow plant crops from the Pueblo Indians, but also hunted and gathered wild plant foods. They dominated and terrorized enormous areas of

Texas, New Mexico, Arizona, and northern Mexico. They were a warlike people often fighting with warlike neighbors or preying upon the peaceful Pueblo Indians, who practiced agriculture much as they are doing now. The marauders obtained much of their subsistence from the Pueblos and later from the Spanish and Anglo settlers. They chose to destroy the smaller Pueblo villages and, as a result, other Pueblos of the vulnerable valley floors were forced to build their villages high up on the cliffs of the mesas.

After the horse was introduced upon this continent by the Spaniards, wild horses spread rapidly and were used by the Indians of the Southwest even before the Spaniards arrived in that area. The horse increased the capacity of the Apache and Navajo to terrorize the other Indian tribes of the Southwest for about two centuries before the arrival of the Anglos. Then they fought long and bloody campaigns against the latter before capitulating (Terrell, 1971).

Today extensive reservations include large areas of good high-mountain summer grazing country and lowland areas in which the Apaches can keep their cattle during the winter. The former hunting and gathering culture of the Apaches was exchanged for a commercial pastoralism in which their love of horsemanship and the open country could be retained unabated. The Navajos have retained their tribal tradition of raising large flocks of sheep and goats and many horses. The ranges have a very low carrying capacity. With the increase of herds and flocks that accompanied the great increase in the Navajo population, there followed much overgrazing, range destruction, and poverty.

The Apache Indians were not particularly noted as agriculturists but in fact had considerable acreages of "Indian corn." In his *Apaches and Longhorns,* Will Barnes (1941) tells of having found a walled-off cave near Fort Apache, Arizona, in 1883. Inside were thirty wicker baskets ranging from one to five feet in height that had been filled with corn "in the common red, green, blue, and yellow colors." Weevils had devoured all but the hulls of the kernels and the cobs in each basket. The plan was to ship the baskets to the Smithsonian Institution in Washington, but it developed that the corn had been a funeral cache placed in the cave years before upon the death of a beloved member of the Apache tribe. The Indians were so outraged that the military at Fort Apache decided it would be expedient to replace the baskets and reseal the cave entrance.

Besides herding livestock, both the Apaches and Navajos now have a considerable acreage of plant crops.

### Indians of the Great Basin

While the Indians of the Great Basin were generally considered to be culturally impoverished, this was no doubt the result of the meager resources of the arid region. Where local circumstances permitted, such as in fertile river valleys, such cultural indicators as irrigated agriculture were evident. Fray Escalante (see following section) noted in his diary on October 14, 1776, that on the Pilar River (now Ash Greek) near its junction with the Virgin River 25 miles below Zion Canyon, he found a well-made platform with a large supply of ears of corn on it. Along the river bank were three small cornfields with very well-made irrigation ditches. Escalante related further:

> From here down the stream and on the mesa and on both sides for a long distance, according to what we learned, these Indians apply themselves to the cultivation of maize and calabashes (probably squashes). In their language they [the Indians] are called Parrusis. [Woodbury, 1950]

A half-century later, Jedediah Smith found Indians raising corn and squashes in southern Utah, along Santa Clara Creek, which in his diary he called "Corn Creek" (Sullivan, 1934).

## EARLY SPANISH EXPLORERS AND MISSIONARIES

Much of Arizona and New Mexico, as well as the Texas Panhandle and southwest Kansas, were penetrated by the expedition of Francisco Vásquez de Coronado, as early as 1540, in vain quest of cities of gold (Hammond and Metz, 1978). Don Juan de Oñate, a wealthy mineowner in Mexico, at his own expense equipped an army of about 400 soldiers and settlers, a supply train of 83 wagons and carts, and 7,000 head of stock, leaving a village in southern Chihuahua, Mexico, on January 30, 1598. Moving up the west side of the Rio Grande, Oñate halted on April 30 to build an altar and proclaim possession of *Nuevo Mexico* for Spain. This was nine years before the English founded Jamestown. In 1599 Oñate designated San Gabriel, near the present town of Chamita, as New Mexico's capital city; Santa Fe, the present capital, became so in 1610. Like Coronado, Oñate found no gold, but was instrumental in establishing the Catholic religion and a mission system in New Mexico. His last journey was across Arizona to the Colorado River and down the Colorado to the Gulf of Mexico, experiencing the same hardships in crossing the desert as American explorers and mountain men were to experience two centuries later (Mann and Harvey, 1955).

Where the Spanish explorer Fray Marcos de Niza first entered the Pima Indian country of southern Arizona in 1534, he found Indians farming the land on both sides of the Gila River. The Indians diverted water from the river with their dams of brush, rocks, and mud and brought water via hand-dug canals through the flat land of the valley (Fey and McNickle, 1959).

An explorer and missionary with great influence on early Arizona agriculture was Father Eusebio Francisco Kino. Born in Austria in 1644, he taught mathematics at the University of Ingolstadt in Bavaria. He visited the first Indian village in what is now Arizona in 1691. He also founded a mission at San Cayetano de Tumacacori, a short distance above the present Mexican border in the spacious Santa Cruz Valley. Kino then gradually worked his way northward as far as the Colorado. He explored, made maps, preached, and gave agricultural instruction to the Pima and Papago Indians. Kino was impressed by the flourishing fields irrigated by the friendly Pimas in the Gila Valley. He was shown a massive tower with ruins of houses around it—the ruin of a prehistoric Hohokam town center. He named it Casa Grande, the name it retains to this day (Peck, 1962). When the Pima Indians obtained wheat and barley in 1697, they immediately recognized their value and began planting them and storing surpluses[2] (Fey and McNickle, 1959). Within fifteen years Kino had brought horses, mules, sheep, and cattle into Arizona. He taught the Indians how to care for animals so that the herds would increase and how to butcher and prepare the meat for eating. He introduced various grains and fruits and showed the Indians how to plant, harvest, store, and cook them. His principal objective was to save souls, but he found that they were easier to save when their possessors were well fed (Faulk, 1970).

[2]During the war with Mexico, when General Stephen Kearny marched his troops from Santa Fe to San Diego, they were fortunate to obtain much-needed food supplies from the friendly Pimas and admired their civilization and good farming. The same was true of Colonel Philip S. Cooke's famous Mormon Battalion, which passed along the Gila River the same year, building the first route for wheeled vehicles to cross southern Arizona (Peck, 1962).

After Kino's death there followed a period of deterioration in relations between the Spanish missionaries and the Indians, exacerbated by the passage of control from the Jesuits to the Franciscans. This difficult period lasted until the arrival of Father Francisco Tomas Garcés in 1768. Garcés rivaled Kino in ability to gain the loyalty and affection of the Pimas. It was Garcés's ability to communicate successfully with the Yuma Indians that ensured the success of the famous 1773 expedition of Captain Juan Bautista de Anza, from Mexico to California, of which more will be related in the following chapter (Peck, 1962; Faux, 1970).

In Arizona, mining was originally a far greater attraction than farming. Nevertheless by 1804, 4,000 cattle, 2,600 sheep, and 1,200 horses grazed at Tubac. There was also at the time considerable native agriculture and cloth was being made from cotton grown by Indians. Then began the era of large land grants, such as Canoa, Sonoita, and San Bernardino (Faulk, 1970). Yet Arizona remained primitive and neglected between two thriving provinces—California and New Mexico. The only towns were Tubac and the walled presidio of Tucson. Starting in 1810, garrisons on the frontier were greatly reduced or abandoned, for the Spaniards needed all available soldiers to fight revolutionaries in Mexico. Then the Apaches emerged from their mountain strongholds along the frontiers of New Mexico, Arizona, and Sonora, destroying the ranches, driving off cattle, and almost exterminating the white population. Tumacacori was abandoned. It has since been preserved as a national monument and the buildings have been repaired as much as possible, using the old records as a guide (Peck, 1962).

Arizona came under Mexican rule in 1821. During the brief Mexican regime, Indian hostilities were resumed from time to time against Tucson and Tubac, but settlers could take refuge within the walls of the presidios and repel attacks by as many as a thousand Apaches. In the 1830s Sonoran officials, who controlled the Arizona area, resorted to the payment of bounties for Apache scalps: 100 pesos (ca. $100 American) for the scalp of an adult male, 50 pesos for the scalp of a female, and 25 pesos for the scalp of a child. However, Indian scalps could not be distinguished from the scalps of Mexicans and entire Mexican villages were depopulated by unscrupulous bounty hunters. By the end of the 1840s the population of whites and Christian Indians was only about 600 and the "wild" Indians were in almost complete control (Faulk, 1970).

The first white men to reach the heart of the Great Basin were Spaniards in the Dominguez-Escalante expedition, which left Santa Fe on July 29, 1776. Father Escalante kept a journal of the expedition's travels that was so vivid, detailed, and accurate as to leave no doubt as to the route taken. The expedition made its way north and northwest through Colorado, then veered westward to cross the Continental Divide at the head of Spanish Fork Canyon, thence to the east shore of Utah Lake. Traveling south through Utah and southeast through northern Arizona, the party reached Santa Fe on January 2, 1777, having been gone over five months (Hunter, 1940).

After the Arze-Garcia expedition to Utah Lake, in 1813, the Spaniards, and later the Mexicans, traded with the Paiute Indians, the principal article of trade being Indian women and children taken south as slaves.[3] There is a strong infusion of Indian blood among modern New Mexicans.

[3]In 1850, when the first Mormon exploring party reached an Indian village on the Santa Clara River, near the present city of St. George in southern Utah, they found almost no women and children; most of them had been sold to Spanish traders. The Mormons legislated against the slave trade and fixed severe penalties for offenders (Hunter, 1940).

## MOUNTAIN MEN

Important in the exploration of the West were the beaver trappers, along with prospectors, traders with Indians, or merely footloose adventurers. These were reckless, audacious, and colorful characters who became known as "mountain men." They were the pathfinders who discovered the routes through the West that were used by later arrivals; in the Southwest they shared this honor with the early Spanish explorers. Some became invaluable as guides for the early immigrant groups.

Most of the mountain men supported themselves by trapping or trading with the Indians for beaver pelts. Beaver fur was prized in European markets for its beauty, durability, and warmth. Much was used in making "beaver hats," while those were in fashion. The fact that the hats went out of style by 1840 probably saved the beaver from extinction in North America. Trappers of beaver, starting with the first New England settlers and the Indians with whom they traded goods, employed the "strip-and-run" method of all the early exploiters of natural resources in this country—exploiters of wild animals, forests, petroleum, minerals, and topsoils—in the belief that the resources of America were inexhaustible, or in what Stewart Udall (1963) called the "myth of superabundance." Many streams that once reached the plains were dried up when the beaver were trapped out.

Mountain men were irresistibly drawn to a life of peril and adventure in the vast, practically unknown western wilderness—a brief life in most cases, recklessly lived and usually ending in violence. Without them the advances of the miners, cattlemen, sheepherders, and homesteaders into the lands beyond the Rockies would have been greatly delayed, as would also the establishment of the great trailways to the West Coast.

The true stories of the mountain men, as researched and interestingly narrated by a number of writers, are beyond the wildest fiction. Americans are likely to connect the fur trade with place names in the northern Rockies such as the Missouri, Yellowstone, Snake, Green, and Columbia rivers, the Wind River Mountains, Jackson's Hole, and Forts Laramie, Bonneville, and Hall. It was left to the historian Robert G. Cleland, in *This Reckless Breed of Men* (1950), to point out that beaver, the principal objective of the trappers, were also found in large numbers in the lower Colorado, Rio Grande, Arkansas, Humboldt, Sacramento, and San Joaquin rivers and their tributaries. Beaver in the desert rivers had a somewhat lighter fur, but their winter pelts were only slightly inferior to those in the rivers farther north. Trappers and traders opened the great trails of the Southwest—the Santa Fe, Gila, and Old Spanish trails, as well as those that penetrated the western wilderness farther north.

Of the mountain men, the most daring and adventurous were the trappers—such individuals as John Colter, Jim Bridger, Christopher (Kit) Carson, Jedediah Smith, William Sublette, "Peg-leg" Smith, George Young, "Old Bill" Williams, Paulino (Pauline) Weaver, Ewing Young, J. J. Warner, Louis Robidoux, George C. Yount, and William Wolfskill. The last six named were among the few who survived the hardships and dangers of the tumultuous 1820-1840 period. Weaver was to become one of the founders of modern Arizona. Young raised cattle in the Willamette Valley of Oregon. Robidoux, Yount, Warner, and Wolfskill became prominent farmers and ranchers in California.

Trappers led a precarious life, menaced by the perils of nature, including man and beast. Some died from cold, thirst, prairie fires, cloudbursts, snowslides, disease, blood poisoning, or tetanus. In case of gangrene, the trapper's affected limb was cut off with a handsaw or butcher's knife by one of his companions and the exposed flesh cauterized with a hot iron. "Peg-leg" Smith amputated his own leg, the bones of which had been shattered with a rifle ball in an Indian fight. He then fashioned a wooden leg out of the limb of an oak tree. The distinguished author, Francis Parkman, Jr., described a trapper with chest marked with the scars of six bullets and arrows, an arm broken by a shot, and one of his knees shattered, but who continued to follow his perilous occupation. Parkman believed that "there is a mysterious, resistless charm in the basilisk eye of danger, and few men perhaps remain long in that wild region without learning to love peril for its own sake, and to laugh carelessly in the face of death" (Parkman, 1857).

The trapper's greatest menace was the Indian, and the blood feud between trappers and Indians took a heavy toll of the lives of both. The trapper had a superb knowledge of woodcraft and self-defense and was remarkably resourceful and skillful with his rifle and tomahawk, these qualities accounting for his survival sufficiently long to be recorded in history. He might do his trapping alone, in small groups, or in extended expeditions—the famous explorations. When alone, he chose as the base of his winter operation a remote and hidden area, where he hoped to be out of sight of roving Indians. His lodge consisted of skins extended over an arched framework of slender poles that were bent in the form of a semicircle and with their extremities in the ground. The dirt floor of the lodge was covered with reeds, dried grass, or small evergreen boughs, on which fur robes and heavy woolen blankets were spread.

Grizzly bears were second only to the Indians as a menace. Grizzlies occurred in what now seem incredible numbers and appeared to be fearless and truculent, at least in those early days. A grizzly was difficult to kill unless shot through the head.

As an example of the energy and ambition of the American trappers is the story of the first expedition of the party headed by Jedediah S. Smith (Cleland, 1950). Smith had barely escaped death at the hands of Indians and was horribly mutilated by a grizzly. The bear left an ear and part of the scalp hanging from Smith's bleeding skull. One of his men stitched up Smith's lacerated skin with needle and thread and Smith was soon on his way again. He organized an expedition of eighteen trappers and adventurers, leaving from near Salt Lake, Utah, on August 26, 1826, to blaze a new "Wilderness Trail" southwestward to the Pacific. Suffering great hardship and subsisting at times on the "dry, tasteless, leather-like meat of their dying horses," the intrepid band of mountain men reached the Franciscan mission of San Gabriel in southern California on November 27, 1826. The historic implications of the trip could hardly have been envisioned by the small band of uncouth, half-starved trappers, devoted brown-robed friars, and some 2,500 Indian neophytes.

Of special interest to Californians was William Wolfskill, who survived the frenetic period of the beaver-fur trade to become the founder of the southern California commercial citrus industry. Wolfskill was the grandson of Joseph Wolfskill (Wolfskehl), the youngest of seven brothers who, because of their height and physical fitness, had been impressed into the service of Frederick the Great as members of his famous Potsdam Regiment. Joseph fled to America in 1742 to escape this duty and to share the perils and rewards of the American frontier. His grandson William, born in Boonesborough, Kentucky, on March 20, 1798, was well schooled in the methods of hunting, trapping, plowing, planting, and the raising of livestock, along with intermittent instruction in reading and writing, in the Missouri wilderness. By the time William was twenty-three, St. Louis had become the home of the principal fur companies and the base for outfitting western expeditions. Wolfskill decided to carry on his family pioneer tradition by following the fur trade into the unexplored areas of the great Southwest (Wilson, 1965).

Wolfskill found it expedient to become a Mexican citizen and a Catholic in order to continue his trading and trapping in Mexican territory. In September 1830 he and

George C. Yount organized an expedition to California. The route they chose was northwest from Santa Fe through southwest Colorado to central Utah, then in a generally southwestern direction through southern Nevada, then over the Mojave Desert, through Cajon Pass, and on to the ranch of Antonio María Lugo, which was reached on February 5, 1831. The party then went on to Mission San Gabriel, where they spent two weeks. Wolfskill's route was the first one feasible for wagon trains for the entire distance from Taos, New Mexico, to the Pacific. It became known as the "Old Spanish Trail" because it was considered to be a continuation of an old trail used by the Spaniards to the Utah Indian country (Wilson, 1965).

In the drama of the American West, the cattlemen, cowboys, sheepherders, prospectors, and homesteaders have their modern counterparts. But the brave and hardy trappers played their exciting and colorful role in the drama and then left the stage without leaving a trace—submerged in the tide of history. The story of the mountain men is like a strange, wild, improbable myth that had become reality for a moment in history and then again became myth, to tantalize the imagination of those of us who love freedom, adventure, and wilderness.

## POWELL SHAPES ARID LANDS POLICY

Although the hardy, fearless mountain men were given credit for having "learned almost everything of importance about the geography of the West" in the period 1820-1840 (Chittenden, 1902), the region's most important river, the Colorado, remained largely unexplored. The lower Green and upper Colorado rivers and a fairly wide strip of territory on either side of them, for a distance of about a thousand miles from northern Utah to western Arizona below Grand Canyon, were designated as "unexplored" in official Land Office and War Department maps. It remained for John Wesley Powell, a largely self-educated geologist, paleontologist, and ethnologist who would eventually excel also in physiography (a science for which he laid the foundation), topographical mapping, hydrology, soil classification, land reform, and land laws, to undertake one of the most daring and hazardous of all the stirring feats of exploration that fill the pages of western history.

"Wes" Powell had four specially designed boats, with watertight compartments fore and aft, constructed in Chicago and shipped to Green River, Wyoming. The Union Pacific Railroad had only shortly before reached this point on its historic race to meet the Central Pacific, which was then moving eastward from California. There Major Powell launched the four boats with a party of ten men on May 24, 1869. Except for three men who had defected and were soon after killed by Indians, Powell's party survived the acute suffering and many accidents of the perilous journey. On August 30 the men reached the mouth of the Virgin River, where they found three Mormons and an Indian who had instructions to watch for any fragments or relics of the party which might float downstream. Powell and his brother left the river at that point; others left it at Fort Mojave or continued down to tidewater.

Powell's river trip led to his theory, at that time revolutionary, that the Green and Colorado rivers had at one time flowed through flat country; then, as the earth was gradually pushed upward by internal stresses, to distances of thousands of feet, the rivers with their high content of silt and sand wore their way down to continue in their old "right-of-way." In the Grand Canyon the Colorado had to cut its way deeply into the hard Archean rock.

Powell's explorations and surveys, and his rapidly increasing prestige as a scientist and man of vision, encouraged him to enter boldly into the dangerous realm of land politics, taking on the congressmen who represented the cattle, timber, and mining kings. In those halcyon years for unscrupulous speculators and grafters, huge tracts of land were taken over by land companies operating through homesteaders. Land company employees filed claim to land under the Homestead Act. On these 160-acre tracts they placed 16 x 24-inch model "homes," or even packing cases, to comply with the provisions of the act that required the homesteader to build a domicile on his land. Then after five years, when the land legally became his, the homesteader sold it to the company for perhaps fifty cents an acre, for without water it was worthless. Large areas of valuable timber and grazing lands and watersheds came into the possession of the unscrupulous speculators. Genuine homesteaders were charged outrageous prices for water monopolized by water companies.

According to the Timber Culture Act of 1873, a homesteader could apply for another 160 acres, and it would be deeded to him, provided he set out and cared for trees on 40 acres, later reduced to 10. The intent of the act was to encourage the growing of trees in areas where none existed. If within ten years the entryman fulfilled all requirements as affirmed by affidavits and witnesses, he could have title. Again speculators and cattlemen used dummy entrymen to file claims for thousands of acres. Of 290,278 timber culture filings made before 1921, only 62,265 went to patent and 7,108 had been commuted, strongly indicating the extent to which filings were made merely to hold the land, but without intention to comply with the law (Gates, 1977).

The Desert Land Act of 1877, lobbied for by stockmen's associations, permitted a settler to obtain 640 acres of grassland, presumably enough to enable him to make a living at stock-raising. He needed only to construct, within three years, some kind of gravity system by which water could be brought to his land. Again entrymen were used to obtain the property. It was easy to get witnesses to swear that they had seen water on the land, but the water may have been poured on with a bucket. Again it was easy to obtain entrymen to file claims and to sell the land to cattlemen for a few cents an acre.

Wealthy speculators could purchase land from the Land Office for $1.25 an acre and sell it at any price they chose. Speculators and cattlemen purchased 140 million acres of such land and a great deal of it was sold to farmers for as much as twenty times what was paid for it. Lumbermen likewise obtained large tracts of forest through direct purchase, or through "dummies," in the same way.

Under the Morrill Land Grant Act, the states were granted some 140 million acres of western lands which they could sell, provided they used the proceeds to train young men to be farmers. Much was sold to speculators for as little as fifty cents an acre.

By the time Powell had begun his campaign to save the West from destruction, exploiters and monopolists had obtained over half a billion acres of public domain: the railroads 181 million, the states 140 million, Land Office sales more than 100 million, and land given to Indians and then taken from them and sold mostly to speculators, also about 100 million.

In an address to the National Academy of Sciences in 1876, Powell emphasized that in the West, government land of 160 acres was practically worthless without water and that wealthy individuals and stock companies had already appropriated all the streams. This kind of talk led to praise from scientists, but bitter criticism from railroad interests, mining and timber combines, cattlemen's associations, bankers, the Indian Bureau, and the Land Office. Newspapers, conscious of Powell's great reputation, were glad to carry his story.

In Powell's Report on the *Lands of the Arid Region of the United States* (1878), generally referred to as "Arid Lands," he emphasized that in the enormous arid region between the 100th meridian and the meandering mountain line in Washington, Oregon, and California, more than two-fifths of the coterminous United States, a large population could be supported only by proper and judicious use of its resources of water, timber, and grasslands. This required revolutionary change in national thinking and drastic revision of virtually every land law in the Federal Code. The basic precept of Powell's land philosophy was simply that "right to use water should inhere in the land to be irrigated, and water rights should go with land titles." Without water the 160-acre homestead, or any multiple of it, was practically worthless. (Powell recommended that the smallest grazing ranch, to be profitable, should comprise at least four sections [2,560 acres] of land.) But sources of water were being acquired by wealthy speculators; homesteaders wishing to use the water had to pay dearly for it. Powell advised that monopolies of water supplies be prevented. Moreover, he said, an extensive and equitable use of irrigation water in the West required a system of dams and canals far in excess of the capacity of private enterprise to develop. Federal intervention and aid were required.

In his courageous fight against the firmly entrenched forces of greed, exploitation, corruption, and graft and their political minions, Powell had the support of a wise and public-spirited Secretary of the Interior, Carl Schurz, who had "Arid Lands" printed and then reprinted and widely circulated on Capitol Hill. Nevertheless, Powell's proposals, bitterly opposed by western interests, were shelved in Congress and two decades were to pass before his progressive ideas were supported by appropriate legislative action. In the meantime thousands of homesteaders were being ruined by droughts, cyclones, extremes of heat and cold, blizzards, floods, locusts, chinch bugs, and other natural calamities. And the kings of timber, cattle, and water continued to prosper.

In all fairness one should also bear in mind that the refusal of Congress to recognize the need for large homesteads (e.g., Powell's recommended 2,560 acres) in arid regions tended to force even those seriously interested in ranching to subterfuge. The Desert Land Act of 1877 allowed for 640 acres, but this was reduced to 320 acres in 1890. Through the Grazing Homestead Act of 1916 the rancher could procure 640 acres "provided he could perjure himself in regard to his irrigation works and other requirements" (Webb, 1931).

At Powell's urging, Congress passed a bill creating a United States Geological Survey, with the man he recommended, the famous geologist Clarence King, as director, and a United States Bureau of Ethnology, to which Powell was appointed as director. However, when Powell's long-time friend James A Garfield became president in 1880, one of the first acts of his administration was to recommend Powell as director of the Geological Survey, a recommendation quickly confirmed by the Senate. He also remained director of the Bureau of Ethnology.

As Powell had predicted, a cycle of dry years, beginning in 1887, followed the cycle of wet years that his opponents said would "follow the plow," and the cry arose for government action on irrigation and water conservation. As director of the Geological Survey, Powell was asked, through a Senate resolution, to make a survey of all lands that could be cultivated through irrigation and to select sites for dams from which these lands could be irrigated. Predictably, speculators rushed in to buy land that might be irrigated and sell it or lease it to settlers at a great profit. Congress then withdrew from settlement all lands that might be irrigated as a result of the new irrigation survey. All claims to land filed after October 2, 1888, were canceled and public domain would not be reopened until after Powell's survey was finished. Powell estimated that out of a desert area of 1,300,000 square miles only 150,000 square miles could be brought under cultivation. He pointed out that with irrigation, however, this would add enormously to the wealth of the United States.

Predictably, powerful western political pressure was brought to bear that led to Congress once more opening the public domain to settlement. The $720,000 Powell had sought for his irrigation survey was cut to $162,500 and in 1894, Powell, at the age of sixty, resigned from the Geological Survey. But before Powell died of a stroke on September 23, 1902, President Theodore Roosevelt had sent a message to Congress urging government to take an active part in irrigation in the West and requested all that Powell had recommended ten years previously. "These things take time," said Powell.

A few months before Powell died, Congress had established a Reclamation Bureau, the first step in many vast government water storage and irrigation projects reaching their apex in Hoover Dam, damming up the river on which Powell and his brave companions had risked their lives thirty-three years before the bureau was established. But before that time many irremediable mistakes had been made in western land policy. Many of the disasters Powell had brilliantly foreseen had become a part of historic record.

Theodore Roosevelt's term as president marked the turning point in reclamation development in the United States. The Reclamation Act of 1902 placed the federal government firmly in the business of constructing and maintaining irrigation projects. In 1907 the administration of the act was placed with the newly established United States Reclamation Service, in the Department of the Interior. In 1923 the service was reorganized and became known as the Bureau of Reclamation. The role played by the federal government in the development of irrigation projects gradually increased until during the depression years of the 1930s it was responsible for practically all development. During the 1930s the cost of irrigation development had exceeded the ability of the land to return the cost and leave the farmer a profit. Then the relative amount of irrigated land developed by the government again decreased. About a fifth of the land now irrigated in the United States was developed by the Bureau of Reclamation.

The primary function of the bureau is the construction, operation, and maintenance of water resource development facilities to aid in the economic growth of the West. Its first major achievement in this respect was the Hoover Dam, constructed during the period 1931-1935, creating the largest artificial reservoir in the world and making possible the eventual cultivation of more than a million acres. Then within a period of three decades following the construction of Hoover Dam, the bureau had completed 252 storage dams, 126 diversion dams, and 48 hydroelectric power plants.

Drainage has long been an important part of land reclamation practice—for instance, in the case of swampland and marshy areas to make waterlogged land suitable for cultivation. In regions of the country where rainfall is heavy and the land has little slope, much farmland is underlain with a system of underground tile drains to draw off excess water. Proper drainage of irrigated land is also important to maintain the groundwater table below the plant root zone and to provide a favorable balance of salt in the soil.

In arid regions, soils commonly have underlying layers with high concentrations of salts. After penetrating to these layers, irrigation water moves upward by capillary action, bringing dissolved salts to the surface. This serious disadvantage is compounded if the irrigation water itself has a high salt content. Without proper natural or artificial drainage, the irrigated fields may have areas of white, salty surface soil and stunted plants. With proper land leveling and good natural or artificial drainage, the salts can be flushed out into the drainage, particularly if the soils are permeable and the water is not too salty (Kellogg and Orvedal, 1969).

## UTAH AND NEVADA

In the entire region of what is now western Utah and Nevada, the Salt Lake Valley of Utah was the only area with potential for irrigated agriculture that was not entirely bypassed in the great migrations of settlers from the Mississippi Valley and the East to California and Oregon in the middle decades of the nineteenth century. It would have been shunned much longer if it were not for the Mormons remaining in this part of the "Great American Desert" in the course of their historic exodus. The Salt Lake Valley had a feature desired by the Mormons; it provided isolation from the world in which they had been persecuted, and noninterference in their efforts to establish an enduring settlement in the West (Arrington and May, 1975). The area was not yet under the control of the United States and it was separated from Spanish influence by the Grand Canyon barrier. This happy state of affairs was soon terminated by the Treaty of Guadalupe Hidalgo, which brought Mormon country under American rule in 1848.

## The Mormon Exodus from the Midwest

The history of Utah, and in fact the history of the entire Great Basin, is intimately related to the history of the Mormons after their arrival in the West. Mormons are members of the Mormon church, properly called the Church of Jesus Christ of Latter-day Saints. Prevented by Gentile opposition and persecution in four attempts, between 1831 and 1846, to build Mormon communities in the Midwest, the Mormons were left with the grim alternative of massacre or exodus. Choosing the latter, they left their beautiful city of Nauvoo, Illinois, and crossed the Mississippi, camping in the open on the Iowa shore with the temperature at -20°F (-28.9°C), then began their painful trek across Iowa and spent the winter in Winter Quarters, now Omaha, Nebraska.

After the murder in Carthage, Illinois, of the Mormons' prophet and leader Joseph Smith, Brigham Young, as President of the Quorum of Twelve Apostles, became the new leader. (He was elected president of the Mormon church the following year.) Young firmly but wisely employed all the great political power and moral influence that theocratic and patriarchal Mormon society bestowed on his position. The primary objective of the Mormons was to find a refuge from the jurisdiction of the United States. Joseph Smith had collected as much information as was readily available on the Rocky Mountain region, and as early as 1842 he had predicted the relocation of the Saints in that region. He had sent several scouting parties to Texas and certain areas of the West. The unclaimed valleys of the western edge of the Wasatch Range had been established as the primary destination of the Saints by the fall of 1845, although California and Vancouver Island also received consideration from time to time (Allen and McLeonard, 1976).

The advance party in the trek to the Salt Lake Valley left its rendezvous on the Elkhorn River, Nebraska, near Winter Quarters on April 17, 1847. In this group there were 148 persons, among them 3 women (including one of Young's wives and one of his brother's) and 2 children. There were 7 prairie schooners, 93 horses, 52 mules, 66 oxen, 19 cows, 17 dogs, a few cats, and some chickens. One of the women discovered that the jolting of the wagons would churn milk. Thereafter, all the Mormon emigrants made butter en route. They dug hollows in the hillsides to serve as ovens in which to bake the dough that they prepared as they drove along the trail (Werner, 1925).

Brigham Young learned of the Salt Lake Valley from reports and maps of Colonel John C. Fremont and long conversations with Father De Smet, the famous French missionary. The advance party reached the valley on July 22, 1847, which they entered via the infamous Hasting's Cutoff from Fort Bridger that had so tragically delayed the California-bound Donner Party a year previously. Young was delayed two days by a severe attack of "mountain fever."[4] When he finally passed through Emigration

[4]Probably Rocky Mountain spotted fever, caused by a microorganism transmitted by the bite of a tick.

Canyon on July 24 and looked upon the Jordan River winding through the Salt Lake Valley and the late sun glittering on the Dead Sea, he said simply, "This is the place." The Great Basin was to be the heaven-directed sanctuary for the Mormons and the Salt Lake Valley the site of the new City of Zion (Whitney, 1892; Werner, 1925; Stegner, 1942).

Brigham Young's goal was an agrarian society. He was rivaled only by Thomas Jefferson in his zeal for a society of small, independent farmers as an American ideal. Now, with an industrial society plagued by pollution and rapidly dwindling resources, and under the threat of nuclear extinction, would it not be appropriate to reexamine the philosophies of these two giants in American history? As might be expected, such reexaminations are being made by savants of such eminence, for example, as Caltech's Harrison Brown, who predicts: "If industrial civilization eventually succumbs to the forces that are relentlessly operating to make its position more precarious, the world as a whole will probably revert to an agrarian existence" (Brown, 1956). Who would have imagined at the turn of the century that in the year 1975 *The New Pioneer's Handbook* (Bohlen, 1975) would be popular reading? Yet the philosophies of Jefferson and Young understandably lacked an important element; neither of the two leaders envisioned the day when population explosion would make either agrarian or industrial society untenable. The aim in the eighteenth and nineteenth centuries was to populate the vast uninhabited areas of the country as rapidly as possible, to ensure its strength, security, and prosperity.

Brigham Young had become the Moses of the "last days." He not only had led his people "out of the land of Egypt and out of the house of bondage" but, like Joshua, he was also destined to rule them in the Promised Land for many years.[5] There appeared to be no limit to the toil and sacrifice that he could call on his loyal followers to endure. One might expect that he had at last exceeded mortal bounds when he called on "handcart brigades" to expedite the great exodus. But no, the faithful looked forward eagerly to any test, however severe, of their suitability to inherit the Kingdom and share in the millennial fulfillment.

Some Mormons had understandable reservations about Salt Lake Valley as the proper place to colonize; they would have preferred to pass on to the fertile valleys of northern California. The sight from Emigrant Pass may have been inspiring, but closer inspection of the sandy desert, barren except for sagebrush and some sunflowers infested with crickets, seemed to some in the pioneer party to hold little promise. The three women broke down and wept, appalled by the loneliness, desolation, and general wretchedness of the area. But Brigham Young, whose word was law, concluded with good reason that this at last was an area the Gentiles would not covet. In any case, survival demanded immediate action; there was little time for melancholy.

## Irrigation

Exploring parties found streams of water in the timbered canyons of the Wasatch Mountains. Irrigation of the parched land was a prerequisite for survival, and the Mormons lost no time in diverting the water of City Creek for this purpose. As previously related, Indians of the Southwest had long before developed extensive irrigation systems. The Spaniards in California and ethnic Germans in Pennsylvania had greatly

[5]Young encouraged the people to come to him for advice even for the most intimate details. On one occasion when a woman asked him whether she should wear red or yellow flannel next to the skin, he solemnly counseled her by all means to wear yellow. When another of his flock tearfully complained that her husband had told her to go to hell, Young said, "Well, don't go; don't go" (Werner, 1925).

increased the productivity of their farms with irrigation. But to the Mormons belongs the distinction of having begun the first extensive white civilization in America "with its economic basis resting almost wholly upon irrigation practice" (Price, 1976).

Before the end of July 1847, land had already been flooded and allowed to dry out enough to support the weight of oxen. Farmers were then able to plow the land; previously it had been too hard to penetrate with a plow. The indefatigable Mormon settler could now see his dream materialize on the parched desert land.

> Oh faith rewarded! Now no idle dream,
> The long sought Canaan before him lies;
> He floods the desert with the mountain stream,
> And low! it leaps transformed to paradise.

Irrigation required cooperation between farmers, particularly in the pioneer stage. Young decided that initially a large common farm would be most successful, and even necessary to prevent starvation. The famed community spirit and high morale of the Mormons were equal to the task. Their survival assured, many believed that their interests would be best served if a large private company would irrigate the entire territory and charge for the use of the water, but Young insisted that each farmer should build his own canals and furrows and pay tribute to no one. He wanted no farmer to own more land than he and his family could care for. Rise in land values resulting from community development could not be appropriated as unearned increment by individual owners. Water rights went along with land ownership. No speculator could buy up water sources and charge exorbitant prices for water use, or deny it altogether. The tragedies that blighted the lives of thousands of settlers on the arid High Plains were avoided by sensible and humane land and water policy.

Parts of the irrigation system that required greater effort than individuals or families could perform were built and owned cooperatively by groups of farmers and their families. Mormon doctrine stressed community and cooperation, combining cooperative endeavor and private enterprise. Poor and mostly illiterate Yankee farmers, English miners and mechanics, and Scandinavian peasants, under the inspiring leadership of Brigham Young and through their own ingenuity, developed the first large-scale Anglo-American irrigation projects in the United States. By 1865 there were over 1,000 miles of irrigation canals in Utah and, by 1946, 8,750 miles.

The Mormon settlers had more going for them than just hard work, perseverance, and an incorruptible land ethic. Snowfall is moderately heavy in Utah's mountains, especially in the north, and forms the chief reservoir of moisture, conveniently close to practically all farming areas. Pacific storms from the west move through northwestern Utah during the wettest winter and spring months. In summer this region is comparatively dry, while eastern Utah receives appreciable rain from summer thundershowers associated with moisture-laden air masses from the Gulf of Mexico (Nuttonson, 1966). The Mormon settlers found that the Wasatch Mountains, benchlands, bottomlands, and the Jordan River paralleled one another along the length of the farming areas of the Salt Lake Valley. Creeks flowed westward to the Jordan at intervals of two or three miles. They were small and nearly always a diversion dam sufficed to turn the water into a canal; the latter rarely needed to be more than two or three miles long in the early days.[6] This was a good bit of fortune for the nearly 16,000 persons camped along the

[6]Even as late as 1865, after considerable agricultural expansion, the average length of 277 canals that had been built in Utah was about 3.7 miles, a very low figure when compared with that of other western irrigation systems.

Missouri River and thousands more in England and Scotland, all awaiting the call from church leaders to come to the Promised Valley (Arrington and May, 1975).

Far more difficult than the simple irrigation engineering system was the task of applying water to planted fields, often on undulating land, so that a given type of soil and seed would get the right amount of water. There also had to be a decision on whether furrow irrigation or flooding would promote maximum yields. Successful irrigation innovations were communicated from farmer to farmer and the church organizational structure spread new and successful ideas throughout the commonwealth. The success of Mormon agriculture soon became known in California and other irrigated areas of the West and in foreign countries. In 1873 Utah fruit won awards in national expositions in Philadelphia and Boston.

The Mormons rapidly demonstrated that irrigated agriculture could form the basis for a complex society of professional men, businessmen, artisans, and farmers. This principle was soon tested in California, where gold was discovered in 1848 and irrigated agriculture was called upon to form the food base for miners and the industry and commerce that developed with mining (Higbee, 1957).

Eventually irrigation systems in other areas surpassed those of Utah in degree of sophistication and more attention began to be paid to the Mormons' success in innovating distinctive patterns of rural community life. They had shown that American civilization could be perpetuated in the previously forsaken trans-Missouri regions. Irrigation permitted the widest possible diversification of crops, abundant self-sufficiency for each family, and virtual independence from the outside world.

Most of the cultivated lands of Utah were at an elevation of between 4,200 and 5,000 feet. Only in the southwestern corner of Utah, in the Rio Virgin Valley, "Utah's Dixie," did cultivated land lie below 4,000 feet. Nevertheless, considerable crop diversification was possible. Utah's mountains had an important influence in preventing frost or freezing temperatures and therefore most of the farming lands were near them. On clear nights, the colder air usually moves down to the valley bottoms, leaving the foothills and bench areas relatively warm. It is on these higher lands at the edges of the valleys where the valuable and delicate fruits, berries, and vegetables were grown, leaving the bottomlands for the hardier grains and vegetables. Nevertheless, in some years there was, and is to this day, considerable damage to fruits from late spring frosts (Nuttonson, 1966).

The important crops in Utah are, in approximate order of importance, alfalfa (for hay or for seed; fig. 29, left bottom), dryland wheat, irrigated grain (barley, wheat, oats; fig. 29, right bottom), corn for silage or grain, potatoes, fruit, and vegetables. Only about 3.5 percent of the land area of the state is devoted to crops. The remainder is desert, used only for winter grazing for sheep and cattle, and mountain lands that serve as watersheds and provide summer grazing for the livestock. The mountains also produce some commercial timber (L. A. Jensen, correspondence).

Farming in Utah became intensive rather than extensive; a family could sustain itself on as little as ten or twenty acres. Farmers could live in villages, rather than isolated farmhouses; thus they could avoid the rural loneliness that had been characteristic of life in the arid West. They had easy access to schools, public libraries, and churches and, initially, defense against Indians.

## Mormon Colonization

During the first two years after their arrival, the Mormons had explored all the areas of the Great Basin that held promise for colonization. They also explored west to San Francisco and southwest to Los Angeles and San Diego. They not only explored these

*Fig. 29. "He floods the desert with the mountain stream / And lo! it leaps transformed to paradise" (canal courtesy David P. Lunt; others courtesy Cleon Kotter).*

regions but they also remained in them. The success with which Young was able to convince his followers of the merits of the Mormon agrarian society is shown by the way they stayed in Mormon country when gold was discovered in California. A large contingent of Mormons had sailed from New York to San Francisco in 1846. Despite the better agricultural potential of California, and despite the lure of gold, irresistible to most emigrants, the Mormons all heeded Young's call for them to come to the Salt Lake Valley. Mormons who were panning and mining for gold left for Salt Lake City and contributed their gold to the building of the new Zion in the wilderness.

Within eight years after the arrival of the Mormon pioneers in Salt Lake Valley, Young and his assistants had sent 500 colonists to San Bernardino, California (in 1851); founded Genoa in Carson Valley, Nevada; founded Lemhi on the Salmon River in Idaho; built Fort Bridger and Fort Supply at the east entrance of the Great Basin; and had small, flourishing communities in most of the valleys of the Great Basin that had any potential for agriculture. Never did any American colony builder so completely leave his stamp on his people and their institutions. One of Young's contemporaries, historian E. W. Tullidge, concluded that the Mormon leader "was perhaps the greatest colonizer the world has ever seen," a judgment that has well stood the test of time (Hunter, 1940).

Brigham Young realized that his dreams of an eccleseosocial and agrarian commonwealth encompassing the vast region his scouts had explored could not be fulfilled with only the 25,000 members of the Mormon church at that time in the United States and Canada. He also realized that a potent factor in his favor was the poverty and dissatisfaction of many of the world's people. Accordingly he established the "Perpetual Emigrating Fund Company" to assist converts financially. Young's active worldwide proselyting campaign resulted in bringing over 70,000 European proselytes to the Mormon empire before his death in 1877. Most of the foreigners came from Great Britain, where proselyting was particularly effective, but many came from bleak, poverty-stricken Scandinavia and from Canada.

### Mormon Indian Policy

The first Mormon exploring party to the Santa Clara valley in southern Utah saw about a hundred acres of land along the river that had been cultivated by the Paiute Indians, principally planted to corn and squashes. In 1855, the Mormon missionaries helped the Indians construct the first dam across the river. The dam was 100 feet long and 14 feet high. Half of the impounded water ran out one side of the dam for the whites and half on the other side for the Indians, arousing great enthusiasm among the latter. This was an example of Mormon Indian policy—to establish missions, cultivate friendship, bestow gifts and goods, and encourage improved agriculture. The policy was in accord with the Book of Mormon, which purports to be a religious history covering a period of about a thousand years of a small group of Israelites from Jerusalem who migrated to the Americas about 600 B.C. According to this history, the Indians were said to be at least in part descended from these people. Through conversion to Christianity they would become again a "white and delightsome people." The Mormons' Indian policy greatly facilitated their early expansion in the Great Basin (Woodbury, 1950).

In later years, after greater Mormon immigration led to a decline in wild game, edible fruits, and seeds upon which the Indians had formerly subsisted, Navajos, Paiutes, and Utes began stealing and raiding. This in turn led to bloody reprisals by the Mormons. It became evident that the Indians would become neither white nor delightsome. In the Black Hawk War against the Utes, from 1865 to 1868, nearly 3,000 Mormons were enlisted, 70 lost their lives, and over a million dollars were spent. Hostility lasted longest with the Navajos, who were notoriously skillful and effective raiders. It was the remarkable Jacob Hamblin, at great risk to himself, who walked into the Navajo camp and laid the foundation for a durable peace. Thereafter Mormons and Indians traded peacefully (Little, 1909).

### Deseret

Brigham Young wished to colonize Mormon country in such a way that his converts would be the original settlers and remain in the majority. He planned to hold the territory by right of colonization. Before the state of California was created, his plan included a "Mormon Corridor"—a continuous line of Mormon settlements from Salt Lake City to the sea. The Mormon commonwealth was to include not only the Great Basin but also the territory drained by the Colorado River and its tributaries and most of southern California, including the seacoast from Los Angeles to the present Mexican border. The California coast was closer to Salt Lake City than the American frontier in the East, and would be open throughout the year, whereas the eastern route was closed by winter snows for nearly half the year. The proposed road followed roughly the course of present Highway 91 from Los Angeles to Salt Lake City.

In the spring of 1849, delegates from the enormous region were called to Salt Lake City for the purpose of writing a constitution and applying for admission into the Union. The proposed state was to be called the state of Deseret. Deseret was a *Book of*

*Mormon* word signifying "honey bee." A beehive is represented on the great seal of Utah, a reminder of the name selected for the state by the first settlers. The grandiose plans for Deseret were nipped in the bud, of course, when the territory of Utah was established by Congress in 1850 and the state in 1896, within the relatively modest boundaries we see on the map today. Mormons resented the loss of land by the state of Deseret—land that had been first broken by them. They knew that they could colonize and make a living on land where other settlers were doomed to failure. Nevertheless, within 30 years Brigham Young had supervised the building of some 40 towns along the Mormon Corridor to the sea as well as 325 other towns and cities in the West. By the time Young died, there were 140,000 Mormons in "Mormon country." Mormons were so disciplined that if Young told a farmer to sell his place and join a group to start a new colony in some distant valley, there was never any argument or hesitation. The Saints accepted the "call" as coming from God (Hunter, 1940). Everywhere the well-planned and attractive towns founded by the Mormons, with their wide streets, characteristically bordered with Lombardy poplars, stood out in striking contrast to their shabby non-Mormon counterparts throughout the western frontier (Stegner, 1942).

### Nevada

Deseret included what is now the huge state of Nevada, where Mormons had considerable influence on early agricultural development. Opportunities for irrigation were limited. Of the nation's fifty states, only the smallest and the largest (Rhode Island and Alaska) rank lower in agricultural production.

There is considerable range country in the northern half of the state and 600,000 cattle comprise its principal agricultural product. Dairy products, hay, potatoes, sheep, alfalfa seed, and grain are other relatively important products (NCLRS, 1977).

## ARIZONA AFTER AMERICAN RULE

In 1848 the war with Mexico ended and the Treaty of Guadalupe Hidalgo was signed on February 1. The United States acquired New Mexico, California, Nevada, Utah, parts of Colorado and Wyoming, and Arizona north of the Gila River. Through the Gadsden Purchase agreement of 1853, for the sum of $10,000,000 the rest of Arizona, down to the present border of Mexico, was acquired. Arizona became a territory in December 1863, with a population of 4,573. The end of the Civil War found the territory still largely the domain of the Indian, with only a few white settlements—Tucson, Tubac, Prescott, and Yuma—"islands of safety in a sea of murder and pillage." The Apaches and Yavapais had not ever been appreciably contained. They lived as they had for centuries, even before the Spanish occupation, by raiding. The Navajos were equally troublesome in the north. Not until 1886 were all Indians on reservations, lands comprising some 27 percent of the Territory of Arizona (Faulk, 1970).

### Cattle and Sheep

Kino and the Jesuits had introduced Longhorn cattle to the missions late in the seventeenth century and soon there were immense herds, as well as many sheep. However, many ranches were later abandoned because of Indian depredations, and by the time of the American occupation there were few herds remaining. Many cattle and sheep were driven through Arizona to supply the demands of the California forty-niners, the cattle bringing $100 a head.

Spanish and Mexican ranchers had varying degrees of success with livestock on the ranges of southern Arizona for they were continually harassed by marauding Apaches.[7] The lone Anglo rancher, Pete Kitchen, was so fearless and such a good shot that the Apaches failed to dislodge him from his ranch at Portrero, just north of the present Mexican border. From his adobe house on a hill above a spring, miles of surrounding country were visible and some of Pete's men were continuously on guard at that superb vantage point. Those working in the fields always had rifles with them. Kitchen continued to raise cattle, hogs, and forage crops and was on hand to welcome other Anglo ranchers following the Gadsden Purchase (Peck, 1962).

Among Texans who were driven west by nesters and barbed wire was a veteran trail boss and Indian fighter by the name of John Slaughter. He happened to be in Tombstone when silver was discovered there, and decided to remain. He founded what was to become the famous Slaughter cattle empire. Grass was said to grow luxuriantly "to a horse's belly" on the rolling rangelands and in the brushy hills and canyons of southern Arizona in those days. Along with Slaughter's cattle came rustlers to add their depredations to those of the Apaches. Rustlers built corrals in remote canyons where there was water. At first they dealt mainly with cattle stolen in Mexico and driven north to their canyon hideouts. Cochise County ranchers did not interfere with the outlaws until the latter began to steal the ranchers' cattle and horses. Then John Slaughter, widely known for his nerve and shooting skill, was appointed sheriff of Cochise County. Four years of hard work by Slaughter and his posse put an end to the rustler menace. Slaughter then purchased the huge old San Bernardino land grant and herded thousands of head of cattle in Mexico as well as Arizona. Living in the style of the Mexican haciendas in a great Spanish-style house surrounded by trees and flowers and artesian wells, the Slaughters became famous for their elegant style of living and their hospitality. The entire region was public domain and other famous cattle kings established their vast empires in the region (Peck, 1962).

Henry Clay Hooker, who got his start by driving 500 turkeys from California to the Nevada mining camps and later by driving cattle from Texas to army posts in northern Arizona, finally established his Sierra Bonita Ranch in Sulphur Springs Valley in southeastern Arizona in 1872. Despite Apache depredations, Hooker developed what was to become the largest and most successful ranch in Arizona Territory. He introduced new cattle breeds—Shorthorns, Durhams, and Herefords—to his 800-square-mile ranch. He kept a dairy herd, poultry, and a large garden, for the benefit of the many guests who stayed on the ranch.[8] He pioneered in breeding and racing horses of great stamina and beauty. Gradually the cattle industry spread west and north. Cattle ranches were developed in the Gila River Valley and in the tallgrass country of Chino Valley on the Prescott plateau. The largest corporate venture in the north was that of the Aztec Land

[7]When viewing the Apache raids from the standpoint of the white settlers, one is apt to forget that white men were the intruders, raiding the ancestral lands of the Indians. Indians were at times forced to steal cattle and sheep to avoid starvation. When the Indians were persuaded to make peaceable arrangements requiring mutual trust and honesty, their efforts might be rewarded by massacre of an entire village, sometimes by the U.S. Army and sometimes by groups of private citizens. Such, for example, was the massacre of 125 unarmed Apaches, mostly women and children, on a Sunday morning in the spring of 1871. These were Apaches who had been given sanctuary on a military reservation near Camp Grant, Arizona. The infamous raid was carried out by a group of prominent Tucson citizens assisted by Papago Indian mercenaries. This was but one of many examples of such senseless brutality against Indians. To compound the shame of such behavior, western newspapers generally praised the perpetrators. In the case of the Camp Grant massacre, they alleged that the Apaches had been conducting raids, but offered no evidence, and in fact the Indians had been living in peace and working hard to develop their own agriculture (Terrell, 1972).

[8]Among prominent guests were Stewart Edward White and his bride. This prolific writer of outdoor-adventure fiction obtained the inspiration and locale for the thrilling stories of his *Arizona Nights* (1907) while on a honeymoon on the famous ranch.

and Cattle Company, with headquarters near Joseph City on the Little Colorado River. This ranch, 40 by 90 miles in extent and with 60,000 cattle, was plagued by absentee ownership and rustlers. It ceased operation about 1900 (Peck, 1962; Faulk, 1970).

The transcontinental railroads brought a boom to cattle-raising in Arizona, resulting in overgrazing. Thousands of homesteaders, with barbed wire and windmills, broke up the open ranges. Cattle ranchers themselves began to drill wells, fence in their land, and introduce new breeds. Modern ranching became well established. It was more enduring, if less romantic, than the open ranges of the old days (Faulk, 1970).

Northern Arizona had been sheep range from the time the Navajos and Hopis began depending on their flocks for most of their livelihood and the Mormons drove sheep down from Utah. Flagstaff became the major center of the sheep industry. Mutton was never popular in the West but the railroads opened up a good market in the East. More important, the price of wool proved to be steadier than the price of cattle. As on the Great Plains, cattlemen and sheepmen contended for grazing rights. Open war erupted when the Tewksburys moved sheep south of the Mogollon Rim into Tonto Basin, the territory of the Grahams. The war lasted five years and sheep were finally driven out of the basin, but twenty-nine men died, including every male Graham and all but one male Tewksbury (Faulk, 1970).

Cattlemen and sheepmen eventually recognized that they had common interests in their struggle against homesteaders, conservationists, and the National Forest movement. Cattlemen finally realized that cattle and sheep could graze on the same land, and started raising some of the detested "woolies" themselves. They benefited from the steady price of wool in years when the price of beef fell abruptly. Cattlemen also lobbied for the creation of the Arizona Rangers, a force of tough peace officers who could ride and shoot, and felt no compunction about crossing the Mexican border to run down and arrest cattle rustlers (Faulk, 1970).

Generation after generation, countless books and movies have kept very much alive the aura of romance of picturesque, colorful, Old Arizona. In Will C. Barnes's *Apaches and Longhorns* (1941), he tells a stirring story of his own life on the Arizona cattle ranges in the old days, first in the Army Signal Corps in the struggle against the Apaches, then as a cattleman with the traditional antagonism against federal range regulation, then as Inspector of Grazing in the Forest Service after his conversion by the wise and farsighted Gifford Pinchot (head of the newly created service), and finally as Assistant Forester and Chief of Grazing. It was Pinchot's policy of at first placing the administration of new grazing regulations in the hands of western men who could "talk the language" of the cattlemen and sheepherders, and understood their problems, that ensured the success of the federal program. Even then it often took great nerve and courage to enforce the grazing regulations. The admirably suited pioneer forest rangers were then gradually replaced by young men educated and trained for the work.

When the Tonto National Forest was created in 1905, the forest rangers prepared a sheep drive through the area along which the herders might drive their flocks from the north for winter feeding in the Salt River Valley. Passing through Tonto National Forest in winter, one may sometimes see mostly Mexican or Basque sheepherders driving their sheep down to the valley for winter feeding, while in the same valley cattle are being fattened in alfalfa pastures.

### Modern Rangelands

The western rangelands make up an area of about 540 million acres—over half of the billion acres of range and pasture in the coterminous forty-eight states. They are in that portion of the seventeen western states that lies roughly west of 100° longitude and comprise most of the area of those states. To the tourist seeing this immense area from the air, it is a region of mountains, deserts, plateaus, dry basins; rocky, shallow, saline soil; and semiarid grasslands with clumps of inedible shrubs and ranging up to impenetrable forests on high-mountain slopes. The tourist might prefer that we utilize this land for its beauty and recreational value and leave its meager forage for the wild animals. However, looks are deceiving, and this region is making substantial contributions to the national wealth, including more than half of the national production of livestock (Love, 1970).

Fortunately or unfortunately, depending on one's point of view, the western rangelands will no more be left to "nature's plan" than any other section of the country. The fate of the West will depend on how well man can manage its immense areas to meet his multiple requirements for food, water, recreation, mining, grazing, and lumbering on a sustained basis.

On Southwest cattle ranges in good condition, the principal forage is grass. Perennial grasses tend to predominate in the higher altitudes and annual ones in the lower. However, production of grasses increases with altitude, because of increasing rainfall, so on most of the Southwest ranges the perennial grasses are the more important. The latter respond in growth from even one heavy rain, whereas annuals require proper follow-up storms for germination, establishment, and growth.

On the grass-shrub ranges, grass species can be maintained satisfactorily when as much as 40 percent of the grass is removed. Proper distribution of cattle, to avoid overgrazing, is essential. Optimal distribution can be attained by fencing, and developing and controlling water to the extent practical, and using salt and supplements to draw animals to otherwise lightly used areas. Over a ten-year period, on the Santa Rita Experimental Range south of Tucson, Arizona, the average yearlong animal units per section (640 acres) required to utilize 40 percent of the perennial grasses were 12-24 at high elevations; 6-11, intermediate; and 3-6, low.

Whether for grazing or for recreational use, the foremost enemy in the western lands is brush. Brush uses up the scarce soil moisture, provides fuel for devastating fires, makes many areas impenetrable, and deprives the land of the potential value of grasses and useful browse. The intermontane country has 115 million acres of sagebrush. In Arizona, New Mexico, and Texas there are 60 million acres of spiny shrub mesquite besides many millions of acres of cactus, burroweed, and snakeweed. They are difficult to eradicate and every year encroach on new areas. Every acre of brush utilizes six more acre-inches of water than herbaceous vegetation. A farmer who converted brushland to grass on a dry range found that streams that once dried in summer began to flow again all year, allowing him to build a reservoir to supply water for irrigated pasture. A livestock operator who burned off 2,400 acres of brush, and seeded the area to grass, found that within six years the area supported 250 cows and many deer where formerly it supported 20 cows and only a few deer (Love, 1970).

Among undesirable range shrubs, the effect of mesquite is best known. Mesquite has greatly increased in the Southwest since the last decades of the nineteenth century. It has spread from its original habitat along drainage channels and arroyos to the uplands (Glendening, 1952). Grazing cattle disseminate the seed, and reduced grass cover makes it easier for the seed to become established (Glendening and Paulson, 1955). Although mesquite furnishes some forage, especially during the season of seed production, it does not produce enough to compensate for the loss of grass (Reynolds and Martin, 1968).

As related in chapter 6, range improvement in grass-shrub rangelands includes shrub suppression. This can involve burning, hand grubbing with mattocks, cabling or chaining, and using low-grade diesel oil or herbicides. Where the range has deteriorated, re-seeding may be desirable, following ground preparation with special equipment.

On the Santa Rita Experimental Range, rodents and rabbits consumed an estimated two-fifths of the forage produced. They do more damage to range vegetation than cattle, because they forage much closer and may even dig up roots during dry periods. Certain species, particularly kangaroo rats, disseminate seeds of unwanted shrubs. Control of rodents and rabbits may be necessary, preferably by a competent rodent-control specialist (Reynolds and Martin, 1968).

## Farming

With the notable exception of the extensive wheat fields of Pete Kitchen north of Nogales, there was little farming in Arizona from the time of Father Kino to the post-Civil War era. Then, as previously related, the homesteaders ("nesters") began to compete with the cattle kings for land, particularly in river valleys. As in California, the large and often extravagant and fraudulent[9] claims based on the old Spanish-Mexican land grants required years of litigation. This did not prevent the homesteaders from rushing in on public domain, stimulated by (1) the Homestead Act of 1862, which provided 160 acres virtually free to anyone who would live on the land five years and make improvements each year; (2) the Timber Culture Act of 1873, providing for an additional 160 acres for any settler who would plant 40 acres in trees; and (3) the Desert Land Act of 1877, by which a settler could gain title to 640 acres if he irrigated the land. The alluvial sandy soil of the river valleys proved to be very productive and the long growing season enabled farmers to grow two crops a year. The entire valley of the Gila River, more than 400 miles long, had good potential for farming and this is where Arizona's largest farming area is now located.

In Arizona, as in California, mining was a stimulus to agriculture. Prescott, the capital of the new territory, could use the products grown in the Verde Valley to the east. However, the fruits of the farmers' labor were often lost to the Apaches, who descended from the craggy cliffs of the Mogollon Rim. The first agricultural activity in the Salt River Valley was the cutting and drying of grass that grew luxuriantly along the remains of the canals of the prehistoric Hohokam Indians. The hay was for the horses of the men at the army posts who kept watch on the Apaches until the Indian troubles finally ceased. A "hay camp" was established at the site of the future town of Phoenix. This camp was visited by prospectors and miners passing through the area, one of whom was Jack Swilling, concerning whom, more later (Peck, 1962).

## Mormons in Arizona

Mormons were by no means the pioneer Anglos of Arizona, but unlike their predecessors, who came to Arizona mainly to prospect or work in the mines, the Mormons came to till the soil and establish settlements. It was the Mormons, from an agricultural background in Utah, who developed the fertile areas of Mesa, Lehi, the Safford-Thatcher-Franklin district, St. David on the San Pedro River, and the many settlements along the Little Colorado and its tributaries in northeastern Arizona, such as St. Joseph (now Joseph City), St. Johns, Snowflake, Show Low, and Springerville.

In 1880, David K. Udall, leader of the St. Johns settlement, drove his herd of cattle

[9]James Addison Reavis, the "Baron of Arizona," filed for 10,467,456 acres, but after ten years of investigation by Land Office agents, Reavis was convicted on conspiracy charges and sentenced to six years in prison (Faulk, 1970).

from Utah to Lees Ferry on the Colorado. To reach the ferry, the Mormons had to lower their wagons down cliffs by means of ropes. Udall then drove his cattle across the river on solid ice, for the river had frozen solid in that unusually cold winter. Then he followed the Mormon wagon road to St. Johns. The Mormons constructed dams on the Little Colorado with brush and rocks. The dams were annually washed out by floods, but were just as regularly rebuilt until eventually they became permanent structures (McClintock, 1921; Udall, 1959; Peck, 1962).

Throughout American history rural villages have with surprising frequency been birthplaces of persons of great stature,[10] and the tiny hamlet of St. Johns contributed more than its share. David K. Udall became the grandfather of Stewart L. Udall, Secretary of the Interior under Presidents Kennedy and Johnson, and author of *The Quiet Crisis* and *National Parks of America,* and of Morris K. Udall, lawyer, author, and congressman. The two brothers were born in St. Johns in 1920 and 1922, respectively.

Another St. Johns pioneer was Miles P. Romney. He was editor and publisher of the pioneer weekly *Orion Era* and one of his three wives, Anna, was the first schoolteacher in a small log schoolhouse built in 1881.[11] When Congress outlawed polygamy in 1885, the Romneys and others in the same plight sought asylum in Mexico. These refugees, and others who came, even after the Mormon church banned the controversial plural marriages in 1890, obtained permission from President Porfirio Díaz to buy lands and establish colonies in Mexico. In Colonia Dublan, one of the eight agricultural colonies established in northwest Mexico,[12] George W. Romney, grandson of Miles P. Romney, was born in 1907. George Romney became general manager of the American Motors Corporation, governor of Michigan, and secretary of the U.S. Department of Housing and Urban Development (Mahoney, 1960). One of the first children born in the neighboring Colonia Juarez, was Carl F. Eyring, who became a famous physicist-chemist at the University of California, Berkeley, and at Princeton, where he made a major breakthrough in the application of the principles of quantum mechanics to the laws of the chemical bonds between the elements. Also from tiny Colonia Juarez came Franklin S. Harris, who became head of the Agronomy Department and director of the Agricultural Experiment Station at Cornell University and later president of Brigham Young University. An astounding percentage of the children who obtained their elementary schooling in the primitive, makeshift schoolhouses of the pioneer Mexican Mormon colonies became prominent in the business and professional world (Romney, 1938), reflecting the zeal for education and the self-discipline, high motivation, perseverance, and dedication that characterized the social milieu in which they were reared.

## A Large Pig Farm

On the Silver Creek tributary of the Little Colorado is Snowflake, at an elevation of 5,640 feet and with a population of about 3,000. The town was named after W. J. Flake, the founder, and Apostle Erastus Snow, an early visitor. Flake was called by the church to lead the first Mormon expedition from Utah to Silver Creek Valley, arriving

[10]Including all the presidents of the United States of the last half-century, with the exception of John F. Kennedy.

[11]Miles Romney was one of the never more than 3 percent of the Mormon men who embraced "plural" marriage.

[12]The Mormons developed a thriving agriculture before they were forced to flee before the contending forces of the Mexican agrarian revolution of 1910. They had to hurriedly leave homes, farms, and places of business which were ransacked, looted, and in many cases completely destroyed (Romney, 1938; Johnson, 1972). Eventually Colonia Dublan and Colonia Juarez were resettled by Mormons. These two colonies are again thriving, with great emphasis on fruit-growing.

*Fig. 30. A portion of Snowflake Pig Farms, Inc., Snowflake, Arizona.*

on May 21, 1878. The Flake expedition experienced the usual agonizing travail of the long Mormon trek from Utah to the northeast Arizona plateau (O. D. Flake, undated; S. E. Flake, 1970).

Nearly all the farmers of the Snowflake area are Mormons, including the principal owner of Snowflake Pig Farms, Inc. (fig. 30) along Highway 77 a few miles north of the town. Starting with twenty-one 40-pound weaner pigs in 1966, Jim Caldwell now has one of the largest pig farms in the nation, a sow inventory of 2,211. Another 750-sow unit is under construction and Caldwell has his sights on 5,000 sows.

Whereas cattlemen in the area are pessimistic concerning the cattle market, Caldwell and his three brother-partners, De Lynn, Bud, and Lemoyne, have no reason to be anything but optimistic about pigs. A sow will average six to eight litters in her lifetime at Snowflake Farms, but some produce ten or twelve. The average number of pigs weaned is 7.95. Caldwell's target is a 225-pound quality meat-type hog obtained in an average of six months from birth to market. Four to five loads of hogs are shipped per week to a market in Stockton, California, in Caldwell's two triple-deck trucks and trailers—an 850-mile ride of 17-18 hours.

Caldwell obtains the widest genetic vigor by a three-way rotational cross-breeding between Hampshire, Yorkshire, and Duroc, breeding the most unrelated boar to the female. He has recently added some English Large Whites and Landraces. Only SPF (specific pathogen free) breeding stock is allowed to come onto the farm. The original stock of this type is obtained by Cesarian operation to obtain pigs free of certain diseases. As a graduate student at the University of Missouri, Jim Caldwell participated in some of the original research on this important procedure.

Caldwell mixes his own feed—milo and soybean meal—in his four-ton mixer. Lemoyne, the "innovator" of the Caldwell team, developed a recycling system that salvages feed from the floors in the "grow houses." Called RAF (reclaimed antibody feed), it is collected in a flush system and settling pond and fed wet to the sows. In addition to a savings of $105,879 per year, Caldwell found that scours (swine dysentery) was completely eliminated. Lemoyne also switched the heat source for heating the floors of the pig farm from LP gas to sawdust, with a saving of close to $16,000 per year, apart from the satisfaction of using a renewable energy source.

Social innovations, involving employee shareholding, a pension program, medical insurance, a profit-sharing trust, and a productivity bonus system are additional interesting and commendable features of Caldwell's pig-farm operation (Anonymous, 1977*c*).

### Irrigation

Jack Swilling, an enterprising Confederate deserter, raised $10,000 and excavated ancient and barely discernible canals of the prehistoric Hohokam Indians in the Salt River Valley and organized the Swilling Irrigation Canal Company in 1867. Supplies for canal construction were hauled fifty-four miles from Wickenburg by eight-mule teams. On the irrigated land, crops of barley and pumpkins were growing by the summer of 1868. Workers in the irrigated fields watched Indian women gather wood in the mesquite thickets while their warriors stood guard with bows and arrows at the top of a ridge. Fortunately for the pioneers of the Salt River Valley, the Pima and Maricopa Indians of the area were peaceable and willing to cooperate with white men when necessary.

By 1870 the 300 settlers of the little town of Pumpkinville changed its name to Phoenix, for the town was springing up Phoenixlike from the ashes of an ancient civilization (Peck, 1962; Faulk, 1970). In 1868, Columbus Gray had been the first to plant citrus and deciduous fruit trees in the Salt River Valley. The citrus trees became the harbingers of an important industry. Maricopa County, of which Phoenix is the county seat, was destined to become Arizona's most important agricultural county.

At the settlement at Tempe, Judge Charles T. Hayden built a ferry to accommodate the Prescott-to-Tucson traffic that crossed the Salt River. He also used the water of the newly constructed Tempe Canal to turn the wheels of a gristmill, where farmers from miles around had their grain ground. This thriving enterprise developed into what today is known as the Hayden Flour Mills. In the 1870s and 1880s fields of wheat, barley, and alfalfa and a few orange groves began to dot the hot, dry, desert land and a few Mormon farmers came down from Utah to lay out farm plots, dig ditches, and build homes in the settlement of Mesa. In those days most of the houses were built of adobe, the coolest type of construction (Peck, 1962).

All the early mining and agricultural development was taking place while stage and coach travel was still difficult and perilous, the sources of hazard being the Indians and the Mexican and American highwaymen and outlaws. Freight was hauled by two or three wagons hitched together and pulled by many pairs of mules from the nearest rail terminals in New Mexico or Texas or from the Colorado River. Passengers and freight came by ships from San Francisco, around the peninsula of Lower California to Puerto Ysabel at the mouth of the Colorado River, thence by river steamers up the Colorado to Ehrenberg or Fort Mojave and overland via stage coaches and freight wagons. Phoenix was not reached by a branch line of the Union Pacific until 1887 (Peck, 1962). Who could have foreseen that in less than a century Arizona would become the fastest growing among the coterminous forty-eight states?

Farming would never reach its full potential without sufficient, dependable, year-round quantities of water. Canal companies were formed—joint-stock companies in which a farmer could either purchase a share or personally help to build the diversion dam and canals. In addition, small fees for the water paid for repairs and for the salary of the overseer (*zanjero*) who distributed the water. From 1890 to 1900 the land under cultivation in Arizona increased from 70,000 to 180,000 acres (Faulk, 1970).

By a congressional act of June 17, 1902, the federal government was authorized to use the money derived from the sale of public lands for the construction of reclamation works. One of the first projects was a dam on the Salt River of Arizona, a tributary of the Gila. The dam cost $10,500,000 but this included power facilities, transmission lines to Phoenix, a road to the Salt River Valley and canals to the valley farms. The dam was formally dedicated on March 18, 1911, by Theodore Roosevelt and named after that fervent advocate of conservation and reclamation. This was only forty-eight years after Arizona had become a territory (in 1863), at which time there had been little besides the tiny settlements of Tucson and Yuma, and a few ruins left at the site of abandoned mines, to distinguish it from the wilderness found by the early Spanish conquistadores. In 1912 Arizona became a state, having acquired in the short period of forty-nine years, "the trappings of civilization, of culture, of polite society" (Faulk, 1970).

It took four years to fill Roosevelt Dam, but the 1.4 million acre-feet of water in the reservoir would be enough to irrigate the valley below for three years. The Salt River Valley Water Users Association was given charge of the operation of the dam and canals in 1917. Revenues totally paid the cost of the dam in thirty-eight years. The Roosevelt Dam was such a great success that other dams throughout the country were inevitable. The only practicable way of obtaining the expensive structures was through government financing. The large, arid areas of Arizona called for further development of dams, and it was obvious that the Colorado River would have to be harnessed.

Arizona, a state with few natural lakes, nevertheless has a considerable number of artificial bodies of water either entirely within the state or shared with neighbors (fig. 31). Entirely within the state is San Carlos Lake created by Coolidge Dam on the Gila River and a string of lakes called Theodore Roosevelt, Canyon, Saguaro, and Apache, created by a series of dams on the Salt River. On the Colorado Glen Canyon Dam created Lake Powell, 200 miles long when full, and shared with Utah; behind Hoover Dam is Lake Mead, 115 miles long, shared with Nevada, as is also Lake Mohave behind Davis Dam; and behind Parker Dam is Havasu Lake, which along with Imperial and Laguna Reservoirs behind dams of corresponding names, is shared with California.

Arizona's present share of Colorado River water was not obtained without a long and bitter dispute with California, beginning in the 1920s. In 1952 Arizona filed suit in the U.S. Supreme Court over the division of the river's annual flow, which resulted, eleven years later, in the Court awarding Arizona 2.8 million acre-feet of the water annually. In 1947 a plan now known as the Central Arizona Project (CAP) was introduced into the U.S. Senate and this bill and its successors won congressional approval in 1968. The project would take water from Havasu Lake. Through an elaborate system of dams, aqueducts, tunnels, pumping stations, reservoirs, and long pipelines, the water would be brought to Phoenix. An aqueduct would serve Tucson.

## MODERN ARIZONA AGRICULTURE

Arizona's economy has been characterized by the five *C*s—cattle, cotton, citrus, copper, and climate. But this leaves out manufacturing, an important and continuously growing segment of Arizona's economy. From the standpoint of cash receipts in 1970, beef cattle were still Arizona's most important single agricultural product (see table 2). Among livestock and livestock products, dairy products rank second, followed by

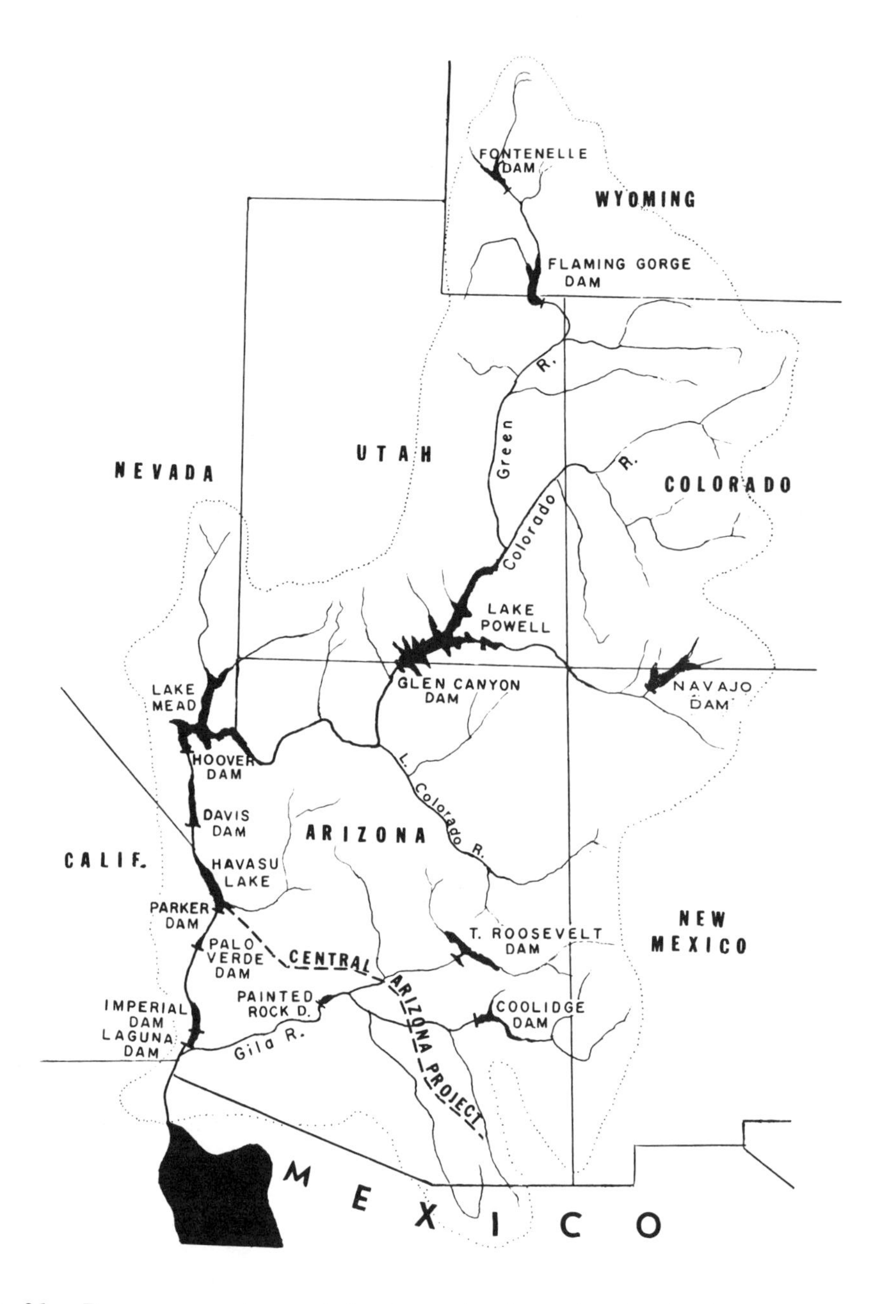


*Fig. 31. Dams, reservoirs, and the Central Arizona Project in the Colorado River Basin (redrawn and modified from Bailey, 1975).*

poultry and eggs, and sheep and lambs. By far the most important plant crop is cotton, accounting for about a third of the total crop value. Among the other important crops are the field crops—hay, sorghum grain, barley, and corn—the vegetables, especially lettuce and cantaloupes, and the citrus fruits.

On the average, total income from plant crops is approximately the same as total income from all livestock and livestock products, each annually contributing over $600 million to Arizona's economy. The leading agricultural counties for production of cotton are Maricopa, Pinal, and Yuma. These are the three leading counties for most plant crops. The number of cartons (ca. 40 pounds per carton) of citrus fruits produced in Arizona in the 1974-1975 season was as follows: lemons, 14,400,200; oranges, 9,940,000; grapefruit, 5,540,000; and tangerines, 1,220,000 (Ariz. Agr. Stat., 1975).

The rangelands of Arizona support a considerable quantity of cattle breeding stock. Cattle are shipped to feedlots in Phoenix, Yuma, and Tucson, fattened in pens, and then shipped out again. In 1975, Pinal, Maricopa, and Yuma counties had 59 percent of the 1,280,000 beef cattle and milk cows on farms. Among range cattle and calves, all of Arizona's fourteen counties contributed substantially (Ariz. Agr. Stat., 1975).

## Dams and the Southwest's Future

The development of the Southwest was intimately bound with the construction of dams, yet many environmentalists take a dim view of them. The grandeur and beauty of the canyons of the Colorado and its tributaries are unsurpassed and could be enjoyed for countless millennia. This loss of natural beauty must be weighed against livelihood for millions of Southwesterners and the beauty of humanized nature: green fields and orchards in bloom, rising Phoenixlike from desert sand. The huge artificial lakes provide water sports and access to rugged, picturesque, colorful canyons for countless thousands. But reservoirs eventually become filled with silt. They serve their purpose for only a fleeting moment even when measured in terms of the relatively brief period of recorded human history. The average annual rate of storage depletion resulting from silting varies from 2.5 percent for small reservoirs (capacities between 1,000 and 10,000 acre-feet) to 0.2 percent for the largest (more than 100,000 acre-feet). Rate of storage depletion tends to be inversely related to reservoir size and is, of course, also influenced by the amount of silt in the river. The utility of most reservoirs is seriously impaired by the time half of their original capacity for water storage is replaced by sediment (Glymph and Storey, 1967). The capacity of Lake Mead behind Hoover Dam is annually being reduced by 137,000 acre-feet by the silt that is washed into it by the Colorado River. At that rate the usefulness of the dam will end in less than two centuries. With the exception of the Grand Canyon, there are no remaining canyons in the Southwest comparable to Boulder Canyon in water-storage potential. As long as twenty years ago the National Research Council pointed out that government sedimentation surveys indicated that 83 percent of all reservoirs in existence at that time would become extinct from siltation in less than 200 years (Higbee, 1957). Serious impairment of the utility of half of the small reservoirs will occur in about 20 years (Dendy et al., 1973). Nationwide the capacity of man-made reservoirs is currently being reduced through siltation at the rate of about a million acre-feet per year (Baker, 1977). And there are other problems. Glen Canyon's spongelike walls are soaking up huge quantities of water and it is not known whether they will become saturated or continue to soak up water indefinitely, or whether the soaked-up water may be recharging usable aquifers.

The huge reservoirs behind dams share with natural underground reservoirs the tendency to increase the population, agriculture, and industry of a region but without the capacity for supplying the expanded need on a sustained basis. To all but the most dedicated cornucopians, the various limits to growth on a finite planet appear to be obvious, but possibly least recognized by the public is the limit set by the availability of fresh water. Yet this may prove to be the most imminent and serious limit, particularly in arid, irrigated regions. As Professor D. F. Peterson (1975), dean of the School of Engineering at Utah State University, has pointed out, technology has contributed greatly toward increasing man's useful water supply, but "the time may come when he may find it wiser, in some instances, to adapt to his environment rather than to modify it." This point of view was given added emphasis by the disastrous drought of 1976-1977.

Between 4.5 and 5 million acre-feet per year of groundwater is pumped in Arizona. Yet annual replenishment, including natural recharge and return seepage from irrigation, is probably not more than 2.0 million. Thus there is a perennial overdraft of groundwater approaching 3.0 million acre-feet per year. The U.S. Geological Survey reports that since 1952 declines in groundwater have caused an estimated 4,500 square miles in Maricopa and Pinal counties, including 120 square miles southeast of Phoenix, to sink an average of seven feet. When CAP is completed, Arizona will have the balance of the water from the Colorado River to which it is legally entitled, but this will amount to only about 1.2 million acre-feet and will be substituted for groundwater on existing croplands. It will replace only a portion of the present groundwater overdraft (DeCook et al., 1975).

Rapid evaporation of impounded water in hot, dry, desert areas, and its recycling for irrigation purposes as it makes its way to the sea, results in its ever-increasing salination. Water becomes essentially valueless for agriculture at a salt concentration of about 4,800 ppm. More water per acre must be used as its quality decreases, and this increases its cost (Pillsbury and Blaney, 1966). Mexico's share of Colorado River water has been increasing in salinity. However, the problem has become somewhat alleviated now that saline water from Arizona's Wellton-Mohawk irrigation district is no longer flowing into the Colorado River. It is being transported to the Gulf of California via a recently completed 51-mile concrete-lined extension of that district's bypass drain. Engineering plans and specifications for a desalting plant are now being made so that water from the irrigation district can be desalinated and can flow into the Colorado, thereby further reducing its salinity. Brine waste from the desalting plant will then be transported to the gulf via the extension of the bypass drain.

Using nonrenewable coal and nonrenewable gas as sources of energy, man is pumping water that is also, for practical purposes, nonrenewable, for as Bowden (1977) has pointed out, "the economies that exploit it cannot abide a low rate of use." It is considered sound economics to deplete an aquifer if by doing so a degree of local affluence is gained that allows purchase of more expensive water from a more distant source (Thomas, 1962). At exponential growth rates, vast regions are depleted with alarming suddenness. An ominous feature of the human predicament is that prosperity is based on using resources that are nonrenewable or, if renewable, at a rate greater than their natural replacement. Because of low rainfall, replacement of water in the aquifers of the arid West is extremely slow (Bowden, 1977).

At the very time when world events are emphasizing the need for increasing agricultural production, there is already talk of changing agricultural to industrial land because of the imbalance of agricultural need and availability of water. But taking irrigated land out of production in areas of high temperature and abundant sunshine removes much more food-producing potential per acre than in other areas. For example, the productivity of corn and alfalfa on irrigated land is about twice as high as on nonirrigated land (Kruse, 1975). Also, where irrigated land is in hot desert areas, crops can be grown throughout the year. Only areas with warm climates can produce the steady supply of fresh vegetables that is now taken for granted throughout the nation during the long winter season. Nevertheless, as water becomes scarcer and increases in cost, economics increasingly favors its use in industry, where income per dollar of water used greatly exceeds that of agriculture. But industry soon brings about the congestion and pollution that negates the area's original lure as a "tropical paradise."

The frightfully cold winters of 1976-1977 and 1977-1978 in the eastern half of the nation will probably drive people from that region at an accelerated rate to sunny Arizona, already the fastest growing state. They will bring the profligate water-use habits of the Midwest and the humid East to an area where water conservation should be of top priority. This comes at a time when analysis of the world's most extensive bank of tree-ring data at the University of Arizona at Tucson indicates that we may have passed through a benign period of higher rainfall than is normal for the Southwest and may be returning to a drought-plagued era that, over the centuries, has been more characteristic of the region. However, decreasing precipitation would only hasten the day of reckoning, made inevitable by the endless waves of immigration. The basic problem is that the sum of all rights to Colorado River water held by Great Basin and Southwest states and Mexico greatly exceeds what the river has to offer. In addition, the Navajo Indians are seeking to establish a right estimated as high as 5 million acre-feet. Fortunately for southern California, the storage capacity of Colorado River dams is so large that this area suffered no deprivation during the 1976 and 1977 drought years. But once Arizona is able to claim its legal share of Colorado River water through CAP, scheduled for completion between 1985 and 1988, southern California will no longer be able to escape the serious consequences of major droughts (Rood, 1977).

Leaving behind them an environment that has demonstrated its ability to heal its wounds and sustain itself, new arrivals from the Midwest and East will share with the already present Southwesterners the transient, seductive amenities of a "fool's paradise"—situated on a time bomb.

## WATER PROBLEMS OF THE WEST

Astronauts who looked back at our beautiful planet, our only haven in endless, inhospitable space, saw principally the brilliant blue and white of water in all its forms—our most precious resource. Ninety-seven percent of this water is in the oceans. Of the earth's freshwater, more than 99 percent is retained in the ice cap and glaciers or is underground; less than 1 percent is in lakes, rivers, and soil. Yet even this amount is tremendous. The average annual rainfall in the United States is approximately 30 inches, depositing about 4.3 trillion gallons per day. But only a third of the water precipitated in the continental United States is west of a north-south line passing along the Kansas-Missouri border and two-thirds of that one-third is precipitated in the Columbia River Basin. About half of the United States is arid or semiarid.

### How Water Is Used

About 70 percent of the annual rainfall is lost to the atmosphere by evaporation or transpiration, and the latter is the portion most important to man, for it is the portion that sustains plant and animal life. About 2,500 pounds or 300 gallons of water are transpired by the wheat plants that produce the grain for 2.5 pounds of bread, the quantity that would be required to sustain life if man could "live by bread alone." Alfalfa transpires 800 pounds of water in producing 1 pound of dry matter. A steer not only drinks about 12 gallons of water a day, but also consumes about 30 pounds of alfalfa, for the production of which 24,000 pounds of water are consumed. If a person eats one pound of meat per day, this represents a consumption of 2,900 gallons of water, about 25 times more than for a pound of wheat (Bradley, 1962; Maxwell, 1965).

The daily per capita use of water in the United States, principally by the indirect means just discussed, is more than 15,000 gallons, 95 percent of which is lost to the atmosphere. At that rate of water use, the nation might not be able to reach a population of 230 million without a change in its way of life. In 1970 it was estimated that 7 percent of the water used in the United States was for municipal and domestic purposes, 38 percent for agriculture, and 48 percent for industry (Pereira, 1973). The limiting restraint on food production may be the water budget rather than available land (Bradley, 1962; Maxwell, 1965). Likewise the water budget may be the limiting factor in the current mindless expansion of metropolitan areas. For example, in the mid-1960s the cities of the eastern seaboard of Pennsylvania, Delaware, and New Jersey were already using 2.9 billion gallons of water per day in a region where only an estimated 3.5 billion gallons per day are available (George and McKinley, 1974).

### Possible New Sources of Water

There is now no practical source of artificially developed energy that could substantially contribute to the work the sun is doing in separating freshwater from the saltwater of the sea and carrying the freshwater far inland. A modest contribution to coastal water supplies might be made by desalting seawater even with our present energy supply; Congress approved a $200 million five-year research and development program toward that end, involving pilot-scale test plants. The four basic desalting processes that are being developed are distillation, electrodialysis, reverse osmosis, and crystallization. The brine and heat would have to be dissipated in some way in the surrounding sea, drastically altering the ecologic balance. If nuclear fission rather than nuclear fusion is employed, tons of highly lethal radioactive waste would have to be disposed of annually—waste that would remain deadly for about a thousand years. The cost of domestic water obtained from the sea and delivered to the home would be 35-40 cents per thousand gallons. The agriculturist probably could not afford to pay more than about 5 cents per thousand except in the case of a few high-value crops (Peterson, 1975).

It is tempting to view the enormous uninhabited desert areas of the Great Basin and Southwest as a potential granary, but there is little factual basis for this belief. Far greater returns per unit of capital and effort invested would derive from increased erosion control and fertility maintenance of the soils of the humid East.

Many people have endless faith in the "technological fix." "It is hard to conceive of a growth-induced disaster," says Richard Zeckhauser (1973), "that could not be averted or deflected by some scientific innovation. Indeed, nothing more than the successful exploitation of fusion energy would splash rosy paint on many black predictions of recent years." But there is no escape from the facts of thermodynamics.

One can imagine a circumstance in which an unlimited source of energy might enable man to develop freshwater from the sea and pump it up to any arable land surface, provided, of course, that the separated brine and the heat evolved in this Brobdingnagian operation would be ecologically tolerable. Even without such a quantum jump in energy requirements, climatologists and physicists have calculated that, at the present rate of increase in energy utilization of 4-6 percent per year, waste heat from fossil and nuclear fuels would result in an intolerable amount of heat pollution (the "thermal barrier") within periods estimated to be as little as 80-180 years in the future.

The most widely accepted hypothesis of future climates is that the cumulative effect of all mankind's environmental modifications, particularly the increase in carbon dioxide from burning fossil fuels, will be predominate over natural processes of climate change before the turn of the century (Mitchell, 1977; Norwine, 1977; Kellogg, 1978). Carbon dioxide is one pollutant that will not be easily controlled, for it is a product of combustion of all hydrocarbon fuels. Its release into the atmosphere has been increasing at an average rate of 4.3 percent per year. Nearly 20 billion tons of the gas are annually released into the atmosphere through combustion. Only about 20 percent of this gets back into the carbon reservoirs, principally investing the oceans or vegetation (Mitchell, 1977). W. W. Kellogg of the National Center for Atmospheric Research estimates that the atmosphere's present carbon dioxide level of about 325 parts per million by volume (ppmv) could be doubled by the year 2050 if the present rate of increase continues (Kellogg, 1978). Carbon dioxide is nearly transparent to sunlight but reflects back to earth a significant part of the terrestrial heat radiation that otherwise would pass through the atmosphere into space. The net effect is similar to that of a layer of glass in a greenhouse. A doubling of the present level of atmospheric carbon dioxide would increase world temperatures by about 4.5°F (2.5°C). A heat increase of about 3-5 times greater than the global average is expected in the polar regions (Manabe and Wetherold, 1975). According to Kellogg (1978), the earth's ice sheets on land contain enough water to raise sea level by 213 feet (65 m), if they were all melted. This would cause a flooding of coastal lowlands.

If a maximum carbon dioxide increase of 50 percent above the preindustrial concentration would be tolerable, its production would have to be reduced rapidly after the beginning of the twenty-first century. We would be able to burn in total, over the next centuries, not much more than 10 percent of the known fossil-fuel reserves (Siegenthaler and Oeschger, 1978). Converting to sources of energy not based upon coal, oil, and gas would appear to be a matter of high priority and one that we can scarcely afford to leave to the next generation (Mitchell, 1977), but nuclear energy does not appear to be the answer, for the possibility of reaching the thermal barrier would not be eliminated (Benoit, 1976; Wilcox, 1975, 1976).

Alvin M. Weinberg, director of the Energy and Development Office of the Federal Energy Administration, believes that we will have to adjust the world's energy policies to take into account the limit imposed by the thermal barrier on energy production in as little as 30-50 years. He proposes establishing one or more institutes of climatology with a long-term commitment to determine the global effect of man's production of energy (Weinberg, 1974).

A large-scale plan for importing Canadian and Alaskan water has been proposed by the North American Water and Power Alliance (NAWAPA). The plan envisions water supply to all of the western states and to the Great Lakes. The technical and economic parts of the plan were made by the water-resources-planning engineers of the Ralph M. Parsons Company, based in Los Angeles, leaving problems of jurisdiction and organization to social and political specialists. The NAWAPA plans are continental in scope, affecting Canada, the United States, and northern Mexico. One obtains an idea of the unprecedented magnitude of the project from the size of just one feature: four dams, on the Fraser, Thompson, Columbia, and Kootenay rivers, would result in inundating the divides between the headwaters of those rivers, creating a freshwater lake 500 miles long and 5-15 miles wide. Mexico, at the lower end of the NAWAPA system, could receive enough water to irrigate seven or eight times more land than Egypt will reclaim as the result of the construction of the Aswan High Dam. A concise popular account of the vast, complex, interrelated system may be found in *The Water Crisis* (1967), written by Senator Frank Moss of Utah. The cost of the NAWAPA development is estimated to be $100 billion over a 30-year period, provoking one critic to call it "a boondoggle visible from Mars." But in 30 years $100 billion would be only about 10 percent of GNP and an expenditure of 3 or 4 billion dollars a year would be only about 10 percent of our present annual defense budget (Peterson, 1975).

At least at this stage, the NAWAPA plan is opposed by many in the water-donor areas as well as by conservationists. In the final analysis the plan will have to be evaluated on the basis of risk versus benefit, just like any other large-scale environmental modification. Quite aside from the many environmental problems with which Canadians are concerned, NAWAPA reflects a basic human dilemma discussed in R. C. Bocking's *Canada's Water: for Sale?* Future water requirements are generally based on projections of probable population, industrial, and agricultural growth of the future. But such projections also remind us that, at present rates of population and industrial growth, in less than two centuries the United States would be completely covered with electrical generating plants or totally paved with freeways and parking lots. So it is currently fashionable to extend projections to the year 2000. "Building for such predicted levels of water use will incorporate such a rate of growth in water consumption that the problems of keeping up with requirements in the year 2001 will be far greater than those encountered in supplying the quantities 'required' prior to 2000" (Bocking, 1972).

Artificial modification of precipitation has a solid scientific basis. There is no doubt that the introduction of condensation nuclei in the form of dry ice or silver iodide crystals causes supercooled fog or mist to freeze into ice, thus permitting precipitation. A National Academy of Sciences Panel presented data indicating that cloud-seeding operations have resulted in precipitation increases of 10-20 percent. Carefully designed engineering research and development efforts are now underway. A criticism of cloud-seeding is that any change in the weather is likely to be detrimental to someone. Seeding cannot be controlled sufficiently to plan the precipitation either as to amount or location. Also, drought or floods could be blamed on seeding even if no proof were provided.

There are many ways of extending scarce water supply in arid regions, ranging from rainwater harvesting from hillslopes and man-made catchments to controlled-environment facilities. The latter are expensive, but require very little water and are currently popular in some Arab countries long on newly acquired wealth and short on water (NAS, 1973).

Where urbanization occurs, municipal wastewaters are available for irrigation. The use of such water is increasing throughout the nation (Godfrey, 1973). With a discharge rate of 100 gallons per day per capita with each increase of 100,000 in urban population about 11,200 acre-feet of usable wastewater per year becomes available, sufficient for irrigation of 2,500 acres of farmland, assuming 3.4 acre-feet delivered per acre and a 25-percent loss in storage and conveyance of water to the land (DeCook and Cluff, 1973). The technology also exists for reconstituting municipal wastewater for its original use. Also, irrigation practice can be greatly modified as a means of conservation.

Water conservation in farm practices and the development of plant varieties that use less water have been suggested as ways of helping to solve the water-shortage problem in arid and semiarid areas. New conservation steps in use of water for domestic and industrial practices would also be helpful. In the refinement of crude petroleum, for example, water requirements now range from as low as 1.73 to as high as 44.5 gallons of water per gallon of petroleum (Peterson, 1975).

## Water and Fossil-Fuels Development

Plans proceed for "development" in the arid West with alarming disregard for limits set by availability of water. The direct and indirect consequences of strip mining in the West may be much more serious in relation to the quantity of water required than in relation to the ability to reclaim mining sites; yet the latter has received most attention. The 176 potential sites for coal gasification can be calculated to require almost 1.5 trillion gallons of water per year, roughly thrice the water than the annual flow of the California Water Project, one of the world's largest. A million-barrel-per-day oil-shale industry would require from 121,000 to 189,000 acre-feet of water a year (Perelman and Gardner, 1975). (There are at least a trillion barrels of shale oil in Colorado alone.)

Such plans for the development of the West could be made only at the cost of decline in western agriculture.

Fortunately, a breath of fresh air is emanating from the nation's capital. Secretary of the Interior Cecil D. Andrus assures us that the BLM (Bureau of Land Management) will no longer stand for Bureau of Livestock and Mining, adding that "The public's lands will be managed in the interests of all the people because they belong to all the people." Andrus has called for a five-year moratorium on new strip-mining ventures on prime agricultural land. The same spirit pervades throughout the new administration. For example, Secretary of Agriculture Bob Bergland, himself the owner of a 600-acre farm in Minnesota, is concerned with the heavy and continuing losses of topsoil from erosion with intensification of food-production pressures on the land and says he will ask the president's science adviser, Frank Press, to form a panel to tell the public the probable results of another 25-30 years of the current rate of land degradation. Many an eyebrow will be raised in some circles at the two questions that Bergland believes call for a larger research effort: (1) the effect of increasing use of nitrogen fertilizer on the atmosphere's ozone layer and its long-term implications for agriculture, and (2) human nutritional needs. He points out that "We know more about the nutritional needs of a dairy cow than about the nutritional needs of children" (Carter, 1977 *a, b, c*).

## DESERTS IN THE COMING SOLAR AGE

Just as wind might be utilized as an abundant source of energy on the Great Plains (chapter 6), sunlight might be utilized for the same purpose in the desert areas of the Southwest. In a prize-winning paper presented at the 1977 meetings of the Woodlands (Texas) Conferences on Growth, Jerome Weingart (1977) suggested a network of solar systems ranging from small localized units to large complexes producing electricity and synthetic fuels with interconnection over thousands of kilometers. Available for large-scale use would be the more than ten million square kilometers of arid sunny wastelands of the globe. (Southwestern United States and northern Mexico have a generous share of such wastelands.) According to Weingart's scenario, energy would be transmitted electrically or as hydrogen, the latter via pipeline or tanker. Technologically advanced but sun-poor nations of Western Europe, for example, would form partnerships with sun-rich neighbors farther south. Hydrogen would be stored for short periods in aquifers and for longer periods (decades to centuries) in other natural formations, including depleted oil and gas fields and similar formations containing water. Weingart believes that indirect forms of solar energy (hydropower, ocean thermal gradients, wind, and waves will contribute in a relatively limited way and that the same is true of biomass, because of competition with other land uses. He believes that an energy supply comparable with the potential from the fast breeder and fusion, can be obtained only with the high-efficiency, direct-conversion solar option: thermal energy for low grade end uses; solar thermal hydrogen from solar-generated, high-temperature heat of 600 to 2500°C; solar thermal electricity; and photovoltaics. However, energy systems could be tailored to suit the needs and structures of a wide variety of communities around the world, amenable to local control and management. A new kind of rural society, not possible with nuclear energy sources, is an exciting possibility. And all within the framework of a stable, sustainable world.

Some authorities believe that all possible sources of energy, including nuclear, must be exploited to fulfill the needs of the post-petroleum world (Weinberg, 1978). If true, this would indicate the degree to which mankind has allowed itself to be painted into a corner. But an ever-increasing number of authorities consider fossil fuel and nuclear energy to be only interim sources until the various solar alternatives have reached their full potential. Denis Hayes, in his *Rays of Hope: The Transition to a Post-Petroleum World* (1977), an in-depth study of energy alternatives, points out that some currently impractical technologies, for example photovoltaic cells, will probably become increasingly competitive with further technical improvement, the advantage of economies of scale, and the rapidly increasing cost of fossil fuels and nuclear energy. Construction costs for nuclear power plants, for example, have more than quadrupled in recent years, and there are also high environmental and social costs. Hayes believes that existing solar technologies could meet most of the energy budget of the United States and Canada.

The President's Council on Environmental Quality reports (CEQ, 1978):

> For the period 2020 and beyond, it is now possible to speak hopefully, and unblushingly, of the United States becoming a solar society. A national goal of providing significantly more than half of our energy from solar sources by the year 2020 should be achievable if our commitment to that goal and to energy conservation is strong. . . . In principle, with an average collection efficiency of only 15 percent—achievable with present technologies—all of our current energy needs could be met by using roughly 1 percent of our land area. As a perspective, approximately 17 percent of our land area is now used for crops.

According to *Science* staff writers A. L. Hammond and W. D. Metz (1978):

> But predictions of any kind are difficult because new collectors are still being designed, new materials are being developed, and the integration of solar collectors into practical energy systems is just beginning. What is certain is that solar collectors are a booming field of technology. If the past few years are prelude to the future, it will be a future energized by sunlight collected in myriad ways.

# 9 California

*The family farm has preserved the values—honesty, dependability, hard work, and faith—which we need to rediscover as a nation. There is no more dramatic success story than the story of the American farm family, and all of us are proud of its achievement.*

—Jimmy Carter

California, 158,693 square miles in area, is exceeded in size in the coterminous United States only by Texas's 267,339 square miles. Its medial length is 780 miles and its width ranges from 150 to 250 miles. Its largest agricultural area is the Central Valley, which ranges from the junction of the Cascade Range and the Klamath Mountains in the north to the Tehachapi Mountains in the south. The latter form a short connecting link between the Coast Ranges and the Sierra Nevada and form the northern rim of the Mojave Desert. The Central Valley is about 400 miles long and an average of 50 miles wide. It is bordered on the east by the Sierra Nevada and on the west by the Coast Ranges. The sources of the many tributaries of the southward-flowing Sacramento and the northward-flowing San Joaquin are in the Sierra Nevada. The two rivers converge northeast of San Francisco, flow through the agriculturally rich delta region, and empty into Suisun Bay, an arm of San Francisco Bay. Other smaller agricultural valleys lie within the Coast Ranges or between them and the sea. They include the Santa Rosa, Napa, Livermore, Santa Clara, Salinas, and San Luis valleys.

In the south below a complex of narrow, generally east-west ranges, including the San Gabriel, San Bernardino, and San Jacinto mountains, are remnants of what were once rich agricultural valleys that included a large acreage of citrus trees (see fig. 6) but great areas of urban sprawl invaded most of this land. East of the southern California ranges, and east and south of the Sierra Nevada, as well as in the rugged northeast corner of the state, lie large areas of mostly desert land located in Region D (fig. 3). Much of this land, particularly in the southeast (Imperial, Coachella, and Palo Verde valleys), is irrigated and produces a year-round abundance of many crops.

But that is getting far ahead of our story. In the sixteenth century California was a faraway legendary land, set off from the rest of the world by desert and mountain wilderness and the Pacific Ocean. Even the name "California" was steeped in legend originating in a novel that was at the height of its popularity when Hernando Cortez was exploring America. California was the name of a mythical island of tall, bronze-colored Amazons ruled by the pagan queen Calafía, who assisted the pagan forces besieging Constantinople. It was the bold empire-building Spaniards of the sixteenth century who removed New World California from the realm of legend.

## CALIFORNIA INDIANS

Bones with carvings depicting animals now long extinct were taken from the La Brea tar pits in Los Angeles. The bones were found by radiocarbon dating to be 15,500 years old. More recent sophisticated dating procedures indicate that man existed in California at least 50,000 years ago (Schiller, 1975). Despite its antiquity, the aboriginal population of California had not reached more than about 275,000 by the time the Spaniards arrived. Yet this was then about 30 percent of the population of the present coterminous United States. Thus at the time of the first Spanish settlement in 1769, the population density of California Indians was about six times the national average. This relatively high population might be attributed to the mild climate, plentiful food, and possibly the peaceful coexistence of the many tribes.

The Indians appeared to have experienced little cultural evolution during the long period of their existence in California. The Spaniards could not fail to notice their primitive culture when compared with that of the Aztecs of Mexico, nor did the pioneer Anglo-Americans fail to notice their primitive culture when compared with that of the Iroquois of the Northeast or of the civilized Indian tribes of the Southeast. Nevertheless, the California Indians had an interesting oral literature of some 135 different languages in 21 or 22 linguistic families (Kroeber, 1939). Usually the highest political unit was the village community. There was nowhere a nation of several thousand people, with a strong and militant national consciousness, as in other parts of North America, so it was impossible to organize any widespread and effective resistance to the white man (Kroeber, 1925; Bean, 1968; Heizer and Whipple, 1971).

California Indians lived a hunting-fishing-gathering type of existence. They raised no crops except in the extreme southeast where the Yuma Indians dropped corn, bean, and pumpkin seeds in the mud left by the annual flood waters along the lower Colorado River.

The valley floors and foothills over much of California contained many oaks, and acorns provided the natives with a substantial part of their diet. Acorns were gathered in season and stored in raised cylindrical cribs. They had a higher caloric value than wheat. By pounding the husked acorns with a stone mortar and pestle and winnowing them by tossing them in a shallow basket, a flour was produced which could then be repeatedly dowsed with water to wash away the tannic acid. Usually the flour was placed in a basket with water and brought to a boil by placing hot stones in it. (Baskets might be caulked with pitch or tar, but often they were so tightly woven that caulking was not necessary to make them hold water.) The resulting porridge was eaten plain or flavored with berries, grass seeds, nuts, or bits of meat or fish. Less often, bread was baked in an earth oven. Deer and small game were abundant and insects were widely accepted as food. Fish were plentiful in streams, and fish and meat were often dried. Along the coast, mounds of shells testify to extensive use of shellfish, some heaps accumulating for periods of up to 3,000 years (Kroeber, 1925; Bean, 1968).

Despite their Stone Age level of existence, California Indians surprised the Spaniards with the facility with which they learned to speak the Spanish language and mastered the mechanical arts. The California missions were built almost entirely by Indian workmen under the direction of the friars. They became skillful carpenters, weavers, and farmers, and were surpassed by none as cattle herders (Rolle, 1963).

California Indians were at first notably docile; easy to preach to and control. Regardless of how honorable the initial intentions of the missionaries may have been, the fact remains that disease brought by white men from the Old World decimated the Indians to such an extent that the Spanish military had to go inland to the coastal ranges and even into interior valleys for more Indian workers. Then the number of fugitives from the missions and the extent to which the Spaniards had to resort to punishment became clear indications of the psychic upset suffered by the forcibly recruited Indians (Borah, 1970).

During the Spanish and Mexican regimes, the military and civil population arriving from Mexico were mostly mixtures of European, Indian, and Negro, who would have been in the lowest social strata in Mexico—laborers and servants. In California the indigenous Indians served as such menials in addition to supplying women as mates for the colonizers. Unlike the later-arriving Anglos,[1] the Ibero-Americans considered intermarriage with indigenous peoples as the natural result of conquest. The *mestizo* became the dominant population. Thus the Indians were considered a useful element of the population.

The gold rush and subsequent economic activity brought a steady stream of Anglos and their families to California. They intended to do their own work and brought with them a bitter, frontier-bred hatred for Indians. The Anglos invaded Sierra foothill and Central Valley areas that had not been settled by Spanish-Americans and Mexicans. Diseases (smallpox, malaria, typhoid fever, measles, tuberculosis, and syphilis), liquor and white military campaigns that generally involved the burning of native villages and destruction of stored food led to an accelerated Indian population decline. By 1880 there were not over 20,000 Indians left in California (Cook, 1970).

## VEGETATION BEFORE THE SPANISH OCCUPATION

When Spanish colonists arrived in California in 1769, they found a pleasant Mediterranean-type climate and an abundance of nutritious forage for the livestock that was to become the mainstay of their economy. Records of Portola's expedition from San Diego to San Francisco in 1769-1770 attest to the high quality of the indigenous forage. Although available records are not adequate for a precise picture of California's pristine grasslands, when combined with knowledge from modern ecological studies of plant distribution and succession they serve to give a fairly reliable composite description.

Burcham (1957, 1961, 1975) estimated the approximate areas, in millions of acres, of plant communities of California at the time of the arrival of the Spanish colonists as follows: prairie, 22.3; oak woodland, 10.6; marsh grass, 0.5; sagebrush, 8.1; chaparral, 9.0; coniferous woodland, 3.2; coniferous forest, 21.5. The remainder was desert. Some 52.7 million acres, about 53 percent of the California total land area, were grazeable.

[1]According to Webster, the word Anglo may be taken to mean "of English origin" or, in the Southwest, "an Anglo-American as distinguished from a Spanish-American or a Mexican." Yet the term is commonly used to refer to any Caucasian not of Spanish-American or Mexican descent. It therefore includes many ethnic groups not of English ancestry and is used in that sense throughout this book.

## Grasses

Prairies of the Central Valley and valleys of the Coast Ranges, as well as the herbaceous cover of the oak woodlands of the foothills, were dominated by perennial bunchgrasses, chiefly needlegrasses (*Stipa* spp.). Only one of the dominant grasses, the beardless wild rye (*Elymus triticoides*), departed from the bunchgrass habit. It was a sod former that covered large areas of the central San Joaquin Valley and some of the larger coastal valleys. In the prairies of the cooler, moister northern coast ranges, oat grasses (*Danthonia*), fescues (*Festuca*), hair grasses (*Deschampsia*), and bent grasses (*Agrostis*) largely replaced the needlegrasses and their associates as dominants.

Associated with the grasses were seasonal societies of broad-leaved herbs known as forbs. When in full bloom, their brightly colored flowers gave the characteristic aspect to the landscape that was described in glowing terms by early travelers in California prairie and woodland areas, such as T. F. Cronese (1868) and John Muir (1911). Many were perennials, but even more were annuals. The most important forbs to the stockmen were the native legumes—many clovers, lupines, trefoils and deervetches (*Lotus*), and related genera. Authorities believe that an extensive livestock industry could have been sustained indefinitely, even under yearlong grazing, under methods of management adjusted to growth characteristics of the plant cover.

Overgrazing resulted in native annual plants of inferior quality, as well as many accidentally introduced species, rapidly taking over. This took place to a much greater extent in California than in areas east of the Rockies. The remains of some introduced species (annual bluegrass, *Poa annua;* mouse barley, *Hordeum leporinum;* and ryegrass, *Lolium multiflorum*) were found in adobe bricks from the oldest portions of the earliest mission buildings. Other introduced species were wild oats (*Avena fatua* and *A. barbata*), black mustard (*Brassica nigra*), filaree (*Erodium*), wild barley (*Hordeum*), nut grass (*Gastridium ventricosum*), and many others. Fortunately some of the alien plants had well-recognized value as forage. These included wild oats, soft cheat (or chess) (*Bromus mollis*), ryegrasses, bur clovers (*Medicago*), filaree, and others. Also some good range grasses have been planted.

The seeds of the herbaceous annual plants germinate after the first "effective rain" (about one-half inch in a single storm) in the fall. The plants then grow slowly through the rainy winter, bringing the luxuriant carpet of green so pleasing to the eastern visitor, along with many showy flowering plants in some areas. The annual vegetation matures by late March or early April in the south and by June in the north, then scatters its seeds and dries up (Burcham, 1957, 1961, 1975).

Prominent among the conspicuous grasses of the valleys and foothills are the wild oats, believed to have been first accidentally introduced, along with livestock, on the early sailing ships from Spain. They grow rapidly after the winter rains start in earnest and create much of the green landscape of winter and spring. In the summer the fields of wild oats change from green to gold, first along the hilltops where the soil is driest, then everywhere, changing eventually to the yellow-brown of summer and fall, set off in hilly areas by the dark green of live oaks. The distinctive sound of the California coastal breezes in fields of wild oats is caused by the tiny metallic glumes, the chaff around the grains, clapping against one another—"a rustle as of stiff petticoats" (Anderson, 1952).

## Forests

Of California's 100 million acres of land, forest land comprises over 42 million acres, of which 17 million are commercial forest, exceeded only by Oregon and Washington.

The timber species in California, in order of volume of raw timber harvested annually, are Douglas-fir, true firs, ponderosa pine, coast redwood, sugar pine, other softwoods, and hardwoods (California Almanac, 1973).[2]

The coast redwood (*Sequoia sempervirons*) is almost entirely confined to California. It is a tall, massive tree, which when mature generally ranges in height from 100 to over 300 feet and in diameter from 2 to 16 feet or more. It is a rapidly growing tree that reproduces from seed or from sprouts that grow out from stumps.

The coast redwood reaches a relatively great age because, like its near relative the giant sequoia (*Sequoiadendron giganteum*) of the Sierra Nevada, it is highly resistant to fire, insects, and fungus disease. While its normal life-span is from 500 to 800 years, the greatest age so far determined by count of growth rings is 2,200 years. This may be compared with 3,200 years for the oldest giant sequoia and 4,600 years for the oldest of California bristlecone pines (*Pinus aristata*). The giant sequoia also exceeds the coast redwood in trunk diameter with a record of 32.2 feet (Adams, 1969).

The early Spaniards, confining themselves to the coastal valleys and foothill grasslands, saw (and utilized) very little of California's rich forest resource.

### Wildlife

After Jedediah Smith and his party of mountain men reached southern California in 1826, recounted in the preceding chapter, they proceeded north, crossed the Tehachapi Mountains and trapped the tree-lined rivers of the wild, unmapped Central Valley of California, finding beaver to be abundant. Deer, antelope, and wild horses roamed by the thousands across the valley floor and in the surrounding hills. Elk were said to be almost as abundant in the valley as the cattle of the Spaniards were on the Coast Ranges. (Twenty years later they were profitably hunted for their hides and tallow.) Brown bear and grizzlies multiplied unmolested and wildfowl were plentiful on lakes and ponds. Indians populated the 300-mile length of the Central Valley and, fortunately for the early white explorers, they were relatively peaceful.

## THE SPANISH REGIME
(Rolle, 1963; Caughey, 1970)

What is now California was discovered by the Portuguese-born navigator Juan Rodriguez Cabrillo, whose two ships entered San Diego Harbor on September 28, 1542, after three months at sea from the port of Navidad on the west coast of Mexico. Cabrillo sailed as far north as a cape near what was later to become known as Fort Ross, seeking the nonexistent Strait of Adrian (believed at the time to be a shortcut from Europe to the Orient) and not seeing such important landmarks as Monterey Bay and the Golden Gate.

The English privateer, Francis Drake, whose vessel was the famous *Golden Hind,* made some landings in California, claimed title for the queen of England, and called the land New Albion. This was destined to hasten Spanish occupation of the area. Subsequent voyages by Spaniards, particularly that of Vizcaíno in 1602, added much to the knowledge about California, but from 1602 to 1769 no further explorations occurred. This Spanish lethargy regarding California was ended by Russians from the north, in seach of furs.

[2]"Forest land" is at least 10% occupied by forest trees, of any size, and is not currently developed for non-forest use (USFS, 1973).

## Portolá and Serra

In 1769 the soldier Don Gaspar de Portolá and a Franciscan friar, Fray Junípero Serra, were sent to Upper California to occupy the area. They were both capable and conscientious men and set about to occupy the ports of San Diego and Monterey and to establish five missions. Serra's cargo included seeds of flowers and vegetables from the Old and New Worlds. The Spaniards, long familiar with irrigated agriculture, were the first to practice the ancient art in California, bringing water via a simple system of canals from the San Diego River to the San Diego Mission. (Subsequent facilities included a large diversion dam and a six-mile conduit to the mission, the remains of which can still be seen.) A herd of 200 cattle was driven into Upper California from the northernmost mission of Baja California. These were the progenitors of California's chief source of wealth for several generations, thanks to the immense areas of native grassland, as previously related. This first colonizing venture involved two expeditions by land and two by sea.

It is fitting to recall that the hardship and suffering endured and the indomitable will demonstrated by the Anglo-American pioneers who crossed the continent from the eastern United States were matched in every respect by the early Spaniards who penetrated California from the south. Those who came by sea were stricken with scurvy and dysentery; those by land with scurvy, hunger, thirst, and exhaustion. Yet among the latter, Portolá selected sixty "skeletons," and leaving Father Serra behind to save heathen souls, the party made a long overland trek north, for which the survivors were treated with the magnificent sight of San Francisco Bay on November 2, 1769.

## De Anza's Overland Route

The routes to Alta California by sea or overland via Baja California proved to be too difficult, hazardous, and time-consuming for bringing families of settlers to the province and adequately supplying them. A land route from Sonora, Mexico, needed to be established. For this formidable task the Spanish viceroy Bucareli made a fortunate choice in selecting Captain Juan Bautista de Anza, who was experienced in Indian control, Apache fighting, and desert campaigning. Anza's expedition in 1773-1774 to explore the trail and in 1775-1776 to bring supplies, livestock, and families of settlers to Alta California, has been compared in length of journey and difficulties with terrain and hostile natives to the Lewis and Clark expedition to the Pacific Northwest and the subsequent migration of pioneer settlers to Oregon via the Oregon Trail, respectively. Perhaps the more remarkable of the two Anza expeditions was the second, which involved transporting 240 persons, 695 horses and mules, and 355 cattle. That expedition eventually arrived at what is now San Francisco, where a presidio was founded on September 17, 1776, and the mission was dedicated on October 9 of the same year.

## Spanish Missions

A chain of missions ranging from San Diego to Sonoma, about thirty miles apart (a day's journey on horseback), was built along *El Camino Real* (The King's Highway). The mission sites were selected for good soil, convenient water supply, and a large local Indian population. The mission buildings, the ruins of which are well known to Californians, were made of adobe brick or cut stones and were slowly built by Indian labor. They were distinctive and well suited for California conditions. The red-tiled roofs resisted fire and the thick walls, sometimes reinforced by heavy buttresses, were a provision for protection against earthquakes. Open courts, long colonnades, and many arches and corridors were characteristic—a combination of Moorish and Roman influences.

Usually there were two friars at a mission and they were virtual rulers of the mission domains, treating the Indians as wards. The elder padre had charge of interior matters and religious instruction, and the younger had charge of agricultural and other outside activity. The Indians were taught various trades and under suitable direction accomplished remarkable feats in constructing buildings, dams, and canals. Agricultural surpluses, such as meal, wine, oil, hemp, hides, and tallow, were shipped down the coast of New Spain, principally to Acapulco, where they were exchanged for clothing, furniture, implements, and tools. By the time of Father Serra's death in 1784, 5,800 natives had been converted to Christianity in the nine missions that he founded. They were busily engaged in spinning, weaving, carpentry, masonry, raising crops, and tending thousands of acres of flocks and herds.

The San Gabriel Mission, where Jedediah Smith and his party rested in August 1826, after their historic trek from northern Utah, had some 30,000 cattle, and thousands of horses, sheep, and swine. Besides grainfields and vineyards, there were orchards of oranges, apples, peaches, and figs. There was a water-driven gristmill and a distillery. Skilled workers produced cloth, blankets, and soap.

After Serra's death, it was principally the capable Fray Fermin de Lasuen who carried on with the expansion of California missions, serving in that capacity for eighteen years (from 1785 to 1803). During that period the number of missions doubled and Serra's nine missions were rebuilt, all in what is now known as the mission style. The Christian Indian population increased to about 20,000. Under Lasuen, stock-raising and farming activities were increased and diversified. Activities other than agriculture were introduced. Artisans from Mexico helped the Indians become good carpenters, masons, and smiths. However, Spain treated California as a mere holding operation, finding no advantage in building up the colony. A proposed second chain of missions in the interior was not authorized.

## Pueblos

Settlers were induced to reside in pueblos, where community life revolved around the central plaza. Each settler was entitled to a house lot, livestock and implements, and an allowance for clothing and supplies amounting to $116 for each of the first two years and $60 for the next three. He was also allowed the use of government land and was exempt from taxes for five years. He was required to sell his surplus agricultural products to the presidio (frontier fortress) and be ready for military service, along with his horse and musket, in any emergency. Each settler was also required to build a house, dig irrigation ditches, cultivate the land, repair his implements, and maintain a specified number of animals. In five years he obtained permanent title to his land.

The Spaniards paid little attention to the interior valley between the Coast Range and the Sierra Nevada although, beginning in 1806, Gabriel Moraga organized many military expeditions against the Indians that sometimes reached into the foothills of the Sierra. Moraga's forays into the Sierra foothills left a heritage of Spanish place names, or their English equivalents. However, missions and pueblos were not established in the Central Valley by the Spaniards or the Mexican governments that were to follow. Thus the vast area was to remain of secondary importance until the hordes of immigrants from the eastern United States arrived.

### The Fur Trade

The Spanish became apprehensive concerning the interest shown by Russia and England in the Northwest, but their troubles began in earnest when the Americans started illicit trading with Californians and engaged in the lucrative trade in sea otter furs with China around the turn of the century. Contraband trade succeeded because of New Spain's neglect of California, reflected in the poverty of its inhabitants, who were eager to obtain a few amenities of civilization. Yankee traders, and particularly the whalers, who might be away from the home port for as long as four years, appreciated the use of several protected ports in California in which to repair their ships and stock them with fresh meat, fruit, grains, and other provisions for the long return trip around the Horn. The whalers in turn had such items as needles, stockings, jewelry, thread, and bolts of cloth, which were comparative luxuries to the Californians. Whalers were instrumental in popularizing California among Easterners with their glowing accounts of that pastoral province of New Spain.

The Russians in Alaska were rich in furs but poor in food supplies and were eager to purchase the latter from Yankee sea captains. The Yankees suggested that the Russians send the skillful native Aleuts of the Aleutian Islands to hunt land and sea otters "on shares." Subsequently the Russian-American Fur Company was formed and sent its own hunters to California. In 1812 the Russians built warehouses at Bodega Bay, about fifty miles north of San Francisco, for the storage of goods. Ships wintered and made repairs at Bodega, but the Russians built a rectangular fort, mounted with cannon, on a plateau about eighteen miles north of Bodega. This came to be known as Fort Ross. The Russians did not leave until 1841, when they sold all movable property to a Swiss immigrant, John Augustus Sutter, for $50,000, and ended their thirty years of occupancy of the outpost. They had threatened to burn it rather than give it away. A few partially preserved buildings and some archival records are all that remain at Fort Ross today.

## THE MEXICAN REGIME
(Rolle, 1963; Caughey, 1970)

The banner of Spain was lowered and the Mexican imperial flag was unfurled at Monterey in September 1822. By 1895, as Mexican control of California became increasingly jeopardized, the governors were authorized to use land grants as compensation for services or to sell lands to pay bills and raise money for defense. During the brief Mexican regime (1821-1848) more than 800 grants were made, totaling about 10 million acres. Although these grants are often referred to as the "Spanish land grants," under Spanish control less than 30 grants were ceded to private persons; the missionaries feared that the Indians would be morally contaminated by settlers.

### The Ranchos

According to a colonization law of 1824, any Mexican of good character, or any foreigners willing to become naturalized and accept the Catholic faith, could petition for as many as 11 square leagues of land, although foreign settlers were usually unable to obtain grants within 10 leagues (about 30 miles) of the coast. A square league comprised 4,438 acres and a ranch of 4 or 5 square leagues was considered small. These large tracts were only roughly measured, and this resulted in much litigation when the land was later acquired by Anglos. The land was free from taxes for five years. Yankee maritime traders began to settle in California and marry into old California families. Spanish and Anglo cultures began to blend long before the first overland parties of American pioneers arrived.

Hides and tallow were much in demand, and many fortunes were made, especially after the beginning of the extensive traffic with English and American vessels. San Diego became the principal depot for the hide business, being the one good port in southern California and having ideal weather for the kind of hide-curing that was required for the long voyage around the Horn. According to one estimate, there were 1,222,000 head of cattle on the ranchos at the height of California's pastoral period. Little cash was exchanged; a dried steer hide, the "California bank note," had a value of about a dollar. Cattle received practically no attention except at branding and butchering time. They were valued mainly for hide and tallow. In seasons of drought, when forage had to be preserved for cattle and sheep, competition from the many wild horses on the range could not be tolerated. Wild-horse hunting became both a diversion and a necessary task.

For plowing, the Indian laborers usually scratched the soil with a rude wooden plow made from the crooked limb of a tree and possessing an iron point. Harrowing consisted of dragging large tree branches over the ground. Yet the virgin earth yielded well. Wheat was harvested with sickles; then it was scattered over a circular piece of ground that had been watered and pounded and became very hard when dry. From 75 to 100 mares were driven over the wheat until the grain was trampled out. For winnowing, the wheat was tossed against the wind. At first the grain was ground by hand, with stone mortars and pestles, and later by water-driven gristmills. However, a more common mechanical method was the *arrastre*. This consisted of a stationary millstone on top of which another was rotated by means of a crossbeam dragged around in a circle by a horse or mule. The grain was crushed between the two stones, as grain had been ground in Spain for centuries.

Each mission and most ranchos raised small flocks of sheep for their mutton and wool. The warm, wiry fibers were woven by Indians on handmade looms into cloth and blankets. Hogs were raised mainly for lard to be used in making soap. Sufficient flax and hemp were raised for making rope. The Spanish and Mexican colonials failed to exploit two of California's great natural sources of wealth, fur-bearing animals and gold. American and Russian hunters reaped the rich harvest of sea otter and seal pelts and gold was discovered later.

On the ranchos as well as in the towns, the houses were generally square or oblong unadorned structures with thick walls of adobe (sun-dried brick). This type of structure was warm in winter and cool in summer. Furnishings were simple and strictly utilitarian, except for the homes of some well-to-do rancheros in which mahogany furniture imported from South America or the Philippines might be seen. The wealthier rancheros might own a home in the pueblo as well as on the rancho. Likewise among the wealthy, costly and strikingly beautiful and picturesque gowns were worn by women at fiesta time, only to be exceeded in flashy brilliance by the ornate costumes of the men and the embroidered trappings of their horses. The most favored pastime was the dance, to the music of guitars, and the jarabe, fandango, and many other dances accompanied practically every public or private event; in fact it was a rare evening when there were no guests at a ranch house to join the family in singing and dancing. Hospitality was most generous; if a traveler arrived with an exhausted horse, a fresh one, saddled and bridled, was provided for his use the next morning. A few coins might be left in a dish in his bedroom, to save the traveler the embarrassment of asking, if there was reason to suspect he was in need. Travelers could also count on a night's lodging at the missions, about a day's trip apart by horseback.

Californians felt even less of a bond with Mexico than with Spain. The great distance, difficulty of communication, and strong local pride led the people of the province to identify themselves with California, not Mexico. Within Mexico there was continual internal revolution. The newly liberated country was financially exhausted, had difficulty in maintaining her own autonomy, and was helpless to protect herself from dismemberment in the Texas revolution. Mexico understandably had little energy to devote to California and neither money nor troops to send there in case of need. As late as 1848 the entire non-Indian population of the province was estimated to be only about 19,000, nearly equally divided between Californians (people of Spanish birth or background who became residents during either the Spanish or Mexican eras) and foreigners. There was increasing acceptance of Russian fur traders and English and American trading enterprises, all of which was pleasing to the rancheros, no longer dependent on the uncertain arrival of government supply ships.

### Secularization of Missions and Weakening of Mexican Rule

In 1833 Mexico promulgated a sweeping decree that secularized the missions of both Upper and Lower California, leaving them in the status of parish churches. Outright looting of mission properties became rampant. The missions began to crumble into dust, and Junípero Serra's dream was soon to become no more than a nostalgic memory.

There followed a succession of Mexican governors, all unpopular in a province where the citizens were beginning to call themselves Californians instead of Mexicans. Prominent native sons, such as Mariano Guadalupe Vallejo, became convinced that Americanization of the province would be a good thing—the best hope for California's future tranquility and progress. Rolle (1963) points out that this growing attitude among Californians in those days is sometimes forgotten by historians.

The man most successful in defying the Mexican authorities was Sutter, "a dreamer with a gifted tongue," who came to California to "escape a debtor's prison and an angry wife." In 1839 Governor Alvarado gave him permission to occupy 50,000 acres of land about two miles from the confluence of the Sacramento and American rivers and also conferred on him Mexican citizenship after he convinced the governor that he would act as a semiofficial representative of the government in the interior of the province. Sutter built a large fort, with an adobe wall 18 feet high by 3 feet thick. He mounted twelve guns on the fort. These he had obtained when he bought Fort Ross, along with a small launch and some horses and cattle, from the Russians. Sutter engaged in many pursuits, including the diversion of water from the American River for irrigation, planting large areas to wheat, grazing large herds of cattle and horses, building a flour mill and distillery, and running his launch on the Sacramento River on a regular service for freight and passengers from his colony, which he called "New Helvetia," to San Francisco. New Helvetia became the nucleus of all political and economic activity in what was then the only settled portion of the interior of California.

## THE INFLUX OF FOREIGNERS

Emigrant parties began the long and hazardous trek to California on horseback and by ox wagon long before the stampede for the gold mines in the late 1840s. Their imaginations had been fired by extravagant stories of "paradise" by such early California boosters as T. J. Farnham who, in his *Life and Adventures in California* (1846), wrote of the fabulously good life that the new land had to offer. At the same time he described the Spanish-speaking inhabitants as lazy, without ambition, and lax and inefficient in administration—a plight that presumably could be soon remedied by vigorous and ambitious Anglos. R. H. Dana's *Two Years Before the Mast* (1840) and Alfred Robinson's *Life in California* (1846) likewise created much interest, the former remaining popular to this day. "In the hands of an enterprising people," said Dana, "what a country this might be!"

The true pioneer settlers from the Midwest and East were farmers who came to California before the gold rush with the aim of making the Mexican province their home. Beginning in 1841, an important factor distinguishing Anglo settlers in California was their coming by an overland route rather than by the sea route. Nearly all were from the Midwest and many brought their families with them. Unlike former foreigners, relatively few married California women.

The first wagon train to cross the formidable barrier of the Sierra Nevada in order to reach California's Central Valley was that of the Bidwell-Bartleson Party in 1841. This party crossed the crest of the range about twelve miles north of Sonora Pass. A diary of the remarkable journey was kept by John Bidwell. In 1890 it was published serially in *Century* magazine and in a book titled *Echoes of the Past,* edited by M. M. Quife (1928). Bidwell struck gold at Bidwell's Bar on the Feather River, purchased the Chico Ranch, and expanded it to 26,000 acres. Besides much livestock and wheat, he eventually had over 115,000 fruit trees and some 200 acres of vineyard. He and his wife, Annie Kennedy Bidwell, donated the land on which Chico State University is now located (Hunt, 1942; Hunt and Adams, 1974).

Ordinarily the pioneer wagon trains, headed for what became known as California's Central Valley, followed the "California Trail." Whether they reached Nevada's Humboldt River from the north via the Raft River or from the east via Hasting's Cutoff and the Great Salt Lake, they followed most of the river's course, for it provided the only practicable east-west route through Nevada's numerous north-to-south ranges. Then the trail divided into a north and south branch, and these in turn rebranched into the nine most commonly followed routes over the Sierra Nevada through passes as far north as the Fandango Pass near the Oregon border and as far south as the Sonora Pass east of Stockton. In a six-year study of these routes, veteran backpackers T. H. Hunt and R. V. H. Adams (1974), armed with notebooks and cameras, retraced them step by step, as closely as possible, guided by old journals and diaries of the emigrants, occasional ruts worn by wagon wheels, resin-coated trail blazes on ancient trees, boulders with names and dates carved or painted on them, trailside graves, a few rusty barrel hoops, corroded nails, or bits of broken pottery. With great empathy and devotion they prepared a photo essay on the California Trail, with a text enriched by excerpts from the emigrant journals and diaries at appropriate places and with maps, sketches, 34 black-and-white and 147 beautiful full-color photographs. Hunt wrote: "You stand there in their tracks now, alone and in almost painfully silent wastes, and if you have any sort of feeling at all for the human species, you can't help feeling awed by the experience—awed, and humble, and proud."

The authors make an eloquent plea for the protection and preservation of areas of beauty and historic interest along the California Trail.

The overland emigrants preferred the inland, unsettled, and agricultural portions of California. Recalling what had happened in Texas, California's Mexican regime was not enthusiastic about the influx of overland immigrants. Fortunately for the latter they enjoyed the unreserved hospitality of Sutter, whose armed garrison at his fort made him immune to any interference by Mexican authorities. Nevertheless, the number of settlers going to California in the early 1840s was probably only about a tenth of the number going to Oregon. The flood tide of immigrants to California was not to come until the gold rush.

In the early 1840s the journey across Utah, Nevada, and the Sierra Nevada was beset with continuous hardship and tragedy. Sometimes emigrants traveled across salt plains for days without water and they were occasionally misled by mirages. Often, in an effort to speed up the journey and reach the Sierra Nevada of California before winter set in, the emigrants abandoned their wagons and much of their baggage, but they were usually inexperienced in balancing loads on animals so that the gear would be held in place. Much of their baggage became scattered over the desert along the line of march. The oxen could not keep up with the faster walking mules and horses and the parties became scattered each day. By the time the weary emigrants reached the Humboldt River in Nevada, many were on foot. They did not know how far the mountains were, or whether the winter snows would come before they reached the "granite hell." If the snows came before they reached the Sierra Nevada, they would not be able to cross and many might die of starvation, as happened to the ill-fated Donner Party (Houghton, 1911; Stewart, 1936; McGlashan, 1940).

Thousands of horses, mules, and oxen died from thirst, hunger, and exhaution along the trails. The putrefying flesh of dead animals was often eaten by starving travelers. Cholera, scurvy, and dysentery took a heavy toll both on land and sea routes in the gold rush days. Accident, thirst, hunger, fatigue, and attack or theft by Indians were additional perils.[3] Despite rescue teams sent from California, one year 1,500 graves were counted between Salt Lake and Sacramento along the Truckee route alone (Cleland, 1923).

## DIVISION OF THE RANCHOS
(Gates, 1967)

The strife between cattle kings and homesteaders on the High Plains had a counterpart in California, where the owners of the huge ranchos permitted their lands to remain relatively undeveloped while harassing and driving off settlers who were looking for land. The large landowners were repeatedly criticized in the editorials and news columns of the weekly *California Farmer.* Some of the huge Mexican grants remained intact for generations and were eventually incorporated into such holdings as those of Miller and Lux, with 304,000 acres and great herds of cattle and sheep as late as 1922, and those of the Kern County Land Company, comprising 390,000 acres in California and 1,525,000 acres in Arizona and New Mexico.

The immense Mexican grants included land that, if used at all, was used only for grazing, yet included some of the best farmland. Disillusioned miners who failed to "strike it rich," and with no other occupation, took up farming, often as squatters, thus coming into conflict with the original claimants of the land and their assignees. Most of the later Mexican grants had been neither located nor developed. Some permitted the grantees or their assignees to locate in areas five or ten times greater than the grant itself, surveyed in such a way as to include all the best land. Understandably, people coming from states where farms generally were 80-160 acres and plantations were

[3]Theft was the more common danger. The scanty resources of the despised "Digger" Indians of the Great Basin allowed them to exist only at an extremely low cultural level. After the destruction of the fragile desert ecology of the region through which the California Trail passed, the Diggers were forced to steal draft animals and livestock from the emigrants in order to survive.

rarely more than 2,000-3,000 acres, were astonished at the size of such empires as those of the Pico family's 700,000 acres, the de la Guerra family's 488,000 acres, the Vallejo family's 291,000 acres, the Castro family's 280,000 acres, and the Yorba family's 218,000 acres.[4]

For some years there was great resistance to subdividing the vast but ill-defined cattle ranches, where cattle raising was still carried on as in the old days, the range having to supply feed through all the seasons without assistance from granary or haystack. There were many conflicts between rancheros and squatters, the latter sometimes banding themselves together to resist dispossession.

The old grants were broken up by the Land Act of 1851. The high cost of the litigation involved has been severely criticized, but Paul W. Gates, who made a thorough investigation of the charges made against the act, concluded that

> it was no more the cost of litigation than the extravagance of the old families, their neglect to pay taxes, their dependence on loans carrying high interest rates, their failure to realize that the day of sparsely grazed *ranchos* had gone, and that their lands called for a heavy investment of capital to make them productive and able to bear taxation, that broke up the large estates.

Despite all the strife and turmoil, by the end of the 1860s agriculture had displaced mining as the principal occupation.

Even after the large claims were divided, the resulting farms were by no means small. "Bonanza farms" of 500-5,000 acres were common, particularly in Marin, Sonoma, Napa, and counties of the Sacramento Valley. In southern California, not until the disastrous drought of 1863-1864 resulted in huge cattle losses was the process of land division begun; many of the ranches remained intact into the twentieth century. Nevertheless, the drought forced the breaking up of the large cattle holdings into small ranches that were sold to the ever-increasing numbers of settlers. Thus the diversified and highly productive forms of agriculture that were to contribute so much to California's prosperity and mode of life had their beginning.

## STATEHOOD AND THE GOLD RUSH

By 1846 the war with Mexico provided the opportunity to terminate the Mexican regime. Probably the continual infiltration of settlers would soon have achieved the same result, as it had in Texas and Oregon, without war, particularly in view of the inadequacy of Mexican control. Caughey (1970) points out that although the Mexican period is sometimes dismissed as a mere interregnum between the Spanish and American regimes, it was of value in providing a period during which secularization and the introduction of such concepts as republicanism, constitutionalism, and representation, as well as initial contacts with non-Spaniards, could take place under Mexican rule, thus avoiding an uncomfortably abrupt transition.

On January 24, 1848, James William Marshall noticed flecks of yellow along the tail race of a sawmill he was constructing at Coloma on the American River. The news of his discovery of gold spread to eastern United States and to Europe and fortune seekers —almost 40,000 in 1849—arrived by way of Cape Horn, Panama, or overland. But most of the "forty-niners" arrived by the overland trails.

[4]The only states with the huge grants of the Spanish system were California, New Mexico, Texas, Louisiana, and Florida. The cattle kingdoms of the Central Plains had no legal basis, but led to similar turmoil.

Although many disillusioned miners left California, an even greater number came and stayed. The result was that by the end of 1849, the non-Indian population of the state exceeded 100,000; in 1852, 224,435; and in 1860, 380,015. In a province that had lain practically dormant for generations, the rapid rise in population clearly presaged far-reaching political, social, and economic changes. As late as 1863, mining still employed more of the state's workers than any other pursuit, but it was also a stimulus to many other industries, including, of course, agriculture.

## THE IMMIGRANT IN CALIFORNIA AGRICULTURE

From 1860 to 1900, when foreign immigrants swelled the growth of America's cities by almost 36 million persons, they increased the farm population by about 9 million. The immigrant influence was particularly great in the west; in 1870 more than half of the men aged twenty-one and over in Utah, Nevada, Arizona, Idaho, and California had been born abroad (Rolle, 1968). The federal census of 1910 and 1920 showed that during that period, the Pacific Coast states showed a gain of more than 14,000 foreign-born farmers, whereas in the nation as a whole the number declined by more than 83,000. Of the 55,397 foreign-born farmers in the Pacific Coast states in 1910, 28,250—more than half—lived in California. In each of the three states, the Germans constituted the largest ethnic group (Saloutos, 1975).

The percentage of farm ownership among the foreign-born whites was a little higher than among native-born whites; in 1920 the percentages were 81.2 and 77.9, respectively. Among the foreign-born farm owners in the Pacific Coast states in 1920, the Germans were followed numerically by the Swedes, Canadians, English, and Italians, but the Swedes had become the largest group in Washington. The percentage of ownership was much lower among Japanese and Chinese, owing to alien land laws. But Japanese comprised almost one-third of all farm tenants in the Pacific Coast states, with the Italians and Portuguese a distant second and third. In California, the Germans ranked first in farm ownership among foreign immigrants, followed by the Italians, Portuguese, Canadians, and Swedes in that order. The Japanese constituted the largest foreign group of farmers, but ranked thirteenth as farm owners, first as tenants, and third as managers (Saloutos, 1975).

The special knowledge and skills that the many ethnic groups brought from their native lands led to rapid growth and great diversification in California agriculture. The choice grape cuttings imported from Italy, France, Hungary, and the Rhine Valley, and the special skills of the immigrants from those regions in grape culture and wine-making, resulted in a rapid rise in the wine industry.

The Frenchman Jean Louis Vignes imported enough vine cuttings from France in the 1830s to start vineyards and was the first person to establish a vineyard and raise oranges in Los Angeles (Carosso, 1951). Italians from San Francisco formed the Italian-Swiss Agricultural Colony in Sonoma Valley in 1881 (Lord et al., 1906; Rolle, 1968). German immigrants, likewise from San Francisco and with no previous experience in agriculture, in 1857 established what was to become the thriving agricultural community of Anaheim in southern California, with special emphasis on viticulture. By 1884 more than 1.25 million gallons of wine were produced by eight wineries. That year Pierce's disease, caused by a bacterium, began to appear, which completely destroyed all Anaheim vineyards within three years. However, walnuts and oranges brought a new period of prosperity (Raup, 1932; MacArthur, 1959).

A Hungarian nobleman, Colonel Agostan Harazthy, came to California in 1849 at the age of 37. His systematic and steady acquisition of imported varieties of grapes resulted in his being considered the "father of California's modern wine industry." With the help of his two sons he planted 80,000 vines in 140 acres of land east of Sonoma in a single year. Harazthy was responsible for Sonoma County—close to the San Francisco market—becoming the foreign vine nursery and source of some of California's finest vines. He demonstrated the success of hillside grape culture. His *Report on Grapes and Wine in California* was an important treatise on practical viticulture that was printed and circulated by the California Agricultural Society in 1859 (Carosso, 1951; Jacobs, 1975) and was later expanded by Harazthy into his *Grape Culture, Wines and Wine-Making* (1862).

The Portuguese specialized in the growing of artichokes and sweet potatoes. They also became prominent in the dairy industry, as did also the Dutch, Swiss, and Danes. In 1895 the Danes organized a cooperative creamery in Fresno, making that town famous for its butter. Armenians were active in the grape, raisin, dried fruit, asparagus, melon, and wholesale fruit and vegetable industries, particularly in agriculturally rich Fresno County. In 1918 Henry Markarian had a 160-acre fig orchard, the largest in the world at that time, and became the first president of the California Fig Growers Association (Saloutos, 1975).

The most spectacular of the familiar rags-to-riches stories in the cattle industry was that of the firm of Miller and Lux. Christened Heinrich Alfred Kreiser, Miller arrived in New York as an indigent steerage passenger from Germany in 1847 at the age of nineteen. In 1850 he had the good fortune to obtain a steamship ticket to California from a friend, Henry Miller, who had changed his mind about making the trip. The ticket was not transferable, and that is how Heinrich Kreiser became Henry Miller. Like many passengers, Miller protracted "Panama fever" during the arduous journey on foot across the Isthmus of Panama. He arrived in San Francisco ill, emaciated, and with six dollars in his pocket. Eschewing the lure of the goldfields, he utilized his skill as a butcher to good advantage, for meat was in great demand (Treadwell, 1931). In 1858, after a trip to the San Joaquin Valley, Miller obtained a loan from a San Francisco banker to finance a cattle empire. He formed a partnership with a fellow countryman, Charles Lux, who attended to the city business. Miller and Lux began an orgy of land and cattle buying that lasted 30 years (Treadwell, 1931). They acquired all or parts of large ranchos, legitimately or with the aid of dummy entrymen, as was the custom among big landowners (Gates, 1975*b*).

Miller constructed levees across low places in San Joaquin River sloughs to spread water over large areas of land and thereby raised luxuriant stands of grass for his cattle. One *zanjero* could water thousands of acres of land. Soon Miller had the best canal system in the West, watering over 150,000 acres. Later he obtained a third of the water from Kern County's Buena Vista reservoir. He extended his cattle operations into Nevada and southeast Oregon, where the cooler climate was more healthful for cattle. These areas became breeding grounds for steers finished in California (Treadwell, 1931). By 1891 Miller and Lux held 750,000 acres in California, Oregon, and Nevada, and through ownership of water they controlled a far larger acreage than this figure indicates. Their gross annual income was as high as $1.5 million from the sale of cattle, sheep, horses, and hogs (Gates, 1975*b*). As late as 1973, Ralph Nader's Study Group estimated that the Miller and Lux firm still owned 93,000 acres in California (Fellmath, 1973).

The cattle kings could hardly have envisioned the day when dairy cows would vie with beef cattle for first place in California agriculture, yet as early as 1857 the Steele Brothers (Edgar, Rennselaer, George, and Isaac) had made an auspicious start in that direction. Beginning with twenty-five cows in 1857, they eventually developed a thriving business, specializing in making cheese (Steele, 1941).

In 1870 the Steele Brothers' dairy herds, located in 15,000 acres of land near Pescadero and 45,000 acres at San Luis Obispo, totaled about 1,400 cows. The Steele Brothers also had over 2,000 steers, calves, and young cattle and many hogs. Their beautiful rolling oak-studded land was well watered with springs and small streams. They produced good crops of grain as well as indigenous grasses.

Isaac Steele, as a friend of Professor Eugene W. Hilgard and staunch advocate of education, worked for the creation of an agricultural college for the University of California. Later Peter J. Shields, raised on a California dairy farm, drafted bills to establish such a college in 1901, 1903, and 1905. The 1905 bill was passed by the state legislature and immediately signed by Governor Pardee. Established at Davis and first called the "University Farm," the unit later became known as a branch of the University's College of Agriculture. A College of Agriculture was established at Davis in 1952. (The Regents declared Davis a general campus of the university in 1959.) The extensive and effective program of research and extension in animal science and veterinary medicine at Davis was a vital factor in the amazing productivity of California's dairy herds, surpassing all others in the nation.

Further details may be appropriate at this point concerning some major ethnic groups not given special attention previously because of their relatively late arrival in America and because their principal influence in agriculture was in California.

### Italians

From 1880 to 1924, the period of greatest Italian immigration, there was a net total immigration of over 6 million out of Italy's population of only 35 million. Perhaps 80 percent of these immigrants came from a farm background, with an excellent agricultural tradition, but no more than 20 percent moved back into agriculture. By the time of the peak Italian immigration, the federal government and the railroads had already dispersed the large, cheap tracts of land. Italians came too late to reap the benefits of the homestead system. During those years, economic opportunities for poor and propertyless persons were more available in cities. Italians wishing to farm had to find bits of land discarded by others, as in the outskirts of cities, and they generally planted truck crops. They believed that with ambition and hard work they could overcome their hardships. As related previously, immigrants tend to congregate in areas that remind them most of their homeland. Among western states, California, in topography and climate the state most like Italy, was the one most attractive to Italians (Rolle, 1968).

Italian immigrants were exploited almost as ruthlessly as other foreigners had been before them. The long ocean voyage was still a harrowing ordeal of lice, scurvy, and seasickness, alleviated somewhat by the knapsacks full of cheese and salami to supplement the meager fare of soup and hardtack doled out to the miserable, crowded, steerage passengers. After arrival, the *padrone,* or boss, recruited immigrant Italian mining, railroading, and agricultural laborers and herded them about in gangs while negotiating contracts for them. The gullible and ignorant immigrants were often bound up for seven-year work periods and were paid whatever the padrone saw fit. Nevertheless the padrone system, like the earlier indenture and redemptioner systems, had positive effects in that many immigrants would have been lost without such help. Honest padroni could serve such useful functions as helping ghetto dwellers get out into smaller communities or farm areas and could perform such services as banking, buying and selling property, and settling estates (Rolle, 1972).

European immigrants were good farmers—industrious and thrifty. In 1920 the leading nationalities in value of owner-operated farms were the Danes, Germans, and Scots, all with farms averaging more than $16,000 in value, compared with $13,484 for all foreign-born and $10,019 for native-born farmers. The Italians had smaller farms, for reasons already explained. Their farms averaged only $7,531 in value, but their value per acre was $80.12 compared with $73.28 for all foreign-born and $60.72 for all native-born (Brunner, 1929).

California wine-making inherited much from the Italians. Some towns in grape-growing regions, like Asti and Lodi, were named after towns in Italy. Secondo Guasti, a peasant from the Italian Piedmont, arrived in 1881 at the age of 21, and during the 1880s he established in southern California the Italian Vineyard Company, "the globe's largest vineyard." Many Swiss-Italians went into dairying from Santa Barbara north along the California coast, aiming to make California the "Argentina of North America" (Rolle, 1972).

Among the early ethnic Italians, the greatest impact on California agriculture was made by the famous banker Amadeo P. Giannini. Even while in partnership with his stepfather in the fruit and vegetable business, Giannini had banking experience that consisted of advancing money to farmers, sometimes tens of thousands of dollars, and without interest—advances against future delivery of commodities. Farmers are in particular need of loans because they must spend so much for seeds, water, fertilizer, pesticides, labor, and equipment before their crops can be harvested and sold.

Giannini started the Bank of Italy in 1904, prompted to do so because of the neglect by other banks of the Italian immigrants of San Francisco's North Beach. Most of the immigrants had never been in a bank before; they hid their surplus cash in such places as under the mattress, in teapots back of the kitchen stove, or on the top shelf of the pantry. The existing banks had no desire to make small loans. When it was necessary for poor people to borrow money, they were at the mercy of loan sharks. Giannini taught them the advantages of putting their money in interest-bearing savings accounts. He believed in making small loans to poor people and getting large numbers of small investors in his bank, for past experience showed that most of these people would be successful and some would be the leaders of a new generation. No one, not even "A. P." himself, was allowed to own more than 100 shares of stock, and Giannini was pleased at the large number of poor people who bought two, three, or four shares apiece. These were the kind of people whose patronage was to become so important for the kind of bank he had in mind.

California agriculture, already rich and highly diversified, enjoyed a great boom during World War I. Much capital was required to finance irrigated agriculture. More than 43 percent of California's farmers paid 8 percent interest on money they borrowed, plus 0.2 percent commission, and some isolated communities paid 12 percent. The Bank of Italy made plenty of money available at 7 percent. Formerly a small farmer was often unable to obtain even a modest loan because prominent merchants, big farmers sitting on the board, or a board's officers first made sure of safeguarding their own financial needs. The Bank of Italy guarded against monopoly of loan services either by branch officers or by advisory boards.

Following the Great Depression, A. P. Giannini, through his bank and some of its affiliates—and with 423 branches in 255 California communities at that time—played a role in the recovery of the farmer that was probably greater than that of any other Californian. During the first full year of the operation of the Farm Credit Administration (FCA) in 1934, federal agencies held 25.67 percent of California's farm-mortgage debt of nearly a half-billion dollars. California banks held 23.24 percent, of which nearly half was financed by the Bank of America, the new name for the old Bank of Italy. The Bank of America's crop-production loans gradually increased in relation to those of FCA. In 1952 they were $159,292,000 compared to $25,208,000 for FCA's. At the peak of its heavy farm involvement, the Bank of America's affiliate at that time, California

Lands, foreclosed on farms representing only 1.7 percent of California's tillable soil.

In 1945 the Bank of America, with assets of over $5 billion, became the world's largest bank. At the time of his death on June 3, 1949, a few weeks after his seventy-ninth birthday, A. P.'s estate was appraised at only $489,278. In 1928 he had turned over $1.5 million due him under the terms of a percentage of profits agreement to the University of California at Berkeley to form the basis of the Giannini Foundation of Agricultural Economics. One of the buildings in the agriculture complex at the university bears his name (James and James, 1954; Rink, 1963).

### Portuguese

Portuguese immigration was small until the decade 1871-80, when 14,082 arrived in the United States, mainly via fishing boats from the western Azores. Immigrants from mainland Portugal were not numerous until after 1910. More Portuguese came to California than to any other section of the country. They tended to become farmers, fishermen, or whalers. The farmers first engaged mainly in truck farming, later turning to dairying. In the San Joaquin Valley, many took up leaseholds of from 120 to 160 acres in the 1880s and operated in companies of 6 to 15 men, usually unmarried (Bohme, 1956).

J. B. Avila, the "father of the sweet potato industry," planted in Merced a small patch of sweet potatoes from the Azores. With cultivation and irrigation the patch grew to 6 or 7 acres the next year. The sweet potato proved to be popular in San Francisco restaurants and hotels and its cultivation spread over large areas of the Central Valley.

Before and after World War I, a group of Portuguese dairy workers were hired as milkers in California. After they had saved sufficient money, they left their Dutch, Swiss, Danish, and other employers and bought land and herds of their own. They now are a major factor in California's dairy industry in various areas of the state from Eureka to southern California. They are particularly prominent in the San Joaquin Valley, where they own stores and operate feed lots as well as dairies. In the lower San Joaquin Valley, Portuguese-Americans own from 70 to 75 percent of the dairy herds and their herds tend to be larger than those of other ethnic groups.

### Japanese

Like the Italians and Portuguese, the Japanese came from a country where the greatest possible sustenance had to be extracted from the smallest piece of land and where population pressed hard against resources. Most Japanese immigrants were sons of farmers and cherished little hope for economic success in their native land. They generally intended to remain in the United States for a few years and then return home with funds to enlarge their family farms. Not until 1891 were there as many as a thousand Japanese in California, but by 1930 there were 97,000—70 percent of the Japanese in the nation. Yet they amounted to only 1.7 percent of California's population (Modell, 1970).

In California the Japanese obtained relatively higher earnings and a better opportunity to eventually become landowners than in their native land. Compared with white migratory workers, they were more efficient and reliable and were willing to accept lower wages and cruder housing accommodations. Bosses of Japanese labor camps could supply farmers with workers from their organized gangs and then direct them elsewhere when jobs became scarce. Eventually these Japanese bosses, speaking for the men they represented, could organize strikes and boycotts at harvesttime and could secure increases in wages that were equal to or even surpassed those of white workers. The Japanese bosses often negotiated to lease land and were instrumental in the transition of the Japanese from wage earners to tenants and even to landowning farmers (Iwata, 1962; Saloutos, 1975).

Small white farmers favored the California Alien Land Act of 1913, which forbade ownership of land by Orientals. They wanted to force the Japanese back into the wage-earning class and to hire them as farm laborers. Instead, the Japanese opted for urban trades and occupations, earning the money to enable them to become thriving tenant farmers. (The Alien Land Act allowed them to lease land.) This pleased the large landowners because the Japanese paid higher rents, worked longer hours, used family labor, employed scientific techniques, produced larger crops, and were satisfied with smaller profits. Leasing to a few Japanese bosses, the large landowners often obtained a monopoly of the most desirable Japanese laborers. Smaller farmers had difficulty in obtaining any (Kawakami, 1921; Matson, 1953). Anti-Japanese agitation was by no means confined to small owners and farm organizations (e.g., the Farm Bureau and Grange). It was just as intense in the ranks of organized labor and small business. Big business strongly opposed anti-Japanese activity, however, fearing possible harm to trade relationships with Japan (Matson, 1953).

After the federal government and some California land companies abandoned rice-growing as a futile project, the Japanese succeeded in the effort. They introduced superior early-ripening varieties from Japan and devised methods of irrigation and cultivation. In the early 1920s about 85 percent of the varieties of rice grown in California were from Japanese stock, but the Japanese operated no more than 29,000 of the 150,000 acres of rice grown in the state at that time (Toyoji, 1922). Eventually the survival of the rice industry depended on mechanization of culture and harvesting. By 1948, when 900 man-hours of labor per acre were required for rice-harvesting in Japan, only 7.5 were required with the highly mechanized system used in California (Bainer, 1975*a, b*).

Like the Italians, the Japanese concentrated on intensive farming on small pieces of land; by using fertilizers and crop rotation they could make a financial success of them. In 1920, when California farms averaged about 200 acres, Japanese farms in the state averaged only 57 acres. Much stoop labor was required on these farms, compared with the extensive farming operations by the Caucasians; the latter required expensive machinery and elaborate equipment. With a shrewd choice of crops requiring the least capital and the most labor, the Japanese were remarkably successful and proved difficult for whites to compete against. Along with successful farming, the Japanese soon entered wholesaling and retailing, completing the vertical integration of the industry (Modell, 1970; Saloutos, 1975).

The Alien Land Law appeared to be too loosely drawn to destroy Japanese agriculture or even reverse its growth. The Immigration Act of 1924 denied to Japan even a normal quota of immigrants, but by that time second-generation Japanese with native-born citizenship, known as nisei, were becoming a competitive factor in the state's agricultural and business life. The relocation of both Japanese immigrants and nisei after Pearl Harbor, through a baseless notion of military necessity, did little to deter the determination of this ethnic minority to succeed in every facet of California economic life. The opening of sections of the armed forces to persons of Japanese extraction, and the proven loyalty of the latter in battle, served to enhance their status. Most Japanese-Americans chose to return to California after the end of their internment in early 1945. Their characteristic response to discrimination and adversity was to recognize that they would have to "try harder" than the white majority (Modell, 1970).

The Japanese-Americans were eminently successful in obtaining the training and education demanded for success in modern life. Comparing California employment statistics for 1940 and 1960, one finds that the percentage of Japanese-American males among "all farm workers" had declined only from 47.1 percent to 31.9 percent, compared to a decline from 12.7 percent to 5.6 percent for all California males. But the percentage of Japanese-American males in the category of "professional, technical, and kindred workers" during the same period increased from 2.7 percent to 15.7 percent, compared to an increase of from 8.9 percent to 14.3 percent for all California males. Within two decades, a substantial part of which was spent behind barbed wire, the Japanese-Americans had reached a proportion in this prestigious category higher than the statewide average! "The case of the Japanese in California," concluded historian John Modell, "is an indication that white racism is not a simple, irreducible, and unchangeable quantity" (1970).

### Okies

The "Okies" were native-born semisubsistence farmers from the southern Great Plains who were driven out by the drought and dust storms of the 1930s. They learned new work habits and discipline in California's "factories in the fields." Contrary to the hopes of farm-labor reformers and the fears of farmers' organizations, Okies were not very responsive to union appeals—much less so than Mexicans and Filipinos—and they did not strike. The Okies had been farmers and hoped someday to again have farms of their own. However, 10 percent of California's agricultural concerns—principally the large-acreage, corporate-owned agribusinesses—employed two-thirds of the agricultural labor. In their wretched "Little Oklahomas" on the outskirts of the farm towns of the Central and Imperial valleys, the Okies found no opportunity for upward mobility. Rescued by the defense industries of World War II, their places in the fields were taken by Mexican *braceros* (Stein, 1975).

### Mexicans

The pioneer agricultural contributions of the Spaniards and Mexicans prior to the mid-nineteenth century have been recounted earlier in this chapter. These pioneers followed a wide variety of agricultural pursuits, but mainly on a subsistence basis and using Indian labor. Beginning with the gold rush, the rapid growth of Anglo agriculture in California brought with it the need for cheap farm labor. The need was filled by Chinese until the Exclusion Act of 1882 terminated this source. The Japanese were used in gang labor like the Chinese, but were small in numbers. Filipinos were used as farm workers, but again not in large numbers. (Beginning in 1924, they increased in number and eventually were exceeded only by the Mexicans as farm laborers.)

After 1910, the greatest source of farm labor in California was Mexico. That was the year of the beginning of the Mexican revolution, when for a decade poverty became more widespread than usual in Mexico and the political and social restraints that had confined people to a particular hacienda or village began to break down. Coincidentally, there was an acute shortage of farm labor in California just at the time that World War I created an increased demand for the state's crops. Although immigrants from Mexico came to the United States primarily from rural areas and were better suited to farm work by tradition and culture pattern than many previous ethnic groups, relatively few became farmers. As farm laborers, however, Mexican-Americans played a dominant role and continue to do so to the present day.

In the southeastern California desert bordering Mexico is the Imperial Valley, where intensive irrigation projects opened much land to cultivation. This was one of the first major agricultural areas in California to become dependent on Mexican labor, a process that was repeated throughout rural areas of the state. By 1928 about a third of the 60,000 permanent residents of Imperial County were people of Mexican origin who came to be known as "Chicanos." Nearly all were field workers and were employed chiefly by individuals or corporations that owned or leased large acreages of farmland. These large landholders needed big work crews and used the services of labor contractors, often of Mexican descent.

In the 1930s the Dust Bowl tragedy and the depression forced many Anglo migrants, the "Okies," into the California unskilled farm-labor market and during that decade there was actually a net outflow of people from the United States to Mexico. With the return of prosperity during World War II, the federal government guaranteed cheap migrant labor from Mexico in the Bracero Program. A bilateral agreement between Mexico and the United States allowed Mexican field laborers to work here for limited periods. The program was to be terminated when the wartime labor shortage ended, but it continued under various guises and was finally reestablished and strengthened in 1951 when Congress passed Public Law 78. The number of braceros increased until during the years 1956 through 1959 over 400,000 per year were contracted. Spanish-speaking farm laborers received much competition from Mexican "wetbacks," a name given to those persons who illegally crossed the Rio Grande to work on the American side of the border. Sometimes they were tolerated because of need for cheap labor. However, there were 1,035,282 separate deportations of wetbacks during the peak year of 1954. The Bracero Program ended in January 1964. In 1965 the first numerical limits were placed on immigrants from nations of the Western Hemisphere (120,000 in 1968) (Scholes, 1966; Wollenberg, 1970).

The decreased flow of laborers from Mexico improved the bargaining position of resident agricultural workers and increased labor costs for California growers. Higher labor costs encouraged California farmers to invest in more laborsaving machines. This reduced the number of unskilled workers needed, but also created opportunities for some workers to move to higher-paying jobs and to shift from seasonal employment to year-round regular work. Nevertheless, during the peak seasons of labor needs in summer and fall, nearly twice as much labor is hired than during the slack season of the year.

In California, wages for farm labor have consistently been higher than the nationwide average by 25-30 percent. The number of hired domestic farm workers in California now remains fairly constant at around 200,000. There, where nearly 80 percent of farm labor is hired, labor costs amount to 20 percent of the total cost of agricultural production compared to 8 percent nationwide (Reed, 1976).

Farm mechanization tends to lessen the differences between industrial and rural work forces. This in turn increases the justification for farm labor to enjoy the benefits and prerogatives of industrial workers as contained in the National Labor Relations Act. In 1974 a sixteen-member Blue Ribbon Committee on the Future of California Agriculture recommended that, pending congressional action in this area, "the state's agricultural industry would benefit from action by the state to establish secret balloting, freedom of organizational choice, unemployment benefits, and the prohibition of secondary boycotts and product boycotts" (ABRC, 1974).

The need of farm workers for an effective and aggressive leader was satisfied by a Chicano farm worker by the name of Cesar Chavez. Born in Yuma, Arizona, in 1927, he grew up in immigrant labor camps and began to organize the National Farm Workers Organization in 1962. In 1965 his union joined Filipino grape pickers in a strike against growers. It lasted several years. The table grape growers held out into the 1970s. Under Chavez's leadership the union changed its name to the United Farm Workers

(UFW), affiliated with the AFL-CIO, and organized boycotts of nonunion farm products, particularly lettuce.

Most adversely affected are the small farmers. Increased farm wages lead to mechanization. At current prices, only the large farmers and corporations can afford this. Small farmers are soon priced out of business. They have a great financial investment, work long hours, have all the worry, and take all the risk, but many are not as well off financially as those who work for them.

In addition to the usual risks and hazards of farming, the farmer has suddenly become a victim of the latest crime wave. The problem is nationwide but is particularly severe in California. Thefts of machinery and crops, as well as vandalism, are reaching epidemic proportions, and are believed to be costing California farmers about $30 million a year.

## Illegal Aliens

During the past decade the flood of illegal aliens into the United States has increased to such proportions as to be described as "out of control" and "a national disaster." The number of illegal aliens apprehended by immigration officials increased from 94,778 in 1967 to 874,492 in 1976. Current estimates of illegal aliens in this country range from 6 to 12 million, most of them Mexican nationals, with about two-thirds of them going to urban areas. When returned to Mexico, these unfortunates rejoin the ranks of the unemployed, about 40-50 percent of that nation's work force. Mexico provides no unemployment benefits for its jobless workers and there is no welfare system to help the poor.

The Green Revolution resulted in Mexico temporarily becoming an exporter of grain, but its population has doubled since those days and Mexico is again importing grain. Faced with the dilemma of either increasing staple crops to feed its people or export crops to increase foreign exchange earnings to pay its astronomical foreign debts, Mexico opted for the latter. In the newly developed, mechanized, agricultural areas of northwest Mexico, the big Mexican growers, in partnership with agricultural interests and banks of the United States, are the main beneficiaries of government programs financing fertilizers, machinery, irrigation, and agricultural research. They have planted almost all their land in fruits and vegetables for export; for example, about 60 percent of the winter vegetables consumed in the United States. Meanwhile, food shortages are affecting both the urban and the rural population. In 1977 Mexico imported 1.5 million tons of corn and 1.4 million tons of wheat. The same thing is occurring throughout most of Latin America, where between 1964 and 1974 per capita production of export crops increased by 27 percent while per capita production of staple crops decreased by 10 percent (Lappé and Collins, 1977; Flynn, 1978).

In Mexico, as in the United States, many small farmers have lost their lands to large expansion-minded agribusinesses. The number of landless Mexican peasants has increased from 1.5 million to over 5 million since 1950. They have the Hobson's choice of either staying on the land as underpaid seasonal workers, under wretched living conditions, or joining the ranks of the unemployed in the teeming slums of the overcrowded cities.

There is obvious need for change in social and economic structures (Lappé and Collins, 1977; Flynn, 1978), but also an urgent need for birth control. In Mexico, with a current population of 62 million and a birth rate of 3.7 percent, the highest ever measured by any country, the population will double every twenty years. Anthropological studies of the past fifty years showed that Mexican village women want fewer children and would practice birth control if the Catholic church would accept it. Fearing church and community censure, they are afraid to practice contraception (Critchfield, 1977).

This amounts to a grim portent not only for Mexico but also its neighbor to the north, which is no longer welcoming the world's "huddled masses," having plenty of its own (Bernstein and Del Olmo, 1977; Kendall, 1977; Meyers, 1977). Three or four million illegals hold jobs in the United States. Illegal aliens cost the American taxpayer $1 billion a year for various benefits and they send an estimated $3 billion out of the country (Chapman, 1976).

## LIVESTOCK FARMING
(Hart, 1946; Dasmann, 1965; Gates, 1967)

The early California rancheros had no incentive to properly maintain or improve their cattle during the Mexican regime, but the great tide of immigration in 1848 and 1849 resulted in an immediate demand for meat and scrawny Longhorn cattle sold for high prices. Cattle that formerly sold for $2-3 a head sold for $25-52 even as late as 1853. Droves of 700 to 1,000 cattle were trailed north via the coast route or via the San Joaquin Valley. Usually many cattle were lost through Indian raids, poisonous plants, exhaustion, dehydration, and insufficient grass. Yet it was only through cattle that the "cow-counties" of southern California could indirectly benefit from the financial boom of the gold rush.

Herds were also trailed to California from Texas and the Midwest. The Texas cattle were not much better than those from the California ranges, but those from the Midwest and the Mississippi Valley were substantially superior, particularly livestock from Ohio and Kentucky, which may have had some Shorthorn blood. Cattle were being trailed to California from all these states in numbers estimated to be as high as 50,000 per year. Likewise, thousands of horses and mules were being driven into California. The number of cattle lost through exhaustion and deprivation, or run off by Indians, on these long and grueling cattle drives, can well be imagined, but the $100-150 a head obtained for the best cattle, much more than obtained for California stock, kept a steady stream of the luckless beasts moving over the plains, deserts, and mountain passes. Some of the better animals were used to improve the California breeds.

Stock was not required to be herded or fenced and did much damage to grain. California ranges became overstocked and overgrazed. Drought afflicted some part of the state almost every year and in the worst years it caused heavy losses. Whereas the drought of 1862 reduced herds in southern California, floods the same year in the Sacramento Valley destroyed 150,000 cattle, 100,000 sheep, 300,000 lambs, and 25,000 horses and mules.[5] Nevertheless by 1863, cattle were increasing faster than they could be sold. Cattlemen were advised to slaughter a third of their cattle three years of age and over for their hides and tallow, spay for two or three seasons an equal number of broodstock under three, and castrate their inferior bulls. Those who could afford it began to import purebred Shorthorn and Devon bulls and cows. Such expensive animals generally arrived via the Panama route. They were driven across the isthmus, of course; the canal was not completed until 1914.

It did not take long for immigrants to discover California was not the Eden they had been led to expect from extravagant reports that had circulated in the East. It was flood

[5]When the pioneers arrived in the Sacramento Valley they found much of it to be an impassible quagmire in winter and spring and an area of sunbaked, cracked earth in summer and fall. The extensive system of levees along the Sacramento River and its tributaries, and the various dams and drainage projects, resulted in the valley becoming one of the richest agricultural areas in the world.

control and irrigation that eventually transformed the state's potentially fertile valleys into the nation's most productive and diversified agricultural region.

Compared with cattle, sheep were of little importance on the lands of the missions and on the early ranchos. Californians did not care for mutton. Before the American period, sheep were small, coarse-wooled, and had lighter fleece than in other states (Wentworth, 1948). The gold rush suddenly raised the price of sheep from twenty cents to twenty-five dollars a head. In the summer of 1849 the first of the great sheep drives got under way when 25,000 sheep from New Mexico started west in ten bands, going by way of Fort Yuma, the Mojave Desert, and Tehachapi Pass to the San Joaquin Valley to reach the mining camps of the central Sierra Nevada. Year after year the great flocks were trailed in, over a half-million from New Mexico alone from 1852 to 1860. As in the case of cattle, increasing interest was shown in obtaining improved breeds, particularly the Merino, from the eastern United States, from Europe, and from Australia. From 1852 to 1860 the number of sheep in California increased from less than ninety thousand to over a million, principally in the coastal counties from Los Angeles County to what is now San Benito County.

Cattle declined to a low of about half a million by 1870; a sixth of the number they had reached in the peak years. Sheep require less water and can be readily herded to follow the forage supply. By 1875 they had reached a peak number of 5.5 million. The day of the migratory Basque sheepherders, immigrants from Spain, had begun.

Unlike cattle, horses brought into California by the Spaniards were of excellent quality. This beautiful Arab breed was not as fast as the Thoroughbred, but was nevertheless fast, surefooted, and capable of long journeys. The wild horses of California and the Rocky Mountain range country were descendants of these horses. In California many of these wild horses were destroyed to conserve forage for meat animals. The larger all-purpose American horses of Thoroughbred blood, trailed in from the East, gradually replaced or were bred with the Spanish horses. Later, draft horses were introduced to replace oxen, which were considered to be too slow. Mules were found to be superior for packing and hauling, and large numbers were raised, beginning in the late 1860s.

Camels from Egypt were introduced into the United States in 1856 and again in 1857. A herd of twenty-eight left San Antonio, Texas, on June 25, 1857, swam the Colorado River, and proceeded to Fort Tejon, familiar today to tourists driving north on the Ridge Route to Bakersfield. Camels possess certain advantages. They can carry 1,000 pounds, travel 35-40 miles a day, and find their own feed. However, they were considered too slow, and men trained to handle horses found them difficult to manage. The Civil War hastened their demise as a domestic animal and by 1864 all that remained were sold to circuses, zoos, and ranches. There were three importations of the Bactrian, a two-humped camel from Asia, starting in 1860, but this venture also failed.

Hogs were brought into California by the Spanish missionaries, but a superior breed (the Berkshire) was not introduced until 1847, by Philip D. Armour, later to become well known as the head of a large packing business. California has never been an important producer of hogs and this is true of the western states in general. In 1972 California marketed a paltry 185,000 of the 76,000,000 hogs marketed in the United States (USDA Agr. Stat., 1977).

### Livestock Feed-By-Products of Industry and Agriculture

It is now difficult to believe that California has been deficient in all types of livestock throughout most of its history. Its present status as an exporter of animal products was more or less forced upon it by new and better field crops and a steady increase in the by-products of industry and agriculture which proved valuable for feeding of the stock. Local production of the animals needed to be encouraged so that feeds could be consumed near the site, avoiding the expense of shipment to the Atlantic Coast or to Europe. Likewise, improved meat quality, as demanded by an increasingly discriminating market, depended on having a variety of feeds, which the highly diversified agriculture of the state could supply in great quantity. Feed producers and processors and livestock raisers and packers recognized their mutual dependence, with great benefit to the entire Pacific Coast area.

The amount of research that was required for the development of the diversity of feed supply that is now more or less taken for granted can be readily imagined. Much by-product feed is furnished by the sugar-beet industry. Whether dry, wet, or as silage, beet pulp is very palatable to livestock. It is high in total digestible nutrients (TDN) per unit of dry matter, but low in protein and phosphorus, so it must be supplemented with other feeds containing these nutrients. Beet tops may also be used as feed (Leonard et al., 1978).

Cottonseed meal is a very important animal feed, but it was not until 1925, after many fruitless endeavors, that success with cotton was ensured by concentrating on a single long-staple variety high enough in yield and quality to encourage extensive plantings. In 1975 California produced 786,100 tons of cottonseed, exceeded only by Texas. After the oil is pressed out, the cottonseed meal that remains is high in protein and phosphorus.

The milling of rice produces another important by-product for stock feed. All of the bran, most of the polish, and some of the grits are used for this purpose. Copra, the meat of the coconut, yields oil when pressed, but the remaining meal is a valuable by-product consumed by beef and dairy cattle, sheep, and poultry in California. Orange and lemon pulp are available in large quantity as by-products of the citrus fruit-processing industry. This material has approximately the feeding value of barley or dried beet pulp. In the production of fish oils and canned fish, the offal was originally disposed of as fertilizer. Its high content of protein, as well as minerals and vitamins, stimulated research on a processing procedure that would render this by-product suitable for feeding to all classes of livestock and poultry.

### Present Status of California's Livestock Industry

The livestock industry is currently the most important sector in the California agricultural economy (table 5). (The data for 1975 are presented in table 5 because that was the year preceding the unprecedented drought of 1976 and 1977. The drought did not decrease total agricultural production but influenced the relative acreage planted to the various crops.) In 1975, 26.7 percent of the gross value of agricultural production in California was provided by livestock. Among forty agricultural commodities evaluated, cattle and calves were the most important, with a value of $1,101,000,000, followed by milk and cream with a value of $997,000,000. Yet over a third of the beef, nearly all of the pork, and over half of the lamb consumed in California are imported (Cothern, 1974). Southern California retained a prominent place in livestock production, although the industry moved inland.

## THE EGG INDUSTRY

Egg production is one of the major facets of the livestock industry. In 1972 there were 302 million hens and pullets of laying age in the United States. These, however,

**Table 5**
**Rank and Farm Value of California's Ten Most Important Agricultural Commodities in 1930 (Reed, 1976) and 1975 (CCLRS, 1975)**

| 1930 | | | 1975 | | |
|---|---|---|---|---|---|
| Rank | Commodity | Value (million dollars) | Rank | Commodity | Value (million dollars) |
| 1 | Oranges | 98 | 1 | Cattle and calves | 1,101 |
| 2 | Milk and cream | 90 | 2 | Milk and cream | 997 |
| 3 | Cattle and calves | 51 | 3 | Cotton, lint & seed | 566 |
| 4 | Eggs | 46 | 4 | Grapes | 479 |
| 5 | Grapes | 34 | 5 | Hay | 459 |
| 6 | Lemons | 28 | 6 | Tomatoes, processing | 454 |
| 7 | Dry edible beans | 22 | 7 | Eggs | 351 |
| 8 | Cotton, lint & seed | 17 | 8 | Nursery products | 294 |
| 9 | Chickens | 16 | 9 | Rice | 256 |
| 10 | Prunes | 15 | 10 | Lettuce | 248 |

were only about 10 percent of the number of broilers produced that year (discussed in chapter 4). But the nation's $1.8 billion gross income from eggs surpassed the $1.6 billion gross income from broilers. California is the nation's leading producer of eggs, with about 40 million laying hens and pullets in 1976. Among regions, the South is the leading producer, but all areas of the country are well represented (USDA Agr. Stat., 1977).

Sonoma County was once the leading poultry-raising county in the United States. In the early days it was favored by climate and access to San Francisco by waterway (Hart, 1946). In the last thirty years the center of poultry production has shifted southward; the leading counties for market eggs are in southern California and the San Joaquin Valley—Riverside, San Bernardino, San Diego, Stanislaus, and San Joaquin counties (McGregor et al., 1977*a*).

## History

The domestic chicken (*Gallus domesticus*) may have been descended from the red jungle fowl (*Gallus gallus*) of Asia, which it most closely resembles and which may still be found in the jungles of India. Spanish explorers brought chickens to the New World in the sixteenth century and English settlers brought them to Jamestown and Plymouth (Wilson and Vohra, 1974).

Artificial incubation and brooding made first real progress in the 1880s. In 1901 dry mash feeding was begun. (Mash is now a meal composed of scientifically determined proportions of cereal or by-products, vegetable and animal protein materials, and mineral and vitamin supplements.) The first college poultry department was started at Connecticut Agricultural College. Baby chick shipments were admitted to the United States mails in 1918 and poultry and egg farm income in the United States first reached a billion dollars in 1917. Under the guidance and coordination of the USDA, the National

Poultry Improvement Plan was put into operation, resulting in remarkable control of pullorum disease and other benefits. Disease control has been an important factor in the development of the poultry industry. Mass vaccination of from ten to thirty thousand chickens by means of aerosols, drinking water, or dust, has been developed, producing high and durable immunity in most cases (Wilson and Vohra, 1974). Ready-to-cook poultry was introduced and egg and poultry merchandising became well developed in the 1940s.

## Breeding

The domestic hen descended from an ancestor that laid from twenty-five to fifty eggs a season, hatched them into chicks, and took care of them as long as needed. The fact that such a bird could evolve into an efficient egg-laying machine able to produce up to 366 eggs a year indicates the amazing plasticity of the species. Early man noticed that by removing eggs from the nest of a hen, the onset of broodiness could be postponed and the periods of egg-laying could be prolonged. Through this procedure alone, it is unlikely that the productivity of the hen could have been increased to much beyond 100, or at most, 150 eggs per year. However, when hens with a poorly developed mothering instinct were given a greater opportunity for reproduction, by long-continued selection, the productivity of the chicken population was increased by gradually eliminating the "broody instinct" (Nalbandov, 1974).

One of the spin-offs from the hybrid corn investigation was the increased interest in using inbreeding and hybridizing techniques to improve livestock. It is of historic interest that Henry B. Wallace, son of the illustrious Henry A. Wallace of corn-breeding (and political) fame, specialized in poultry, employing the same techniques that had made possible hybrid corn. After completing his college work, young Wallace assumed entire direction of the poultry-inbreeding program that had become a project of the Pioneer Hi-Bred Corn Company organized by his father in 1926. The first of the new hybrid chickens were produced in 1942 and demonstrated that egg production could be increased by 40-50 percent. Entered in the 1945 Illinois Egg Laying Contest, Wallace's new chickens outlaid the best standard-bred hens by a wide margin (Crabb, 1947).

The most widely distributed egg breed in the developed countries of the world is the familiar white Leghorn, imported into America from Italy in 1835. It matures early and is an excellent layer of chalk-white eggs. Its small size results in it being an efficient converter of feed into eggs. The pullets commence laying at six months of age or earlier. It is, however, inferior as a meat bird.

The Rhode Island Red is second in importance to the Leghorn as an egg-laying breed in America and Europe. It was developed mainly in Rhode Island and Massachusetts from 1850 to 1890. It lays a medium-brown-shelled egg (Nordskog, 1974).

## Management

Two systems of management are employed in poultry houses. The birds may be kept in built-up litter or slatted floors or they may be confined in cages with one or two birds per cage or up to twenty-five or thirty birds in wire-floor colony pens. The floor space allowed per layer is 0.5-0.7 square feet per layer in cages and 2-3 square feet on the floor (Wilson and Vohra, 1974). Most commonly, adult layers are kept in individual wire cages and are watered in troughs suspended outside the cage. Droppings fall through a wire floor, and the eggs roll forward on the inclined floor to a tray. Egg production is stimulated during fall and winter by artificial lighting. Eggs typically receive no processing after they leave the farm, being simply sorted according to size and grade, so the farmer's share of the retail price, about 70 percent, is higher than for most agricultural products.

In their second adult year, hens lay fewer eggs than in their first year, so many farmers dispose of them as they enter their second winter. These hens are then used mainly for making canned chicken soup. Egg production per hen or pullet rose from an average of about 120 in 1930 to 218 in the 1960s, and some can now average an egg per day.

Today's egg factory, the most integrated and mechanized of any industry engaged in food production, is a far cry from the daily tedium of watering, feeding, cleaning, and gathering eggs in pioneer days—traditionally the work of women and children and the source of their "pin money."

### Manure Disposal

Livestock in the United States produce over two billion tons of liquid and solid wastes per year, about ten times the amount produced by the human population. The problem of waste disposal is magnified by the modern tendency to concentrate large numbers of livestock, and so are the problems of water pollution, odors, and the breeding of pest flies.

Poultry ranches in particular are generally not far away from towns and cities whose inhabitants frequently complain of flies and odors. Caged layers provide the more difficult problem because the manure is more liquid than that from deep litter houses, the latter being diluted with litter material. The manure is disposed of through anaerobic decomposition, city sewers, composting, dehydration, as feed supplement for ruminants (Harmon et al., 1975; Federal Register, 1977), incineration lagoons, or as fertilizer spread on land (Wilson and Vohra, 1974). In any case, proper disposal procedure can minimize the number of pest flies. In the case of caged laying hens, a concrete base beneath the cages catches the droppings and prevents house flies from pupating in the soil. Parasites and predators can then destroy a large proportion of the immature flies in the manure. During the season when flies are a problem, only the top portion of manure is removed, and a 6- to 8-inch pad remains on the concrete base, or all manure should be removed on an alternate row basis. In either case, parasites and predators from the manure that remains can quickly invade adjacent fresh droppings. If all manure is removed at one time, a period of 6-12 months is required to reestablish the natural enemies of the flies (UCAE, 1971-72).

## TURKEYS

The modern domestic turkey is believed to be descended from a wild subspecies found in Mexico and Central America and a larger species, native to the United States, and with a characteristic bronze plumage. The modern strains of bronze turkeys closely resemble the latter species, which probably had the predominant influence on the development of domestic types. The wild turkey still occurs in the United States, particularly in the deep woods and borders of swamplands in the eastern half of the country. It has been successfully reintroduced in some areas where it had been hunted to extinction, as in some parts of the Ozarks.

Of the 140 million turkeys raised in the United States in 1976, California raised 17.5 million, being exceeded by Minnesota, with 24 million. Other states with heavy production were North Carolina, Arkansas, Missouri, and Texas (USDA Agr. Stat. 1977). Of the forty-six varieties of turkeys raised for the commercial market in this country, forty are white and only six are bronze. The white varieties carry a recessive white gene that almost completely prevents the appearance of pigment in the plumage. White turkeys seem to withstand hot summer sunshine better than colored varieties. They also grade better because their pin feathers are less conspicuous.

As with broilers, the growth rate of turkeys responds readily to selection. Conformation, a function of both live structure and fleshing, is of particular importance in turkeys and was an important accomplishment of modern breeders. An undesirable side effect has been the poor reproductive performance common to most large strains of turkeys (Nordskog, 1974).

At one time the birds were fed in green grainfields in the spring and early summer, then on the stubble in summer. Feed was supplied for only a few weeks after hatching and before marketing. As in the case of chickens, this era was followed by confinement in small yards, improvements in artificial incubation and brooding, and improved rations, especially protein and vitamin supplements.

Only about two pounds of feed are required to produce one pound of turkey, which may be compared with four pounds for a pound of pork and eight for a pound of beef. Besides, turkey is second only to fish in percent of protein. After a highly efficient processing of the bird in a modern plant, feathers and offal are sold for processing into poultry and pet food and heads and other "trim" are used as food for mink ranches (McGuinness, 1974).

## GRAIN-FARMING (Gates, 1967)

Grain-farming first flourished in the counties bordering the San Francisco Bay and in Santa Cruz and Sacramento counties. After the first winter rains had wet the soil to a depth of six inches, it was easily plowed and prepared for seeding, which was done by hand or by drills. Yields came to thirty, forty, or fifty bushels and in a few areas even eighty bushels per acre—more than anywhere else in the United States. The grain was harvested in the dry season and there was no worry about getting crops in before the anticipated rain or hail, as was the case in the Midwest and Great Plains.

There was intensive activity in the development of reapers and threshers in California because the fields were large and wheat straw was long and heavy, yielding from twenty-five to sixty bushels of heavy grain per acre, whereas in the Great Plains the straw was short and light and yielded only from sixteen to twenty bushels. At first horse-powered treadmills worked the threshing machines on the big ranches, but they were soon replaced by steam power. At least one steam-powered threshing machine was said to be able to thresh between 1,000 and 1,500 bushels of wheat a day at the expense of three-quarters of a cord of wood and with a staff of twelve to seventeen men. Four- to six-ton wagons hauled the grain to San Francisco for shipment.

The combine, considered to be a failure in Michigan, was shipped around Cape Horn to San Francisco in time to cut 600 acres of wheat in California in 1854. A more suitable climate, larger acreages, and the "California love for big machinery" started the combine on the way to success. Soon various California models appeared and factory production began in that state in 1880; by 1889 there were about 500 machines at work. A combine, with four men and with a team of twenty-four or more horses or mules to pull the heavy machine, could harvest twenty-five acres a day. Horses and mules were gradually replaced by steam-engine tractors in the mid-1880s. By 1912 gasoline power began to replace steam in California for pulling the combine and operating its mechanism. During the World War I period of labor shortage, combines, powered by gasoline tractors and auxiliary engines, began to be employed also east of the Rocky Mountains, but were still used only for small grains and in semiarid climates. Then smaller combines began to be built for family farms and for a variety of crops. As previously related, the self-propelled (SP) combine gradually replaced the "pull-type" machines hitched to tractors (Gittins, 1950; Bainer, 1975*a, b*).

The flour-milling industry followed the wheat-raising boom. The hard, dry California wheat was milled into flour that was excellent for shipping, even through the tropics. By the end of the 1850s California had over 200 flour mills, several able to mill a thousand barrels a day. They supplied the mining camps of California and the Rockies with flour, and large shipments were also made to Japan, China, the British Isles, and continental Europe.

California's wheat bonanza was to be of short duration. As land was sown to wheat year after year, yields declined and pests and diseases accumulated. Smut, rust, and blight became prevalent. Some years grasshoppers were very destructive, and the Hessian fly was a severe pest in several areas. Weeds such as mustard, wild oats, wild clover, and volunteer wheat and barley were not killed during the winter, as they were east of the Rockies, and their seeds became mixed with the grain when harvested. In those days equipment was not adequate to separate weed seeds from the grain; the wheat and flour made from it brought low prices (Davis, 1868). Farmers eventually had to either abandon growing grain (wheat and barley) or use a rotation system in which the crop was cultivated on the same land only one or two years in five.

Although the gross value of wheat grown in California in 1975 was about $225 million, this was only 3.6 percent of the value of the state's plant crops and 2.7 percent of the value of all agricultural commodities. Rice, with a value of $265 million, was a more important crop. Nevertheless, wheat ranked twelfth in value among California's 200 agricultural commodities, ninth among plant crops, and exceeded such important crops as oranges, cut flowers, barley, peaches, potatoes, and almonds (McGregor et al., 1976*a, b*).

## CALIFORNIA'S CITRUS INDUSTRY

Gold, grain, lumber, and cattle contributed greatly to California's economic development, but only cattle could be claimed as a substantial contribution in southern California in the early days. But the benighted region was to experience a "gold rush" of its own, with golden "nuggets" growing on trees, just as Goethe found them in sunny Italy:

> Gibt es ein Land wo die Zitronen blühen,
> In dunkler Laub die gold Orangen glühen.

All but three of the chain of twenty-one Spanish missions established from San Diego (1769) to Sonoma (1822) had fruit orchards where oranges, as well as figs, olives, and grapes were grown. An orange orchard with about 44 seedling trees was planted on the San Gabriel Mission grounds, probably in 1804 or 1805, and proved that climate and soil conditions in southern California were suitable for citrus culture. Probably no fruit was sold from this orchard, but trees were obtained from it for planting elsewhere, including thirty-five that were transplanted on the property of Don Luis Vignes on Aliso Street in Los Angeles, as well as the trees planted by erstwhile mountain man William Wolfskill.

## A Mountain Man Turned Orange Grower

Not all the famed mountain men succumbed to the hazards of the traplines and the perilous exploratory expeditions in the western wilderness. The veteran trapper and explorer William Wolfskill lived to pioneer again, but in a more genteel and less hazardous occupation—growing orange trees and grapevines in sunny southern California.

After Wolfskill and George C. Yount blazed the Old Spanish Trail from Santa Fe to the Pacific, already related, Wolfskill chose to remain in southern California. An account of his career not only as a mountain man but also as a pioneer agriculturist and leading citizen was presented for the first time, as a continuous story, in Iris Higbie Wilson's *William Wolfskill, Frontier Trapper to California Ranchero* (1965). After an unsuccessful venture into sea-otter hunting, he bought a small tract of land with grapevines near Los Angeles. Then began a period of experimentation with different ways of planting grapevines, distilling brandy, producing wine commercially, and cultivating many kinds of plants, especially citrus trees. In this enterprise Wolfskill was joined by his brother John in 1838.[6] The Wolfskills planted 32,000 grapevines and eventually were among the leading vineyardists.

In 1841 Wolfskill married Magdalena, the niece of Don Antonio María Lugo, whose ranch had been Wolfskill's first stopping place at the end of his historic trip to California. The couple had three daughters and two sons who grew up on the ranch and became influential members of the Wolfskill community.

Wolfskill planted many varieties of fruit and nut trees, but was always particularly intrigued by the possibilities of oranges. In 1841 he planted the nurseries of what was to become his famous orange orchard, obtaining his trees from the San Gabriel Mission. This was a two-acre orchard planted next to his large, luxuriously furnished adobe dwelling. The sale of his fruit proved to be so lucrative that he increased the size of his orchard to twenty-eight acres in the early 1850s and eventually to seventy (Coit, 1915), including a considerable spread of lemons (Lorenz, 1949). He had four wine cellars with a total capacity of up to 100,000 gallons. The Los Angeles *Star* reported his 1857 vintage as "750 tons of grapes, 12,000 gallons of wine, 2,000 gallons of angelica[7] and 300 gallons of brandy."

There was a considerable increase in orange-planting in southern California after 1850 and Wolfskill's profits were believed to be the principal stimulating factor (Spaulding, 1922).

In 1868, 2,200 boxes of oranges were shipped by boat from Los Angeles to San Francisco, where the fruit competed with citrus fruits shipped from Hawaii and Sicily. For historic perspective, it is of passing interest that at that time deer, antelope, bighorn sheep, and grizzly bears still abounded in parts of Los Angeles and Ventura counties. In the Santa Clara Valley, about fifty miles northwest of Los Angeles, grizzlies raided apiaries and sheepfolds and sometimes killed full-grown cattle and occasionally a farmer, during the 1850s and 1860s (Cleland, 1940).

The last crop of oranges during William Wolfskill's lifetime, in 1866, sold on the trees for $25,000 (Coit, 1915). His son Joseph shipped the first trainload of oranges to

[6]John Wolfskill later developed a ranch along both sides of Putah Creek in the Sacramento Valley west of the city of Sacramento. He grew good crops of corn, wheat, and barley and also had a vineyard and fruit trees. He founded the town of Winters on a site where he had once slept in a tree to avoid grizzly bears. Most of the original Rio de los Putos grant was passed down to his descendants and in 1936 Frances Wolfskill Taylor left a large tract of land at Winters to the University of California. It has been used by the College of Agriculture at Davis for orchard and vineyard research.

[7]Angelica was made by adding one gallon of grape brandy to three gallons of unfermented grape juice, and was said to be "a most palatable and agreeable drink, but woe to him who drinks too deeply" (Wilson, 1965).

eastern markets in 1877, via the recently completed Southern Pacific Railroad. The oranges were individually wrapped in paper, placed in specially designed boxes with dividing centerpieces, and the cars were refrigerated with ice.

## The Insect Menace

Joseph W. Wolfskill should also be given credit for his cooperation with the Department of Agriculture in its investigation of the cottonycushion scale, *Icerya purchasi,* a pest native to Australia, which appears to have been introduced into California on acacia at Menlo Park in 1868 or 1869 and within a period of about ten years caused great damage to citrus orchards in the south. This insect extracts plant sap, causing defoliation, fruit drop, and decrease in the vitality of the tree. Its most conspicuous feature is its large, elongated egg sac, twice as long as the body of the adult female and cottony in appearance. This insect became so abundant on citrus trees that they appeared from a distance to be covered with snow. Growers were preparing to have their trees removed, and the very existence of the citrus industry in California was threatened.

Joseph Wolfskill was the first to cover an orange tree with a tent, using heat under it to kill the black scale (*Saissetia oleae*). Later he used carbon disulfide gas. A USDA entomologist, D. W. Coquillett, was so impressed with the method of treatment that in September 1866 he began experimenting in the Wolfskill orchard and eventually found that hydrocyanic acid (HCN) was the most effective gas for killing cottonycushion scale as well as all other scale insects.

The classic achievement in the suppression of an insect pest by biological control also got its start in the Wolfskill orchard. Another USDA entomologist, Albert Koebele, was sent to Australia to search for possible natural enemies of the cottonycushion scale. On November 23, 1888, Coquillett received the first shipment of predators and these were placed in Wolfskill's orchard. The predator was a small red-and-black lady beetle called the vedalia (*Rodolia cardinalis*). The vedalia thrusts its bright-red eggs under the scale or attaches them to the large egg sac. The orange-red larvae enter the egg mass from beneath and feed on the eggs, and later on all stages of the host. The scales are attacked by the adult beetles in the same manner, but neither larvae nor adults feed on other species of insects. The cottony-cushion scale was practically eliminated from California orchards within eighteen months, marking the first successful introduction of a beneficial insect into any country to destroy another insect.

The control of the cottonycushion scale was followed by other striking successes in the biological control of citrus pests. In 1923 Professor Harry S. Smith joined the staff of the University of California's Citrus Experiment Station (now Citrus Research Center)[8] at Riverside to head a Department of Biological Control. Under his direction a large staff of experts was gradually built up, and a considerable number eventually became located at the university's facilities at Berkeley and Richmond. One success after another rewarded these groups of investigators in bringing about the complete or partial control, by natural means, of many important pests, including weeds. Meanwhile entomological literature was enriched by the results of their basic research on problems relating to insect populations, particularly the complicated biology and taxonomy of the parasitic hymenoptera, which proved to be the most important of the various types of natural enemies (DeBach, 1964). Also important, of course, was the training received by students.

[8]The Citrus Experiment Station was established in 1907 near Mount Rubidoux and was moved to its present location east of Riverside in 1917. A College of Letters and Sciences was established in the same location in 1954 and a School of Agriculture in 1960 (now the College of Natural and Agricultural Sciences).

As related in chapter 3, modern pest control aims to use pesticides, when necessary, in conjunction with cultural and bioenvironmental control measures in such a way as to least interfere with the work of natural enemies. Among the pesticides developed in connection with citrus pest control, the most important and durable were the refined petroleum spray oils, which have played an important role in the control of citrus pests since the 1920s. Apparently insects cannot become resistant to the suffocation and/or desiccation to which they are subjected by a very thin film of spray oil. Spray oils also have the advantage of leaving no poisonous residues on the treated crops and have relatively little adverse effect on natural enemies.

Fortunately for the citrus industry, the Citrus Experiment Station had a collection of 600 citrus varieties in the 1930s when tristeza, a disease caused by the "quick decline" virus, was threatening the orange industry; it eventually destroyed 50,000 acres of oranges. The virus is transmitted by aphids, but it affects only trees budded on sour-orange rootstock. In the 1930s they comprised two-thirds of the California orange acreage. Sour-orange rootstock was replaced with tristeza-tolerant stocks. Troyer citrange, one of the varieties in the university gene bank, became the favored stock.

## Budded Citrus Varieties

The first orange trees planted in California, were seedlings. Unlike some fruits, the offspring from orange seedlings are very good, indicating that the produce was quite acceptable when it first attracted the attention of primitive peoples, and that subsequently the crude selection of the seeds of the best fruits, for propagation, had been going on for many centuries before oranges began to be cultivated in Europe (Webber, 1943).

Budded varieties were introduced in the 1870s. The U.S. Department of Agriculture obtained twelve small trees from Bahia, Brazil, in 1870, which were planted and further propagated by budding. The fruit was said to be seedless. Two of the young trees were sent to Mrs. Eliza Tibbets in Riverside in 1873, and these marked the beginning of the great California navel orange industry. It was not until the fourth year that the two trees were allowed to bear fruit, but then they proved to skeptics in the Riverside settlement that the fruits were indeed seedless and of superb quality. L. C. Tibbets refused an offer of $10,000 apiece for the two trees, but sales of budwood from them rose to about $600 a month. By 1880 an orchard of 75 acres had been budded to Washington Navels, as Tibbets called them. Land that could be purchased in 1880 for $20 an acre sold for $1,000 an acre 4 years later (Holbrook, 1950).

The 1870s also marked the introduction of the Valencia orange into California. A. B. Chapman imported a number of orange varieties from Thomas Rivers, a London nurseryman who imported orange trees to be grown in English greenhouses. One of Chapman's imported varieties, the name of which had been lost, he called "Rivers Late." This was later identified with the variety called *La Naranja Tarde de Valencia* in Spain, and thus derived its present name. It is believed to have originated in the Azores.

The completion of the Southern Pacific Railroad "valley line" in 1876 and "southern line" to New Orleans in 1883 opened up new and better fruit regions and greatly stimulated the development of the orange industry. Then in 1885 the opening of the Atchison, Topeka and Santa Fe lines initiated a great boom in orange-planting, particularly of the navel orange, which was recognized for its superior flavor after its exhibition at a citrus fair held in Riverside in 1879. Returns of $800-$1,000 per acre were common and as much as $3,000 per acre was reported—an incredibly lucrative business in those days (Lelong, 1902).

## Lemons

In the United States, the lemon industry originated in California; that state still produced 86 percent of the nation's crop of 17,820,000 boxes in 1975-1976, the rest being produced in Arizona. (The previous year the nation's production had been 29,400,000 boxes, of which California produced 76 percent (USDA Agr. Stat., 1977). The principal varieties are the Lisbon and the Eureka. Nursery stock of the Lisbon was imported from Australia and the Eureka originated in Los Angeles, the progeny of a tree grown from lemon seed obtained in Sicily and producing fruit superior to that of the parent and sister trees (MacCurdy, 1925).

Urbanization has eliminated nearly all the formerly large lemon acreage in Los Angeles and Orange counties. The coastal plains of Ventura and Santa Barbara counties have become the most important lemon-growing areas, although lemons have also been planted extensively in Coachella and other desert valleys. Of California's 64,494 acres of lemons, 35,314 are Eureka, 27,328 Lisbon, and 1,852 other varieties (McGregor et al., 1977*a, b*).

## Grapefruit

In California, grapefruit is grown principally in desert areas, especially the Coachella Valley, where the fruit matures in winter and early spring. In 1976, 4,100,000 boxes of grapefruit were grown in California desert valleys and 3,100,000 boxes in "other areas," mainly the San Joaquin Valley. The total of 7,200,000 boxes is only about 13 percent of the nation's grapefruit crop, the bulk of which is grown in Florida, with some also grown in Texas and Arizona (USDA Agr. Stat., 1977).

The principal California grapefruit variety is the white-fleshed Marsh seedless, although small quantities of Red Ruby and other pink-fleshed varieties are also grown. The winter desert climate attracts retired people and others who find it interesting and profitable to invest in new grapefruit orchards. These are then operated by local management firms (Johnston and Dean, 1969).

## The Chaffey Brothers and Irrigation Policy

Much credit for the rapid expansion and striking success of the citrus industry in southern California belongs to George and William B. Chaffey. They were Canadian immigrants with commendable ambition and foresight, combined with engineering skill and business acumen, who established the first of their famous irrigation colonies fourteen miles west of San Bernardino, naming it Etiwanda. Along with L. M. Holt, they developed a simple solution to the previously insoluble problems posed by the old California riparian law, inherited from England, but not adaptable to irrigation in the arid West. Beginning in 1881, the Holt-Chaffey Mutual Water Company gave each purchaser one share in the water company for every acre of land he purchased from the company. The colony was divided into ten-acre lots. Water from streams in the San Gabriel Mountains was conducted under pressure, through concrete pipes, to the highest corner of each lot—the first such system in the west (Alexander, 1928).

The success of the Etiwanda project encouraged the Chaffey brothers to plan another settlement in the winter of 1882, about six miles west of Etiwanda and thirty-eight miles east of Los Angeles, where they purchased land and developed the irrigation colony they named Ontario. It was irrigated by surface and ground water from San Antonio Canyon. They developed a settlement unequaled in beauty and attraction for farmers and home seekers. They were the first to include hydroelectric power plants as a part of the irrigation system, recognizing power, whenever its development was possible, to be one of the best "cash crops" of an irrigation project.

The Chaffey brothers then developed similar irrigation colonies (Mildura and Renmark) along the Murray River in Australia. George Chaffey returned to California and developed additional water sources for Ontario. Then on April 3, 1900, he became president, chief engineer, and one of the directors of what was at the time the nearly defunct California Development Company, financing and revitalizing with his engineering genius a renewed effort to reclaim the million-acre desert area he was later to name the Imperial Valley.

A great range of sand dunes made it necessary to construct a canal through about fifty miles of Mexican territory, following an ancient, dry, overflow channel of the Colorado before it turned north and reentered the United States near where the twin cities of Mexicali and Calexico, founded and named by Chaffey, are now located. The main canal was then extended twelve miles north to what is now the town of Imperial. Water arrived on May 14, 1901. Within another year 400 miles of irrigation ditches were dug and settlers were pouring in to farm the typically fertile desert land (Alexander, 1928). The huge concrete-lined All-American canal, just north of the border, was not completed until 1940.

## Citrus-Fruit Cooperatives

Although farmers are generally considered to be proudly independent people, cooperation has been an important factor in farming from the very beginning. The colonial farmers cooperated informally in erecting buildings, constructing roads, importing purebred cattle, harvesting their crops, making quilts, and fighting Indians. Later cooperation became formalized in the formation of cooperatives for the processing and sale of farm products. Starting in 1867, local and state Granges were developed and finally national cooperatives were sponsored throughout the nation under their auspices. Already as early as 1920 there were more than 12,000 active cooperative enterprises for the cooperative marketing of most farm products in sufficient quantity to permit carload shipment. One of the best-known cooperatives got its start in the early citrus industry of southern California.

In the early days of the orange industry, prospective buyers would visit an orchard and make a "Yankee guess" at the quantity of fruit on the trees and bid a lump sum for the crop. The buyer assumed all responsibility for picking, packing, shipping, and marketing. After a few years this practice was modified, the buyer paying a stated price per packed box and doing the packing and shipping, and frequently the picking. As orange production increased, it became apparent that there would have to be some understanding between buyers as to distribution of the fruit among eastern markets to prevent glutting in some areas and undersupply in others (Lelong, 1902).

In the fall of 1885, at a Los Angeles meeting of growers lasting for several days, "The Orange Growers Protective Union of Southern California" was organized. Under the direction of the executive committee, five men were sent east to sell, regulate, distribute, and do all services as required by them under regulation of the committee or the board of directors. During the 1885-1886 season 1,969 carloads of oranges and lemons were shipped from southern California. The union was followed by other organizations in a movement that, not without strife and setbacks, eventually led to the famous Southern California Fruit Growers Exchange, organized in 1895. Its name was changed to California Fruit Growers Exchange in 1905 and to Sunkist Growers in 1952.

Early efforts in marketing lemons in the East did not meet with success and the reason was that the fruit was carelessly packed and did not keep well. This led to the organization of the lemon growers of Ontario and Cucamonga in order that their fruit might be cured under a uniform procedure, graded systematically, and shipped under a common brand name. The lemons reaching eastern markets were then as readily accepted as lemons from Sicily, which for years had been considered to be the best obtainable. Soon lemon growers in other districts were similarly organized and in 1896 they joined the orange growers in the Southern California Fruit Exchange. Fruit sold by the exchange bore the Sunkist trademark. Among the important functions of the organization was advertisement and the education of the American public about the healthful qualities of citrus fruits. The California Fruit Growers Exchange was only one of many examples of the superb ability of American farmers to cooperate effectively when matters of mutual concern and benefit are involved.

The story of California citrus (and walnut) cooperatives is inseparable from that of Charles C. Teague. Born in northern Maine in 1873, he started his career in southern California as a young ranchhand near Santa Paula. His father's death when Teague was twenty years old left him and his mother and two sisters with a paid-up insurance policy of $8,500. With a striking display of hard work and the traditional Yankee ingenuity and business acumen, young Teague eventually parlayed this modest sum into the famed Limoneira Ranch—1,600 acres of citrus and avocados.

Teague served many years, without compensation, on the board of directors of the California Fruit Growers Exchange (1911-1926) and then as president from 1926 to the time of his death in 1960. He also served as president of the California Walnut Growers Association from 1912 to 1942. He served on many committees and on the Federal Farm Board at the urgent request of President Herbert Hoover, who considered him "the most outstanding representative of the western cooperative movement."

Although Teague was an ardent advocate of the private enterprise system, he had little patience with those who criticized cooperatives as "a sort of socialistic movement." In his *Fifty Years as a Rancher* (1944) Teague considered cooperatives "just as much a part of the business enterprise system as is manufacturing or any other form of industry." Another point made by Teague is apt to be overlooked in the current "small is beautiful" sentiment:

> I am a believer in the development of the family size farm in America to the greatest possible extent. However, I am also confident that small farm ownership in California would not have developed to its present proportions if some of the early pioneers had not been able, through the expenditure of considerable capital, to blaze the trail and prove the practical and economic value of producing agricultural crops in various regions.

The leadership and guidance of the big growers was especially valuable in the pioneer stages of the various agricultural industries—before the days of cooperatives and the considerable contributions of governmental research and extension. The small grower, regardless of his ability, when compelled to work from daylight to dark, could not achieve enough economic independence to devote his time and energy to such tasks as the creation of marketing cooperatives.

In 1924 the University of California conferred on Charles Teague the honorary degree of Doctor of Law.

## Modern Cooperatives

Sunkist Growers demonstrated what cooperation could do for farmers by bringing stability and prosperity to thousands of citrus growers through coordinating their processing and marketing needs. The local packing associations of the organization not only pack the fruit, but also supply picking crews and hauling facilities and undertake the work of pruning and pest control. This large organization of 7,500 members is now a federation of about 65 local and regional cooperative associations. Sunkist packs fruit either at its own plants or in plants contracted to it. It advertises and sells the fruit domestically and abroad, handling 70 percent of all western fresh citrus fruit. For the year ending October 31, 1977, estimated revenues were more than $500 million (R. Smith, 1977).

Nationwide, there are over 7,500 farmer organizations such as Sunkist Growers, and they do about $57 billion in business, handling 27 percent of all farm products. The Capper-Volstead Act of 1922 provided farmer cooperatives with antitrust immunity, permitting their members to jointly market their products so as to have the bargaining power to compete with large corporate farms. Local co-ops have joined others to form regional organizations, which in turn have joined interregional and federated co-ops. Such co-ops control 80 percent of the nation's wholesale dairy markets, 40 percent of the cotton crop, 42 percent of the grain, and 52 percent of the fruit and vegetables. Supply co-ops control 40 percent of the agricultural chemical, fertilizer, and farm petroleum products markets, thus taking advantage of economies of scale that enable them to compete with such agribusiness giants as Ralston Purina, Kraft, W. R. Grace, and Dow Chemical (R. Smith, 1977).

Co-ops receive nonprofit tax treatment but they do not have the opportunity to raise equity in the stock market, so growth is limited to what members can provide or borrow. Therefore co-ops are still small when compared with some of their agribusiness competitors. Intense public and congressional sensitivity to food prices has drawn the attention of the Federal Trade Commission (FTC) to what it alleges to be the tactics of some co-ops in stifling competition in various ways. Nevertheless the FTC admits that, as a whole, co-ops provide a desirable degree of "good hard competition" to the corporate giants. Kenneth Naden, president of the National Council of Farmer Cooperatives in Washington, D.C., contends that the need for Capper-Volstead continues undiminished, not only to protect large co-ops but the 90 percent with annual sales below $5 million. He believes that farmers need the ability provided by the co-ops to withhold products at times to even out agriculture's typical boom-and-bust cycles. He points out that co-ops cannot control how much farmers plant and therefore they are still subject to laws of supply and demand. The USDA Farmers Cooperative Service believes co-ops to be the main bulwark of the family farm (R. Smith, 1977).

### Orchard Heaters

Southern California residents who believe that present-day smog is the worst thing that could happen in the way of air pollution probably have not experienced a period of freezing weather during the years when the citrus industry had reached its maximum acreage and extended in a continuous belt across the entire Los Angeles Basin. In those days the idea was prevalent that the smoke from orchard heaters (smudge pots) burning a low grade of fuel oil, as well as the heat generated, protected the trees from the cold. An agricultural engineer, F. A. Brooks, from the University of California, Davis, demonstrated that only the heat was a protective factor—the smoke contributed nothing. Under pressure from irate southern California citizen committees, the university hired a combustion engineer, Arthur Leonard, on loan from the Research Department of the Standard Oil Company, for eighteen months. Leonard developed the return stack, which recirculated about 15 percent of the inert gases, formerly lost, back into the smudge-pot bowl to serve as a diluent to separate the carbon particles, thus ensuring more complete combustion. This engineering innovation resulted in a practically smokeless heater. Four million smudge pots, however, could not be quickly exchanged for the new heaters. Leonard was again engaged to work on "conversion units" to convert old orchard heaters to make them nonpollutive. He was again successful and stayed with the university until he retired.

Another UC Davis engineer had developed a "wind machine," a large power-driven fan mounted on a tower, that could bring warm air down from above and circulate it through the orchard. One wind machine with a 75-hp motor could circulate air for 10 acres. Such machines could raise temperatures as much as 8 degrees under ideal conditions, but did not suffice when there was a moderate air drift or during severe freezes or exceptionally heavy frosts. However, with a wind machine the citrus grower could get as much protection with fifteen heaters per acre as he used to get with forty-five (Young, 1929; Kepner, 1951; Adams, 1952; Bainer, 1975*b*).

### No-Tillage

Citrus growers were the first fruit growers to practice no-tillage in California and probably about 80 percent of the citrus acreage is now under some form of it. Also around 50 percent of California's deciduous acreage is now estimated to be under no-tillage and about 60 percent of almond orchards are under some minimum tillage system. Among the factors favoring no-tillage are herbicides that will give season-long control of weeds with one application, compared with six or more applications in past years. Benefits include less erosion, less soil compaction, less destruction of surface feeder roots, less breakage of limbs in close plantings, and lower costs for equipment, fuel, and labor (Anonymous, 1978).

### Mechanical Pruning

People driving past citrus orchards at present are likely to notice the manicured appearance of the trees in some orchards. Large mechanical pruning equipment consisting of giant saws on rotating arms has been developed. When mounted vertically, the equipment will cut off the sides of the trees; when mounted horizontally, it will cut off the tops.

### Present Status of the California Citrus Industry

Urbanization took a large acreage of citrus out of production, but then new plantings in the San Joaquin Valley and the Southeast desert were developed so rapidly that California now has more citrus acreage than it had when the large acreages of coastal southern California were at their peak, about 1930. In 1976 California had 123,050 acres of navel oranges, 80,722 acres of Valencia oranges, 820 acres of other orange varieties, 64,494 acres of lemons, 26,419 acres of grapefruit, 5,774 acres of tangelos, and 4,990 acres of tangerines—a total of 306,269 acres of citrus trees (McGregor et al., 1977*b*).

Navel oranges ripen in winter and are shipped mainly from November through April. They are used primarily for fresh consumption, so competition from processed oranges from Florida has had less impact on California navels than on its Valencias. In 1976-1976, California produced 28,300,000 boxes of navel oranges, valued at $88,296,000 (McGregor et al., 1977*a*).

Before World War II, about 85 percent of California's Valencias were shipped for fresh consumption. Then after the war, frozen concentrate became widely accepted as a year-round substitute for fresh orange juice, sharply reducing the seasonal advantages of California Valencias. Nevertheless, about 60 percent of the California Valencias are still consumed fresh. On the other hand, as stated in chapter 4, 92.5 percent of Florida's huge orange crop is processed. In 1975-1976, California produced 24,000,000 boxes of Valencia oranges, valued at $66,480,000 (McGregor et al., 1977*a*).

After the development of frozen lemonade concentrate in 1950 and expanded utilization of single-strength lemon juice, California's high-quality lemons for fresh consumption lost some of their competitive advantage and imports of lemon juice increased. About 40 percent of the California lemon crop is processed.

Perhaps 60 percent of California's grapefruit crop is sold fresh, compared with 41.5 percent of Florida's. Prices for fresh grapefruit are two to five times higher than for processed.

## OTHER SUBTROPICAL FRUITS

### Avocados

Although the avocado was seen by the Spaniards in Colombia as early as 1519, it was not until they reached Mexico and Guatemala that they learned the importance of the fruit, for it was then, and still is, one of the staple foods there, being much more nutritious than other fruits. The fleshy edible part is of a buttery consistency, yellow or greenish-yellow, and has a rich nutty flavor in the best varieties. The fruit is highly nourishing, with an oil content averaging about 17 percent.

Paleobotanic studies have revealed that close relatives of present-day avocados, and in the same genus (*Persea*), occurred throughout much of California in prehistoric times when the flora of the area appeared to be dense and semitropical (Schroeder, 1968). The avocados now cultivated in the United States are derived from two species: *Persea americana* and *P. drymifolia.* Among 28 cultivated plants recovered from excavations in Tehuacán, in southern Mexico, *Persea americana* was among the most ancient, believed to have come into cultivation about 9000 B.C. (Smith, 1965).

Avocados are now grown commercially in California from San Diego to Santa Barbara counties. The Fuerte, a fall and winter variety, is declining in popularity because of its variable and unreliable bearing habits. The Haas, a spring and summer fruit, has become the dominant variety. The Fuerte must be grown in localities with sufficient heat during blossoming to assure a crop set. The Haas may be satisfactorily grown in cool coastal areas where climate is unfavorable for setting Fuerte fruit. The latter develops a skin blemish when subjected to moderate to high levels of air pollution (Johnston and Dean, 1969).

The California avocado industry was of no significance as late as 1925, then grew rapidly. In 1976 there were 22,051 acres of Haas, 10,617 acres of Fuertes, 4,758 acres of Bacon, 3,385 acres of Zutanos, and 2,927 acres of other varieties in California, for a total acreage of 43,738—a greater acreage than for such traditional California crops as apples, apricots, cherries, figs, grapefruit, freestone peaches, and plums (McGregor et al., 1977*b*). The fact that 33.6 percent of the acreage was devoted to young, nonbearing trees in 1976 indicates that the avocado industry continues to expand. In the five-year period from 1972 to 1977, inclusive, California averaged 67,000 tons of avocados per year while Florida averaged 25,400 tons (USDA Agr. Stat., 1977).

## Dates

Date palms were once grown in California mission gardens, but the cool coastal climate prevented ripening of the fruit. In 1889 and 1890 the USDA obtained offshoots of Algerian, Egyptian, and Arabian date varieties to be planted in California, Arizona, and New Mexico. In 1904 the famous Deglet Noor variety was introduced and a new Government Date Garden was established in Indio, in the Coachella Valley of California. During the 1920s the seedling acreage was virtually eliminated and standard varieties were planted. The date palm scale (*Parlatoria blanchardi*), accidentally imported from North Africa, was eradicated in a joint federal-state campaign terminating in 1936.

In commercial date gardens the flower clusters are pollinated artificially to improve fruit set. Also the fruit is thinned to increase its size. After the palms are 10-15 years old, ladders and platforms are permanently attached to the trunks of the trees to facilitate harvesting. In California and Arizona, citrus trees are commonly grown between the rows of the tall date palms.

In 1975 there were 3,272 acres of Deglet Noor dates in California and 996 acres of other varieties, nearly all in the Coachella Valley (McGregor et al., 1977*b*).

## Figs

The Mission fig, common in the early mission gardens, was the only variety known in California for about eighty-five years and is still among the five varieties of commercial importance in the state. From about 1850 to 1880, varieties were introduced from the east, from England, and from the Mediterranean region, and the fig became a common dooryard and garden tree in the Sacramento and San Joaquin valleys. These included the Lob Injir or Smyrna fig, later called the Calimyrna. This was known to be the best drying variety, but the fruit failed to set. It had been rumored that pollination was required for Smyrna fig production, but this was believed to be a peasant superstition. As a result, the White Adriatic fig became the most widely planted variety for two decades; however, the dried product lacked the high quality of the figs imported from Smyrna.

Pollination of Smyrna figs is accomplished by a small, black, wasplike insect, *Blastophaga psenes*. In order to go through its life history, this insect must develop in a variety known as caprifig. Generally, blastophaga-containing figs are distributed at intervals of four days over the required period. A 16-foot tree requires two baskets, each containing two figs, every four days, or a total of sixteen figs per season.

The Calimyrna proved to be superior for drying, the Kadota for canning, and these varieties plus the White Adriatic, Mission, and Turkey for shipment as fresh figs.

In the United States, figs are grown only in California. In 1976 there were 10,754 acres of the Calimyrna variety; 3,351 acres of White Adriatic; 2,348 acres of Black Mission; 1,439 acres of Kadota; these plus other varieties brought the total acreage to 18,754 (McGregor et al., 1977*a, b*). In 1970, 10,070 tons of figs were dried and 1,900 tons were sold fresh (USDA Agr. Stat., 1977). Nearly all figs that are to be dried are allowed to drop to the ground and are then picked up mechanically.

## Olives

The olive first arrived in California in the form of cuttings shipped, along with crude farm implements, from New Spain (Mexico). The San Diego Mission alone had more than 500 trees by the year 1800. The olive became generally distributed in California as a dooryard tree by 1850, and by 1875 there were close to 15,000 olive trees in the state. California olive oil soon came to be recognized as being equal to the best European product. Quantity and quality were superior in the hot interior valleys. By 1897 there were about 2.5 million olive trees in the state.

By the turn of the century it was generally realized that olive oil could not be produced cheaply enough in California to compete with the European product or with other vegetable oils produced in the United States. However, the large ripe olives grown in California, when properly processed, proved to be palatable, nutritious, and altogether different from the European green pickled olives. Thereafter the production of large-fruited varieties for processing became the main objective, with oil production as a secondary and salvage operation.

The University of California did much work with variety trials and tree management and pruning; showed that early harvesting reduced alternate bearing; and developed control measures for diseases and insect pests, including the introduction from foreign lands of important parasites of the olive scale.

Conventional harvesting of olives is very expensive. For mechanical harvesting, an olive tree must be considerably restructured over a period of several years by pruning. It is better to prune the tree for mechanical harvesting while it is young. A few growers now harvest their olives mechanically, using tree shakers and mechanized, roll-out, canvas catching frames. Mechanical harvesting is likely to increase in the future (Hartmann and Opitz, 1977).

In the United States, olives are grown commercially only in California's Central Valley, where there were 42,840 acres in 1976, 72 percent in bearing trees that produced 80,000 tons (McGregor et al., 1977*a, b*).

## DECIDUOUS FRUITS

The mission fathers encouraged the planting of fruit trees along with other agricultural crops. Nearly all the kinds of fruits and nuts grown in California today, including almonds, apples, apricots, cherries, citrus fruits, dates, grapes, figs, olives, peaches, pears, plums, pomegranates, prunes, and walnuts were grown in mission orchards. However, the first extensive commercial development of fruits and nuts was done by Anglo settlers and was stimulated by the completion of the overland railroad in 1869. The arid climate, with many subtropical and temperate climatic zones, and the availability of irrigation water, made possible the growing of a great variety of fruits and nuts of high quality and in great quantity. California's extensive wheatfields gave way to thousands of orchards and vineyards, and an enormous tonnage of fresh, canned, and dried fruit was shipped to other parts of the country and abroad.

### Role of the University of California

Although the University of California was established in 1868, it was not until 1874 that E. W. Hilgard, a German immigrant, was appointed Professor of Agriculture. Although Professor Hilgard was primarily a chemist and soil specialist, he possessed a fortunate combination of broad European training in the natural sciences and a sympathetic interest in agriculture. He fostered the idea that the agriculture faculty should assume responsibility for research in the laboratory, and follow the results of their research in the field, as a part of their regular duties. He further pioneered the publication of Experiment Station bulletins. Hilgard Hall on the Berkeley campus reminds us of his lasting influence on California agriculture.

The extensive variety collections of the University of California at Berkeley and four substations throughout the state served to correct the nomenclature of fruits already in cultivation, to supply scion wood and plants for the rapidly growing industry, and to introduce new varieties. After 1908 this work was all done at the University Farm at Davis, which by 1926 had about 1,350 named varieties of fruits and nuts. Extensive studies of pollination requirements, beginning with the creation of the Division of Pomology in 1913, would have been greatly handicapped without the university's extensive variety collections. Cooperative work on a general fruit-breeding program began in 1936 in the Wolfskill Experimental Orchards at Winters in northern California and at the university campus at Davis in 1941. Also, pruning investigations by the university resulted in less severe cutting than practiced by the early orchardists, who brought with them the traditions developed in the intensive cultivation of fruits in southern European gardens, much of it of the espalier and cordon type, rather than the extensive cultivation of standard trees as eventually developed in California. This led to increased yield per acre (Tufts, 1946).

The university also greatly aided the grape industry, developing improved methods of pruning, thinning, girdling (for greater production), maturity standards, packing, precooling, pest control, prevention of spoilage in transit by the use of sulfur dioxide, as well as the curing of raisins. The university's agricultural engineers have made the major contribution toward the mechanical harvesting of fruit crops.

As J. B. Kendrick, director of the University of California Agricultural Experiment Station and Cooperative Extension recently emphasized, California's highly technological agricultural enterprise depends heavily on readily available supplies of energy and water resources, and there should be no delay in initiating some new research strategies. "Future yields and cropping systems," he says, "will depend on research underlying the conservation, development, and efficient use of our resources; how resources can be replenished, reused, and substituted for; and on research to provide a better basis for resource planning and policy formulation" (Kendrick, 1978).

## Refrigeration

Cold-storage plants have played an important role in the fruit industry, particularly for fruits that can be stored for long periods, such as apples and citrus fruits. The addition of 10 percent carbon dioxide to the atmosphere resulted in fruit being held in good condition at 38-40°F (3.3-4.4°C) for as long a period as at 32°F (0°C) when the gas was not present. This was particularly important in apple storage because at 32°F a storage disease of Pajaro Valley Yellow Newton apples was usually severe.

In the first transcontinental shipments of deciduous fruits there was serious loss of the softer and more perishable fruits. Then in 1887 a Chicago fruit dealer interested the western railroads in using several of his refrigerated cars. The first refrigerated carload of deciduous fruit, consisting of ripe cherries and apricots, reached eastern markets in June 1888, and the three transcontinental railway lines soon were constructing insulated cars with large ice bunkers at either end, each holding several tons of ice. Also icehouses were constructed along the railway lines at intervals of twenty-four hours in traveling time for reicing the cars in transit.

## Apples

Among the 207 apple trees grown by the Russians at Fort Ross in Sonoma County, two of the trees had apples identical to the Gravenstein variety as now grown there.

That variety is the first important apple of the season in the United States, maturing in August. Its skin is a greenish-yellow overlain with broken stripes of light and dark red.

Nursery stock shipped around the Horn from New York State was planted in Napa Valley in 1850. By 1860 a nurseryman in San Jose had 263 varieties and during the remainder of the century apple-growing became widely dispersed in California. But California's 21,230 bearing acres (in 1976) are now mainly concentrated in the Pajaro Valley or Watsonville district (winter apples) and in Sonoma and Napa counties (a summer apple—the Gravenstein—and a few other varieties).

A small percentage of the apple crop is grown at elevations of 2,800-4,000 feet in the Sierra foothills and in northern California at elevations of 4,000-5,200 feet. At Oak Glen in the San Bernardino Mountains, growers take advantage of their mountain scenery and proximity to a huge population center by selling most of the crop at roadside stands in the district. Apple growers do the same at Julian, in a mountain valley in central San Diego County.

The principal California varieties, in order of importance, are Gravenstein, Red Delicious, Golden Delicious, Newtons, and Rome Beauty. California processes more of its apples than any other state—up to 70 percent of the total production. The processed products are canned (applesauce), dried, made into juice, and frozen (Johnston and Dean, 1967). Only about 10 percent of the apples—those used for cider, vinegar, wine, jam, and sauce—are machine-harvested.

In 1976 California's 27,290 acres of apples (bearing and nonbearing) produced 240,000 tons, about 7.5 percent of the nation's apple crop, being exceeded by Washington and New York (McGregor et al., 1977*a, b;* USDA Agr. Stat., 1973).

## Apricots

The quality and commercial importance of the apricot in California is said to be unmatched anywhere else in the world. Although grown by the mission fathers, the fruit did not gain much attention until the gold rush. New varieties were imported after 1851, including most of the present commercial varieties. Apricot production in the United States is almost entirely confined to California, which had 29,604 acres and produced 124,000 tons in 1976. Only 9.2 percent was consumed fresh, the remainder being canned, dried, or frozen (McGregor et al., 1977*a, b*). Washington and Utah have a small acreage (USDA Agr. Stat., 1977).

The apricot is a very long-lived tree when compared with some other stone fruits, and Tufts (1946) observed some eighty-year-old trees that were still yielding profitable commercial crops. During dormancy the apricot can stand as much cold as the peach, but it is the flowers and young fruits that are especially susceptible to harm from frosts; thus climate is an important factor delimiting the areas where the fruit can be grown. Most of the apricot acreage is found in coastal areas from the Santa Clara Valley south to the Hollister district and in certain areas in the Sacramento and San Joaquin valleys (Johnston and Dean, 1969). The most important varieties of apricots are the Royal or Blenheim, hardly distinguishable. About 10 percent of the crop is machine-harvested.

## Cherries

The two major centers of cherry production in California are the Santa Clara Valley in the central coast region and the Lodi-Linden district in the San Joaquin Valley. The Santa Clara Valley is steadily declining as a fruit-producing area because of urban expansion. The Bing and Royal Ann are the principal varieties. California had 11,889 acres in 1976 (McGregor et al., 1976*a*). It produced 46,700 tons, exceeded only by Washington's 54,300 tons. The three Pacific Coast states produced 85 percent of the nation's 164,000-ton sweet-cherry crop (USDA Agr. Stat., 1977).

About half of California's cherries are sold fresh and half are processed. Those to be sold fresh mature early. Most are shipped before mid-June and marketed before supplies from other states become abundant (Johnston and Dean, 1969). That 5 percent of the crop that is machine-harvested is generally used for maraschino cherries.

## Grapes

The wine grape (*Vitis vinifera*) was apparently domesticated in southwestern Asia well before 5000 B.C., for that is when it was brought into Palestine from the north. It was known to have been grown in Egypt as early as 2375 B.C. (Schery, 1972). As related in chapter 3, several species of grapes are indigenous to North America. Although used by the Indians, American grapes had neither the size nor quality of those introduced from Europe.

The planting of wine grapes was high on the list of priorities of Spanish padres, who presumably heeded the advice given to Timothy by Apostle Paul: "Drink no longer water, but use a little wine for thy stomach's sake and thine often infirmities." The Mission variety was planted at Missions San Gabriel, Santa Barbara, and San Luis Obispo and bore grapes for more than a century. Such vines furnished the cuttings for many of California's vineyards. This variety, probably a seedling type the mission fathers developed, is still grown in California and produces some of the best wines of the sweet, or dessert, types (the Mission fig and Mission olive are other legacies from the Spanish missions still grown commercially in California). One mission vine in Santa Barbara County, when it was sixty-five years of age, covered 12,000 square feet and produced 10,000 pounds of grapes. Another vine in Carpenteria, planted in 1845, yielded 16,000 pounds of grapes and had a trunk reported to be 9 feet in circumference.

The grape industry was threatened by the grape phylloxera, *Phylloxera vitifoliae,* an insect similar to the aphids, which attacks the roots of grapevines. The insect occurred in wild species of vines east of the Rockies, where it was indigenous. From there it was introduced into France in 1863 and into California in 1873. It caused great destruction in both regions before being controlled by the use of resistant American rootstock.

The vines, 400-600 per acre, trained to standardized trellises, are often sprayed with gibberellins, which are plant-growth regulators, to reduce fruit set and increase the size of the grape berries.

Grapes may be divided into three general varietal classes—raisin, wine, and table, although some varieties of the raisin and table classes may go into two or all three outlets, depending on relative supplies and prices. The Thompson Seedless is the main raisin variety, but nearly half the crop is used for wine and for fresh use. Most of the raisins are of the natural sun-dried type. Of the strictly wine-type grapes, Zinfandel, Carignane, Cabernet Sauvignon, French Colombard, Barbera, and Grenache are the leading varieties among the more than seventy varieties grown. Among strictly table grapes, Emperor, Flame Tokay, Ribier, and Perlette are the leaders (McGregor et al., 1977*b*).

Most of California's grape acreage is on the east side of the San Joaquin Valley. Some high-quality wine grapes are grown in central Sonoma County and the Napa—St. Helena districts of the central coast area, in Santa Clara and San Benito counties, and near Ontario in San Bernardino County. Some of the Thompson Seedless acreage is in the hot desert of Coachella Valley in southern California, where the grapes are harvested in June; whereas in the San Joaquin Valley they are harvested in July to September. The early-maturing desert grapes bring premium prices for table use.

In 1976 California had 322,650 acres of wine grapes, 243,011 acres of raisin-type grapes, and 65,640 acres of table-type grapes—a total of 631,301 acres (McGregor et al., 1977*b*). California produced 3.6 million tons of grapes—90 percent of the nation's total (USDA Agr. Stat., 1977). As shown in table 5, in 1975 grapes ranked fourth among California's agricultural commodities, with a value of $479 million.

### Nectarines

The nectarine tree is identical to the peach in tree and bud characteristics. Its fruit is similar to the peach except in having a smooth skin. Nectarine trees have grown from peach seeds and vice versa. Nectarine branches have appeared as bud mutants on peach trees. By crossing some of the older nectarines with very large fruited peaches such as J. H. Hale, some large, delicious nectarine varieties have been obtained. Their round form and bright red color over golden yellow result in their being among the most strikingly beautiful fruits to be seen in the market (Chandler, 1957).

There were 19,613 acres of nectarines in California in 1976, of which 32.4 percent were nonbearing, rivaled among fruit crops only by the avocado's 33.6 percent, and indicating increasing interest in nectarines in recent years. The San Joaquin Valley has 99.4 percent of California's nectarine acreage, with Fresno and Tulare the leading counties (McGregor et al., 1977*a, b*).

### Peaches

The peach is one of the most important of stone fruits. It appears to have originated in China, but was introduced into Persia before Christian times and then was distributed throughout Europe by the Romans. After the introduction of some European varieties into Florida by the Spaniards, there were plantings for domestic use—first by the Indians and then by the white settlers—but there was no commercial production in the United States until the nineteenth century (Schery, 1972).

Peaches are grouped as freestone or clingstone, depending on whether the flesh separates readily from the pit. California is the nation's only important area of commercial clingstone peach production with its 51,164 harvested acres yielding 748,000 tons in 1976 (McGregor et al., 1977*a, b*). Nearly the entire crop is canned. The cling peach acreage lies almost entirely in interior valleys. About two-thirds of the bearing acreage is in the Stockton, Modesto-Merced, and Kingsburg-Visalia-Lindsay districts of the San Joaquin Valley. Most of the remaining acreage is in the Sacramento Valley centering in the Marysville-Yuba City district (Johnston and Dean, 1969).

Increasing cost and limited availability of good "hard-harvest" labor is necessitating harvest by machine. As with other fruit crops, a mechanical device is clamped onto main branches or the trunk, delivering an impact or a sustained shake. In 1975 a detailed survey made by the Canning Peach Association revealed that 18 percent of the cling peach crops were mechanically harvested. It appeared in 1977 that this percentage had increased to about 25 percent (J. A. Beutel, correspondence).

California is also the nation's leading producer of freestone peaches, with 21,556 acres yielding 232,000 tons in 1976 (McGregor et al., 1977*a, b*). Yields per acre are about double those in the southeast. South Carolina, Georgia, Pennsylvania, and New Jersey are also important producers of freestone peaches. The major variety in California is the dual-purpose Elberta, which can be sold either fresh or canned. Also, earlier and later varieties are grown to extend the market. In their order of ripening, the Babcock, Red Top, July Elberta, Suncrest, and Rio Oso Gem are among the many other California freestone peaches destined for the fresh fruit market. California sells only about a third of its freestone peaches fresh, whereas in the rest of the nation practically the entire crop is sold that way.

### Pears

California, on its 37,620 acres of bearing trees, produced 353,500 tons, 43 percent of the nation's pear crop, in 1976. Washington, with 235,000 tons and Oregon with 207,000 tons, were the other major producers. The Bartlett variety comprised 97.6 percent of California's pear crop, but only 50 percent of the crop that was grown in the Pacific Northwest (McGregor et al., 1977*b;* USDA Agr. Stat., 1977). Pear acreage has been increasing in Lake and Mendocino counties, in the Sacramento Valley near Courtland and Yuba City, and the northern San Joaquin Valley. On the other hand, acreage has been decreasing in the mountain counties of Placer and El Dorado because of the malady known as "pear decline" and in the Santa Clara Valley because of new housing and industrial subdivisions.

In California, 72 percent of the pear crop is canned, compared with 53 percent in Washington and 23 percent in Oregon. California's fresh pears mature earlier and are marketed before those of the Northwest. A new variety, "California," reaches maturity at the same time or a few days earlier than the Bartlett and resembles the Comice in size and form more than it does the Bartlett. Like the Comice, it will not ripen properly without a period of cold storage. Its attractive appearance (ground color yellowish-green to yellow with overcolor of bright red), fine flavor, and long shelf life should make the California a desirable pear for the fresh-fruit market. Qualified nurserymen may obtain commercial licenses for propagating and selling the California pear from the University of California Board of Patents (Griggs and Iwakiri, 1974).

### Plums

In 1976, California's 24,600 acres of bearing plum trees produced 115,000 tons (McGregor et al., 1977*a, b*), over 90 percent of the nation's production, with Michigan producing most of the remainder.

Since the mid 1930s, bearing acreage of plums has declined in the mountain area (Placer County) and the Yuba City area in the Sacramento Valley and has increased in the San Joaquin Valley, particularly in the "fruit belt" from Clovis to Porterville, in the Linden District of the San Joaquin County and the Arvin area of Kern County, south of Bakersfield. Statewide, the leading varieties are Santa Rosa, Casselman, Laroda, late Santa Rosa, and El Dorado. These varieties also have the greatest areas of new plantings (McGregor et al., 1977*a, b*).

### Prunes

A prune is a plum that can be dried or that has been dried without fermentation even though the pit is not removed. Dried prunes are well suited to international trade and much of California's crop is exported. California has produced as much as 205,000 tons (in 1973), and generally grows about 90 percent of the nation's crop, with Oregon, Washington, and Idaho producing the rest. In 1976, California produced 145,000 tons of dried prunes valued at $62 million (USDA Agr. Stat., 1977).

Prune production is relatively highly mechanized, featuring mechanical harvesting of about 95 percent of the fruit. This has been a factor favoring the planting of prunes in recent years. In 1976 about 10 percent of California's prune acreage was in young, nonbearing trees.

Although the prune requires adequate winter chilling for normal blossoming and leaf-bud emergence (as do most deciduous fruits), it thrives best where the spring and summer are warm, dry, and clear. Some of the coastal valleys where fog is not excessive are satisfactory, and until about 1950 possibly three-fourths of the acreage was in Santa Clara, Sonora, and Napa counties. Since then, most new planting has taken place in several areas of the Sacramento Valley and Tulare and Fresno counties in the San Joaquin Valley. In all areas, the French prune is by far the most important variety.

## TREE NUTS

### Almonds

With 336,037 bearing and nonbearing acres in 1976, almonds exceeded all California tree crops in acreage and were exceeded only by grapes, with 631,301 acres, among fruit and nut crops. The 233,000 tons produced were valued at $166 million (McGregor et al., 1977*a, b*). Almonds also led by far all fruit and nut crops in new acreage planted; 23.6 percent of the acreage was devoted to young nonbearing trees. New plantings continue as new lands come under cultivation, particularly in the western San Joaquin Valley, as the result of the Bureau of Reclamation's giant Central Valley Project canal (fig. 32). By the late 1970s, about 65 percent of California's total almond production is expected to be in the San Joaquin Valley, with the remainder principally in the Sacramento Valley. Acreage in the central coast area (mainly in Contra Costa and San Luis Obispo counties) is declining. A factor favoring the almond industry is that it is largely mechanized, including the mechanical shakers that grab and shake the tree limbs, causing the nuts to fall to the ground where mechanical sweepers pick them up.

California produces practically all the almonds grown in the United States, 233,000 tons in 1976, amounting to over half of the world's supply. More than half of the state's production is exported, Germany being the biggest customer, with Japan, Britain, Sweden, France, Canada, and the Soviet Union also serving as prime markets. Exports increased ninefold in a decade and now exceed the domestic demand.

About two-thirds of the almond acreage is planted to the Nonpareil variety, but every third row is planted to Mission or Merced or some other complementary variety to provide the necessary cross-pollination, for almonds are self-sterile. Most growers rent bees for the pollination period—as many as three hives to the acre, some brought from as far away as Montana and North Dakota.

A leaf-scorch disease, referred to as "golden death" or "almond decline" since its discovery in 1958, has become widespread in California's almond-growing areas. Portions of otherwise healthy, green, full-sized leaves suddenly desiccate from either the edges or tips, with the tissue between the green and scorched areas appearing as a bright yellow band. The diseased trees do not recover. Bacteria in affected almond leaves are markedly similar in appearance to those that cause Pierce's disease in grapevines. They can be transmitted to grapevines by means of insects (leafhoppers) causing typical symptoms. Also the bacteria causing the disease in grapevines can be transmitted to almond, inducing leaf scorch (Sanborn et al., 1975).

*Fig. 32. Sections of canal of the Bureau of Reclamation's Central Valley Project in California, San Luis Unit, on the west side of the San Joaquin Valley (above, in Fresno County, and below, in Merced County, near Los Banos; courtesy U.S. Bureau of Reclamation).*

## Walnuts

By 1850 a few orchards of Persian (English) walnuts had been planted in California, ranging from San Diego to Napa, and by 1880 there were walnut orchards as far north as Winters and Chico in the Sacramento Valley. However, these were the small, round, hard-shelled inferior types introduced by the mission fathers. In 1868 some walnuts, believed to have come from Chile, were planted in Goleta, Santa Barbara County, giving rise to the greatly superior Santa Barbara soft-shell seedling type of nut that became the foundation of the southern California walnut industry. In the 1860s, French varieties planted at Nevada City, Nevada County, became the basis for the walnut industry in central and northern California and in Oregon. Like almonds, walnuts are mechanically harvested.

In southern California the Placentia and Eureka varieties were discovered between 1890 and 1905 and the planting of seedling trees was discontinued. The walnut industry developed later in the north than in southern California because of the mistaken idea that the equitable climate in the south was required for safe commercial production. However, when once begun the industry in central and northern California expanded rapidly, with the Franquette and Payne varieties predominating, whereas in the south the acreage declined because of the rising cost of pest control for codling moth and walnut husk fly; the high cost of land, irrigation water, and taxes; the reduced yields some years because of "delayed foliation" after the warmer winters; and the displacement of walnuts by Valencia oranges and lemons.

In 1976 California's 205,609 acres of walnuts produced 183,000 tons, valued at $116,000,000. California produces over 98 percent of the nation's walnut crop, with a small acreage in Oregon (McGregor et al., 1977*a, b;* USDA Agr. Stat., 1977).

### Pistachios

The edible pistachio nut (*Pistacia vera*), a native of Asia, was introduced into California but created little interest until the 1960s. Inadequate nursery stock, alternate bearing (heavy one year and light the next), and high harvesting cost were among the factors retarding development of the industry. Satisfactory production in experimental plantings at the former USDA Plant Introduction Station at Chico and successful commercial use of machine harvesters have revived interest. Certain areas in the San Joaquin and Sacramento valleys are suitable for commercial production. In 1976 California had 31,041 acres of pistachios, but only 833 acres of bearing trees. Thus the acreage of young, nonbearing pistachio trees is exceeded only by that of the almond among California's fruit and nut crops and is an indication of the phenomenal rate of growth of the pistachio industry.

Male and female flowers of the pistachio are borne on separate trees and wind carries the pollen. It is necessary to have enough male trees to ensure adequate production, at least one pollinator in ten, with twice the number on windward border rows (Opitz, 1975).

## VEGETABLE CROPS

California ranks first in the nation in such major vegetable crops as asparagus, lima beans, broccoli, brussels sprouts, carrots, cauliflower, celery, garlic, lettuce, cantaloupe and honeydew melons, spinach, and tomatoes. In 1975 the value of the state's vegetable crops was $1.6 billion. From 70 to 75 percent of the vegetable produce is utilized fresh, although California supplies about a third of the national processed vegetable market. Only the two most important of California's twenty-seven vegetable crops of commercial value will be discussed.

### Tomatoes

In 1975 California harvested 326,000 acres of tomatoes producing 7,690,000 tons valued at $566,546,000 (McGregor et al., 1976*a, b*). In 1975, the state produced 33 percent of the tomatoes grown in the United States for the fresh market while Florida produced 45 percent. California produced 83 percent of the nation's processed tomatoes (USDA Agr. Stat., 1977).

Fresh tomatoes are available to consumers in the United States every month of the year. In California, peak harvest occurs from May through October, although some production finds outlets to local and eastern markets throughout the year. Florida and Mexico are the large producers of winter tomatoes. Most of California's early spring tomatoes for the fresh market are grown in the Imperial Valley. Early summer and early fall types for the fresh market are grown in the San Joaquin Valley and in Monterey, Ventura, and San Diego counties.

Processing tomatoes comprise 95.8 percent of the total tomato crop and form California's leading vegetable type. They are grown principally in the Central Valley. Growers have successfully adopted yield-increasing and cost-reducing innovations, including mechanized growing and harvesting, that have made California competitive in eastern markets despite the distinct freight disadvantage.

The mechanized harvesting of relatively nonperishable crops such as cereals, potatoes, sugar beets, and cotton was undertaken long before being applied to soft fruits and vegetables. The need for such harvesting was long felt, however, because of the large and highly seasonal labor force required, greatly adding to the cost of production. Devising mechanical harvesters proved to be particularly difficult. Close cooperation was required between engineers and the geneticists, plant physiologists, and botanists who developed new crop strains suitable for mechanical harvesting. California is fortunate in not having summer rains. Tomatoes cannot be mechanically harvested when the ground is muddy. That is one of the reasons why most of the tomatoes grown for canning or other processing are grown in California (Kelly, 1967; Rasmussen, 1968; Bainer, 1975*a, b*).

Harvesting machines normally have their least potential on crops intended for the fresh market, for the latter require several selective pickings, or are easily bruised. Crops that are processed spend less time in storage and therefore are less likely to develop rot where they are punctured or bruised in the mechanical harvesting process. Also they are subject to peeling, slicing, or other operations that can eliminate small bruises with little if any loss.

The principal contributions toward the tomato harvester (fig. 33) were made by a plant breeder, G. C. (Frank) Hanna and an agricultural engineer, Coby Lorenzen, at the University of California Agricultural Experiment Station at Davis. The plant breeder is nearly always involved in research on mechanical food-crop harvesting, but this was probably the first time that a plant breeder and an agricultural engineer were asked to work as a team on such a project from the very beginning. This was because it was so obvious that a very different type of tomato would have to be developed before there could be any hope of mechanization. Hanna bred tomato plants designed for machine handling. The plants produced fruits that were of uniform size and ripened at about the same time so that the vines could be removed when the fruit was harvested. (Previously developed varieties had to be picked three or four times because of uneven ripening.) The fruit had to be easily detached, yet not drop off the vine prematurely. It had to have a skin tough enough to withstand mechanical handling, and a jointless stem or pedicel to prevent puncture of the fruit in handling.

In the meantime, Lorenzen worked on the design of the harvester. This machine cuts the plant off at ground level and lifts it to a vibrator that separates the fruit from the plant, then deposits the fruit in a bin in which it is hauled to the processing plant (Lorenzen and Hanna, 1962), Lorenzen and others had a prototype machine in operation in 1959. The University of California then patented the machine and licensed it to a manufacturer to undertake commercial construction. The concern had fifteen machines by 1960 and twenty-five by 1961 in the hands of growers. By 1963 about 1.5 percent of California's processed tomatoes were harvested by machine. Then the termination of Public Law 78 on December 31, 1964, drastically reduced the importation of foreign

*Fig. 33. Mechanical tomato harvester in action; as harvester moves along, the conveyor belt drops tomatoes into a giant truck trailer (courtesy USDA).*

labor. (Previously, about 85 percent of the tomatoes for processing were picked by Mexican nationals.) This led to the rapid adoption of mechanical tomato harvesters, for processors were already considering the shift of their operations to Mexico, where labor was plentiful. Rapid mechanization could not have happened, however, if the University of California had not already had fourteen years of interdisciplinary research experience in mechanized tomato-harvesting in cooperation with farmers, agricultural equipment manufacturers, and processors who were willing to risk their economic future on mechanization. Within five years about 1,300 mechanical tomato harvesters harvested approximately 95 percent of the 240,000 acres of tomatoes in California at that time. A half-billion dollar industry was retained in California that would have been lost to the state without mechanical harvesting of the tomato crop (Lorenzen, 1969; Bainer, 1975*a*).

The tomato harvester can harvest ten tons of tomatoes an hour, and the question arose as to how this amazing quantity of fruit was going to be boxed and hauled to the cannery. A UC Davis engineer, Michael O'Brien, had already shown that peaches, plums, and prunes could be handled in 4 x 4 x 4-foot boxes holding a half-ton of fruit and hauled hundreds of miles without injury to the fruit. These boxes proved to be satisfactory for tomatoes also. The huge containers were loaded with forklifts. The next step was to dispense with boxes entirely and fill the entire truck bed with tomatoes, haul them to the cannery, and dump them into water (Bainer, 1975*b*). It was a surprise to everyone that so much fruit could be piled into a truck bed without causing damage. One cannot drive through the Central Valley during the harvest season without seeing many huge trucks and trailers with either the familiar 4 x 4 x 4 boxes of fruit or with their beds heaped high with tomatoes.

As previously explained, mechanization in agriculture caused the exodus of millions of farmers and farm laborers to the cities where they were ill equipped to adopt and find jobs. Many who chose to remain in farming were reduced to poverty because they were not financially able to obtain the expensive new equipment and expand their acreage sufficiently to make mechanization profitable to them. There is obviously some social process at fault, but should mechanization per se be blamed? On the credit side, the latter is responsible for the alleviation of the drudgery so long associated with farm work. It has brought food in abundance to the American table at a lower percentage of the consumer's income than in any other nation. What is needed is not resistance to change but possibly the help of our present crop of futurist "think tanks" to develop a new level of competence in the management of change—to develop techniques for harmonizing the future.

As stated by James B. Kendrick (1977), director of California's Agricultural Experiment Station and Cooperative Extension, "The real challenge to all of us is to adapt innovation to the good of all people rather than to fear the exploration of the unknown." He pointed out that the tomato harvester, while in the short run reducing labor requirements, also resulted in an increase in tomato acreage from 143,000 to 233,000 acres. The concomitant increase in processing, transportation, and other activities associated with increased production has actually increased total employment in the industry. Moreover, stoop labor, possibly the most excruciating and debilitating of all forms, was eliminated, which could be added to the credit side as a social gain.

To reinforce his argument, Kendrick calls attention to California's processed asparagus industry, in which mechanization of harvesting has so far been unsuccessful. High production costs have resulted in loss of markets to foreign producers. The state's asparagus acreage decreased from 74,000 to 34,000 acres, along with a corresponding reduction of jobs in production and processing.

It is the low-income families that gain most from increased efficiency in agricultural productivity, for food purchases comprise a far greater portion of their expendable income than in the case of families with higher incomes.

### Lettuce

Among vegetables, lettuce is exceeded only by tomatoes in acreage, production, and value. In 1975 California's 155,900 acres of lettuce produced 1,912,500 tons valued at $248,401,000 (McGregor et al., 1976*a, b*). In 1976 California produced 73 percent and its nearest rival, Arizona, produced 14 percent of the lettuce grown in the United States (USDA Agr. Stat., 1977).

In California the lettuce season begins with winter production in the Imperial and Palo Verde valleys (fig. 34) then shifts north to the Salinas Valley and other areas in the spring through the fall seasons. Lettuce is consumed fresh, and the industry has developed so as to provide supplies throughout the year. As labor costs rise, California and Arizona will realize benefits from mechanization. Efficiency of mechanical methods depends on high and stable yields such as are best realized in irrigated areas.

One of the most sophisticated of agricultural machines is the lettuce harvester. This machine goes over the field more than once, picking only mature heads, selected on the basis of size and firmness. The fewer the pickings the more economical the harvesting. Here again varieties are sought that will mature uniformly, and cultural practices are developed that will induce uniform maturation.

*Fig. 34. Lettuce being irrigated in the Palo Verde Valley near Blythe, California (U.S. Bureau of Reclamation photo by E. E. Hertzog).*

## FIELD CROPS

The importance of field crops in California agriculture is indicated by the fact that cotton, hay, and rice are among the first ten commodities in value (table 5). An extension of table 5 would have shown that sugar beets and wheat were eleventh and twelfth, respectively, in value in 1975. Space permits only a brief discussion of the principal field crops.

## Cotton

A relatively latecomer among California field crops was cotton. In the early 1870s only about 2,000 acres were planted to cotton and 80 acres of this was within the Los Angeles city limits. It was not until the new irrigation projects in the Imperial and southern San Joaquin valleys were developed that the industry began to boom—138,000 acres in the Imperial Valley by 1918 and 90,000 acres in the San Joaquin Valley by 1929. California now has 875,000 harvested acres, with major production in the lower San Joaquin Valley. There the industry is unique in that, by law, only the high-quality, longer-staple Acala variety of cotton has been grown since 1925.

In 1975 California produced 1,954,000 bales of cotton, exceeded in production only by Texas, which produced 2,382,000 bales (480-pound net weight bales). California's harvested acreage was 875,000 compared with Texas' 3,900,000. Thus with only 22.5 percent as much acreage California produced 82 percent as much cotton as Texas (USDA Agr. Stat., 1977). High yields and mechanization have favored expansion of cotton production in California and other irrigated areas such as Arizona and New Mexico.

## Alfalfa

Much California land—about 1.9 million acres—is devoted to raising hay and about two-thirds of it is planted to alfalfa. Alfalfa is grown commercially in almost every county. A substantial proportion is produced as a cash crop for shipment to the state's dairy-producing and livestock-feeding centers, the dairies in the northwest using substantial quantities from the Sacramento and northern San Joaquin valleys, and those of Los Angeles and vicinity drawing heavy shipments from the Imperial Valley, the Mojave Desert area, and the southern San Joaquin Valley. Cattle-feeding operations in some of the alfalfa-growing areas, such as the Imperial and Palo Verde valleys, use large tonnages of alfalfa hay.

There is also a substantial domestic and foreign demand for alfalfa meal pellets and 1¼-inch alfalfa hay cubes. These dehydrated products are fed to poultry, hogs, and cattle.

California is the leading state in the production of alfalfa seed, having produced 28 million pounds in 1976—35.6 percent of the nation's total. Washington and Idaho are also big producers (USDA Agr. Stat., 1977).

## Barley

Barley is grown on 20 percent of the acreage devoted to field crops in California, compared with 18 percent for alfalfa. The state was exceeded only by North Dakota and Montana in barley production. Most of the California crop goes to livestock and feed-manufacturing industries, unlike the Midwest where much barley is used by breweries. Leading counties in production of barley are Fresno, Kings, Tulare, Kern, and San Luis Obispo.

## Rice

In 1975 rice was ninth in value among California's agricultural commodities. There were 525,000 acres and the 1.5-million-ton crop was valued at $256 million (McGregor et al., 1976*a, b*). California produced 22.9 percent of the nation's rice crop in 1975, exceeded only by Arkansas (USDA Agr. Stat., 1977). Most of California's rice acreage is in the lower Sacramento Valley, where soil, climate, and abundance of water favor rice culture.

Combines modified from those first used to harvest and thresh wheat, along with equipment for artificial drying, enable growers to harvest rice when its moisture content is 22-25 percent, instead of the 14-15 percent necessary if it were to dry in the field. Therefore a rice grower can harvest his crop earlier and reduce the risk of loss from rainstorms. He also obtains a higher yield of whole kernels (head rice) than was previously possible.

As early as 1930, some combines in California were mounted on a track-layer chassis and the header was mounted in front of the tracks, with central delivery from the header into the combine. When the header was attached in front, the rice was cut ahead of the tracks, avoiding the necessity for running over uncut rice. This is a feature still retained by most rice-harvesting combines today. By 1948, when 900 man-hours of labor per acre were required for rice-harvesting in Japan, only 7.5 were required under the highly mechanized system used in California (Bainer, 1975*a, b*).

Probably with no other crop has the role of the airplane been as important as with rice. The airplane is the seeder, fertilizer spreader, insecticide or desiccant applicator, and herbicide dispenser. In the latter role it is a substitute for the cultivator.

### Sugar Beets

The ancestor of the sugar beet (*Beta vulgaris*) apparently grew wild in parts of Asia and the Mediterranean. At an early time it was cultivated in southern Europe and Egypt. By 1854 sugar was being extracted from beets on a large scale in Europe. The first successful beet-sugar factory in the United States was established in California in 1879. The sugar-beet industry was once faced with total failure because of "curly top," a virus disease transmitted by leafhoppers. In 1934 over 85 percent of the planted sugar-beet acreage was abandoned. Plant breeders then developed hybrids that were not only resistant to curly top (and Cercospora leaf spot), but possessed greatly superior agronomic qualities. By 1941, 97 percent of the planted acreage was harvested.

The sugar content of the sugar-beet root often exceeds 20 percent of the root weight. The plant is highly tolerant to salinity, which favored the development of the sugar-beet industry in the arid West. Because the cost of transporting the beets is an important factor, sugar beets are grown in areas not far removed from a processing plant (Hughes and Metcalfe, 1972; Pound, 1975).

In 1975 California's 326,000 acres produced 8.5 million tons of sugar beets, valued at $232 million. The leading counties in sugar-beet production were Imperial, San Joaquin, Kern, Yolo, and Solano. California was the nation's principal sugar-beet producing state with 31 percent of the total, followed by Idaho, Colorado, and Washington.

The beet industry is a good example of survival through technology and mechanization. First came beet planters and thinners, the former made possible by Roy Bainer of the University of California, Davis, who found a way of grinding down the beet's polygerm seedball to approximately single-germ units and blowing out the chaff. This served its purpose for eighteen years until plant breeders were able to develop a variety with monogerm seed. Then came mechanical harvesters just in time to save the beet industry from threatened collapse because of acute labor shortage in World War II. Basic contributions were made by the University of California's agricultural engineers such as John B. Powers, who solved the difficult problem of topping the beets mechanically (Bainer, 1948; Powers, 1948; Walker, 1948).

The first machines were built by Blackwelder Manufacturing Company, the firm that also built the first tomato harvester. The company was noted for its initiative and foresight in undertaking the commercial construction of new types of harvesters while they were still in a semiexperimental stage of development and before the larger equipment companies manifested any interest. Because of the war-induced shortage of labor, Ernie Blackwelder recruited people from banks, schools, stores, and the post office to work four or five hours in the early part of the evening as a patriotic duty to help assemble the beet harvesters (Bainer, 1975*a, b*).

## ORNAMENTAL HORTICULTURE

The early Spanish and Mexican settlers had contributed little to the art of ornamental gardening, but the Anglos found it a challenge and delight in the unfamiliar climate of California. There were many plant importations from areas of the world with similar climate even before 1848, the year gold was discovered. The gold rush brought to California people who had been familiar with ornamental plants in many parts of the world. The wealth from the mines stimulated and facilitated luxuries, including ornamental gardens. To take care of large estates, workers as well as plants were imported. Some experienced gardeners who began by growing ornamentals for pleasure later established nurseries, the earliest on record being located near Sacramento in 1848. Australian ornamentals such as the acacia, eucalyptus, and araucaria were particularly well represented among the early introductions and, in fact, in the late 1870s the university at Berkeley was criticized for having so much Australian vegetation on its campus. At a flower show in San Francisco in 1854 the wide variety of ornamentals displayed would be creditable even today.

About 100,000 eucalyptus trees were sold in Oakland along in 1875, and one nurseryman advertised a million of them. They were planted for windbreaks, for timber, but also as ornamentals. Ornamental horticulture developed later in southern California than farther north because of lower population, but the banana, strawberry guava, canary palms, date palms, fan palms, and various Japanese plants were introduced. Southern California also developed an enormous cut-flower industry.

Credit for much of the progress in ornamental horticulture belongs to city parks and botanical gardens such as the Golden Gate Park in San Francisco, Balboa Park in San Diego, and the Huntington Gardens at San Marino, Los Angeles County. Many horticultural societies were formed in the state.

The early importers of plants appeared to be unaware of the danger of importing from other parts of the world pests that were not present in the United States. As previously related, the notorious cottonycushion scale, *Icerya purchasi,* was introduced on acacia at Menlo Park, San Mateo County, in 1868 or 1869. In about ten years it was causing great damage to citrus trees in southern California. Also in that period the San Jose scale was introduced on ornamental peach trees and the araucaria scale on trees of the same name. These accidental introductions had much to do with the passage of the first plant-quarantine laws in 1899.

California's mild climate and rainless summer favored the development of a flower-seed industry and the state now supplies much of such seed to the world. Major concentrations of the industry are at Santa Maria and Lompoc in Santa Barbara County, where the vast fields of solid color are a spectacular sight. Some of the leading breeders of iris, dahlias, chrysanthemums, roses, fuchsias, and other flowers are located in California. They usually distribute them as bulbs, tubers, or plants rather than as seeds.

## PRESENT STATUS OF CALIFORNIA AGRICULTURE

California's massive and diverse agriculture is its biggest business enterprise. In its 100.2 million acres, California had 36 million acres in farms in 1975. Of its farmland, 9.1 million acres were devoted to cropland: 72.7 percent in field crops, 17.3 percent in fruit and nuts and 10.0 percent in vegetables and melons, with gross values of $2.7 billion, $1.9 billion, and $1.6 billion, respectively. Cattle and calves were valued at $1.1 billion and milk and cream at close to $1 billion (table 5). Fresno County, the state's and the nation's leading agricultural county, raised agricultural products valued in excess of $1 billion. California farmers raised 200 crops and in 49 of these it led the nation in production.

Table 5 shows the great change in the relative importance of major crops in the forty-five-year period from 1930 to 1975 and also the enormous increase in value, even when inflation is taken into account. Oranges, the leading crop of 1930, dropped out of the first ten. However, oranges and lemons combined ranked citrus fruits, even without grapefruit and tangerines, eighth on the 1975 list. Other crops that dropped from the top ten were lemons, dry edible beans, chickens, and prunes. Entering the list were hay (mostly alfalfa, reflecting the importance of beef and dairy cattle), processing tomatoes, nursery products, rice, and lettuce.

Along with its large farms, averaging over 1,000 acres, California has the highest degree of farm mechanization, most of it developed in the state. Combines harvest wheat, corn, rice, and sugar beets. Tree shakers drop almonds, walnuts, olives, prunes, and other tree crops on the ground or into cloth receptacles. Plant breeders have developed tomatoes, potatoes, melons, and lettuce especially for mechanical harvesting. Citrus fruits are sucked from trees with pneumatic tubes (Bainer, 1932, 1948, 1975*a, b;* Bainer et al., 1955; Barmington and McBirney, 1952; Fridley and Adrian, 1966; Fridley et al., 1966; Gittins, 1950; Goss et al., 1955; Lorenzen, 1969; Lorenzen and Hanna, 1962; Powers, 1948; Rasmussen, 1967, 1968; Walker, 1948). Nevertheless, much traditional stoop labor is required for such jobs as pruning and harvesting where mechanical harvesters have not been perfected or are for some reason not being used.

## NEW RESEARCH NEEDS

The undeniable accomplishments of agricultural research in increasing crop productivity should not be construed to imply that a continuation of its traditional orientation will continue to be productive. S. H. Wittwer (1978), well-known plant physiologist and director of the Michigan State University Agricultural Experiment Station, has issued a warning that the productivity of major food crops has plateaued in the United States and elsewhere. The striking increases in crop production following World War II and up to approximately 1970 were the results of technologies requiring massive inputs of fossil energy in the form of fertilizer, pesticides, irrigation, mechanization, and new seeds. With the possible exception of genetic improvements, these external inputs are becoming increasingly costly, less available, and subject to increasing constraints. Some are from nonrenewable sources. Meanwhile soil erosion continues unabated nationally and globally. In the United States no more than 25 percent of the farmlands are committed to approved conservation practices after forty years of a federal soil-conservation program. Soil organic matter is being reduced and soil compaction is increasing as the result of excess and untimely tillage.

There has been a twelve-year erosion of federal investment in agricultural research, in manpower and in new equipment and facilities. Although enrollments in colleges of agriculture have tripled in ten years, there has been little, if any, increase in faculty. There has been no increase in scientist years in support of agricultural research since 1966.

Yet there are technologies, according to Wittwer,

> that will result in stable food and fiber production at high levels, are nonpolluting, will add to rather than diminish the earth's resources, be sparing of capital, management, and nonrenewable resources, and are scale neutral.... It is research that will address the problems of enabling plants and animals to more effectively utilize present environmental resources through (i) greater photosynthetic efficiency; (ii) improved biological nitrogen fixation; (iii) new techniques for genetic improvement; (iv) more efficient nutrient and water uptake and utilization, and reduced losses of nitrogen fertilizer from nitrification and denitrification; and (v) more resistance to competing biological systems and environmental stresses.

These are grossly underfunded fields of research not only vital to American agriculture but also the development of Third World nations. They would lead to technologies that "would be economically, socially, and ecologically sound and would ease the inevitable transition we must make from nonrenewable to renewable resources" (Wittwer, 1978).

Besides the problem of increasing crop productivity, there is the problem of reducing the enormous losses of food crops in storage. The world's total food supply could be increased by 25-30 percent if postharvest losses could be avoided. The exposure of foods to low-intensity gamma radiation slows down the growth rate of spoilage microorganisms. Among the many benefits is the extension of the period fruits remain fresh after harvest. Insects, insect eggs, and fungi that may have got into grain, fruits, or vegetables can be killed in the same process, and the sprouting of bulb vegetables such as potatoes and onions can be inhibited. Exposure to high-intensity radiation destroys all the microorganisms in food, making possible the preparation of certain meat, poultry, and marine products for long-term storage (about sixteen months) or shipment without refrigeration. Unfortunately, food irradiation is liable to the same type of regulations as those imposed on the use of chemical additives, even though it is a *physical* process like heating, freezing, or dehydrating. This has greatly delayed and increased the cost of the commercial development of this very promising modern technology. Yet no harmful effects in animals that have consumed irradiated food have ever been detected in more than twenty-five years of intensive research (Libby and Black, 1978).

Mechanization tends to make urban and rural areas more alike with respect to social problems. Because farm mechanization came sooner and in larger measure to California than elsewhere, it is no surprise that the controversy between unionized farm workers and attendant labor versus management should first assume major proportions in the state. At a "listening conference" at Asilomar, California, in October 1977, the state university was criticized by activists from various segments of society, including worker and consumer groups, for not keeping up with social change, not addressing itself to current human aspirations, and not concerning itself with the social impact of its research. In reply, agricultural economist Chester McCorkle (1977) agreed that the university had fallen behind times in shifting and broadening its program emphasis:

> We should give more attention to environment concerns and to the impact of research on the labor force, farm sizes, income distribution patterns, and concentration of economic power. If we do not turn some research attention to these kinds

> of issues or to these aspects of the problems we research, we fail to tackle some of the most important issues of our time in California, as elsewhere in the nation. . . . We can shift some research emphasis toward limited resource farmers, intermediate size technology, energy and capital requirements for intensive agriculture, and studies on impact analysis.

Another indication of changing times is that the Food and Agriculture Act of 1977 "supports expanded research and extension for the benefit of small farmers."

The basic problem derives from the inability of the economic system to employ all our citizens, especially youth. Unemployment among men and women in their late teens and twenties ranges from 15 to 20 percent in some sections of the United States to as high as 45 percent among urban minority youth. In highly industrialized nations, automation has resulted in a steady increase in the number of unemployed or marginally employed young people, despite the once-prevalent belief that in the long run the market takes care of unemployment in the free-enterprise system. Page Smith, emeritus professor of history at UCLA and UC Santa Barbara, suggests that we should recognize that modern industrial technology has been so efficient that it has produced a surplus of human energy that should be utilized to perform tasks urgently needed to conserve and enhance the natural and human environment. Smith urges that we should

> stop viewing unemployment as evidence of personal or social economic failure and recognize it as the most brilliant achievement of modern technology. . . . We have freed a vast army of young men and women from routine labors and given them an opportunity to perform useful, constructive and rewarding work for their country or, more broadly, for humanity. . . . We are already, for better or worse, committed to not allowing people to starve in the most flourishing economy in the history of the world. We spend or waste countless billions on armaments too destructive to use and that will be obsolete tomorrow. We dissipate other billions in overbureaucratized, poorly conceived stopgap measures. We paid the necessary price to salvage our children in the depth of a devastating depression; in times of almost euphoric prosperity, surely we can do no less. (P. Smith, 1977)

Ill fares the land, to hastening ills a prey
Where wealth accumulates and men decay.
*—Oliver Goldsmith*

## CALIFORNIA LAND CONSERVATION ACT OF 1965

An important adverse effect of population explosion is the constant conversion of agricultural land to other uses. Of California's original 8.7 million acres of prime agricultural land, over 2 million had already been urbanized by 1972. Another million acres were expected to be subdivided by 1980, resulting in a total loss of about one-third of the best agricultural land (Heller, 1972). Recognition of this threat led to the passage of the California Land Conservation Act of 1965 (CLCA), also called the Williamson Act. One of its objectives was to preserve a maximum amount of available agricultural land to maintain California's agricultural economy and ensure an adequate food supply for the nation's future. The act provides that, in return for use-value assessment for property taxes, typically lower than for market value, the landowner agrees to restrict his or her land to agricultural or related use for at least ten years. Contracts are automatically renewed each year unless one party gives notice of nonrenewal.

By 1975-1976 about 45.9 percent of available farmland in 47 counties was represented in CLCA. However, results have been somewhat disappointing. Much land remote from incorporated areas and in little or no danger of being converted is under contract and much considered by owners as having development potential is not under contract. Thus, under this voluntary program, substantial amounts of California's best agricultural land will continue to be subject to development. And the public continues to make a significant investment in the program through subvention payments to school districts and local governments to partially offset the fiscal impact of the act (Carman and Heaton, 1977).

## CALIFORNIA'S WATER SUPPLY

In California about 8.8 million acres of land are now irrigated. Such acreage is expected to reach 9.7 million by the year 2000. This is all semiarid land with less than 15 inches of rainfall per year. As in other western states, the area of irrigable land depends principally on the amount of water available as runoff from the mountains which, in California, amounts to 71 million acre-feet per year. (Imperial and other desert valleys obtain their water from the Colorado River, as previously recounted.) Approximately two-thirds of the water originating in California is from the north, as a result of heavier rainfall and snow, but two-thirds of the need for water is in the area south of Sacramento. Leading the list of irrigation water users from the state's own watersheds, in order of consumption, are Fresno, Kern, Tulare, San Joaquin, and Kings counties (ABRC, 1974).

### Canal Systems

The canal systems (fig. 35) that carry water from areas of excess to areas of deficit are the State Water Project (SWP) administered by the state's Department of Water Resources, and the Central Valley Project (CVP) administered by the U.S. Bureau of Reclamation. The two systems represent a total investment of nearly $7 billion and constitute the nation's most extensive and sophisticated water storage and delivery system. The two projects include 37 dams and reservoirs, 1,400 miles of canals and tunnels, 10 power plants, and 25 pumping plants.

Federal participation began in 1935 when the state failed to sell the $170 million in bonds required to start the project. The principal reservoir is Shasta Lake behind the 602-foot-high Shasta Dam (fig. 36) on the Sacramento River north of Redding, completed in 1944. Water stored in this and other reservoirs flows down the Sacramento River and into the Sacramento-San Joaquin delta. At the south end of the delta, the Tracy Pumping Plant of the CVP pumps water into the 117-mile-long Delta-Mendota Canal to serve farmers in the San Joaquin Valley (fig. 32). Another major unit of the CVP is Friant Dam and Millerton Lake on the San Joaquin River 25 miles northeast of Fresno. Most of the water from Millerton Lake flows south down the 152-mile Friant-Kern Canal to farmers as far away as the outskirts of Bakersfield, but some also flows north in the 36-mile Madera Canal (fig. 35).

The SWP parallels the CVP in many respects and the two operate some joint facilities, such as the San Luis Reservoir near Los Banos. SWP construction began in the early 1960s. Its biggest project is the 742-foot-high Oroville Dam on the Feather River, the major tributary of the Sacramento. At the time it was completed in 1969, it was the largest earth-filled dam in the world, 16 feet higher than Hoover Dam on the Colorado. Like water from Shasta Lake, the water from Oroville Lake flows down the Sacramento system into the delta and is pumped out at the state's Delta Pumping Plant two

*Fig. 35. California's aqueduct and canal system (redrawn and modified from Rood and Stall, 1976*a*).*

miles from the federal facility at Tracy. Here it is lifted 244 feet into the California Aqueduct. South of the San Luis Reservoir near Los Banos, this major canal transports the water southward. Here it is 257 feet wide at the top and 110 feet wide at the bottom and has a water depth of 36 feet. The canal can handle 13,100 cubic feet per second. The water flows down the 444-mile aqueduct along the west side of the San Joaquin Valley and over the Tehachapi Mountains through a series of pumps and 11 miles of tunnels. The greatest pumping complex of them all, in Kern County, lifts 20 million gallons per hour 1,926 feet in the Tehachapis to reach an elevation of 3,165 feet above sea level. A west branch of the aqueduct then carries water bound for Los Angeles to Pyramid and Castaic lakes while the main aqueduct skirts the desert to the north of the San Gabriel Mountains and on to Perris Reservoir in Riverside County.

*Fig. 36. Shasta Lake on September 30, 1977, after two years of drought. The actual color of the exposed shoreline is a light rusty shade of red (photo by Jack K. Clark and William E. Wildman).*

The CVP and the SWP are the two largest systems. The system of the Metropolitan Water District of Southern California, bringing water from the Colorado River, was described in chapter 2. Los Angeles also obtains water from Owens Valley, east of the Sierra, via the 240-mile Los Angeles Aqueduct, the intake of which was placed on the Owens River just north of Owens Lake. The aqueduct was completed in 1913. It begins at an elevation of 4,000 feet and operates entirely by gravity, producing a substantial amount of power by means of generating plants along its route.

## Groundwater

Subterranean aquifers have long been an important source of water for irrigation and domestic use in California. The Los Angeles Aqueduct, for example, is supplemented with 108,000 acre-feet of Owens Valley groundwater to bring the total to 500,000 acre-feet per year. Extensive use of such water began in 1865, when former governor John G. Downey bored into an artesian aquifer near Compton, triggering a real estate boom that attracted thousands of settlers interested in growing oranges and starting the first extensive use of groundwater in the United States. By 1890 more than 2,000 wells were supplying water to 38,378 acres, and by 1899 there were 152,506 acres of fruit lands irrigated by aquifers in California—89.9 percent of all land in the United

States that was irrigated from wells. By 1959 about 13 million acres were irrigated from groundwater sources in the West, more than a third of it in Texas (Dunbar, 1977).

California has an underground storage capacity of 143 million acre-feet of water, according to an estimate by the state's Department of Water Resources (DWR), compared with 34 million acre-feet in surface reservoirs. This water typically moves no more than 10-1,000 feet a year, so it can be readily utilized before it has a chance to escape. Underground basins receive the surplus water that percolates down during wet years. Good water management would involve a balance in the use of surface and underground water. Surplus water can be recharged into the underground supply by artificial means during wet years, thus increasing the amount of water available for pumping during dry years. Southern California has far surpassed other areas of the state in its utilization of this procedure. Much more needs to be done in the often drought-plagued Central Valley, which has nearly ten times greater groundwater resources (Todd, 1977).

### The Peripheral Canal Controversy

All of California's water systems have been beset with bitterness and controversy from the very beginning, with no letup to the present day. Space does not permit a discussion of the controversies attending each project, but those pertaining to the giant CVP and SWP projects are particularly crucial at this time. Central to the current issue are the winds of change that have blown over the land. Former Governor Edmund G. ("Pat") Brown, generally credited with providing the political impetus to get the SWP started with passage of the Burns-Porter Act in 1959, in later years recalled, "I was a builder. I loved to build freeways, bridges.... I wanted to build that goddamn water project." This is in sharp contrast with the philosophy of his son, Governor Edmund G. ("Jerry") Brown, Jr., who says, "The values of one generation differ from those of another. We have a limit on our resources. There is a very serious question about how we grow and in what respect" (Rood and Stall, 1976*a*). The "era of limits" is anathema to the Bureau of Reclamation and most water engineers and farmers. Controversy over water use is exacerbated by the governor's appointment of a team of environmentally oriented officials to manage his Resources Agency and Department of Water Resources, charged with building and operating the SWP. With their newfound legal and political influence, the environmentalists have already caused a lengthy delay in the construction of the controversial peripheral canal around the Sacramento-San Joaquin delta (Rood and Stall, 1977*a, b*).

The delta was originally an important swampland habitat for fish and fowl. Levees were built to reclaim the swamps for farming, which resulted in 700 miles of interconnecting waterways and the creation of 60 islands. Leaving the delta, the water enters shallow Suisun Bay and flows on to San Francisco Bay and the Pacific Ocean. Agricultural crops consumed only about one-fourth to one-third as much water as the natural plant life (tules and cattails) they replaced. About 550,000 acres of farmland had been reclaimed in the delta by 1930, making the area one of the state's leading agricultural producers. It is also the vital link between the ocean and freshwater spawning grounds for the state's most important commercial and sport fishery. It is the principal stop along the Pacific Flyway for millions of migratory birds (Jackson and Paterson, 1977).

Increased pumping of water from the delta canals for export to the south has severely damaged the immediate environment, protected by California law. It has reversed the normal seaward flows of some delta channels, thereby luring salmon away from their natural spawning ground. The migration patterns of salmon and steelhead are, in part, directed by water currents. Likewise, the striped bass population is only half what it was when the pumping began. Saltwater from San Francisco Bay is intruding where freshwater once sustained crops and plant life supplying food for waterfowl.

Further intrusion of saltwater would be disastrous to local communities, industry, and agriculture. Much more water than engineers originally said would be needed will have to be released to repel the saltwater pressure. The state's DWR contends that the CVP is obligated to furnish up to 75 percent of the flows needed to keep saltwater out of the delta, but the federal Bureau of Reclamation has claimed that the CVP is not legally required to lease its reservoir waters to protect the delta.[9] The bureau contends that as deliveries to CVP customers increase, it will have even less water available to maintain delta outflows. Yet, a state law provides that only water beyond that needed to protect the delta may be imported for use elsewhere. Water agencies and delta interests disagree on how much water is required to protect the delta.

A proposed 42-mile peripheral canal (fig. 35) would bring Sacramento River water around the delta to state and federal pumps. Proponents say the canal would provide better outflows into the delta system, correct the flow reversal now damaging the fishery, and meet various water-quality objectives at a net water savings of nearly a million acre-feet a year. Delta interests and environmentalists contend the canal would provide high-quality water to the SWP and CVP pumps without regard to what happens locally. The entire state has a stake in the outcome of the peripheral canal and other state-federal controversies. By the year 2020 southern California is expected to receive 48 percent of its water via the canal; the San Francisco Bay area, 30 percent; the San Joaquin Valley, 35 percent; and the central coast, 18 percent, to say nothing of the problem of replenishing ever-receding aquifers. The north coast has at stake the fate of the Eel, Klamath, and Trinity rivers. Regional and legal opposition to their "development" is formidable. The same may be said of the opposition to bringing water south from the Columbia River or its tributaries.

In Los Angeles, with its charm dimmed by overpopulation, boosterism is being replaced by disillusionment in the minds of an increasing number of its citizens as they accommodate themselves to the new "era of limits." No longer seeking far-off water sources, the Los Angeles DWP is encouraging conservation by its customers and wastewater reclamation—much more costly than existing sources but a possible solution to the problem.

## Conservation of Irrigation Water

Agriculture consumes 32 million acre-feet of California's water in a normal year—about 85 percent of the total supply. The Irvine Ranch, south of Los Angeles, has reduced its water demand 25-30 percent by the use of drip irrigation[10] in its citrus orchards, sprinkler systems on other crops, and extensive wastewater reclamation. The reclaimed water costs about $100 per acre-foot. Farmers in other parts of the state point out, however, that high water costs and high-profit crops at Irvine provide a special incentive for water conservation. They claim that farmers in the Central or Imperial valleys could not afford to duplicate Irvine's efforts. The DWR agrees, stating that if all the available conservation techniques were used throughout the state, only about a 3.5 percent savings of water would be effected, even if large holding reservoirs were con-

[9]In August 1978, the U.S. Supreme Court upheld California's legal right to impose conditions on the operation of federal water projects (Rood, 1978).

[10]In drip or trickle irrigation, thin plastic tubes are laid along rows of plants and trees to deliver a metered flow of water through outlets near the roots. This method can reduce water use by 30-50 percent compared to flood irrigation, but is expensive to install.

structed in the Central Valley. Most agricultural water is already used many times before it reaches the ocean. Water applied in excess of what is demanded by crops either percolates down to recharge aquifers or recirculates into the irrigation canal and in either case is reversible. (If water is reversed too often, it becomes excessively saline and should be returned to the ocean.) Also, excess water from the Central Valley eventually flows into the Sacramento delta to prevent seawater from pushing inland. It is true, however, that most farmers could reduce water use, without reduced yields, by watering exactly when the plants need water and in the exact amounts. There are expert consultants available to advise on these matters (Jones, 1977).

## Two Years of Drought

By the fall of 1977, California had suffered two successive years of the worst drought in its history. Precipitation had been less than half of normal, and by October 1, 1977, reservoirs had less than 15 percent of capacity. Shasta Lake (fig. 36) for example, reached its lowest point on September 13, 1977, when it was storing 562,600 acre-feet, or 12.4 percent of total capacity. Precipitation in the winter of 1977-1978, however, was sufficient to end the agricultural water shortage and allow normal allocations of water for 1978.

Southern California suffered the least, for its agricultural areas get most of their water from the Colorado River, whose immense reservoirs could be tapped. Some water normally routed to southern California from the northern part of the state could be diverted instead to the Central Valley. The SWP and the CVP cut water supplies to farmers by 60 percent and 75 percent, respectively. Some farmers were left virtually without surface water supplies. Nevertheless, most of the state's agricultural areas remained green and productive. Gross farm income was expected to be about $9.6 billion, only 1 percent less than in 1976, but about $800 million below what it would have been without the drought (Bulkeley, 1977).

The most important step taken by farmers in combating the drought was to drill more irrigation wells. In 1977, 10,000 wells were drilled by the month of August and many others were reopened or deepened. Many of the new wells were required because heavy pumping had lowered water tables below well depths or caused land to subside and crush well casings (Bulkeley, 1977).

To conserve water, some farmers recirculated runoff water from the foot to the head of their field. Others went to great expense to substitute sprinkler or drip irrigation systems for flood irrigations, or they ran their water in furrows instead of flooding the land. However, by using less water, farmers run the risk of not sufficiently leaching excess salts from their soil. Crops that use much water, such as rice, were reduced in acreage, while crops using less water, such as cotton, were increased. Growers of fruit and nut trees have a special problem, for they may have an investment of 20-30 years or more in their trees and must get water to them at almost any cost.

In the San Joaquin Valley, the water table is dropping an average of 4.5 feet a year, and many shallow wells have already gone dry. Many others will be destroyed by increasing land subsidence. In any case, groundwater is not as satisfactory for irrigation as water from mountain streams, for it is higher in salt concentration.

## Postponing the Day of Reckoning

The saddest aspect of the California water crisis is that water reclamation and engineering, with due respect for their past achievements, are only temporary palliatives. The same may be said of water conservation and the lowering of our demand for the amenities of the "good life"—washing machines, dishwashers, lawns, swimming pools, and so on—praiseworthy as these may be. These measures serve only to enable a still larger population to bring the crisis to a new and higher level—postponing the day of reckoning. We are asked to develop a water system "adequate to our demands." But according to the feedback principles operating in our society, demand *breeds* demand. Already over 1,200 dams have been built in this state without satisfying the burgeoning demands of exponential growth.

The basic fact is that California suffers not from too little water but from too many people, exceeding the carrying capacity of the land. A stopgap solution to the problem might be to confine water delivery to present agricultural and industrial users and not to new consumers, thus curbing further immigration (and subverting the programs of the many chambers of commerce). The long-term solution, nationwide, must be more universal and effective education, and the development of the intellectual climate and political, social, and legal instrumentalities for an effective program of birth control (and an end to illegal immigration). Equal emphasis should be placed on ending wasteful consumerism. These are the fields in which technology could be utilized to confer its greatest blessing toward improving the human prospect.

UCLA's popular zoologist-ecologist, the late Professor Raymond B. Cowles (1976), had this to say after more than a half-century of observation of the environmental degradation of his native South Africa and his enthusiastically adopted California: "Perhaps the most reasonable question today is not 'when shall we commence a rigorous program of population and resource control management?' but rather, 'when should we have begun?' " This is a point of view that has long been commonplace on university campuses, but now that it is also beginning to be expressed in government circles, for example, by California's Secretary of Resources, Huey Johnson (Gendlin, 1978), perhaps there is reason for cautious optimism.

There is much of California's charm and beauty that remains to be rescued. One can hope that universal education and improved technology will stem the tide of population explosion within a reasonable time frame. Yet, as native-son conservationist Raymond Dasmann (1965) emphasized, there will still be needed the determination not to provide areas for the industrial expansion that encourages immigration. Immigration is encouraged by citizens who are still mesmerized by the concepts of "growth" and "progress," even while decrying the consequences. The fact that cities *can* control rate of growth has already been demonstrated by some California cities such as Santa Barbara and Carmel.

## THE FARM-SIZE CONTROVERSY

In California, 79 percent of the land in farms was in units of more than 1,000 acres in 1975, compared to only 4 percent in Iowa and 6 percent in Illinois. Not only economic circumstances but also early land policies and their administration account for the large percentage of big farms (Gates, 1975*b*). As in other states, the high cost of mechanization put an end to the era of mobility up the "agricultural ladder" and an end also to Thomas Jefferson's dream of numerous yeoman farmers as a bulwark of a free republican polity. Large family farms are now predominantly those that are passed down from one generation to the next.

Large and thriving corporate farms in California can be accounted for mainly by the huge areas owned by corporations prior to their great rise in value after water was delivered to them at taxpayers' expense. However, as in other areas of the semiarid West, the size of an individual farm in California is likely to be affected within the next few years by the U.S. Reclamation Act of 1902, which limits to 160 acres per individual the land that receives federally subsidized water. If the law were enforced, a man and wife could legally own 320 acres. The law also requires that farmers receiving federal water live on or near their land. The intention of the law was to ensure the survival of the American family farm.

It is generally recognized that the 160-acre limitation of the 1902 act is out of date. As President Carter has said, "with massive development and large machinery, a larger acreage is necessary for an economically viable farm operation" Amendments to the 1902 law will be proposed to Congress (Shaw, 1978). Governor Brown's administration favors a 640-acre limit on the amount of land for an individual or family of any size receiving federal water.

The enormous Westlands Irrigation District that comprises the western San Joaquin Valley is practically a demographic desert. The rural villages and farmsteads of René Dubos's "humanized nature" are conspicuously absent. According to the Department of the Interior, the Southern Pacific Land Company alone owns more than 100,000 of the district's 600,000 acres. Yet, no doubt many a family farmer would have been glad to farm there if the land had been available at the prices prevailing before the development of the huge federally constructed canals.

## HISTORY OF A CALIFORNIA FAMILY FARM

The history of a family farm at Winters, in Yolo County, could be used to illustrate, in principle, the evolution of farming in California's great Central Valley wherever it has been practiced on a successful and sustained basis. The farm has been in the possession of one family for four generations. Richard E. Rominger (1975), one of the owners of the farm and director of the California Department of Food and Agriculture, told the story at the Symposium on Agriculture in the Development of the Far West held at Davis, California in June 1974.

Rominger's great-grandfather and grandfather grew dryland barley and raised sheep. Fertility of the soil was maintained by a rotation of barley, sheep pasture, and fallow. Electricity and irrigation arrived on the scene, and Rominger's father grew irrigated alfalfa to feed an increased number of sheep. Then rapid changes in agricultural machinery took place. Following college, the two sons returned to the farm and went into partnership with their father. Land was leveled for irrigation and more land was leased. By 1974 ten crops were raised on the Rominger farm. Listed in the order in which they were added were barley, wheat, oats, alfalfa, rice, sugar beets, milo, beans, safflower, and honeydew melons.

A Caterpillar tractor was purchased by the Romingers in 1946 for $6,700. In 1974 the same type of tractor, with the same horsepower, would have cost four times as much. But the price of barley remained about the same. There was no way to stay in business except to risk expanding the operation, adding new crops and adopting new technology. The Romingers now have about 5,000 acres; approximately 2,000 are irrigated and the rest dryland-farmed in the foothills. They own about half of the land and lease the balance. The farm supports three families.

Rominger described the many legal restraints on his farming operations imposed by conservation legislation, but he understands the necessity for them. "Our challenge," he said, "is to find the balance that will provide the food and preserve the environment." For example, the Romingers and other farmers, with the help of resource conservation districts, have planted trees and shrubs to provide feed and cover for wildlife. Rominger pointed out that his father had neither the restraints nor the opportunities of the present-day farmer. He concluded his talk with this eloquent statement of his philosophy:

> I have changed as new techniques and new regulations came along. My children are growing up with this concern for the total environment and are able, I think, to put agriculture in perspective in all its relationships with man and his environment. They have a different feeling for our ranch and agriculture. Whereas I saw it as productive soil and a "good life," they see that plus the beautiful microcosms and complex interrelationships in agriculture and nature and have a greater appreciation of the total environment. Two of them are Sierra Club members (but don't tell their uncle who raises sheep). Perhaps this coming generation will always be able to see the bigger environmental picture.

## OCEAN FARMS

For all at last return to the sea—to Oceanus, the ocean river, like the ever-flowing stream of time, the beginning and the end (Carson, 1961).

The oceans cover 71 percent of the earth. Evidently most of the 2.5 x $10^{21}$ Btu of solar energy annually reaching the earth's surface must be falling on them. It is not surprising that research is now being directed toward harnessing this enormous energy for raising prodigious quantities of biomass as a source of food and energy. Most of this research is centered in California. The urgency of such research is prompted by the knowledge that most of the world's productive arable land in climatically suitable areas is already under cultivation. Much of it is annually lost to urban encroachment. Large areas formerly with rainfall above the level classified as semiarid have become deserts through deforestation, overgrazing, or injudicious farming practices. Siltation, salination, and waterlogging have ruined many irrigated areas, including some of once fabulous productivity. Some people look to the vast tropical jungles as areas for agricultural exploitation, but such areas appear to have little potential for sustained use in this respect. Attempts to farm them extensively with anything other than the traditional shifting (slash-and-burn) system could lead to widespread ecological disaster (Eckholm, 1976, 1977; Friedman, 1977).

### Ocean Food and Energy Farm Project

A project, originally based in San Diego, California, was begun by the United States Navy and is now administered by the General Electric Company. The project is discussed by Wilcox (1975). He presents a realistic and ecologically viable plan for food and energy for the world's peoples on a *sustained* basis, rather than offering just a temporary palliative. He believes that the best way to directly or indirectly use solar energy falling on the ocean is to generate huge new quantities of food, the latter being "the most precious large-scale form of energy known to man." He points out that ocean farm harvests can also be converted into synthetic natural gas, alcohol, gasoline, fertilizers, and plastics; in fact, all the products now obtained from the petroleum-petrochemical industries.

When compared with land-farming, the oceans provide not only a far greater area but seaweeds can be more efficient converters of solar energy than land plants. No irrigation is needed and there is no danger of freezing. Nearly everywhere there is deep nutrient-rich seawater that can be upwelled (by wave-powered pumps) for the sustenance of the crops.

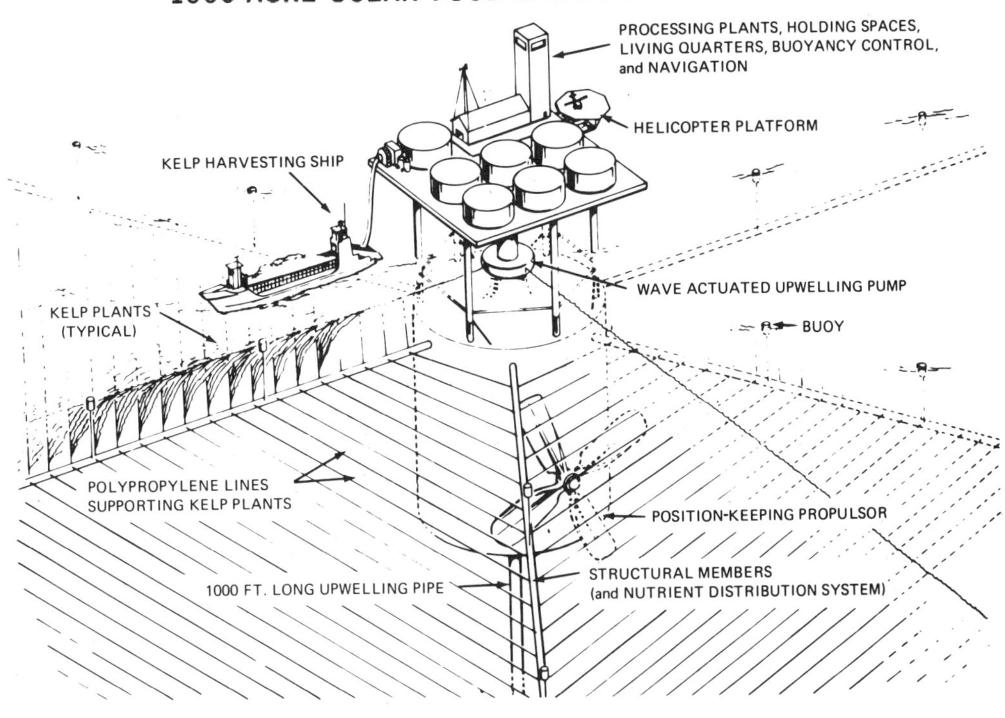


*Fig. 37. Artist's conception of a future ocean farm unit (from Wilcox, 1975).*

Most marine plants use their "roots" (holdfasts) to anchor themselves to the ocean's bottom. Where the ocean is too deep for this, artificial bottoms can be built consisting of far-flung networks of nylonlike lines submerged 40-80 feet. The plants can be attached to these and their surface canopies harvested with ships outfitted with undersea clippers (fig. 37). For farms located far out at sea, the harvested vegetation can probably be processed on giant platforms floating nearby. Some of the vegetation can be fed to cattle, sheep, and marine organisms and the rest given to chemical processing plants housed on the same platforms.

Plants most suitable for the various water temperatures must be investigated. Hybridization and other manipulations of basic genetic qualities are suggested as a means of increasing yield and resistance to diseases and parasites. The initial investigations were with giant California kelp, *Macrocystis pyrifera.* In tests made at the UC Davis, sheep were found to be able to digest this kelp about as well as they can alfalfa. In fact, certain seaweeds can be eaten directly by man, as in Japan and Ireland.[11]

In a report on the progress of the Ocean Farms Project, Wilcox (1977) stated that

[11]In Leon Uris's *Trinity,* young Seamus O'Neill wistfully recalls kelp harvests on the coast of northern Ireland during the months of hunger before the fall harvest season. "Some of it was used for animal fodder, some for making iodine and some for fertilizer. There was edible seaweed that my ma mixed with potatoes and another type that could be jellied to thicken the milk and butter." While the kelp was drying on a long stone wall and was being separated by variety and use, the thousands of trapped cockles and mussles were picked out as a welcome bonus (Uris, 1976).

three small experimental farms have already provided valuable field data and operating experience. Preliminary studies involving many assumptions have led to the following conclusions:

> (1) Large systems—8000 hectares (20,000 acres) or larger—will be economically preferable to small units, (2) dynamically positioned farm systems will be economically preferable to anchored systems for all but the shallowest waters, (3) on-site structures may cost less than $7000 per hectare ($3000/acre) of cultivated ocean, (4) harvesting ship costs will possibly amount to some $3300 per hectare ($1300/acre), (5) associated on-shore processing facility investments may amount to about $4000 per hectare ($1600/acre) of cultivated ocean area, (6) synthetic natural gas costs may range from $2 or less to a high of about $7 per million Btu, depending mainly on credit values assumed for foods and other products.

The most attractive aspect of sea-farming as envisioned by Wilcox (1975) is that it utilizes solar energy, thus obviating the "thermal pollution barrier" which he and other scientists consider to be an ominous threat to mankind if we continue to pursue current policies of developing principally fossil fuels and nuclear energy. He warns,

> But even the solar energy is limited in its rate of supply. Therefore, though the utilization of ocean farms and other types of solar energy converters can temporarily palliate and postpone the present problems generated by man's spiralling growth, it cannot long resolve them. If man does not use this century and the next to develop a no-growth life style, the final and unavoidable hothouse catastrophe will terminate man's present civilization.

## Chromalloy Ocean Resources Project

A project with objectives similar to those of the San Diego Sea Farms is being undertaken by Chromalloy Ocean Resources with offices in Gardena, California. Technical Director, Edward N. Hall (1977) divides human history into three eras: Pre-Agricultural, Careless, and Steady State, the second one now drawing to a close. He considers marine algae, raised on continental shelves, to be the most likely means, among those he has studied, of developing the enormous biomass required to generate the energy and food to redress the consequences of man's unbalanced activities on land and sea. Continental shelves, however, represent only 0.1 percent of the earth's ocean area. The Chromalloy Project does not include artificially induced upwelling of subthermocline ocean waters to ensure nutrition of kelp and differs in this respect from the project begun by Wilcox.

Ocean-farming extends some of the current goals of the land type. Among the many new goals of the federal Food and Agricultural Act of 1977 is the provision for a solar energy research and development program, including the establishment of solar energy demonstration farms within each state, and the establishment of three to five regional solar energy research, development, and demonstration centers.

Solar energy is renewable, nonpollutive, and is the only form that would not add to the amount of heat in the atmosphere, unless it were transmitted via shortwaves to earth from space satellites. In a thoughtful and objective appraisal of the world's energy-food crisis W. J. Chancellor and J. R. Goss, professors of agricultural engineering at the University of California, Davis, concluded that "the key to the system of balance is stopping population growth, and the necessary goal must be a solar powered world. The time to use this key and set out toward this goal must be now—otherwise there may be no such thing as a calamity-free balance between energy and food" (Chancellor and Goss, 1976).

## THE HOLISTIC APPROACH

The development of solar energy is but one facet of the holistic approach—recognition of the whole as something greater than the summation of its parts. That approach, in turn, is an important feature of the qualitative growth to which mankind should now aspire to replace the quantitative growth or "growthmania" that has been far too long associated with progress (Watt et al., 1977).

The human race has the knowledge, technology, and resources to develop a sane and humane world in which basic precepts for wholesome and sustainable life systems will not be violated. We can derive some hope from the fact that ever-increasing numbers of people, including many leading industrialists, are becoming aware that, once we discontinue our obsolete obsession with quantitative growth, we will be free to focus our efforts on a culturally more sophisticated and sustainable way of life. For good practical reasons, society is becoming more concerned with the "enduring things" formerly considered to be exclusively the concern of the poet (Paarlberg, 1976). Signs of the new spirit are increasingly apparent—restless stirrings at the grass-roots level—but, inured to a winter of discontent, we sometimes fail to recognize the first breath of spring.

# Literature Cited


Abrams, H. L., Jr. 1978. A diachronic preview of wheat in hominid nutrition. J. Appl. Nutrition 30:41-55.

Abelson, P. H., and A. L. Hammond. 1977. The electronics revolution. Science 195:1087-1091.

ABRC. 1974. *The Future of California's Agriculture 1974-2000.* Report of Blue Ribbon Committee on the Future of California Agriculture, C. B. Christensen, Chmn., Director, Dept. Food and Agriculture, Sacramento. Processed.

Adams, K. 1969. *The Redwoods.* New York: Popular Library.

Adams, R. L. 1952. Protecting citrus groves from frost. Calif. Agr. Exp. Sta. Bull. 730.

Adams, R. M. 1960. Origin of cities. Sci. Am. 203(3):153-168.

AFI. 1976. Young trees: our most efficient air purifier. In *Forests USA,* p. 10. Am. Forest Inst., Washington, D.C.

Albaugh, R. 1976. Bulls vs. steers. Univ. Calif. Agr. Ext. Leaflet 2310.

Albaugh, R., and H. T. Strong. 1972. Breeding yearling beef heifers. Calif. Agr. Exp. Sta. Circ. 433.

Alexander, A. J. 1928. *The Life of George Chaffey.* New York, London: Macmillan.

All, J. N., and R. N. Gallaher. 1977. Detrimental impact of no-tillage corn cropping systems involving insecticides, hybrids, and irrigation in lesser cornstalk borer infestations. J. Econ. Entomol. 70:361-365.

Allen, D. L. 1967. *The Life of Prairies and Plains.* New York: McGraw Hill.

Allen, J. B., and G. McLeonard. 1976. *The Story of the Latter Day Saints.* Salt Lake City: Deseret Book Co.

Anderson, E. 1952. *Plants, Man and Life.* Boston: Little, Brown.

Anderson, G. B., Jr. 1977. Oregon's forest conservation laws. Part I. Am. Forests 83(3):16-19, 52-56.

Anderson, T. L. 1977. Bonanza farmers and subsistence: a response. Agr. Hist. 51:104-108.

Anonymous. 1966. A national program for research in agriculture. Joint report by Assoc. of State Universities and Land Grant Colleges and the U.S. Dept. Agr. Mimeographed.

——— 1971. Focus on Michigan's bean industry. Mich. Agr. Exp. Sta., Mich. Science in Action Ser., no. 16.

——— 1974. Continuous cropping: it's best for the Palouse. Washington State University, Col. Agr., Ext. Circ. 391.

——— 1975*a*. Herbicides stimulate protein content. Agr. Research 23(9):10.

——— 1975*b*. New York orchard and vineyard survey, 1975. N.Y. Crop Reporting Serv. Release no. 31, Albany.
——— 1977*a*. Whatever happened to high-lysine corn? Farm J. 101(2):C6.
——— 1977*b*. Drilled soybeans: up to 50% higher yields possible. Successful Farming 75(6):38.
——— 1977*c*. "Lots of records" guide Snowflake pig farms. Progressive Farmer 92(4):8-90.
——— 1977*d*. Sorghum breakthrough doubles hybrid possibilities. Progressive Farmer 92(4): 107.
——— 1978. No-till takes a firm hold in California orchards. Agrichemical Age 21(1):20-21.
App, B. A., and G. R. Manglitz. 1972. Insects and related pests. In *Alfalfa Science and Technology*. Ed. C. H. Hanson, pp. 527-554. Am. Soc. Agron., Madison, Wis.
Arizona agricultural statistics. 1975. Arizona Crop and Livestock Reporting Service. University of Arizona Coop. Ext. Serv., Tucson.
Armstrong, F. H. 1977. Jobs for youth. American Forests 83(11):30-33, 42-44.
Arrington, L. J., and D. May. 1975. A different mode of life: irrigation and society in nineteenth-century Utah. Agr. Hist. 49:3-20.
Atherton, L. E., 1961. *The Cattle Kings*. Bloomington: Indiana University Press.
Austin, L. 1977. Caltech professor says answer to energy crisis could be written on the wind. *Los Angeles Times,* April 21.
Austin, M. E. 1965. *Land Resource Regions and Major Land Resource Areas of the United States*. USDA, Soil Conserv. Serv. Agr. Handbook 296.
Bailey, G. 1975. Project independence. *Cry California* 10(1):22-30.
Bainer, R. 1932. Harvesting and drying rough rice in California. Calif. Agr. Exp. Sta. Bull. 541.
——— 1948. The processing of sugar beet seed. Agr. Eng. 29:477-479.
——— 1975*a*. Science and technology in western agriculture. Agr. Hist. 49:56-82.
——— 1975*b*. *The Engineering of Abundance*. Davis, Calif.: Roy Bainer.
Bainer, R., J. R. Gass, R. G. Curley, and D. G. Smeltzer. 1955. Combine used in corn. Calif. Agr. 9(7):12-13.
Bakeless, J. E. 1950. *The Eyes of Discovery*. Philadelphia: Lippincott.
Baker, R. G. 1977. Natural productivity masks the problem. Prairie Farmer 149(2):101.
Ballagh, J. C. 1895. White servitude in the colony of Virginia. In *Studies in Historical and Political Science,* XIII, no. 6-7. Baltimore: John Hopkins University.
Barmington, R. A., and S. W. McBirney. 1952. Mechanizing the production of sugar beets. Colo. Agr. Exp. Sta. Bull. 420-A.
Barnes, W. C. 1941. *Apaches and Longhorns*. Los Angeles: Ward Richie Press.
BARR. 1975. *World Food and Nutrition Study*. Report of Board of Agricultural and Renewable Resources, Nat. Research Council, Washington, D.C.
Barrons, K. C. 1975. *The Food in Your Future*. New York: Van Nostrand Reinhold.
Bayley, N.D. 1975. Cattle. Encyclopedia Americana, pp. 63-82. New York: Americana Corp.
Beadle, B. W. 1972. The mystery of maize. Field Mus. Nat. Hist. Bull. 43:2-11.
Bean, W. 1968. *California, an Interpretive History*. New York: McGraw-Hill.
Beard, C. A., and M. R. Beard. 1944. *The Beards' Basic History of the United States*. Garden City, N.Y.: Doubleday, Doran.
Beasley, R. P. 1972. *Erosion and Sediment Pollution Control*. Ames: Iowa State University Press.
Beeler, D. 1977. Solving the Coastal Plains' big hard-pan problem. Agrichemical Age 21(7):8-9, 24.
Bender, B. 1975. *Farming in Prehistory*. New York: St. Martin's Press.
Bennett, H. H. 1939. *Soil Conservation*. New York: McGraw-Hill.
Bennett, H. H., and W. R. Chapline. 1928. Soil erosion a national menace. USDA Circ. 33.
Benoit, E. 1976. The coming age of shortages. Bull. Atom. Sci. 32(1):6-16.
Benson, A. B. 1937. *Peter Kalm's Travels in North America*. 2 volumes. New York: Wilson-Erickson.
Bernstein, H., and F. Del Olmo. 1977. Economy of Mexico fuels influx of aliens. *Los Angeles Times,* Sept. 18.
Beveridge, A. J. 1928. *Abraham Lincoln, 1809-1858*. Volume I. Boston: Houghton Mifflin.
Bidwell, J. 1928. *Echoes of the Past*. Ed., M. M. Quife. Chicago: Donnelley.
Bidwell, P. W., and J. I. Falconer. 1925. *History of Agriculture in the Northern United States, 1620-1860*. Carnegie Inst. of Washington.
Binford, L. R. 1968. Post-Pleistocene adaptations. In *New Perspectives in Archeology*. Ed. S. R. Binford and L. R. Binford, pp. 313-431. Chicago: Aldine.
Bird, H. R. 1968. Chicken in every pot—the broiler bonanza. USDA Yearbook, pp. 37-41.

Birdsell, J. B. 1968. Some predictions for the Pleistocene based on equilibrium systems among recent hunter-gatherers. In *Man the Hunter.* Ed. R. B. Lee and I. DeVore, pp. 229-240. Chicago: Aldine.
Blegen, T. C. 1963. *Minnesota: A History of the State.* Minneapolis: University of Minnesota Press.
Blevins, R. L., D. Cook, S. H. Phillips, and R. E. Phillips. 1971. Influence of no-tillage on soil moisture. Agr. J. 63:593-596.
Bocking. R. C. 1972. *Canada's Water: For Sale?* Copyright, R. C. Bocking. Toronto: James Lewis and Samuel.
Bohlen, J. 1975. *The New Pioneer's Handbook. Getting Back to the Land in an Energy-Scarce World.* New York: Schocken Books.
Bohlen, J. M., R. C. Powers, and J. A. Wallize. 1975. Main Street pokes along while urban areas boom. USDA Yearbook, pp. 55-65.
Bohme, F. G. 1956. The Portuguese in California. Calif. Hist. Soc. Quart. 35:233-252.
Bolen, J. S., B. F. Cargill, and J. H. Levin. 1970. Mechanized harvest systems for red tart cherries. Michigan State University, Ext. Bull E-660.
Borah, W. W. 1970. The California mission. In *Ethnic Conflict in California History.* Ed. C. Wollenberg, pp. 1-22. Los Angeles: Tinnon-Brown.
Bordley, J. B. 1801. *Essays and Notes on Husbandry and Rural Affairs.* 2nd ed. Philadelphia.
Bottrell, D. G. and P. L. Adkisson. 1977. Cotton insect management. Ann. Rev. Entomol. 22: 451-81.
Bowden, C. 1977. *Killing the Hidden Waters.* Austin: University of Texas Press.
Bradley, C. C. 1962. Human water needs and water use in America. Science 138:489-491.
Braidwood, R. J. 1960. Agricultural revolution. Sci. Am. 203(3):130-148.
Breimyer, H. F. 1975. Where money in farming comes from. Econ. Market. Inf. for Mo. Agr. 18, Sept.
Bremer, F. 1853. *The Homes of the New World: Impressions of America.* Trans. by M. Howitt. 2 volumes. New York: Harper.
Brink, R. A., J. W. Densmore, and G. A. Hill. 1977. Soil deterioration and the growing world demand for food. Science 197:625-630.
Brodie, F. M. 1945. *No Man Knows My History. The Life of Joseph Smith.* New York: Knopf.
Bronowski, J. 1973. *The Ascent of Man.* London: British Broadcasting Corp.
Brown, A., ed. 1890. *The Genesis of the United States.* 2 volumes. Boston and New York: Houghton Mifflin.
Brown, H. 1956. *The Challenge to Man's Future.* New York: Viking Press.
Bruce, P. A. 1896. *The Economic History of Virginia in the Seventeenth Century.* 2 volumes. New York and London: Macmillan.
Brumfield, K. 1968. *This Was Wheat Farming.* 1st ed. Seattle: Superior Publishing.
Brunner, E. DeS. 1929. *Immigrant Farmers and Their Children.* Garden City, N.Y.: Doubleday.
Bulkeley, W. M. 1977. California crop yields are surprisingly good despite long drought. *Wall Street Journal,* Aug. 27.
Burcham, L. T. 1957. *California Range Land, an Historico-Ecological Study of the Range Resource of California.* Div. Forestry, Dept. Nat. Res., State of California.
——— 1961. Cattle and range forage in California: 1770-1880. Agr. Hist. 35:140-149.
——— 1975. Climate, structure, and history of California's annual grassland ecosystem. In the California Annual Grassland Ecosystem, p. 16-34. Davis: University of California.
Burroughs, J. 1922. *My Boyhood.* Garden City, N.Y.: Doubleday, Page.
Cain, M., and J. C. Seibert. 1976. *California Fruit and Nut Acreage,* 1975. Calif. Crop and Livestock Reporting Serv., Sacramento.
Cairnes, J. E. 1863. *The Slave Power: Its Character, Career and Probable Designs.* London: Macmillan.
Calder, N. 1967. *Eden Was No Garden.* New York: Holt, Reinhart and Winston.
California Almanac. 1973. California Forests. In *California Past-Present-Future,* pp. 281-290. Lakewood, Calif.: California Almanac Co.
Camp, A. F. 1947. *Citrus Industry of Florida.* Fla. Dept. Agr.
Cargill, B. F., G. McManus, J. S. Bolen, and R. T. Whittenberger. 1970. Cooling stations. Michigan State University, Coop. Ext. Serv., Bull. 659.
Carman, H. F., and C. Heaton. 1977. Use-value assessment and land conservation. Calif. Agr. 31(3):12-14.
Carosso, V. P. 1951. *The California Wine Industry.* Berkeley and Los Angeles: University of California Press.

Carrier, L. 1923. *The Beginnings of Agriculture in America.* New York: McGraw-Hill.
Carrier, L., and K. Bort. 1916. The history of Kentucky bluegrass and white clover in the United States. Soc. Agr. 8:256-266.
Carriere, B. G. 1976. Prescription for erodible southern soils. Agr. Research 24(12):5.
Carson, R. L. 1961. *The Sea Around Us.* Rev. ed. New York: Oxford University Press.
Carter, L. J. 1974. *The Florida Experience.* Baltimore: Johns Hopkins Press.
——— 1977*a*. Soil erosion: the problem persists despite the billions spent on it. Science 196:409-411.
——— 1977*b*. Bergland to redirect USDA research? Science 196:506.
——— 1977*c*. Interior department: Andrus promises "sweeping changes." Science 196:507-510.
Casson, H. N. 1909. *Cyrus Hall McCormick: His Life and Work.* Chicago: A. C. McClurg.
Castetter, E. F., and W. H. Bell. 1942. *Pima and Papago Indian Agriculture.* Albuquerque: University of New Mexico Press.
Cather, W. S. 1913. *O Pioneers!* Boston: Houghton Mifflin.
——— 1918. *My Ántonia.* Boston: Houghton Mifflin.
Caudill, H. 1975. Appalachia. In *The People's Land.* Ed. P. Barnes, pp. 33-37. Emmaus, Pa.: Rodale Press.
Caughey, J. W. 1970. *California: A Remarkable State's Life History.* 3rd ed. Englewood Cliffs, N.J.: Prentice-Hall.
CCLRS. 1975. *Gross Values of California's Agricultural Production.* Calif. Crop and Livestock Reporting Serv., Sacramento.
Ceci, L. 1975. Fish fertilizer: a native North American practice? Science 188:26-30.
CEQ. 1978. *Solar Energy—Progress and Promise.* Council on Environmental Quality. Washington, D.C.: Gov. Print. Off.
Chancellor, W. J., and J. R. Goss. 1976. Balancing energy and food production, 1975-2000. Science 192:213-218.
Chandler, W. H. 1957. *Deciduous Orchards.* Philadelphia: Lea and Febiger.
Channing, E. 1905. *A History of the United States.* 6 vols., 1905-1925. New York: Macmillan.
Chapman, L. F., Jr. 1976. Illegal aliens: time to call a halt! Reader's Digest, Oct. 1976, pp. 188-192.
Childe, V. G. 1956. *Man Makes Himself.* London: Watts.
Chittenden, H. M. 1902. *The American Fur Trade of the Far West.* 2 volumes. Academic Reprints, Stanford University, 1954.
Clark, T. D., and A. D. Kirwan. 1967. *The South Since Appomattox.* New York: Oxford University Press.
Clark, W. H. 1945. *Farms and Farmers. The Story of American Agriculture.* Boston: L. C. Page.
Clawson, M. 1972. *America's Land and Its Uses.* Baltimore: Johns Hopkins Press.
Cleland, R. G. 1923. *A History of California: The American Period.* New York: Macmillan.
——— 1940. *The Place Called Sespe. A History of a California Ranch.* Privately printed.
——— 1950. *This Reckless Breed of Men.* New York: Knopf.
CLRS. 1977. *Gross Values of California's Agricultural Production by Counties and Commodity Groups, 1975 and 1976.* Compiled from the County Agricultural Commissioners' Reports. Calif. Crop and Livestock Reporting Serv., Sacramento.
Cohen, M. N. 1977. *The Food Crisis in Prehistory: Overpopulation and the Origins of Agriculture.* New Haven: Yale University Press.
Coit, J. E. 1915. *Citrus Fruits.* New York: Macmillan.
Coman, K. 1918. *The Industrial History of the United States.* New York: Macmillan.
Commoner, B. 1976. *The Poverty of Power.* New York: Knopf.
Cook, S. F. 1970. The California Indian and Anglo-American culture. In *Ethnic Conflict in California History.* Ed. C. Wollenberg, pp. 23-42. Los Angeles: Tinnon-Brown.
Cordtz, D. 1972. Corporate farming a tough row to hoe. Fortune 86(2):134-139, 172-175.
Cowles, R. B. 1976. *Desert Journal. A Naturalist Reflects on Arid California.* Berkeley, Los Angeles, London: University of California Press.
CRA. 1976. *Wheels of Fortune. A Report on the Impact of Center Pivot Irrigation on the Ownership of Land in Nebraska.* Center for Rural Affairs, Walthill, Nebraska.
Crabb, A. R. 1947. *The Hybrid-Corn Makers: Prophets of Plenty.* New Brunswick, N.J.: Rutgers University Press.
Crane, T. H. 1976. Pyrolysis of agricultural wastes for resource recovery. In *Agricultural Residues as a Source of Energy,* pp. 22-33. A compendium of papers presented at the Rural Electric Conference, University of California, Davis.

Critchfield, R. 1977. While Mexico's villagers look north with hope. *Los Angeles Times,* Oct. 9.
Cronese, T. F. 1968. *The Natural Wealth of California.* San Francisco: H. H. Bancroft.
Cundiff, L. V. 1975. Beef—from trail drives to America's main course. USDA Yearbook, pp. 116-124.
Curwen, E., and G. Hatt. 1953. *Plough and Pasture: The History of Farming.* New York: Collier.
Dale, T., and V. G. Carter. 1955. *Topsoil and Civilization.* Norman: University of Oklahoma Press.
Dalton, J. D., G. C. Russell, and D. H. Sieling. 1952. Effect of organic matter on phosphate availability. Soil Sci. 73:173-181.
Dana, R. H. 1840. *Two Years Before the Mast.* New York: Harper.
Dana, S. T. 1956. *Forest and Range Policy. Its Development in the United States.* New York, Toronto, London: McGraw Hill.
Dasmann, F. R. 1965. *The Destruction of California.* New York: Macmillan.
David, P. A., H. G. Gutman, R. Sutch, T. Temin, and G. Wright. 1976. *Reckoning with Slavery.* New York: Oxford University Press.
Davidson, J. H., and K. C. Barrons. 1959. Chemical seedbed preparation—new approach to soil conservation. Down to Earth, Winter. Michigan: Dow Chem. Co.
Davis, H. 1868. Wheat in California. Overland Monthly 1:448.
Day, S. 1843. *Historical Collections of the State of Pennsylvania.* Philadelphia: G. W. Gorton.
DeBach, P., ed. 1964. *Biological Control of Fruit Pests and Weeds.* New York: Reinhold.
DeBach, P. 1974. *Biological Control by Natural Enemies.* London: Cambridge University Press.
DeCook, K. J., C. B. Cluff, and M. Asce. 1975. Trends in Arizona cropland irrigation. In *Contribution of Irrigation and Drainage to the World Food Supply.* Am. Soc. Civil. Eng., New York.
Deevy, E. S. 1960. The human population. Sci. Am. 203(3):194-204.
——— 1968. Pleistocene family planning. In *Man the Hunter.* Ed. R. B. Lee and L. DeVore, pp. 248-249. Chicago: Aldine Publishing Co.
Dendy, F. E., W. A. Champion, and R. B. Wilson. 1973. Reservoir resedimentation surveys in the United States. Geophys. Mgt. Ser. 17:349-357.
Dethier, V. G. 1976. *Man's Plague?: Insects and Agriculture.* Princeton, N.J.: Darwin Press.
DeVoto, B., ed. 1953. *The Journals of Lewis and Clark.* Boston: Houghton Mifflin.
Dicke, F. F. 1977. The most important corn insects. In *Corn and Corn Improvement.* Ed. G. F. Sprague, pp. 501-590. Madison, Wisc.: Amer. Soc. Agron.
Dimbleby, G. W. 1967. *Plants and Archeology.* London: Baker.
Dobie, J. F. 1941. *The Longhorns.* Boston: Little, Brown.
Dodge, R. I. 1877. *The Hunting Grounds of the Great West.* London: Chatto and Windus.
Donahue, R. L. 1970. *Our Soils and Their Management.* Danville, Ill.: Interstate.
Drache, H. M. 1976. Midwest agriculture: changing with technology. Agr. Hist. 50:290-302.
——— 1977. Thomas D. Campbell—plower of the plains. Agr. Hist. 51:78-91.
DuBois, W. E. B. 1866. *Black Reconstruction in America.* London: Case.
Dubos, R. J. 1972*a*. *A God Within.* New York: Scribner.
——— 1972*b*. 'Replenish the earth and subdue it.' Human touch often improves the land. Smithsonian 3(9):18-28.
——— 1973. Humanizing the earth. Science 179:769-772.
Dumond, D. E. 1975. The limitation of human population: a natural history. Science 187:713-721.
Dunaway, W. F. 1935. *A History of Pennsylvania.* New York: Prentice-Hall.
Dunbar, J. B. 1880. The Pawnee Indians: their history and ethnology. Mag. Am. Hist. 4:241-281.
Dunbar, R. G. 1977. The adaptation of groundwater-control institutions to the arid west. Agr. Hist. 51:662-680.
Durost, D. D. 1969. Where food originates: the farmer and his farm. USDA Yearbook, pp. 8-14.
Easly, J. D. 1977. Industry scientists and objectivity. Nat. Agr. Chem. Assn. Newsletter, Feb. 23.
Easton, C. 1976. A touch of innocence. Westways 68(12):27-29, 60.
Eckholm, E. P. 1976. Losing ground. Environment 18(3):6-11.
——— 1977. *Losing Ground: Environmental Stress and World Food Prospects.* New York: Norton.
Eichers, T., P. Andrilenas, H. Blake, R. Jenkins, and A. Fox. 1970. Quantities of pesticides used by farmers in 1966. U.S. Dept. Agr., Econ. Res. Serv., Agr. Econ. Rept. 179.

Einsett, J., K. Kimball and J. Watson. 1973. Grape varieties for New York State. New York State Educ. Agr. and Life Sci. Inf. Bull. 66.

Eliot, W. A. 1948. *Forest Trees of the Pacific Coast.* Revised. New York: Putnam.

Elliott, L. 1966. *George Washington Carver: The Man Who Overcame.* Englewood Cliffs, N.J.: Prentice-Hall.

——— 1976. The legend that was Daniel Boone. Reader's Digest 108(647):186-219.

Ellis, D. M. 1946. *Landlords and Farmers in the Hudson-Mohawk Region.* Ithaca, N.Y.: Cornell University Press.

Ellis, F., and S. Evans. 1883. *History of Lancaster County, Pennsylvania.* Philadelphia: Everts and Peck.

Ellis, M. 1978. There is a lot of bull in beefalo. *San Francisco Chronicle,* January 11.

Ellis, W. N. 1977. A. T.: the quiet revolution. Bull. Atom. Sci. 33(9)24-35.

Elton, C. A. 1815. *Hesiodus.* 2d ed. London: Baldwin, Craddock and Joy.

Emerson, A. I., and C. M. Weed. 1936. *Our Trees—How to Know Them.* Philadelphia: Lippincott.

Englund, E. 1938. Farmers. In *Swedes in America 1638-1938.* Ed. A. B. Benson and N. Hedin, pp. 75-91. New Haven, Conn.: Yale University Press.

Epstein, E. 1977. The role of roots in the chemical economy of life on earth. BioScience 27:783-787.

ESCOP. 1962. *State Agricultural Experiment Stations: A History of Research Policy and Procedure.* USDA Misc. Bull. 904.

Ezell, J. S. 1963. *South Since 1865.* New York: Macmillan.

Farnham, T. J. 1846. *Life and Adventure in California.* New York: W. H. Graham.

Federal Register. 1977. Recycled animal waste. Federal Register 42(248):64662-64675.

Faulk, O. B. 1970. *Arizona: A Short History.* Norman: University of Oklahoma Press.

Faust, A. B. 1909. *The German Element in the United States.* Volume 1. Boston: Houghton Mifflin.

Faux, G. 1975. New England. In *The People's Land.* Ed. P. Barnes, pp. 52-55. Emmaus, Pa.: Rodale Press.

Fehrenbach, T. R. 1974. *Comanches: The Destruction of a People.* New York: Knopf.

Fellmath, R. C. 1973. *Ralph Nader's Study Group Report on Land Use in California: Politics of Land.* New York: Grossman Publishers.

Ferber, A. E. 1969. Windbreaks for conservation. USDA Soil Conserv. Serv., Agr. Inf. Bull. 339.

Fergurson, E. B. 1975. The subsidy lingers on. *The Los Angeles Times,* Oct. 17.

Fergusson, E. 1964. *New Mexico, a Pageant of Three Peoples.* 2d ed. New York: Knopf.

Fernow, B. E. 1896. The relation of forests to farms. USDA Yearbook, 1895, pp. 333-340.

Fetterman, J. 1967. *Stinking Creek.* New York: Dutton.

Fey, H. E. and D. McNickle. 1959. *Indians and Other Americans: Two Ways of Life Meet.* New York: Harper and Row.

Flake, O. D. undated. *William J. Flake: Pioneer-Colonizer.* Privately printed.

Flake, S. E. 1970. *James Madison Flake: Pioneer, Leader, Missionary.* Bountiful, Utah: Wasatch Press.

Flannery, K. V. 1968. Archeological systems theory and early Mesoamerica. In *Anthropology and Archaeology in the Americas.* Ed. B. J. Meggers, pp. 67-86. Anthrop. Soc. Wash., Washington, D.C.

Fleming, T. 1975. The "real" Uncle Tom. Reader's Digest 106(635):124-128.

Fletcher, A. C., and F. LaFlesche. 1970. *The Omaha Tribe.* (Reprint of 1911 ed.) New York: Johnson Reprint.

Fletcher, S. W. 1950. *Pennsylvania Agriculture and Country Life, 1640-1840.* Harrisburg: Penn. Hist. and Mus. Comm.

Flynn, P. 1978. U.S. agribusiness is devouring the Third World. *Los Angeles Times,* July 16.

Fogel, R. W. and S. L. Engerman. 1974. *Time on the Cross.* Boston: Little, Brown.

Forsythe, J. L. 1977. Environmental considerations in the settlement of Ellis County, Kansas. Agr. Hist. 51:38-50.

Frame, K. 1692. A short description of Pennsylvania. In *Narratives of Early Pennsylvania, West New Jersey, and Delaware 1630-1707.* Ed. A. C. Meyers. New York: Scribner's Sons.

Franklin, T. B. 1948. *A History of Agriculture.* London: G. Bell.

Fridley, R. B., and P. A. Adrian. 1966. Mechanical harvesting equipment for deciduous tree fruits. Calif. Agr. Exp. Sta. Bull. 825.

Fridley, R. B., P. A. Adrian, and C. Lorenzen. 1966. Harvesting perishable crops. Sci. J., Aug.
Friedman, I. 1977. The Amazon Basin, another Sahel? Science 197:7.
Friggens, P. 1975. Triticale: world's first man-made crop. Reader's Digest, Dec., pp. 33-36.
Frink, M., W. T. Jackson, and A. W. Spring. 1956. *When Grass Was King.* Boulder: University of Colorado Press.
Fussell, G. E. 1976. *Farms, Farmers and Society.* Lawrence, Kansas: Coronado Press.
Gage, C. E. 1942. American tobacco types, uses, and markets. USDA Circ. 249.
Galinat, W. C. 1965. Evolution of corn and culture in North America. Econ. Botany 19:350-357.
——— 1971. The origin of maize. Ann. Rev. Genet. 5:447-478.
——— 1977. The origin of corn. In *Corn and Corn Improvement.* Ed. G. F. Sprague, pp. 1-47. Madison, Wisc.: Amer. Soc. Agron.
Gall, M. 1968. The insect destroyer-portrait of a scientist. USDA Yearbook, pp. 54-57.
Gard, W. 1954. *The Chisholm Trail.* Norman: University of Oklahoma Press.
——— 1959. *The Great Buffalo Hunt.* New York: Knopf.
Garner, W. W. 1946. *The Production of Tobacco.* Philadelphia: Blakiston.
Gaston, L. K., R. S. Kaae, H. H. Shorey, and D. Sellers. 1977. Controlling the pink bollworm by disrupting sex pheromone communication between adult moths. Science 196:904-905.
Gates, P. W. 1961. Charles Lewis Fleischmann: German-American agricultural authority. Agr. Hist. 35:13-23.
——— 1967. *California Ranchos and Farms, 1846-1862.* State Hist. Soc. Madison, Wis.
——— 1975*a*. The homestead law in an incongruous land system. In *The People's Land.* Ed. P. Barnes, pp. 7-11. Emmaus, Pa.: Rodale Press.
——— 1975*b*. Public land disposal in California. Agr. Hist. 49:158-181.
——— 1977. Homesteading on the High Plains. Agr. Hist. 51:109-133.
Gendlin, F. 1978. A talk with Huey Johnson. Sierra Club Bull. 63(5):21-25.
George, C. J., and D. McKinley. 1974. *Urban Ecology: In Search for an Asphalt Rose.* New York: McGraw-Hill.
George, E. J. 1966. *Shelterbelts for the Northern Great Plains.* USDA Farmers' Bull. 2109.
Georgescu-Roegen. 1971. *The Entropy Law and the Economic Process.* Cambridge: Harvard University Press.
Giese, R. L., R. M. Peart, and R. T. Huber. 1975. Pest management. Science 187:1045-1052.
Giles, D. 1940. *Singing Valleys—The Story of Corn.* New York: Random House.
Gill, H. B., Jr. 1978. Wheat culture in colonial Virginia. Agr. Hist. 52: 380-393.
Gillette, R. 1977. Cancer—industrial pollution study now called erroneous. *Los Angeles Times,* Nov. 6.
Gittins, B. S. 1950. *Land of Plenty.* Farm Equip. Inst., Chicago.
Gleason, H. A., and A. Cronquist. 1964. *The Natural Geography of Plants.* New York: Columbia University Press.
Glendening, G. E. 1952. Some quantitative data on the increase of mesquite and cactus in desert grassland range in southern Arizona. Ecology 33:319-328.
Glendening, G. E., and H. A. Paulsen, Jr. 1955. Reproduction and establishment of velvet mesquite as related to invasion of semidesert grasslands. USDA Tech. Bull. 1127.
Glymph, L. M., and H. C. Storey. 1967. Sediment—its consequences and control. In *Agriculture and the Quality of our Environment.* Ed. N.C. Brady, pp. 205-220. Amer. Assoc. Adv. Sci., Washington, D.C.
Godfrey, H. A., Jr., ed. 1973. Land treatment of municipal sewage. In *Civil Engineering.* New York: Amer. Soc. Civil Engin., Sept.
Goldschmidt, W. R. 1946. Small business and the community. A study in central valley of California on effects of scale of farm operations. In *Hearings Before the Subcommittee on Monopoly of the Select Committee on Small Business, United States Senate,* pp. 4465-4616. Washington, D.C.: U.S. Govt. Print. Off.
——— 1947. *As You Sow.* New York: Harcourt, Brace.
——— 1975. A tale of two towns. In *Food for People, Not for Profit.* Ed. C. Lerza and M. Jackson, pp. 70-73. New York: Ballantine Books.
——— 1978. *As You Sow: Three Studies in the Social Consequences of Agribusiness.* Montclair, N.J.: Allanhold, Osmun.
Goldsmith, E. 1974. The ecology of unemployment. Ecologist 4:64-67.
Goss, J. R., R. Bainer, and P. G. Curley. 1955. Field tests of combines in corn. Agri. Eng. 36: 794-796.
GPC. 1936. *The Picture of the Great Plains.* U.S. Great Plains Committee, Washington, D.C.

Graham, A. 1973. *The Gardeners of Eden.* London: George Allen and Unwin.
Graham, F. D. 1944. Ethnic and other factors in the American economic ethic. In *Foreign Influences in American Life.* Ed. D. F. Bowers, pp. 67-83. Princeton: Princeton University Press.
Grant, K. E. 1975. Erosion in 1973-74: the record and the challenge. J. Soil and Water Conserv. 30(1):29-32.
Gray, L. C. 1935. *History of Agriculture in the Southern United States to 1860.* 2 volumes. Carnegie Instit. of Washington, Publ. no. 430.
Gray, W. H. 1975. Agricultural land use in Washington: conversion or preservation. Washington State University, Col. Agr., E. M. 3935.
Green, L. 1978. Illinois farmer races time and weather. *Los Angeles Times,* May 15.
Gregory, W. W., and G. J. Musick. 1976. Insect management in reduced tillage systems. Bull. Entomol. Soc. of Am. 22:302-304.
Griggs, W. H., and B. T. Iwakiri. 1974. "California," a new market year. Calif. Agr. 28(12):8-9.
Grivetti, L. E. 1978. Culture, diet, and nutrition: related themes and topics. BioScience 28:171-177.
Guilford, M. C. 1976. Crop with a future. Agr. Research 25(4):10-11.
Gustafson, A. F. 1937. *Conservation of the Soil.* New York: McGraw-Hill.
Gustafson, R. A., and R. N. Van Arsdall. 1970. Cattle feeding in the United States. USDA Agr. Econ. Rpt. no. 186.
Hagan, W. T. 1961. *American Indians.* Chicago: University of Chicago Press.
Hagood, M. A. 1972. *Center pivot sprinkler systems.* Washington State University, Col. Agr., E. M. 3550.
Hall, E. N. 1977. The ocean resource challenge. Mimeographed. Gardena, Calif.: Chromalloy American Corp.
Hall, J. L., J. R. Dunbar, and M. Ronning. 1975. Cubed alfalfa hay for livestock. Univ. Calif. Agr. Exp. Sta. Bull. 1874.
Halpern, D. 1975. Oklahoma's granite parkland. Sierra Club Bull. 60(9):4-7.
Hamilton, D. 1974. *In No Time at All.* Ames: Iowa State University Press.
Hammond, A. L. and W. D. Metz. 1977. Solar energy research: making solar after the nuclear model? Science 197:241-244.
——— 1978. Capturing sunlight: a revolution in collector design. Science 201:36-39.
Hammond, G. P., and A. Rey. 1940. *Narratives of the Coronado Expedition, 1540-1542.* Albuquerque: University of New Mexico Press.
Handler, P. 1975. On the state of man. BioScience 25:425-432.
Harazthy, A. 1862. *Grape Culture, Wines, and Wine-making. With Notes upon Agriculture and Horticulture.* New York: Harper.
Hardin, G. 1968. The tragedy of the commons. Science 162:1243-1248.
Harding, T. S. 1947. *Two Blades of Grass.* Norman: University of Oklahoma Press.
Hariot, T. 1588. A brief and true report of the new found land of Virginia. London.
Harlan, J. R. 1967. A wild wheat harvest in Turkey. Archeology 20:197-201.
——— 1971. Agricultural origins: centers and noncenters. Science 174:468-474.
——— 1975. *Crops and Man.* Amer. Soc. Agron., Madison, Wis.
——— 1976. The plants and animals that nourish man. Sci. Am. 235(3):89-97.
Harmon, B. W., J. P. Fontenot, and K. E. Webb, Jr. 1975. Ensiled broiler litter and corn forage. I. Fermentation characteristics. J. Animal Sci. 40:144-155.
Harris, D. R. 1972. The origins of agriculture in the tropics. Am. Scientist 60:180-193.
Harris, M. 1977. *Cannibals and Kings.* New York: Random House.
Hart, G. H. 1946. Wealth pyramiding in the production of livestock. In *California Agriculture.* Ed., C. B. Hutchison, pp. 51-112. Berkeley: University of California Press.
Hartmann, H. T., and K. W. Opitz. 1977. Olive production in California. University of California Div. Agr. Sci., Leaflet 2474 (rev.).
Harvey, S. 1975. Shelling out for surplus of peanuts. *Los Angeles Times,* July 3.
Haury, E. W. 1976. *The Hohokam: Desert Farmers and Craftsmen.* Tucson: University of Arizona Press.
Hawkes, J. G. 1970. The origin of agriculture. Econ. Botany 24:131-133.
Hawthorne, N. 1887. *The American Notebooks.* Ed. R. Stewart, 1932, based upon the original manuscript. New Haven, Conn.: Yale University Press.
Hayes, D. 1975. Solar power in the Middle East. Science 188:1261.
——— 1977. *Rays of Hope: The Transition to a Post-Petroleum World.* New York: Norton.
Heady, E. O. 1976. The agriculture of the U.S. Sci. Am. 235(3):107-127.

Hedrick, A. S. 1933. *A History of Agriculture in the State of New York.* New York Agr. Soc., Albany, N.Y.
Heichel, G. H. 1976. Agricultural production and energy resources. Am. Scientist. 64:64-72.
Heiser, C. B., Jr. 1973. *Seed to Civilization. The Story of Man's Food.* San Francisco: W. H. Freeman.
Heizer, R. F. and M. A. Whipple, eds. 1971. *The California Indians.* Berkeley, Los Angeles, London: University of California Press.
Heller, A. 1972. *California Tomorrow Plan.* Los Altos, Calif.: William Kaufmann.
Helper, H. R. 1857. *The Impending Crisis in the South: How to Meet It.* New York: Burdick Bros.
Higbee, E. 1957. *The American Oasis. The Land and Its Uses.* New York: Knopf.
——— 1958. *American Agriculture.* New York: John Wiley.
Hightower, J. 1973. The case for the family farmer. Washington Monthly, Sept.
Hilliard, S. B. 1972. The dynamics of power: recent trends in mechanization on the American farm. Technol. and Culture 13:1-24.
Hillinger, C. 1974. Sunflower gains in crop status. *Los Angeles Times,* Oct. 25.
Hillman, D. 1964. Urea corn silage for dairy cattle. Michigan State University, Coop. Ext. Serv. # 446.
Hodgson, H. 1974. We don't need to eliminate beef cattle. Crops and Soils Magazine, Nov., pp. 9-11.
Hodgson, H. J. 1976. Forage crops. Sci. Am. 234(2):61-68, 74-75.
Hoffman, M. B. 1970. Making a better apple, or puzzles and pomology. USDA Yearbook, pp. 47-52.
Hoglund, C. R. 1975. The U.S. dairy industry today and tomorrow. Mich. Agr. Exp. Sta. Res. Rpt. 275.
Holbrook, S. H. 1950. *The Yankee Exodus: An Account of Migration from New England.* New York: Macmillan.
Hooker, A. L. 1972. Southern leaf blight of corn—present status and future prospects. J. Environ. Quality 1:244-249.
Hostetler, J. A. 1974. *Hutterite Society.* Baltimore: Johns Hopkins Press.
Houghton, E. P. D. 1911. *The Expedition of the Donner Party.* Chicago: McClurg.
Hubert, T. A. 1626. *A Relation of Some Years of Travaille.* London.
Hughes, H. D., and D. S. Metcalfe. 1972. *Crop Production.* New York: Macmillan.
Hughes, L. 1954. *Famous American Negroes.* New York: Dodd Mead.
Hull, J., R. L. Andersen, G. M. Kessler, and R. F. Carlson, 1975. Tree fruit varieties for Michigan. Michigan State University, Coop. Ext. Serv., Ext. Bull. E-881.
Hunt, R. D. 1942. *John Bidwell: Prince of California Pioneers.* Caldwell, Idaho: Caxton Printers.
Hunt, T. H., and R. V. H. Adams. 1974. *Ghost Trails to California.* Palo Alto, Calif.: American West.
Hunter, M. R. 1940. *Brigham Young the Colonizer.* Salt Lake City: Deseret News Press.
Inglis, D. R. 1975. Wind power now! Bull. Atom. Sci. 31(8):20-26.
Iwata, M. 1962. The Japanese immigrants in California agriculture. Agr. Hist. 36:25-37.
Jablow, J. 1950. *The Cheyenne in Plains Indian Trade Relations, 1795-1840.* Locust Valley, N.Y.: Augustin.
Jacks, G. V. 1962. Man: the fertility maker. J. Soil and Water Conserv. 17:147-148, 176.
Jackson, W. T. and A. M. Paterson. 1977. The Sacramento-San Joaquin delta: The evolution and implementation of water policy. Calif. Water Resources Center, Davis, Calif., Contribution no. 163.
Jacobs, J. L. 1975. California's pioneer wine families. Calif. Hist. Soc. Quart. 54:139-174.
James, M., and B. R. James. 1954. *Biography of a Bank.* New York: Harper.
Jensen, R. 1975. The vets rave over beef and milk, and the bacon. USDA Yearbook, pp. 75-84.
Jepson, W. L. 1925. *A Manual of Flowering Plants of California.* University of California, Berkeley: Student Store.
Johnson, A. 1972. *Heartbeats of Colonia Diaz.* Salt Lake City: Publishers Press.
Johnson, B. L., and J. G. Waines. 1977. Use of wild-wheat resources. Calif. Agr. 31(9):8-9.
Johnston, W. E., and G. W. Dean. 1969. California crop trends: yields, acreages, and production areas. Calif. Agr. Exp. Sta. Ext. Serv. Circ. 551.
Jones, E. L. 1974. Creative disruptions in American agriculture 1620-1820. Agr. Hist. 48:510-528.
Jones, R. A. 1977. Farm water issue heats up. *Los Angeles Times,* May 2.

Kalm, P. 1770. *The American of 1750.* Rev. 1937. New York: Wilson-Erickson.
Kaniuoka, R. P. 1976. Unturned furrows. Agr. Research 24(8):2.
Katz, S. H. 1973. Genetic adaptation in twentieth-century man. In *Methods and Theories of Anthropological Genetics.* Ed. M. H. Crawford and P. L. Workman, pp. 403-422. Albuquerque: University of New Mexico Press.
Katz, S. H., M. L. Hediger, and L. A. Valleroy. 1974. Traditional maize processing techniques in the New World. Science 184:765-773.
Kawahami, K. K. 1921. *The Real Japanese Question.* New York: Macmillan.
Kelly, C. F. 1967. Mechanical harvesting. Sci. Am. 217(2):50-59.
Kellogg, C. E., and A. C. Orvedal. 1969. Potentially arable soils of the world and critical measures of their use. Advances in Agronomy 21:109-170.
Kellogg, W. W. 1978. Is mankind warming the earth? Bull. Atom. Sci. 34(2):10-19.
Kendall, J. 1977. Influx of illegal aliens termed 'out of control.' *Los Angeles Times,* Jan. 9.
Kendrick, J. B., Jr. 1977. Social impact of agricultural research. Calif. Agr. 31(6):2.
——— 1978. Resources could change California's agricultural future. Calif. Agr. 32(3):2.
Kepner, R. A. 1951. Effectiveness of orchard heaters. Calif. Agr. Exp. Sta. Bull. 723.
Kickingbird, K., and K. Ducheneaux. 1975. Indian lands. In *The People's Land.* Ed. P. Barnes, pp. 56-61. Emmaus, Pa.: Rodale Press.
Kik, M. C., and R. R. Williams. 1945. The nutritional improvement of white rice. Nat. Res. Council Bull. 112.
Kilgore, W. W., and R. L. Doutt. 1967. *Pest Control.* New York: Academic Press.
Klose, N. 1950. *America's Crop Heritage.* Ames: Iowa State College Press.
Knipling, E. F. 1960*a.* Use of insects for their own destruction. J. Econ. Entomol. 53:415-420.
——— 1960*b.* The eradication of the screw-worm fly. Sci. Am. 203(4):54-61.
Knoblauch, H. C., E. M. Law, and W. P. Meyer, 1962. *State Agricultural Experimental Stations: A History of Research and Procedure.* USDA Misc. Publ. no. 904.
Kraenzel, C. F. 1955. *The Great Plains in Transition.* Norman: University of Oklahoma Press.
Kramer, A. 1973. *Food and the Consumer.* Westport, Conn.: Avi Publishing.
Kroeber, A. L. 1925. *Handbook of the Indians of California.* Smiths. Inst. Bur. Am. Ethnol., Bull. 78.
Kruse, E. G. 1975. Irrigation research to increase production without environmental damage. In *Contribution of Irrigation and Drainage to the World Food Supply.* New York: Center for Civil Eng.
Kyle, L. R., W. B. Sunqerist, and H. D. Guither. 1975. Who controls agriculture now?—the trends underway. In *Who Will Control U.S. Agriculture?* North Central Regional Publ. 32, pp. 3-12.
Lamb, D. 1975. U.S. digs at Maine root of spud crop. *Los Angeles Times,* Oct. 21.
Lappé, F. M., and J. Collins. 1977. *Food First. Beyond the Myth of Scarcity.* Boston: Houghton Mifflin Co.
Larsen, R. P. 1975. The quiet revolution in the apple orchard. USDA Yearbook, pp. 158-168.
Larson, J. S. 1972. Man and wildlife in the modern northeastern landscape. Agr. Sci. Rev. 10(1): 1-6.
Larson, O. F., and T. B. Jones. 1976. The unpublished data from Roosevelt's commission on country life. Agr. Hist. 50:583-599.
Lee, R. B. 1968. What hunters do for a living, or how to make out on scarce resources. In *Man the Hunter.* Ed. R. B. Lee and I. DeVore, pp. 30-48. Chicago: Aldine.
——— 1969. !Kung Bushman subsistence: an input-output analysis. In *Ecological Studies in Cultural Anthropology.* Ed. A. P. Vayda, pp. 47-49. New York: Natural History Press.
——— 1972. Work effort, group structure and land use in contemporary hunter-gatherers. In *Man, Settlement and Urbanism.* Ed. P. J. Ucko, R. Tringham, and G. W. Dimbleby, pp. 177-185. London: Duckworth.
Lelong, B. M., 1902. *Culture of the Citrus in California.* Revised by the State Board of Horticulture. Sacramento: Supt. State Printing.
Lent, H. B. 1968. *Agriculture U.S.A.* New York: Dutton.
Leonard, R. O., M. E. Stanley, and D. L. Bath. 1978. Unusual feedstocks in livestock rations. Univ. Calif. Agr. Ext. Leaflet 21014.
Lessiter, F. 1975. Another million acres of no-till coming. No-till Farmer 3(1):4-5.
Lewis, W. M. 1976. Principles of field crop production with reduced tillage systems. Bull. Entomol. Soc. Am. 22:291-293.
Li, H. L. 1970. The origin of cultivated plants in southeast Asia. Econ. Botany 24:3-19.

Libby, W. F., and E. F. Black. 1978. Food irradiation: an unused weapon against hunger. Bull. Atom. Sci. 34(2):50-55.
Lilienthal, D. E. 1977. Lost megawatts flow over nation's myriad spillways. Smithsonian 8(6): 83-89.
Little, J. A., ed. 1909. *Autobiography of Jacob Hamblin*. Salt Lake City: The Deseret News.
Logan, R. F. 1961. Post-Columbian developments in the arid regions of the United States of America. In *A History of Land Use in Arid Regions*. Ed. L. D. Stamp, pp. 277-297. Paris: UNESCO.
Long, A., Jr. 1972. *The Pennsylvania German Family Farm*. The Pennsylvania German Society, Breinigsville, Pa.
Looper, J. D. 1970. Who should pay for conservation: USDA Yearbook, pp. 236-242.
Lord, E., J. J. D. Trenor, and S. J. Barrows. 1906. *The Italians in America*. New York: B. F. Buck.
Lord, R. 1962. *The Care of the Earth. A History of American Husbandry*. New York: Thomas Nelson.
Lorenz, A. J. 1949. *Centennial of the California Lemon*. Los Angeles.
Lorenzen, C. 1969. The mechanical growing of vegetable crops in the West. Horticultural Science 4:238-239.
Lorenzen, C., and G. C. Hanna. 1962. Mechanical harvesting of tomatoes. Agr. Eng. 43:16-19.
Los Angeles Times. 1976. Buffalo back on the landscape. May 30.
Love, R. M. 1970. The rangelands of the western U.S. Sci. Am. 222(2):89-96.
Lowie, R. H. 1940. *An introduction to Cultural Anthropology*, rev. ed. New York: Farrar and Rinehart.
Luckmann, W., and R. L. Metcalf. 1975. The pest management concept. In *Introduction to Insect Pest Management*. Ed. R. L. Metcalf and W. Luckmann, pp. 3-35. New York: Wiley.
Luginbill, P., Jr. 1969. *Developing Resistant Plants—the Ideal Method of Controlling Insects*. USDA, ARS Prod. Res. Rep. no. 111.
MacArthur, M. Y. 1959. Anaheim, "The Mother Colony." Los Angeles: Ward Ritchie Press.
McCaull, J. 1976. Storing the sun. Environment 18(5):9-15.
McClintock, J. H. 1921. *Mormon Settlement in Arizona*. Copyright, J. H. McClintock, Phoenix, Ariz. Manufacturing Stationers Inc.
McCormick, C. 1931. *The Century of the Reaper*. Boston: Houghton Mifflin.
McCracken, R. D. 1971. Lactase deficiency: an example of dietary evolution. Current Anthropology 12:497-517.
MacCurdy, R. M. 1925. *The History of the California Fruit Growers Exchange*. Los Angeles: G. Rice and Sons.
McDonald, A. 1975. The family farm is the most efficient unit of production. In *The People's Land*. Ed. P. Barnes, pp. 86-88. Emmaus, Pa: Rodale Press.
McDowell, L. L., and E. H. Grissinger. 1976. Erosion and water quality. Twenty-Third National Watershed Congress, Biloxi, Miss. pp. 40-56.
McGlashan, C. F. 1940. *History of the Donner Party*. 4th ed. (First edition published in 1881.) Stanford, Calif.: Stanford University Press.
McGregor, R. A., M. Cain, and J. C. Seibert. 1977*a*. *California Fruit and Nut Acreage*. Calif. Crop and Livestock Reporting Serv., Sacramento.
McGregor, R. A., M. Cain, R. Albert, N. Ciancio, R. Peak, and J. Seibert. 1977. *California Fruit and Nut Statistics, 1975-76*. Calif. Crop and Livestock, Reporting Serv., Sacramento.
McGuinness, F. 1974. The witless, wonderful turkey. Reader's Digest, Nov., pp. 185-188.
McHugh, T. 1972. *The Time of the Plains*. New York: Knopf.
Mackintosh, B. 1977. George Washington Carver and the peanut. Am. Heritage 28(5):66-73.
McKnight, W. J. 1905. *A Pioneer Outline History of Northwestern Pennsylvania*. Philadelphia: Lippincott.
McKorkle, C. O. 1977. Summary comments on the Asilomar Conference, Oct. 2-3. Mimeograph.
McLeod, D. 1975. It was the farmer who built the United States. *Los Angeles Times*, Dec. 19.
McMillen, W. 1966. *The Farmer*. New York: Potomac Books.
——— 1974. *Ohio Farm*. Columbus: Ohio State University Press.
McNeil, R. J. 1974. Deer in New York State. New York State Col. Agr. and Life Sci. Ext. Bull. 1189.
MacNeish, R. S. 1964. The origins of New World civilization. Sci. Am. 211(5):29-37.
McWilliams, C. 1946. *Southern California Country*. New York: Duell, Sloan and Pearce.

Magness, J. R. 1957. Pear growing in the Pacific Coast States. USDA Farmers' Bull. no. 1739.
Mahoney, T. 1960. *The Story of George Romney.* New York: Harper.
Malthus, T. R. 1806. *An Essay on the Principle of Population.* 3rd ed. London: T. Bently.
Manabe, S., and R. T. Wetherald. 1975. The effects of doubling the $CO_2$ concentration on climate of a general circulation model. J. Atom. Sci. 32:3-15.
Mangelsdorf, P. C. 1950. The mystery of corn. Sci. Am. 183(1):20-24.
Mangelsdorf, P. C., R. S. MacNeish, and W. C. Galinat. 1964. Domestication of corn. Science 143:538-545.
Mann, E. G., and F. E. Harvey. 1955. *New Mexico, Land of Enchantment.* East Lansing: Michigan State University Press.
Markham, G. 1620. *Farewell to Husbandry.* London: I. B. Jackson.
Mason, J. S. 1964. *The Ancient Civilizations of Peru.* Baltimore: Penguin.
Matson, F. W. 1953. *The Anti-Japanese Movement in California, 1890-1942.* Master's Thesis. University of California, Berkeley.
Maugh, T. H. 1978. The fatter calf (II): The concrete truth about beef. Science 199:413.
Maxwell, J. C. 1965. Will there be enough water? Am. Scientist 53:97-103.
Maxwell, F. G., N. N. Jenkins, and W. L. Parrott. 1972. Resistance of plants to insects. Advances in Agronomy 24:187-265.
Medsger, O. P. 1939. *Edible Wild Plants.* New York: Macmillan.
Meeker, E. 1912. *Personal Experiences on the Oregon Trail Sixty Years Ago.* St. Louis: McAdoo Printing.
Meinig, O. W. 1968. *The Great Columbia Plain. A Historical Geography, 1805-1910.* Seattle: University of Washington Press.
Melham, T., and F. Grehan. 1976. *John Muir's Wild America.* Nat. Geog. Soc., Washington, D.C.
Memolo, M. M. 1978. We did it! Agr. Research 26(9):2.
Mesarovic, M., and E. Pestel. 1974. *Mankind at the Turning Point. The Second Report to the Club of Rome.* New York: Dutton.
Metcalf, R. L., and W. Luckmann. 1975. *Introduction to Insect Pest Management.* New York: Wiley.
Metz, W. D. 1977. Wind energy: large and small systems competing. Science 197:971-973.
Meyer, L. L. 1975. *Shadow of a Continent.* Palo Alto, Calif.: American West.
Meyers, R. M. 1977. Population problems. BioScience 27:4-5.
Michener, J. A. 1974. *Centennial.* New York: Random House.
Middleton, J. T. 1973. Air pollution. Encyclopedia Americana 1:385-393.
Miller, M. F. 1902. *The Evolution of Reaping Machines.* USDA Bull. 103.
Mingay, G. E. 1963. The "agricultural revolution" in English history: a reconsideration. Agr. Hist. 37(3):123-133.
Mitchell, J. M., Jr. 1977. Carbon dioxide and future climate. EDS, March, 1977, pp. 3-9.
Mitre Corp. 1975. *Wind Machines.* Mitre Corporation of Westgate Research Park, McLean, Va.
Modell, J. 1970. Japanese-Americans: some costs of group achievement. In *Ethnic Conflict in California History.* Ed. C. Wollenberg, pp. 101-119. Los Angeles: Tinnon-Brown.
Moloney, F. X. 1967. *The Fur Trade in New England, 1610-1676.* Hamden, Conn.: Archon Books.
Morgan, D. 1975. Family farm faces threat in Nebraska. *Washington Post,* April 13.
Morgan, E. S. 1975. *American Slavery, American Freedom: The Ordeal of Colonial Virginia.* New York: Norton.
Morgan, M. 1955. *The Last Wilderness.* New York: Viking Press.
Morrison, B. Y. 1945. American plants for the Americas. In *New Crops for the New World.* Ed. C. M. Wilson, pp. 261-271. New York: Macmillan.
Moss, F. E. 1967. *The Water Crisis.* New York: Praeger.
Muir, J. 1911. *My First Summer in the Sierra.* Boston: Houghton Mifflin.
Mustachi, P. 1971. Cesare Bressa (1785-1836) on dirt-eating in Louisiana: a critical analysis of his unpublished manuscript *De la Dissolution Scorbutique.* J. Am. Med. Assoc. 218:229-232.
Nabors, M. W. 1976. Using spontaneously occurring and induced mutations to obtain agriculturally useful plants. BioScience 26:761-768.
NACRP. 1967. *The People Left Behind.* President's National Advisory Commission on Rural Poverty, Washington, D.C.
Nalbandov, A. V. 1974. Egg laying. In *Animal Agriculture.* Ed. H. H. Cole and M. Ronning, pp. 394-408. San Francisco: W. H. Freeman.

NAS. 1973. *More Water for Arid Lands. Promising Technologies and Research Opportunities.* Nat. Acad. Sci., Washington, D.C.
NCLRS. 1977. *Nevada Agricultural Statistics for 1976.* Nevada Crop and Livestock Reporting Serv., Reno.
Neill, E. D. 1669. *History of the Virginia Company of London with Letters to and from the First Colony Never Before Printed.* Albany, N.Y.: J. Munsell.
Nelson, B. 1975*a*. Drive launched to save Great Plains windbreak trees, called key resource. *Los Angeles Times,* Aug. 28.
——— 1975*b*. Hill college: each student has to have job at Berea. *Los Angeles Times,* Nov. 14.
Nelson, G. 1977. The tightening squeeze on the farmer. *Los Angeles Times,* Dec. 12.
Newell, F. H. 1896. Irrigation in the Great Plains. USDA Yearbook, pp. 167-196.
Nicholas, M. E. 1977. Another source of roughage. Agr. Research 25(6):7.
Nordskog, A. W. 1974. Breeding for eggs and poultry meats. In *Animal Agriculture.* Ed. H. H. Cole and M. Ronning, pp. 319-333. San Francisco: H. Freeman.
Norwine, J. 1977. A question of climate. Environment 19(8):6-13, 25-27.
NRC. 1974. *Productive Agriculture and a Quality Environment.* Report of the National Research Council Committee on Agriculture and Environment, Nat. Acad. Sci., Washington, D.C.
Nuttonson, M. Y. 1966. *The Physical Environment and Agriculture of Eastern Washington, Idaho, and Utah.* Amer. Inst. Crop. Ecol., Washington, D.C.
O'Callaghan, E. B. 1850. *Documentary History of New York.* Volume 8. Albany, N.Y.: Weed, Parsons and Co.
Ochse, J. J., M. J. Soule, Jr., M. J. Dijhman, and C. Wehlburg. 1961. *Tropical and Subtropical Agriculture.* Volume 2. New York: Macmillan.
Odle, J. 1977. Big blue gave him more beef per acre. Progressive Farmer 92(4):101.
O'Kane, W. C. 1950. *Sun in the Sky.* Norman: University of Oklahoma Press.
Oliphant, J. O. 1968. *On the Cattle Ranges of the Oregon Country.* Seattle: University of Washington Press.
——— 1946. The eastward movement of cattle from Oregon Country. Agr. Hist. 20:19-43.
Olivers, O. 1971. *Natural Reserve Conservation: An Ecological Approach.* New York: Macmillan.
Olmsted, F. L. 1856. A journey in the seaboard slave states. In *The Slave States.* Ed. H. Wish, pp. 40-125. New York: Putnam.
——— 1859. Journey through Texas. In *The Slave States.* Ed. H. Wish, pp. 128-160. New York: Putnam.
——— 1860. A Journey in the back country. In *The Slave States.* Ed. H. Wish, pp. 168-232. New York: Putnam.
——— 1861. The cotton kingdom. In *The Slave States.* Ed. H. Wish, pp. 234-284. New York: Putnam.
Opitz, K. 1975. The pistachio nut. University of California Div. Agr. Sci., Leaflet 2279.
Osborn, J. E. 1975. Agricultural economy and irrigation under declining water table conditions. In *Contribution of Irrigation and Drainage to the World Food Supply.* Amer. Soc. Civil. Eng., New York.
Paarlberg, D. 1976. Agriculture two hundred years from now. Agr. Hist. 50:303-309.
Paddock, W., and E. Paddock, 1973. *We Don't Know How.* Ames: Iowa State University Press.
Painter, R. H. 1958. Resistance of plants to insects. Ann. Rev. Entomol. 3:267-290.
——— 1968. Crops that resist insects provide a way to increase world food supply. Kansas Agr. Exp. Sta. Bull. 520.
Parkman, F., Jr. 1857. *The Oregon Trail.* 3rd ed. Columbus, Ohio: Miller.
Payen, W. 1970. The Negro land-grant colleges. Civil Rights Digest 3:12-17.
Pearce, J. 1978. Cigarettes: Is there a plot to keep you hooked? Family Health, June, pp. 20-23, 54.
Peck, A. M. 1972. *The March of Arizona History.* Tucson: Arizona Silhouettes.
Peffer, E. L. 1951. *The Closing of the Public Domain and Disposal and Reservation Policies, 1900-1950.* Stanford, Calif.: Stanford University Press.
Peirce, N. R. 1976. Angry Northeast wants a better break. *Los Angeles Times,* Dec. 5.
Pereira, H. C. 1973. *Land Use and Water Resources in Temperate and Tropical Climates.* Cambridge Univ. Press.
Perelman, M., and H. Gardner. 1975. Hidden dimensions of the energy crisis. In *The People's Land.* Ed. P. Barnes, pp. 122-124. Emmaus, Pa.: Rodale Press.
Peterson, D. F. 1975. Man and his water resources. In *Selected Works in Water Resources.* Ed. A. K. Biswas, pp. 1-39. Champaign, Ill.: Intern. Water Resources Assn.

Pfeiffer, J. E. 1977. *The Emergence of Society. A Prehistory of the Establishment.* New York: McGraw Hill.
Pillsbury, A. F., and H. F. Blaney. 1966. Salinity problems and management in river systems. J. Irrig. Drainage Div., Pro. Am. Soc. Civil. Eng. 92 (IR. 1):77-90.
Pimentel, D. 1973. Extent of pesticide use, food supply, and pollution. J. N. Y. Entomol. Soc. 81:13-33.
——— 1976. World food crisis: Energy and pests. Bull. Entomol. Soc. Am. 22:20-26.
Pimentel, D., W. Dritschilo, J. Krummel, and J. Kutzman. 1975. Energy and land constraints in food protein production. Science 190:754-761.
Pimentel, D., L. E. Hurd, A. C. Bellotti, M. J. Forester, I. N. Oka, O. D. Sholes, and R. J. Whitman. 1973. Food production and the energy crisis. Science 182:443-449.
Pimentel, D., E. C. Terhune, R. Dyson-Hudson, S. Rochereau, R. Samis, E. A. Smith, D. Denman, D. Reifschneider, and M. Shepard. 1976. Land degradation: effect on food and energy resources. Science 194:149-155.
Pinkerton, J. 1819. *Travels.* London: Longman.
PINM. 1972. *Facts From Our Environment.* Potash Institute of North America, Atlanta, Ga.
Pinthus, M. J. 1973. Lodging in wheat, barley, and oats: the phenomenon, its causes, and preventive measures. Advances in Agronomy 25:209-263.
Piper, C. V., and K. S. Bort. 1915. Early agricultural history of timothy. J. Am. Soc. Agron. 7:1-14.
Potter, B. G. 1976. The "dirty thirties" shelterbelt project. Am. Forests 82(1):36-38.
Pound, G. S. 1975. Plant disease toll is cut with resistant varieties. USDA Yearbook, pp. 66-74.
Powell, J. W. 1878. *Lands of the Arid Region of the United States.* Washington, D.C.: Govt. Print. Off. 2nd ed., 1879, Harvard University Press, 1962, Ed. W. Stegner.
Powers, J. B. 1948. The development of a new sugar beet harvester. Agr. Eng. 29:347-351, 354.
Prentice, E. P. 1939. *Hunger and History.* New York: Harper.
Price, R. 1976. History and importance of watersheds and their management in the Mormon West. In *Agriculture: Food and Man—Century of Progress.* Col. Biol. Agr. Sci., Brigham Young University, Provo, Utah.
Price, W. A. 1939. *Nutrition and Physical Deterioration: A Comparison of Primitive and Modern Diets and Their Effects.* New York: Harper.
Pritchard, A. 1976. *Lessons in aging.* NRTA Journal 27:8-10.
PSAC. 1967. *The World Food Problems.* A report of the President's Science Advisory Committee, Washington, D.C.
Putnam, P. A., and E. J. Warwick. 1975. Beef cattle breeds. USDA Farmers' Bull. No. 2228.
Quisenberry, K. S., and L. R. Reitz. 1974. Turkey wheat: the cornerstone of an empire. Agr. Hist. 48:98-114.
Raskin, E. 1971. *World Food.* New York: McGraw Hill.
Rasmussen, W. D. 1960. *Readings in the History of American Agriculture.* Urbana: University of Illinois Press.
——— 1967. Technological change in sugar beet production. Agr. Hist. 41:31-35.
——— 1968. Advances in American agriculture: The mechanical tomato harvester as a case study. Technol. and Culture 8:531-543.
——— 1975*a*. *Agriculture in the United States: A Documentary History.* 4 volumes. New York: Random House.
——— 1975*b*. Jefferson, Washington and other farmers. USDA Yearbook, pp. 15-22.
Raup, H. F. 1932. The German colonization of Anaheim, California. University of California Publ. Geog. 6:123-146.
Raup, P. M. 1973. Corporate farming in the United States. J. Econ. Hist. 33:274-290.
Reed, A. D. 1976. Facts about California agriculture. University of California Div. Agr. Sci. Leaflet 2290.
Reynolds, H. G., and S. C. Martin. 1968. Managing grass-shrub cattle ranges in the Southwest. USDA Agr. Handbook no. 162 (rev. 1968).
Reynolds, H. T., P. L. Adkisson, and R. F. Smith. 1975. Cotton insect pest management. In *Introduction to Insect Pest Management.* Ed. R. L. Metcalf and W. Luckmann, pp. 379-443. New York: Wiley.
Rhodes, V. J., and L. R. Kyle. 1973. A corporate agriculture. In *Who Will Control U.S. Agriculture?* University of Illinois Col. Agr. Coop. Ext. Serv. Special Publ. 28.
Rink, P. 1963. *A. P. Giannini.* Chicago: Encyclopaedia Brittanica Press.
Robinson, A. 1846. *Life in California.* New York: Putnam.

Roe, F. B. 1951. *The North American Buffalo*. Toronto: University of Toronto Press.
Rolle, A. F. 1963. *California: A History*. New York: Crowell.
——— 1968. *The Immigrant Upraised*. Norman: University of Oklahoma Press.
——— 1972. *The American Italians: Their History and Culture*. Belmont, Calif.: Wadsworth.
Rølvaag, O. E. 1927. *Giants in the Earth*. New York: Harper.
Romans, B. 1775. *A Concise Natural History of East and West Florida Containing an Account of the Natural Produce of all the Southern Part of British America, in the Three Kingdoms of Nature, Particularly the Animal and Vegetable*. Facs. reprod., Gainesville: University of Florida Press.
Rominger, R. E. 1975. Environmental concerns of four generations of farmers. In *Agriculture in the Development of the Far West*. Ed. J. H. Shideler, pp. 243-246. Agr. Hist. Soc., Washington, D.C.
Romney, T. C. 1938. *The Mormon Colonies in Mexico*. Salt Lake City: Deseret Book Co.
Rood, W. B. 1977. Colorado River salvation for southland—for now. *Los Angeles Times*, Mar. 21.
——— 1978. More water to be available in drought years. *Los Angeles Times*, Aug. 17.
Rood, W. B., and B. Stall, 1976*a*. Who'll get water? U.S. and state at odds. *Los Angeles Times*, Oct. 3.
——— 1976*b*. Sacramento delta becomes battleground in water war. *Los Angeles Times*, Oct. 4.
Roosevelt, T. 1889. *The Winning of the West*, Vol. 1. New York: Putnam.
Rosenberg, H. and A. N. Feldzamen. 1974. *The Doctor's Book on Vitamin Therapy*. New York: Putnam.
Rothstein, M. 1975. West coast farmers and the tyranny of distance: agriculture on the fringes of the world market. Agr. Hist. 49:272-280.
Roy, E. P. 1963. *Contract Farming*, U.S.A. Danville, Ill.: Interstate Printers and Publishers.
Rush, B. 1789. *The German Inhabitants of Pennsylvania*. Reprinted in Penn. German Soc., Proc. and Addresses XIX (1910), pp. 1-21.
Russell, P. F. 1959. Insects and the epidemiology of malaria. Ann. Rev. Entomol. 4:415-434.
Salisbury, H. E. 1976. Travels through America. Esquire, Feb.
Saloutos, T. 1960. *Farmer Movements in the South, 1865-1933*. Berkeley and Los Angeles: University of California Press.
——— 1964. *The Greeks in the United States*. Cambridge, Mass.: Harvard University Press.
——— 1975. The Immigrant in Pacific Coast Agriculture, 1880-1950. Agr. Hist. 49:182-201.
——— 1976. The immigrant contribution to American agriculture. Agr. Hist. 50:45-67.
Saloutos, T., and J. D. Hicks. 1951. *Agricultural Discontent in the Middle West, 1900-1939*. Madison: University of Wisconsin Press.
Sanborn, R. R., S. M. Mircetich, G. Nyland, and W. J. Moller. 1974. "Golden Death," a new leaf scorch threat to almond growers. Calif. Agr. 28(12):4-5.
Sauer, C. O. 1952. *Agricultural Origins and Dispersals*. New York: American Geographical Society.
Schery, R. W. 1972. *Plants for Man*. Englewood Cliffs, N.J.: Prentice-Hall.
Schiller, R. 1975. When *did* "civilization" begin? Readers Digest, May, pp. 119-122.
Schlebecker, J. T. 1963. *Cattle Raising on the Plains, 1900-1961*. Lincoln: University of Nebraska Press.
——— 1975. *Whereby We Thrive, A History of American Farming, 1607-1972*. Ames: Iowa State University Press.
Schlebecker, J. T., and G. E. Peterson. 1972. *Living Historical Farms Handbook*. Washington, D.C.: Smithsonian Institution.
Schneider, S. H. 1976. *The Genesis Strategy. Climate and Global Survival*. New York: Plenum Press.
Schoepf, J. D. 1789. *Travels in the Confederation*, 1783-1784. Trans. and ed. A. J. Morrison, Philadelphia: 1911, I, 168.
Scholes, W. E. 1966. The migrant worker. In *La Raza*. Ed. J. Samora, pp. 63-91. Notre Dame, Ind.: University of Notre Dame Press.
Schrader, W. D., H. P. Johnson, and J. F. Timmons. 1963. Applying erosion control principles. J. Soil Water Conserv. 18:195-199.
Schroeder, C. A. 1968. Prehistoric avocados in California. Calif. Avocado Soc. Yearbook 52: 29-34.
Schroeder, H. A. 1971. Losses of vitamins and trace minerals resulting from processing and preservation of foods. J. Clinical Nutrition 24:562-573.

Schumacher, E. F. 1973. *Small is Beautiful.* New York: Harper and Row.
Schwab, G. O., R. K. Frevert, T. W. Edminster, and K. K. Barnes. 1966. *Soil and Water Conservation Engineering.* New York: Wiley.
Schwartz, H. 1945. *Seasonal Farm Labor in the United States.* New York: Columbia University Press.
Scott, D., and J. Blomquist. 1978. A crossroads for Northwest energy. Sierra Club Bull. 63(4): 38-39.
Scott, J. D. 1976. Praise the Potato! Readers Digest 109(656):205-208, 212.
SCSB. 1973. *The Role of Giant Corporations in the American and World Economies.* Hearings before the Subcommittee on Monopoly of the Select Committee on Small Business, United States Senate, 92nd Congress.
Sears, O. H. 1939. Soybeans: their effect on soil productivity. Ill. Agr. Exp. Sta. Bull. 456.
Sears, P. 1971. An empire of dust. In *Patient Earth.* Ed. J. Harte and R. H. Socelow, pp. 2-15. New York: Holt, Rinehart and Winston.
Seidenbaum, A. 1975. Americans head back to the farm. *Los Angeles Times,* Dec. 1.
Sellers, W. D. 1973. A new global model. J. Appl. Meterol. 12:241-254.
Shannon, F. A. 1934. *Economic History of the People of the United States.* New York: Macmillan.
Shantz, H. L., and R. Zon. 1924. Natural vegetation of the United States. In *Atlas of American Agriculture.* Ed. O. E. Baker. Washington, D.C.: Govt. Print. Off.
Shaw, G. 1978. U.S. decides not to appeal ruling on farm water. *Los Angeles Times,* Jan. 7.
Shell, E. W. 1975. A fish story pans out, and the world is better fed. USDA Yearbook, pp. 149-156.
Shepard, P. 1973. *The Tender Carnivore and the Sacred Game.* New York: Scribner.
Shorey, H. H., R. S. Kaae, and L. Gaston. 1974. Sex pheromones of Lepidoptera. Development of a method for pheromonal control of *Pectinophora gossypiella* in cotton. J. Econ. Entomol. 67:347-350.
Shover, J. L. 1976. *First Majority—Last Minority.* DeKalb: Northern Illinois University Press.
Sidey, H. 1976. More powerful than atom bombs. *Time,* Jan. 12.
Siegenthaler, U., and H. Oeschger. 1978. Predicting future atmosphere carbon dioxide levels. Science 199:388-395.
Simoons, F. J. 1978. Traditional use and avoidance of foods of animal origin: a culture historical view. BioScience 28:178-184.
Sloan, C., and J. B. Frantz. 1961. *Life on the XIT Ranch.* Austin: University of Texas Press.
Smith, C. E., Jr. 1965. Plant fibers and civilization—cotton, a case in point. Econ. Botany 19: 71-82.
Smith, H. N. 1950. *Virgin Lands. The American West as Symbol and Myth.* New York: Vintage Books, Random House.
Smith, J. 1608. *A True Relation of Such Occurrences and Accidents of Note as Hath Hapned in Virginia since the First Planting of that Colony.* Reprinted in *Works of Captain John Smith,* Arber edit. 1884: Birmingham, England.
Smith, P. 1977. America's young jobless should be put to the task of needed public works. *Los Angeles Times,* June 15.
Smith, R. 1977. Are ag co-ops overly healthy? *Los Angeles Times,* Dec. 18.
Smith, R. F., and W. W. Allen. 1954. Insect control and the balance of nature. Sci. Am. 190(6): 38-42.
Smith, T., and F. L. Kilbourne. 1893. *Investigations into the Nature, Causation, and Prevention of Texas or Southern Cattle Fever.* USDA Bur. Animal Indust. Bull. 1.
South, P., and R. Rangi. 1974. NRC's vertical wind turbine. Agr. Eng. 55:14-16.
Spaulding, J. 1974. Solar energy now. Sierra Club Bull. 59(5):5-9.
Spaulding, W. A. 1922. Early chapters in the history of California citrus culture. Calif. Citrograph 7(3):66, 103, 150.
Splinter, W. E. 1976. Center-pivot irrigation. Sci. Am. 234(6):90-99.
Steele, C. B. 1941. The Steele Brothers. Pioneers in California's great dairy industry. Calif. Hist. Soc. 20:259-273.
Stegner, W. 1942. *Mormon Country.* New York: Duell, Sloane and Pearce.
Stein, W. 1975. The "Okie" as a farm laborer. In *Agriculture in the Development of the Far West.* Ed. J. H. Shideler, pp. 202-215. Agr. Hist. Soc., Washington, D.C.
Steinhart, J. S. and C. E. Steinhart. 1974. Energy use in the U.S. food system. Science 184: 307-16.

Stephens, L. D. 1976. Farish Furman's formula: scientific farming and the "New South." Agr. Hist. 50:377-390.
Stephens, S. G. 1976. The origin of Sea Island cotton. Agr. Hist. 50:391-399.
Sternitzke, H. S., and T. C. Nelson. 1970. The southern pines of the United States. Econ. Botany 24:142-150.
Steward, J. H. 1933. Ethnography of the Owens Valley Paiutes. University of California Publ. in Amer. Archeol. and Ethnol. 33:233.
Stewart, G. R. 1936. *Ordeal by Hunger.* New York: Pocket Books.
Steyn, R. 1974. Primitive strains may improve U.S. soybean varieties. Wallace's Farmer 19(17): 110-111.
——— 1975. Agricultural research: Impact on soybeans. Iowa Agr. Exp. Sta., Spec. Rep. Sci. Tech. 77.
Stickney, G. P. 1896. Indian uses of wild rice. Am. Anthropol. 9:115-121.
Stidd, C. K. 1975. Irrigation increases rainfall. Science 188:279-281.
Stone, A. H. 1905. The Italian cotton grower: the Negro's problem. South Atlantic Quart. 4: 42-47.
——— 1907. Italian cotton growers in Arkansas. Rev. of Rev. 35:209-213.
Street, J. H. 1957. *The New Revolution in the Cotton Economy: Mechanization and its Consequences.* Chapel Hill: University of North Carolina Press.
Strickland, W. 1801. *Observations on the Agriculture of the United States.* London: W. Bulmer.
Sullivan, M. 1934. *The Travels of Jedediah Smith.* Santa Ana, Calif.: Fine Arts Press.
Tarver, S. 1976. *A Simple Solution to the Energy Problem.* 3rd ed. Gillette, Wyo.: Wyoming Specialties.
Teague, C. C. 1944. *Fifty Years a Rancher.* Los Angeles: Ward Ritchie Press.
Terrell, J. A. 1971. *American Indian Almanac.* Cleveland: World Publishing.
——— 1972. *Land Grab.* New York: Dial Press.
Thomas, H. E. 1962. *Water and the Southwest: What is the Future?* U.S. Geolog. Surv. Circ. 469.
Thomson, J. 1976. Agripolitics. Prairie Farmer 148(21):12-13.
Todd, D. K. 1977. Is the remedy for future droughts under our feet? *Los Angeles Times,* June 26.
Toffler, A. 1970. *Future Shock.* New York: Random House.
Tollett, J. T. 1975. Protecting protein from rumen microbes by the use of aldehydes. Down to Earth 30(4):13-16.
Tooke, W. 1800. *The Life of Catherine II, Empress of Russia,* volume 1. London: Longman and Rees.
Torrey, V. 1976. *Wind-Catchers: American Windmills of Yesterday and Tomorrow.* Brattleboro, Vt.: Stephen Green Press.
Toyoji, C. 1922. Truth about Japanese farming in California. In *California and the Oriental,* Report of the State Board of Central California, Sacramento.
Treadwell, E. F. 1931. *The Cattle King: A Dramatized Biography.* New York: Macmillan.
Trenkle, A., and R. L. Willham. 1977. Beef production efficiency. Science 198:1009-1015.
Triplett, G. B., Jr. 1976. History of reduced tillage systems. Bull. Entomol. Soc. Am. 22:289-291.
Tufts, W. R. 1946. The rich pattern of California crops. In *California Agriculture.* Ed. C. B. Hutchison, pp. 113-238. Berkeley: University of California Press.
Tunley, R. 1977. Comeback of the small town. Reader's Digest, Oct., pp. 143-147.
Turney, O. A. 1929. *Prehistoric Irrigation.* State Historian, Phoenix, Ariz.
UCAE. 1971-72. Fly control on poultry ranches. University of California Agr. Ext. AXT-72.
Udall, D. K. 1959. *Arizona Pioneer Mormon: David King Udall—His Story and His Family, 1851-1938.* Written in collaboration with his daughter, Pearl Udall Nelson. Tucson, Ariz.: Arizona Silhouettes.
Udall, S. L. 1963. *The Quiet Crisis.* 1st ed. New York: Holt, Reinhart, and Winston.
Ullstrup, A. J. 1977. Diseases of corn. In *Corn and Corn Improvement.* Ed. G. F. Sprague, pp. 391-500. Madison, Wisc.: Am. Soc. Agron.
U.N. 1972. *Pesticides in the Modern World.* Rpt. Coop. Prog. Agro-Allied Ind. with FAO and the U.N. Org.
Updike, W. 1847. *History of the Narragansett Church.* Volume 1. Boston.
Uris, L. 1976. *Trinity,* pp. 266-267. Garden City, N.Y.: Doubleday.
USDA. 1968*a. A National Program of Research for Environmental Quality—Pollution in Relation to Agriculture and Forestry.* U.S. Dept. Agr., Washington, D.C.

——— 1968*b*. Breeds of swine. U.S. Dept. Agr. Farmers' Bulletin no. 1263.
——— 1970*a*. Changes in farm production and efficiency: a summary report. U.S. Dept. Agr. Stat., Bull. 233, pp. 16-17.
——— 1970*b*. An expanded Great Plains conservation program for the 1970s. USDA, Soil Cons. Serv. PA-960.
——— 1971. Farm income situation. U.S. Dept. Agr. Econ. Res. Serv., July.
——— 1972. *Fact Book of U.S. Agriculture*. USDA Misc. Publ. 1063. Washington, D.C.: U.S. Govt. Print. Off.
——— 1973*a*. What is a farm conservation plan? U.S. Dept. Agr. Conserv. Serv. PA-629.
——— 1973*b*. The food and fiber system—how it works. Agr. Inform. Bull. 383. Washington, D.C.
——— 1974. Our land and water resources. USDA, ERS, Misc. Publ. 1290.
——— 1975*a*. New wheat provides valuable protein. Agr. Research 24(4):5.
——— 1975*b*. *Nutrition Value of American Foods*. U.S. Dept. Agr., Handbook 456.
——— 1977. *Agricultural Statistics*, 1977. Washington, D.C.: U.S. Govt. Print. Off.
USDC. 1971. *Federal Plan for a National Agricultural Weather Service*. U.S. Dept. of Commerce, Washington, D.C.
U.S. Department of the Interior. 1968. *Answers to Your Questions about American Indians*. Washington, D.C.
USFEA (U.S. Federal Energy Commission). 1976. Comparison of energy consumption between West Germany and the United States. Conservation Paper no. 33, Washington, D.C.
USFS. 1973. The outlook for timber in the United States. Forest Resource Rpt. No. 20, Forest Service, USDA, Washington, D.C.
Van Arsdall, R. N., and M. D. Skold. 1973. Cattle raising in the United States. USDA Agr. Econ. Rpt. no. 235.
Van Bavel, C. H. M. 1977. Soil and oil. Science 197:213.
van den Bosch, R., and P. S. Messenger. 1973. *Biological Control*. New York: Intext Educational Publishers.
van Wagenen, J. 1953. *The Golden Age of Homespun*. Ithaca, N.Y.: Cornell University Press.
Vavilov, N. I. 1951. *The Origin, Variation, Immunity, and Breeding of Cultivated Plants*. New York: Ronald Press.
Veblen, T. 1899. *The Theory of the Leisure Class*. New York: Macmillan.
Virtanen, A. I. 1966. Milk production of cows on protein-free feed. Science 153:1603-1614.
Vogel, V. J. 1972. *This Country Was Ours*. New York: Harper and Row.
von Hippel, F., and R. H. Williams. 1975. Solar technologies. Bull. Atom. Sci. 31(9):25-31.
——— 1977. Toward a solar civilization. Bull. Atom. Sci. 33(8):12-13.
Wadleigh, C. H. 1968. *Waste in Relation to Agriculture and Forestry*. U.S. Dept. Agr., Misc. Publ. no. 1065.
Walden, H. T. 1966. *Native Inheritance*. The Story of Corn in America. New York: Harper and Row.
Walker, H. B. 1948. A resume of sixteen years of research in sugar beet mechanization. Agr. Eng. 29:425-430.
Watt, K. E. F., L. F. Molloy, C. K. Varshney, D. Weeks, and S. Wirosardjono. 1977. *The Unsteady State: Environmental Problems, Growth, and Culture*. Honolulu: East-West Center.
Way, R. D. 1973. Apple varieties in New York State. New York State Agr. Ext. Inf. Bull. 63.
Weatherwax, P. 1954. *Indian Corn in Old America*. New York: Macmillan.
Webb, W. P. 1931. *The Great Plains*. Berkeley: University of California Press. Boston: Ginn.
Webber, H. J. 1943. History and development of the citrus industry. In *Citrus Industry*, Volume 1. Ed. H. J. Webber and L. D. Batchelor. Rev. ed., 1967, pp. 1-39.
Weber, D. F., and G. E. Leggett. 1966. Relation of rhizobia to alfalfa sickness in Eastern Washington. U.S. Dept. Agr., ARS 41-117.
Weinberg, A. M. 1974. Global effects of man's production of energy. Science 186:205.
——— 1978. Reflections on the energy wars. Am. Scientist 66:153-158.
Weingart, J. M. 1977. *The Helios Strategy—A Heretical View of the Role of Solar Energy in the Future of a Small Planet*. Copyright, The Woodlands Conference. Soc. Intern. Develop., 1346 Connecticut Ave., N.W., Washington, D.C. 20036.
Wengert, N. I. 1952. *Valley of Tomorrow: The TVA and Agriculture*. University of Tennessee Ext. Ser., Volume 28, no. 1.
Wentworth, E. N. 1948. *America's Sheep Trails: History, Personalities*. Ames: Iowa University Press.
Werner, M. R. 1925. *Brigham Young*. New York: Harcourt, Brace.

Wessel, T. R. 1969. *Roosevelt and the Great Plains Shelterbelt.* Great Plains J., Great Plains Hist. Assoc., Lawton, Okla.
White, L., jr. 1962. *Medieval Technology and Social Change.* New York: Oxford University Press.
White, L. A. 1959. *The Evolution of Culture.* New York: McGraw Hill.
White, S. E. 1907. *Arizona Nights.* New York: McClure.
White, Z. W. 1976. Southern forestry is growing up. Am. Forests 82(10):17-19, 58.
White-Stevens, R. H. 1975. Antibiotics curb diseases in livestock, boost growth. USDA Yearbook, pp. 85-98.
Whitney, O. F. 1892. *History of Utah.* Volume 1. Salt Lake City: Cannon and Sons.
Wik, R. M. 1975. Some interpretations of the mechanization of agriculture in the Far West. Agr. Hist. 49:73-83.
Wilcox, H. A. 1975. *Hothouse Earth.* New York: Praeger.
——— 1976. The ocean food and energy farm project. Supplement to Calypso Log (Cousteau Society) 3(2):1-5. (From a paper presented at the Intern. Conf. Marine Tech. Assn., Monte Carlo, Monaco, Oct. 26, 1975.)
——— 1977. *The U.S. Navy's Ocean Food and Energy Farm Project.* Naval Ocean Systems Center, San Diego, Calif. Mimeographed.
Wilkes, G. 1977. The world's crop plant germplasm—an endangered resource. Bull. Atom. Sci., Feb. 1977, pp. 8-16.
Will, B. F., and G. E. Hyde. 1964. *Corn Among the Indians of the Upper Missouri.* (Copyright 1917 by the authors.) Lincoln: University of Nebraska.
Wilson, I. H. 1965. *William Wolfskill, 1798-1866. Frontier Trapper to California Ranchero.* Glendale, Calif.: A. H. Clark.
Wilson, W. O. and P. Vohra. 1974. Poultry Management. In *Animal Agriculture.* Ed. H. H. Cole and M. Ronning, pp. 622-636. San Francisco: Freeman.
Winstanley, D., B. Emmett, and G. Winstanley. 1975. Climatic changes and world food supply. In *Contributions of Irrigation and Drainage to the World Food Supply.* Am. Soc. Civil Eng., New York.
Wittwer, S. H. 1970. Research and technology on the United States food supply. In *Research for the World Food Crisis.* Ed. D. C. Aldrich, Jr., pp. 77-121. Am. Assoc. Adv. Sci., Publ. 92, Washington, D.C.
——— 1976. Food production, people and the future. In *Agriculture: Food and Man—a Century of Progress.* Col. Biol. Agric. Sci., Brigham Young University, Provo, Utah.
——— 1978. The next generation of agricultural research. Science 199:375.
Witzel, H. D. and B. F. Vogelaar. 1955. Engineering the hillside combine. Agr. Engin. 36:522-525, 528.
Wolff, A. 1975. World progress report: vegetable oil. *Saturday Review,* Nov. 1.
Wollenberg, C. 1970. Conflict in the fields: Mexican workers in California Agribusiness. In *Ethnic Conflict in California History.* Ed. C. Wollenberg, pp. 135-152. Los Angeles: Tinnon-Brown.
Wood, W. 1634. *New England's Prospect,* pp. 105-108. London: T. Cotes.
Woodbury, A. M. 1950. *The History of Southern Utah and Its National Parks.* Copyrighted by A. M. Woodbury. Salt Lake City.
Yates, R. G., A. D. Griggs, A. E. Weissenborn, F. T. Hidaka, G. V. Schirk, and T. P. Lynott. 1966. In *Mineral and Water Resources of Washington.* Report for the Committee on Interior and Insular Affairs, United States Senate, 89th Congress, 2nd Session, U.S. Govt. Print. Off.
Yermanos, D. M. 1974. Agronomic survey of jojoba in California. Econ. Botany 28:160-174.
Young, F. D. 1929. Frost and the prevention of frost damage. Washington, D.C.: Govt. Print. Off.
Young, H. M. 1973. 'No-tillage' farming in the United States—its profit and potential. Outlook in Agr. 7:143-148.
Yudkin, J. 1972. *Sweet and Dangerous.* New York: Bantam Books.
Zeckhauser, R. 1973. The risks of growth. Daedalus 102(4):103-118.
Zohary, D., and P. Spiegel-Roy. 1975. Beginnings of fruit growing in the Old World. Science 187:319-27.

# Index